Proceedings of the St. Petersburg Mathematical Society, Volume XV

Advances in Mathematical Analysis of Partial Differential Equations

Photo by Vitali Ivanovich Ochkur

Olga Aleksandrovna Ladyzhenskaya
7 March 1922 – 12 January 2004

American Mathematical Society

TRANSLATIONS

Series 2 • Volume 232

Proceedings of the St. Petersburg Mathematical Society, Volume XV

Advances in Mathematical Analysis of Partial Differential Equations

Darya Apushkinskaya
Alexander I. Nazarov
Editors

American Mathematical Society
Providence, Rhode Island

2010 *Mathematics Subject Classification.* Primary 35B05, 35B65, 35J60, 35K20, 35K41, 35Q35, 35R35, 76D03, 76D05.

ISBN-13: 978-1-4704-1551-8
ISBN-10: 1-4704-1551-8
ISSN 0065-9290

Printed in the United States of America.

∞ The paper used in this book is acid-free and falls within the guidelines
established to ensure permanence and durability.
Visit the AMS home page at http://www.ams.org/

10 9 8 7 6 5 4 3 2 1   19 18 17 16 15 14

Dedicated to the memory of
Olga Aleksandrovna Ladyzhenskaya

Contents

Preface

This is a volume in memory of the outstanding Russian mathematician Olga Alexandrovna Ladyzhenskaya (1922–2004). She was one of the founders of the famous Leningrad (St. Petersburg) school of PDE's and Mathematical Physics. Headed by Olga Alexandrovna, this school has made many fundamental contributions to the theory of PDE's as we know it today.

As a genuine classic, Olga Alexandrovna had a great variety of topics in PDE's and in Mathematical Physics to study on her own or together with her numerous students. Among the most important ones were problems from the spectral theory for differential operators, the convergence of finite difference methods, the calculus of variations, the theory of quasilinear equations, the mathematical fluid dynamics, and the theory of attractors for PDE's. The results she achieved were spectacular and many of them still remain on the top of those areas. Olga Alexandrovna, together with her student Nina Nikolaevna Uraltseva, presented a complete solution to Hilbert's 19th and 20th problems for a wide class of second order PDE's.

It is not a secret that Olga Alexandrovna's main mathematical "love" was a rigorous theory of fluid dynamics. Her very influential book *The Mathematical Theory of Viscous Incompressible Flow*, published in 1961, was translated into many languages and has become a classic in the field. Up to the present, it is an excellent introduction to the mathematical foundations of hydrodynamics.

Some of Olga Alexandrovna's achievements related to the mathematical theory of the Navier–Stokes equations will be remembered the longest. The global well posedness for those equations in dimension 2 proved in 1950'ies is one of her great results of this kind. As it has been shown by her in collaboration with A. A. Kiselev, a similar result is valid in dimension 3 as well but holds only on a finite interval of time. It is worthy to notice that the global well posedness for the three-dimensional problem remains open up to this day. Moreover, the issue of existence and uniqueness of physically reasonable solutions to the NSEs in three dimensions has been chosen as one of the seven Millenium "million dollar" Prize Problems of the Clay Mathematical Institute.

Thinking on turbulence, she proved thex existence of an invariant manifold attracting any bounded set of initial data for the 2D Navier-Stokes equations in the pioneering paper written in 1972. This result came to the attention not only of mathematicians, but of theoretical physicists as well. Moreover, this was a starting point for her to develop the theory of attractors of infinite-dimensional dynamical systems.

Olga Alexandrovna Ladyzhenskaya published more than 250 articles and authored or coauthored 7 monographs and a textbook. She has received many honors and awards, including the highest award of the Russian Academy of Sciences, the Lomonosov Gold Medal.

The present volume contains the proceedings of the international Workshop "Advances in Mathematical Analysis of Partial Differential Equations" held in the Institute Mittag-Leffler, Stockholm, Sweden, July 9–13, 2012. The workshop was partially supported by the European Mathematical Society and dedicated to the 90th anniversary of the O. A. Ladyzhenskaya birthday. The workshop has been attended by well-known experts in the field of partial differential equations from 10 countries. They came to honor Olga Alexandrovna Ladyzhenskaya's memory and to pay tribute to her scientific legacy.

Darya Apushkinskaya
Alexander I. Nazarov
Editors

Amer. Math. Soc. Transl.
Volume **232**, 2014

On singular parabolic p-Laplacian systems under nonsmooth external forces. Regularity up to the boundary

H. Beirão da Veiga

To the memory of Olga A. Ladyzhenskaya, with warm recollections of friendship

ABSTRACT. We study the regularity of the solutions to an initial-boundary value problem for a system of the p-Laplacian type in $n \geq 3$ space variables. External forces are square-integrable in the space-time cylinder. Under this natural assumption, the gradient of the solutions may be unbounded. As a consequence, roughly speaking, the ellipticity coefficient of the linearized parabolic equation may be not bounded from below by a positive constant. We show suitable integrability in the space-time cylinder, up to the boundary, for the second order space derivatives. The singular case is also covered.

1. Introduction and main result

The aim of this paper is to introduce a simple idea in the simplest form. Sometimes, to avoid technicalities, we will assume that solutions (which exist and are unique) are sufficiently smooth. So we left to the reader additional details concerning approximation theory.

In the sequel we consider the evolution problem

$$
\begin{cases} \partial_t u - \nabla \cdot \big((\mu + | \nabla u|^2)^{\frac{p-2}{2}} \nabla u \big) = f(t, x) , & \text{in } (0, T) \times \Omega , \\ u = 0 \ \text{ on } (0, T) \times \partial\Omega , \\ u(0) = u_0 \ \text{ in } \Omega , \end{cases} \tag{1.1}
$$

where $p \in (1, 2]$ (see (1.6) below), $T \in (0, \infty]$, and $\mu \geq 0$ are constants. Here u is an N-dimensional vector field, $N \geq 1$, defined in $Q_T \equiv (0, T) \times \Omega$, where $\Omega \subset \mathbb{R}^n$, $n \geq 3$, is a regular, bounded open set, which we assume is of class C^2.

We recall that scalar multiplication of both sides of (1.1) by u, followed by classical manipulations, will lead to the well-known a priori estimate

$$
\begin{aligned} \| u \|^2_{L^\infty(0, T; L^2(\Omega))} + \| u \|^p_{L^p(0, T; W^{1,p}(\Omega))} \\ \leq c \big(\| u_0 \|_2^2 + \| f \|^{p'}_{L^{p'}(0, T; W^{-1,p'}(\Omega))} \big). \end{aligned} \tag{1.2}
$$

2010 *Mathematics Subject Classification.* Primary 35B65, 35J57.
Key words and phrases. Initial-boundary value problems, p-Laplacian parabolic singular systems, regularity up to the boundary.

So we assume in the sequel that

$$u_0 \in L^2(\Omega) \quad \text{and} \quad f \in L^{p'}(0, T; W^{-1,p'}(\Omega)), \tag{1.3}$$

where p' is the exponent conjugate to p. Further, by appealing to the first equation (1.1) together with (1.3), we get

$$\partial_t u \in L^{p'}(0, T; W^{-1,p'}(\Omega)). \tag{1.4}$$

We call u a weak solution of problem (1.1) if

$$u \in L^\infty(0, T; L^2(\Omega)) \cap L^p(0, T; W_0^{1,p}(\Omega)),$$

if $\partial_t u$ satisfies (1.4), if $u(0) = u_0$, and if (for almost all $t \in (0, T)$)

$$\langle \partial_t u, w\rangle + \langle (\mu + |\nabla u|^2)^{\frac{p-2}{2}} \nabla u, w\rangle = \langle f, w\rangle, \quad \forall w \in W_0^{1,p}(\Omega).$$

Here $\langle \cdot, \cdot \rangle$ denotes the natural duality pairings.

For the existence and uniqueness ($p \geq \frac{2n}{n+2}$) of weak solutions under our hypotheses, we refer the reader to [**21**, Theorem 6.2.1] (see also [**30**, Chap. II, Theorem 1.1], where $p \geq 2$). In reference [**21**] it is assumed that $f = 0$. However, as remarked in [**19**], the proof continuous to hold without this condition. For instance, the above assumptions on f are sufficient.

In the following a main point is that the external force f is assumed to be square-integrable in Q_T. This low integrability prevents boundedness of $\nabla u(t, x)$ (which holds, for instance, if $f \in L^q(Q_T)$, $q > n + 2$). So, the ellipticity coefficient $(\mu + |\nabla u|^2)^{\frac{p-2}{2}}$ does not remain bounded away from zero. Actually, we also assume that f satisfies (1.3). However, this does not invalidate the substance of the above remark (in particular for values of p near 2).

Our aim is to prove that, if $u(x, t)$ denotes the solution of problem (1.1), then

$$f \in L^2(0, T; L^2(\Omega)) \quad \Longrightarrow \quad u \in L^{2(p-1)}(0, T; W^{2,q}(\Omega)), \tag{1.5}$$

where q is given by (1.7) below.

As a rule, we are generous in assumptions in order to avoid complicating remarks. So, we assume that

$$\begin{cases} \frac{2n}{n+2} < p \leq 2, & \text{if } n > 3, \\ \frac{5}{4} < p \leq 2, & \text{if } n = 3. \end{cases} \tag{1.6}$$

In particular, the inclusion $L^2(\Omega) \subset W^{-1,p'}(\Omega)$ holds. We define

$$q = q(p) = \frac{2n(p-1)}{n - 2(2-p)}. \tag{1.7}$$

The central role of this exponent will be clear in the following. Our assumptions on p imply that $q \in (1, 2]$, and also that the immersion $W^{2,q}(\Omega) \subset W^{1,p}(\Omega)$ is compact.

Next, we recall the well-known inequality

$$\|D^2 v\|_q \leq C_0(q)\, \|\Delta v\|_q \tag{1.8}$$

for $v \in W^{2,q}(\Omega) \cap W_0^{1,q}(\Omega)$. Moreover, there is a constant K, independent of q, such that

$$C_0(q) \leq K\, q, \tag{1.9}$$

at least for $q > \frac{2n}{n+2}$ (see [**32**]).

Our main result is the following; note that $\mu = 0$ is allowed. As in [**10**], Ω is assumed to be of class C^2.

THEOREM 1.1. *Let be* $\mu \geq 0$. *Assume that* p *satisfies* (1.6), *define* q *by* (1.7), *and assume that*

$$(2-p)\,C_0(q) < 1\,, \tag{1.10}$$

where $C_0(q)$ *is defined by* (1.8). *Let* $u_0 \in W_0^{1,\,p}(\Omega)$, *and let*

$$f \in L^2(0,\,T;\,L^2(\Omega)) \tag{1.11}$$

for some $T \in]\,0,\,+\infty\,]$. *Define* $\mathcal{C} = \mathcal{C}(u_0,\,f,\,\mu,\,T)$ *by*

$$\mathcal{C}^2 = \mu^{\frac{p}{2}}\,|\,\Omega\,| + \|\,\nabla\,u_0\,\|_p^p + \frac{2}{p}\,\|\,f\,\|^2_{L^2(0,\,T;\,L^2(\Omega))}\,. \tag{1.12}$$

Let u *be the weak solution of problem* (1.1). *Then* $u \in L^{2(p-1)}(0,\,T;\,W^{2,\,q}(\Omega))$. *Moreover, the estimate*

$$\|\,u\,\|^{2(p-1)}_{L^{2(p-1)}(\,0,\,T;\,W^{2,\,q}(\Omega))} \leq C\,(\,T^{2-p}\,\mathcal{C}^{2\,(p-1)} + \mathcal{C}^2) \tag{1.13}$$

holds. Furthermore, the estimates (3.13), (3.14), *and* (3.15) *are satisfied.*

Note that if $n = 3$ and $p > \frac{3}{2}$, then

$$u \in L^{2\,(p-1)}(0,\;T\,;\,C^{0,\,\alpha}(\overline{\Omega}))\,,$$

where $\alpha = \frac{p-\frac{3}{2}}{p-1}$.

In Section 3, the extension of the above result (in particular the regularity result $L^{2(p-1)}(0,\,T;W^{2,\,q}(\Omega))$) to systems of the "same family" will be proposed.

REMARK 1.1. Due to (1.9), condition (1.10) holds if

$$(2-\,p)\,q < \frac{1}{K}\,. \tag{1.14}$$

Furthermore, by appealing to the definition of q, we show that (1.14) holds if

$$2 - \frac{n}{2nK+2} < p \leq 2\,, \tag{1.15}$$

where the constant $K = K(\Omega)$ is independent of p and q. Hence, (1.15) is a sufficient condition on p to guarantee the results claimed in Theorem 1.1.

It is worth noting that, for $p = 2$, the above estimates turn into the classical "heat equation" estimates. For instance, the main estimate (3.17) leads to

$$\|\,u\,\|^2_{L^2(\,0,\,T;\,W^{2,\,2}(\Omega))} \leq C\,\big(\,\|\,\nabla\,u_0\,\|_2^2 + \|\,f\,\|^2_{L^2(0,\,T;\,L^2(\Omega))}\big)\,. \tag{1.16}$$

We may also consider the nonhomogeneous evolution boundary value problem, by appealing to the extension of the results of reference [**10**] to the above problem, which is shown in [**6**]. Other possible extensions are to consider slip boundary conditions, and also to replace, in the first equation (1.1), the gradient $\nabla\,u$ by the symmetric gradient $\nabla\,u + \nabla^T\,u$. In the elliptic case these extensions are considered in the forthcoming paper [**8**].

In references [**15**], [**16**] (see [**13**, chapters IX, X]) local Hölder continuity in $(0,\,T) \times \Omega$ of the space gradient of local weak solutions is proved. Regularity results, *up to the boundary*, are stated in [**13**, chapter X] (see the Theorems 1.1 and 1.2 therein). In particular, in Theorem 1.2, Hölder continuity up to the parabolic boundary (where $u = 0$) of the spatial gradient of weak solutions u is proved. However, regularity results, up to the boundary, for second order space derivatives in $L^q(\Omega)$ spaces and for solutions to the parabolic singular case, seem not known in the literature. Actually, the two types of estimates are not comparable. It is worth noting that in the elliptic case (see [**10**]) the $W^{2,q}(\Omega)$ estimates imply $C^{1,\alpha}(\overline{\Omega})$ regularity, since $q > n$ is admissible.

A very interesting related subject concerns linear elliptic and parabolic equations with discontinuous coefficients (in particular by appealing to BMO and VMO functional spaces). See, for instance, [**20**], [**23**], [**24**], [**25**], and references therein. Results in this direction may be useful to try to extend the regularity results for second order space derivatives without assumptions like (1.10).

Another classical related subject (here $N = 1$) are Harnack's inequalities. See references and results in the recent monograph [**14**]. The first parabolic versions go back to Hadamard [**22**] (see also [**31**]).

For related results we also refer, in addition to the above references, to the monographs [**26**], [**30**], and to references [**1**], [**2**], [**3**], [**4**], [**5**], [**11**], [**12**], [**13**], [**17**], [**18**], [**27**], [**28**], [**29**].

The results stated here were announced in [**7**].

2. An auxiliary result concerning the stationary problem

By $L^p(\Omega)$ and $W^{m,p}(\Omega)$, m a nonnegative integer, and $p \in (1, +\infty)$, we denote the usual Lebesgue and Sobolev spaces with the standard norms $\|\cdot\|_p$ and $\|\cdot\|_{m,p}$. We set $\|\cdot\| = \|\cdot\|_2$. We denote by $W_0^{1,p}(\Omega)$ the closure of $C_0^\infty(\Omega)$ in $W^{1,p}(\Omega)$, and by $W^{-1,p'}(\Omega)$, $p' = p/(p-1)$, the strong dual of $W_0^{1,p}(\Omega)$ with norm $\|\cdot\|_{-1,p'}$. We set $\partial_t u = \frac{\partial u}{\partial t}$.

The symbols c, c_0, c_1, etc., denote positive constants that may depend on μ; by capital letters, C, we denote positive constants independent of $\mu \geq 0$ (for convenience, assume the range of μ bounded from above). The same symbol c or C may denote different constants, even in the same equation.

The proof of Theorem 1.1 shown below strongly appeals to the following regularity result.

PROPOSITION 2.1. *Let the hypothesis assumed in Theorem* 1.1 *concerning* Ω, p, q, *and* μ *hold, and assume that* $f \in L^2(\Omega)$. *Further, let* u *be the unique weak solution to the stationary problem*

$$\begin{cases} -\nabla \cdot \left((\mu + |\nabla u|^2)^{\frac{p-2}{2}} \nabla u \right) = f \quad \textit{in } \Omega, \\ u = 0 \quad \textit{on } \partial\Omega. \end{cases} \tag{2.1}$$

Then u *belongs to* $W^{2,q}(\Omega)$. *Moreover, the following estimate holds:*

$$\|u\|_{2,q} \leq C \left(\|f\|_q + \|f\|_2^{\frac{1}{p-1}} \right). \tag{2.2}$$

This result is essentially a particular case of a more general result, namely, Theorem 2.1 in reference [**10**] (see also [**9**]). So, for the proof, we refer the reader to this last paper. However a small (actually obvious) adaptation is needed. For completeness, we discuss this point in the Appendix.

3. Proof of Theorem 1.1

Besides proving Theorem 1.1, we also want to show that the main devices in our argument easily apply to other systems of equations of the form

$$\begin{cases} \partial_t u - \nabla \cdot S(\nabla u)0 = f(t, x) \text{ in } (0, T) \times \Omega, \\ u = 0 \text{ on } (0, T) \times \partial\Omega, \\ u(0) = u_0 \text{ in } \Omega, \end{cases} \tag{3.1}$$

where $S(\cdot)$ is given by

$$S(\nabla u) := A(|\nabla u|^2)\nabla u, \tag{3.2}$$

and $A(y)$, $y \geq 0$, satisfies suitable p-growth conditions. This will be discussed briefly in the next section. In this section we treat system (1.1) by assuming that

$$A(y) = (\mu + y)^{\frac{p-2}{2}}. \tag{3.3}$$

We define

$$G(y) := \int A(y)\, dy, \tag{3.4}$$

for $y \geq 0$. Hence, in this section,

$$G(y) = \frac{2}{p}(\mu + y)^{\frac{p}{2}}. \tag{3.5}$$

Proof of Theorem 1.1. By integration by parts we show that

$$\begin{aligned} -\int_\Omega (\partial_t u) \cdot \nabla \cdot \big(A(|\nabla u|^2)\nabla u\big)\, dx = \frac{1}{2}\int_\Omega A(|\nabla u|^2)\,\partial_t |\nabla u|^2\, dx \\ -\int_{\partial\Omega} A(|\nabla u|^2)(\partial_t u) \cdot (\partial_n u)\, dS. \end{aligned} \tag{3.6}$$

So, by scalar multiplication of both sides of the first equation (3.1) by $-\nabla \cdot \big(A(|\nabla u|^2)\nabla u\big)$, and by taking into account that the boundary integral vanishes, one gets

$$\begin{aligned} \frac{1}{2}\int_\Omega A(|\nabla u|^2)\,\partial_t |\nabla u|^2\, dx + \int_\Omega |\nabla \cdot S(\nabla u)|^2\, dx \\ = -\int_\Omega f \cdot (\nabla \cdot S(\nabla u))\, dx. \end{aligned} \tag{3.7}$$

By appealing to (3.5), we show from (3.6) that

$$\begin{aligned} \frac{1}{2}\frac{d}{dt}\int_\Omega G(|\nabla u|^2)\, dx + \int_\Omega |\nabla \cdot S(\nabla u)|^2\, dx \\ = -\int_\Omega f \cdot (\nabla \cdot S(\nabla u))\, dx. \end{aligned} \tag{3.8}$$

Hence,

$$\frac{d}{dt}\int_\Omega G(|\nabla u|^2)\, dx + \int_\Omega \left|\nabla \cdot \big(A(|\nabla u|^2)\nabla u\big)\right|^2 dx \leq \int_\Omega |f|^2\, dx. \tag{3.9}$$

By integration with respect to t, one gets

$$\int_\Omega G(|\nabla u(t)|^2)dx + \int_0^t \|\nabla\cdot S(\nabla u(s))\|_2^2\,ds \leq \int_\Omega G(|\nabla u_0|^2)dx + \int_0^t \|f(s)\|_2^2\,ds. \tag{3.10}$$

Further, by appealing to the inequality

$$(\mu + |\nabla u|^2)^{\frac{p}{2}} \leq \mu^{\frac{p}{2}} + |\nabla u|^p, \tag{3.11}$$

we obtain (for $0 < t \leq T$)

$$\|\nabla u(t)\|_p^p + \frac{2}{p}\int_0^t \|\nabla\cdot S(\nabla u(s))\|_2^2\,ds \leq \mathcal{C}^2, \tag{3.12}$$

where $\mathcal{C}^2$ is defined by (1.12). We have used the inequality $(a+b)^\alpha \leq a^\alpha + b^\alpha$, for nonnegative a and b, and $0 < \alpha < 1$. From (3.10), standard manipulations, prove that

$$\|\nabla u\|^p_{L^\infty(0,T;\,L^p(\Omega))} \leq \mathcal{C}^2 \tag{3.13}$$

and that

$$\|\nabla\cdot((\mu + |\nabla u|^2)^{\frac{p-2}{2}}\nabla u)\|^2_{L^2(0,T;\,L^2(\Omega))} \leq \mathcal{C}^2. \tag{3.14}$$

Furthermore, the estimate

$$\|\partial_t u\|^2_{L^2(0,T;\,L^2(\Omega))} \leq 2\mathcal{C}^2 \tag{3.15}$$

holds since

$$\partial_t u = \nabla\cdot S(\nabla u) + f(t,x).$$

Finally, we prove the main estimate (1.13) by applying to the stationary equation

$$\nabla\cdot S(\nabla u) = \partial_t u - f(t,x),$$

for a.a. $t \in (0,T)$, the regularity result stated in Proposition 2.1. So, by appealing to (2.2), it follows that

$$\|u(t)\|_{2,q}^{2(p-1)} \leq C\left(\|\partial_t u - f\|_q^{2(p-1)} + \|\partial_t u - f\|_2^2\right), \tag{3.16}$$

for a.a. $t \in (0,T)$. Further, by integration in $(0,T)$, we show that

$$\begin{aligned}\|u\|^{2(p-1)}_{L^{2(p-1)}(0,T;\,W^{2,q}(\Omega))} &\leq C\Big(\|f\|^{2(p-1)}_{L^{2(p-1)}(0,T;\,L^2(\Omega))} + \|\partial_t u\|^{2(p-1)}_{L^{2(p-1)}(0,T;\,L^2(\Omega))} \\ &\qquad + \|f\|^2_{L^2(0,T;\,L^2(\Omega))} + \|\partial_t u\|^2_{L^2(0,T;\,L^2(\Omega))}\Big),\end{aligned} \tag{3.17}$$

where, for brevity, we have appealed to $q \leq 2$. Finally, by appealing to (1.12), (3.15), and the Hölder inequality, straightforward calculations show that the right-hand side of (3.17) is bounded by $C(T^{2-p}\mathcal{C}^{2(p-1)} + \mathcal{C}^2)$. Hence, the main estimate (1.13) holds. This completes the proof of Theorem 1.1.

Note that from (3.8) it follows that the quantity

$$\int_\Omega G(|\nabla u(t)|^2)\,dx$$

is decreasing with respect to time, if $f = 0$. Hence the norm $\|\nabla u(t)\|_p$ decreases with time.

Professor S. Antontsev kindly informed the author that the device leading to equation (3.9) was previously used by him. We refer, for instance, to the argument around equation (6.4) in reference [**2**]. We are grateful to Professor Antontsev for leading our attention to his wide and deep work on the subject.

4. A more general setting

The extension of the results stated in Theorem 1.1 to systems of equations of the form (3.1), where $S(\cdot)$ is given by (3.2), looks not difficult, provided that a corresponding extension of Proposition 2.1 to the related stationary system

$$\begin{cases} -\nabla\cdot\big(A(|\nabla u|^2)\nabla u\big) = f \quad \text{in } \Omega\,, \\ u = 0 \quad \text{on } \partial\Omega\,, \end{cases} \tag{4.1}$$

holds. This last possibility, claimed in reference [**10**], seems not difficult to prove. Roughly speaking, if the "stationary" extension holds, then the argument followed in the previous section applies to the evolution problem (3.1). Clearly, suitable p-growth conditions on $A(y^2)$ are necessary. Concerning the calculations developed in the previous section, the existence of positive constants c_0 and c_1 such that

$$c_0\, y^p - c_1 \leq G(y^2) \leq \widetilde{c}_0\, y^p + \widetilde{c}_1\,, \tag{4.2}$$

for $y \geq 0$, would be sufficient. For instance, if $(\mu + |\nabla u|^2)^{\frac{p-2}{2}}$ is replaced by $(\mu + |\nabla u|)^{p-2}$, the estimates (4.2) hold by setting

$$c_0 = \frac{1}{p}\,, \quad c_1 = \frac{2^p\,\mu^2}{p\,(p-1)}\,, \quad \widetilde{c}_0 = \widetilde{c}_1 = \frac{2^p}{p}\,.$$

For more detail we refer to [**7**].

5. Appendix

In this section we give some specifications concerning the proof of Proposition 2.1 as a corollary of Theorem 2.1 in reference [**10**] (to which the interested reader is necessarily referred). It is worth noting that the points raised in the sequel are minor points in the context of the complete proof. But they are just those for which a small adaptation is necessary (actually, the reader may easily do this task by himself). To avoid repetition of large parts of the original proof presented in [**10**], we assume that the reader, while reading the original proof, decided to take advantage of the remarks made below.

In [**10**] the authors proved the following result (see [**10**], Theorem 2.1).

THEOREM 5.1. *Let $p \in (1,2]$, and let $q \geq 2$, $q \neq n$, be given. Assume that $(2-p)\,C_0(q) < 1$, where $C_0(q)$ satisfies* (1.8). *Further, assume that $\mu \geq 0$. Let $f \in L^{r(q)}(\Omega)$, where*

$$r(q) = \begin{cases} \dfrac{nq}{n(p-1) + q(2-p)} & \text{if } q \in [2, n]\,, \\ q & \text{if } q \geq n\,, \end{cases} \tag{5.1}$$

and let u be the unique weak solution of problem (2.1). *Then u belongs to $W^{2,q}(\Omega)$. Moreover, the following estimate holds:*

$$\|u\|_{2,q} \leq C\left(\|f\|_q + \|f\|_{r(q)}^{\frac{1}{p-1}}\right). \tag{5.2}$$

Our first remark is that, in reference [**10**], the authors consider the case in which the term $(\mu + |\nabla u|^2)^{\frac{p-2}{2}}$ is replaced by $(\mu + |\nabla u|)^{p-2}$. As expected, the proof applies exactly by following the same way.

The second remark is the following. Assumption (1.11), namely, $f(t) \in L^2(\Omega)$, a.e. in t, leads us to consider the above theorem for the particular value q for which $r(q) = 2$. This value of q, given by (1.7), is less then 2. This is in contrast to Theorem 5.1, where it is assumed that $q \geq 2$. However, as already claimed in [**10**], the proof given in this last reference immediately applies to a range of values which includes $q < 2$, by making a couple of small changes in the original proof (shown below just for completeness). We necessarily assume that the reader has in hand the proof presented in reference [**10**]. Notation is that used in this last reference.

Actually, the following is the unique real modification needed to adapt the proof to our value of q. In [**10**, Sec. 3, third to last row], the authors claim the convergence result

$$f\,(\mu + |\nabla v^m|)^{2-p} \to f\,(\mu + |\nabla v|)^{2-p} \tag{5.3}$$

in the $L^{\frac{q}{2}}$-norm, where v^m denotes auxiliary "technical" and not significant, suitable subsequences (see [**10**]). The above convergence result is not suitable here, since $\frac{q}{2} < 1$. However, as already remarked in [**10**], convergence in the distributional sense is sufficient here. As shown in [**10**], one has

$$|\,(\mu + |\nabla v^m|\,)^{2-p} - (\mu + |\nabla v|\,)^{2-p}\,| \leq \frac{2-p}{\mu^{p-1}}\,|\nabla v^m - \nabla v|\,.$$

In particular, it follows that (5.3) holds a.e. in Ω, since, as in [**10**], ∇v^m converges a.e. in Ω to ∇v.

On the other hand, $(\mu + |\nabla v^m|)^{2-p}$ is bounded in L^t, where $t := \frac{q*}{2-p} > \frac{n}{2}$, where $q^* = \frac{q\,n}{n-q}$. So, it follows from [**30**, Chap. 1, Lemma 1.3] that

$$(\mu + |\nabla v^m|)^{2-p} \to (\mu + |\nabla v|)^{2-p}$$

weakly in L^t. Moreover, (1.6) implies that $(\frac{n}{2})' \leq 2$. Consequently, $f \in L^{(\frac{n}{2})'}$. It readily follows that (5.3) holds in the distributional sense.

We end with two remarks:

i) The set

$$\mathbb{K} = \{v \in W^{2,q}(\Omega) \,:\, \|\Delta v\|_q \leq R\,,\, v = 0 \text{ on } \partial\Omega\}\,,$$

introduced in [**10**, Section 3], is still contained in $W_0^{1,p}(\Omega)$, since $p > \frac{2n}{n+2}$.

ii) As at the very beginning of [**10**, Section 4], here one also has $p < q^*$. So, as in [**10**], we show that the solutions u^μ of problem (1.1) for $\mu > 0$ converge in $W^{1,p}(\Omega)$ to the solution u of (1.1) for $\mu = 0$.

References

[1] S. Antontsev and S. Shmarev, *Elliptic equations and systems with nonstandard growth conditions: Existence, uniqueness and localization properties for solutions*, Nonlinear Analysis, **65** (2006), 728–761. MR2232679 (2007f:35072)

[2] S. Antontsev and S. Shmarev, *Anisotropic parabolic equations with variable nonlinearity*, Publ. Mat., **53** (2009), 355–399. MR2543856 (2010m:35258)

[3] S. Antontsev and S. Shmarev, *Vanishing solutions of anisotropic parabolic equations with variable nonlinearity*, J. Math. Anal. Appl., **361** (2010), 371-391. MR2568702 (2010i:35196)

[4] S. Antontsev and S. Shmarev, *Blow-up of solutions to parabolic equations with nonstandard growth conditions*, J. Comp. Appl. Math., **234** (2010), 2633-2645. MR2652114 (2011g:35205)

[5] D. E. Apushkinskaya, A. I. Nazarov, *The elliptic Dirichlet problem in weighted spaces*, J. Math. Sci., **123** (2004), no. 6, 4527–4538. MR1923543 (2003e:35073)

[6] H. Beirão da Veiga, *On the global regularity for singular p-systems under non-homogeneous Dirichlet boundary conditions*, J. Math. Anal. Appl., **398** (2013), 527–533. doi: 10.1016/j.jmaa.2012.08.058. MR2990077

[7] H. Beirão da Veiga, *Singular parabolic p-Laplacian systems under non-smooth external forces. Regularity up to the boundary*, Online arXiv:1206.1808v1 [math.AP], 8 Jun 2012.

[8] H. Beirão da Veiga, *On the singular p-Laplacian system under Navier slip type boundary conditions. The gradient-symmetric case*, to appear.

[9] H. Beirão da Veiga and F. Crispo, *On the global regularity for nonlinear systems of the p-Laplacian type*, to appear in Discrete and Continuous Dynamical Systems—Series S., Online arXiv:1008.3262v1 [math.AP], 19 Aug 2010. MR3039691

[10] H. Beirão da Veiga and F. Crispo, *On the global $W^{2,q}$ regularity for nonlinear $N-$systems of the p-Laplacian type in n space variables*, Nonlinear Analysis-TMA , **75** (2012), 4346–4354. DOI: 10.1016/j.na.2012.03.021 MR2921994

[11] Y. Z. Chen, *Hölder continuity of the gradients of solutions of non-linear degenerate parabolic systems*, Acta Math. Sinica, **2** (1986), 309–331. MR900477 (89m:35036)

[12] H. Choe, *Hölder continuity of solutions of certain degenerate parabolic systems*, Non-linear Anal. **8** (1992), 235–243. MR1148287 (92m:35147)

[13] E. DiBenedetto, *Degenerate parabolic equations*, Springer, Berlin, 1993. MR1230384 (94h:35130)

[14] E. DiBenedetto, U. Gianazza and V. Vespri, *Harnack's inequality for degenerate and singular parabolic equations*, Springer Monographs in Math., Springer, New York, 2012. MR2865434

[15] E. DiBenedetto and A. Friedman, *Regularity of solutions of non-linear degenerate parabolic systems*, J. Reine Angew. Math., **349**, (1984), 83–128. MR743967 (85j:35089)

[16] E. DiBenedetto and A. Friedman, *Hölder estimates for non-linear degenerate parabolic systems*, J. Reine Angew. Math., **357**, (1985), 1–22. MR783531 (87f:35134a)

[17] E. DiBenedetto, Y.C. Kwong and V. Vespri, *Local space analycity of solutions of certain singular parabolic equations*, Indiana Univ. Math. J., **40**, (1991), 741–765. MR1119195 (93b:35014)

[18] E. DiBenedetto and Y. C. Kwong, *Intrinsic Harnack estimates and extinction profile for certain singular parabolic equations*, Trans. AMS., **330**, (1992), 783–811. MR1076615 (92f:35033)

[19] L. Diening, C. Ebmeyer, M. Růžička, *Optimal convergence for the implicit space-time discretization of parabolic systems with p-structure*, SIAM J. Numer. Anal., **45**, (2007), 457–472. MR2300281 (2008b:65118)

[20] H. Dong and D. Kim, *On the L_p-solvability of higher order parabolic and elliptic systems with BMO coefficients*, Arch. Rat. Mech. Anal., **199** (2011), 889-941. MR2771670 (2012h:35152)

[21] H. Gajewsky, K. Gröger and K. Zacharias, *Nichtlineare Operatorgleichungen und Operatordifferentialgleichungen*, Akademie-Verlag, Berlin, 1974. MR0636412 (58:30524a)

[22] J. Hadamard, *Extension à l'équation de la chaleur d'un theoreme de A. Harnack*, Rend. Circ. Mat. Palermo, **23** (1954), 337–346. MR0068713 (16:930a)

[23] D. Kim and N. V. Krylov, *Parabolic equations with measurable coefficients*, Potential Anal., **26**, (2007), 345-361. MR2300337 (2008f:35161)

[24] V. Kozlov and A. Nazarov, *The Dirichlet problem for non-divergence parabolic equations with discontinuous in time coefficients*, Math. Nachr., **282**, (2009), 1220–1241. MR2561181 (2011b:35189)

[25] N. V. Krylov, *Parabolic and elliptic equations with VMO coefficients*, Comm. in P.D.E., **32**, (2007), 453–475. MR2304157 (2008a:35125)

[26] O. A. Ladyzhenskaya, V. A. Solonnikov and N. N. Ural'tseva, *Linear and quasilinear equations of parabolic type*. Trans. Math. Mono., **23** AMS, Providence, RI 1968. MR0241822 (39:3159b)

[27] G. M. Lieberman, *A new regularity estimate for solutions of singular parabolic equations*, Discrete and Continuous Dynamical Systems—Supplement volume, 2005, 605–610. MR2192719 (2006i:35159)

[28] G. M. Lieberman, *The first initial-boundary value problem for quasilinear second order parabolic equations*, Ann. Sc. Norm. Sup. Pisa **13** (1986), 347–387. MR881097 (88e:35108)

[29] G.M. Lieberman, *Boundary and initial regularity for solutions of degenerate value parabolic equations*, Nonlinear Anal. TMA **20** (1993), 551–570. MR1207530 (94e:35041)

[30] J.-L. Lions, *Quelques méthodes de résolution des problèmes aux limites non linéaires*, Dunod; Gauthier-Villars, Paris, 1969. MR0259693 (41:4326)

[31] B. Pini, *Sulla soluzione generalizzata di Wiener per il primo problema di valori di contorno nel caso parabolico*, Rend. Sem. Mat, Univ. Padova, **23** (1954), 422–434. MR0065794 (16:485c)

[32] V. I. Yudovic, *Some estimates connected with integral operators and with solutions of elliptic equations*, Dokl. Akad. Nauk SSSR, **138**(1961), 805–808; English translation in Soviet Math. Doklady **2** (1961), 746–749. MR0140822 (25:4236)

Dipartimento di Matematica, Università di Pisa, Largo Bruno Pontecorvo 5, 56127 Pisa, Italy

E-mail address: bveiga@dma.unipi.it

Amer. Math. Soc. Transl.
Volume **232**, 2014

On the fundamental matrix solution for higher-order parabolic systems

S. Boccia and N. V. Krylov

In memory of O. A. Ladyzhenskaya

ABSTRACT. We establish existence and some regularity properties of fundamental matrix solutions for higher-order parabolic systems in the whole space with coefficients depending only on time variables in a measurable way. Applications to Tikhonov-type uniqueness results are given. Using potential-based techniques, we also investigate the solvability in spaces with mixed norms.

1. Introduction

In this paper, we are concerned with the fundamental matrix solution of higher-order parabolic systems in $\mathbb{R}^{d+1}$. To be more precise, we deal with the system

$$\partial_t u(t,x) = Lu(t,x) - \lambda u(t,x) + f(t,x), \tag{1.1}$$

where $\lambda \geq 0$ is a number,

$$Lu(t,x) = \sum_{|\gamma|\leq 2m} A_t^\gamma D^\gamma u(t,x),$$

$t \in \mathbb{R}$, $x \in \mathbb{R}^d$, m is an integer ≥ 1, u and f are complex vector-valued functions, and A_t^γ are complex matrix-valued functions of t. We shall prove that such parabolic systems have fundamental matrix solutions in the whole space without imposing the continuity condition on the coefficients of L. We only require that the coefficients be bounded measurable functions of t satisfying the Legendre–Hadamard condition which is used in a great number of papers. This condition is stronger than parabolicity in the sense of Petrovskii, which is also widely used; see, for instance, Friedman [**4**] and Eidelman [**3**]. Solonnikov [**12**] considers parabolic systems more general than the ones satisfying the Petrovskii definition. In these references the coefficients are assumed to be continuous in t. We relax this condition on the account of the stronger requirement of parabolicity.

Our first result bears on the fundamental matrix solution $\Psi_{s,t}(x)$ of (1.1) and generalizes the corresponding result in Lee [**9**], where $m = 1$. We prove the existence of $\Psi_{s,t}(x)$ and also derive the following standard global Gaussian estimate (see

2010 *Mathematics Subject Classification.* Primary 35K41, 35A08.
Key words and phrases. Parabolic systems, fundamental matrix, mixed norms.
The second author was partially supported by NSF Grant DMS-0653121.

Theorem 2.1). If $A_t^\gamma \equiv 0$ for all γ such that $|\gamma| \le 2m-1$, then there exists a constant N such that for any $s,t \in \mathbb{R}$ with $s<t$ and any $x \in \mathbb{R}^d$,

$$|\Psi_{s,t}(x)| \le N(t-s)^{-d/2m} \exp\left[-N\left(\frac{|x|^{2m}}{t-s}\right)^{1/(2m-1)}\right].$$

In order to obtain the estimates of the fundamental matrix solution and its derivatives, we follow the method which was probably first introduced in the framework of PDEs by Ladyzhenskaya in [**7**] in 1950 and then used in many articles and books (see, for instance, Eidelman [**3**], Friedman [**4**], and the references therein). However, the fact that we allow A_t^γ to be measurable only in t brings about many technical difficulties. Having established existence of $\Psi_{s,t}$ and investigated the corresponding potentials ("heat" potentials) allows us to generalize on more results of Ladyzhenskaya [**7**], who proved a Tikhonov-type uniqueness theorem for single higher-order parabolic equations with coefficients depending only on t in a continuous way (see Theorem 2.2).

Next, we turn our attention to proving a priori estimates for solutions of equation (1.1) and proving its solvability. We work in the spaces $L_p(\mathbb{R}, C^{2m+\alpha}(\mathbb{R}^d))$, $p \in (1,\infty]$, of functions $u(t,x)$, $t \in \mathbb{R}$, $x \in \mathbb{R}^d$, with mixed norms

$$\left(\int_{\mathbb{R}} \|u(t,\cdot)\|^p_{C^{2m+\alpha}(\mathbb{R}^d)}dt\right)^{1/p}.$$

An unusual consequence of our a priori estimates is given in Remark 2.12. The technique used is similar to that developed in [**5**] to obtain analogous estimates for solutions of second-order parabolic equations in the whole space. It is based on estimates of the heat potentials, proved first, and on the well-known properties of the Hardy–Littlewood maximal function.

Our results can be applied to constructing fundamental solutions for systems with coefficients which are Hölder continuous in x and measurable in t. This would generalize some results of [**11**] for systems of equations and could be done by standard methods as in [**3**], [**4**], and Ladyzhenskaya et al. in [**8**]. One also can add first-order terms with coefficients depending on (t,x) and growing with respect to x as in [**6**].

The organization of the paper is as follows. In Section 2 we introduce some notation and state the main results. Section 3 concerns the existence and properties of the fundamental matrix of our system. In Section 4 we study the resolvent operator and prove a Tikhonov-type theorem. In Section 5 we establish a priori estimates in the spaces with mixed norms.

2. Main results

We first introduce some notation used throughout the paper. A point in $\mathbb{R}^d$ is denoted by $x=(x_1,\dots,x_d)$. A point in

$$\mathbb{R}^{d+1} = \mathbb{R}\times\mathbb{R}^d = \{(t,x) : t\in\mathbb{R}, x\in\mathbb{R}^d\}$$

is denoted by (t,x). By γ we denote a multi-index that is

$$\gamma=(\gamma_1,\dots,\gamma_d), \quad \gamma_i \in \{0,1,\dots\} \quad \forall i=1,\dots,d, \quad |\gamma| = \gamma_1+\dots+\gamma_d.$$

Introduce

$$D^\gamma = D_1^{\gamma_1}\cdots D_d^{\gamma_d}, \quad D_i^{\gamma_i}u = \frac{\partial^{\gamma_i}u}{(\partial x_i)^{\gamma_i}} \quad \forall i=1,\dots,d, \quad \partial_t u = \frac{\partial u}{\partial t},$$

and for $\xi \in \mathbb{R}^d$ denote

$$\xi^\gamma = \xi_1^{\gamma_1} \cdot \ldots \cdot \xi_d^{\gamma_d}.$$

Let n be a fixed integer ≥ 1, and let $A^\gamma = A_t^\gamma$ be an $n \times n$ complex matrix-valued function depending only on t in a measurable way and given for all multi-indices γ with $|\gamma| \leq 2m$.

ASSUMPTION 2.1 (Legendre–Hadamard condition). There is a constant $\delta > 0$ such that $|A_t^\gamma| \leq \delta^{-1}$ for all t and γ and for all t, $\theta \in \mathbb{C}^n$, and $\xi \in \mathbb{R}^d$

$$(-1)^m \sum_{|\gamma|=2m} \Re(\xi^\gamma A_t^\gamma \theta, \bar{\theta}) \leq -\delta |\xi|^{2m} |\theta|^2, \tag{2.1}$$

where, for two complex-valued vectors a and b of the same dimension, we use the notation

$$(a, b) = \sum_k a_k b_k,$$

$\Re(z)$ denotes the real part of $z \in \mathbb{C}$, and for matrices C by $|C|$ we mean $\left(\sum_{i,j} |C_{ij}|^2\right)^{1/2}$.

Here is our first main result, which is proved in Section 3.

THEOREM 2.1. *For $t > s$ and $x \in \mathbb{R}^d$ the fundamental matrix solution $\Psi_{s,t}(x)$ of system (1.1) with $\lambda = 0$ is well defined and is infinitely differentiable with respect to x. Furthermore, for any multi-index γ,*

$$|D^\gamma \Psi_{s,t}(x)| \leq N \frac{e^{N^*(t-s)}}{(t-s)^{(|\gamma|+d)/(2m)}} \exp\left[-\varepsilon \left(\frac{|x|^{2m}}{t-s}\right)^{1/(2m-1)}\right]. \tag{2.2}$$

Finally, if $A_t^\gamma \equiv 0$ for $|\gamma| \leq 2m-1$, then estimate (2.2) holds with $N^* = 0$. The above constants N and N^* depend only on δ, n, m, d, and $|\gamma|$, and $\varepsilon > 0$ depends only on d, n, m, and δ.

Ladyzhenskaya in [**7**] proved the following Tikhonov-type theorem for single equations with coefficients depending only on t in a *continuous* way.

THEOREM 2.2. *Let $T \in (0, \infty)$, and let $u(t, x)$ be a $\mathbb{C}^n$-valued continuous function on $[0, T] \times \mathbb{R}^d$, which satisfies the equation*

$$\partial_t u = Lu \tag{2.3}$$

in $(0, T) \times \mathbb{R}^d$ in a generalized sense; that is, for any $\zeta \in C_0^\infty(\mathbb{R}^d)$ and any $0 \leq s \leq t \leq T$, we have

$$\begin{aligned} &\int_{\mathbb{R}^d} \zeta(x) u(t, x)\, dx - \int_{\mathbb{R}^d} \zeta(x) u(s, x)\, dx \\ &\quad = \sum_{|\gamma| \leq 2m} \int_s^t A_r^\gamma (-1)^{|\gamma|} \int_{\mathbb{R}^d} u(r, x) D^\gamma \zeta(x)\, dx dr. \end{aligned} \tag{2.4}$$

Finally, assume that $u(0, x) \equiv 0$ and there is a constant N such that $|u(t, x)| \leq N \exp(N|x|^{2m/(2m-1)})$ on $[0, T] \times \mathbb{R}^d$. Then $u \equiv 0$ on $[0, T] \times \mathbb{R}^d$.

This theorem is proved in Section 4.

To state our results about solvability of (1.1), we need to explain how the equation is understood.

DEFINITION 2.3. For a time interval $I \subset \mathbb{R}$ (open or closed, finite or infinite) and functions $u(t)$ and $g(t)$ defined on I, we write

$$\partial_t u = g \quad \text{on} \quad I \tag{2.5}$$

if g is Lebesgue-measurable and for any finite $S, T \in I$, such that $S < T$ we have

$$\int_S^T |g(t)|\,dt < \infty \quad \text{and} \quad u(T) - u(S) = \int_S^T g(t)\,dt. \tag{2.6}$$

Obviously g is the generalized derivative of u with respect to t. It should be noted that notation (2.5) is somewhat ambiguous because one can change g on the set of measure zero and preserve (2.6). Therefore, the rigorous meaning of the object $\partial_t u$ is that it is a *class* of equivalent functions g for each of which (2.6) holds. However, following a common abuse of notation we will still write (2.5) instead of $g \in \partial_t u$.

DEFINITION 2.4. For a time interval $I \subset \mathbb{R}$ (open or closed, finite or infinite) and functions $u(t,x)$ and $f(t,x)$ defined on $I \times \mathbb{R}^d$, we say that (1.1) holds on $I \times \mathbb{R}^d$ if, for any $t \in I$, the right-hand side of (1.1) is well defined (in the classical sense) and, for any $x \in \mathbb{R}^d$, (1.1) holds in the sense of Definition 2.3.

As usual, for $\alpha \in (0,1)$ we set

$$[u]_\alpha = \sup_{\substack{x,y \in \mathbb{R}^d \\ x \neq y}} \frac{|u(x) - u(y)|}{|x - y|^\alpha}, \qquad |u|_0 = \sup_{x \in \mathbb{R}^d} |u(x)|,$$

$$|u|_\alpha = |u|_0 + [u]_\alpha.$$

As usual, by C^α we mean the space of $\mathbb{C}^n$-valued functions u on $\mathbb{R}^d$ with $|u|_\alpha < \infty$. For integers $k \geq 1$ by C^k and $C^{k+\alpha}$, we denote the space of k times continuously differentiable $\mathbb{C}^n$-valued functions u on $\mathbb{R}^d$ with finite norms

$$|u|_k = \sum_{i \leq k} [u]_i \quad \text{and} \quad |u|_{k+\alpha} = |u|_k + [u]_{k+\alpha},$$

respectively, where

$$[u]_i = \sum_{|\gamma| = i} |D^\gamma u|_0, \quad [u]_{k+\alpha} = \sum_{|\gamma| = k} [D^\gamma u]_\alpha.$$

DEFINITION 2.5. For $p \in [1, \infty]$ and $\alpha \in (0,1)$, the space $\mathcal{C}_p^\alpha$ is the set of $\mathbb{C}^n$-valued measurable functions $g(t,x)$ defined on $\mathbb{R}^{d+1}$ such that $g(t,\cdot) \in C^\alpha$ for almost all t and the function $|g(t,\cdot)|_\alpha$ is in $L_p(\mathbb{R})$ (essentially bounded if $p = \infty$).

DEFINITION 2.6. For $p \in [1, \infty]$ and $\alpha \in (0,1)$, the space $\mathcal{C}_p^{2m+\alpha}$ is the set of $\mathbb{C}^n$-valued functions $u(t,x)$ defined on $\mathbb{R}^{d+1}$ such that

(i) the function u is continuous in $\mathbb{R}^{d+1}$;

(ii) for almost all t, we have $u(t,\cdot) \in C^{2m+\alpha}$ and the function $|u(t,\cdot)|_{2m+\alpha}$ is in $L_p(\mathbb{R})$ (essentially bounded if $p = \infty$);

(iii) there is a $g \in \mathcal{C}_p^\alpha$ such that, for any $x \in \mathbb{R}^d$, $\partial_t u(t,x) = g(t,x)$ (a.e.) on $\mathbb{R}$.

REMARK 2.7. If $u \in \mathcal{C}_\infty^{2m+\alpha}$, then $u(t,\cdot) \in C^{2m+\alpha}$ for all t rather than only for almost all and $|u(t,\cdot)|_{2m+\alpha}$ is bounded rather than only essentially bounded. This follows from the continuity of u, the fact that $|u(t,\cdot)|_{2m+\alpha}$ is essentially bounded, and a simple fact from calculus that if $u_n(x) \to u(x)$ for all x and the norms $|u_n|_{2m+\alpha}$ are bounded, then $u \in C^{2m+\alpha}$ and $|u|_{2m+\alpha} \leq \underline{\lim}\, |u_n|_{2m+\alpha}$.

It is also worth noting that for $u \in \mathcal{C}^{2m+\alpha}_{\infty}$, one can always make $|\partial_t u(t,\cdot)|_\alpha$ to be bounded rather than only essentially bounded. This follows from the fact that the functions coinciding for almost all t have the same integrals with respect to t.

In the following we set

$$L^0 = \sum_{|\gamma|=2m} A_t^\gamma D^\gamma,$$

we denote by $\Psi^0_{s,t}$ the fundamental solution corresponding to L^0, and for constants λ we introduce the operator

$$\begin{aligned}\mathcal{G}^0_\lambda f(t,x) &= \int_{-\infty}^{t} \left(\Psi^0_{s,t}(\cdot) * f(s,\cdot)\right)(x) e^{-\lambda(t-s)} ds \\ &= \int_{-\infty}^{t} \int_{\mathbb{R}^d} \Psi^0_{s,t}(x-y) f(s,y) e^{-\lambda(t-s)}\, dyds,\end{aligned}$$

whenever the last expression makes sense. Below $\|\,[f]_\alpha\|_{L_p(\mathbb{R})}$ means the $L_p(\mathbb{R})$-norm of $[f(t,\cdot)]_\alpha$. Similar meaning has $\|\,|f|_0\|_{L_p(\mathbb{R})}$ and $\|\,[u]_{k+\alpha}\|_{L_p(\mathbb{R})}$. For Borel $\Gamma \subset \mathbb{R}$ by $|\Gamma|$, we denote the Lebesgue measure of Γ. All remaining results in this section are proved in Section 4.

THEOREM 2.8. *Let $p \in (1,\infty]$, $\alpha \in (0,1)$, $f \in \mathcal{C}^\alpha_p$, and $\lambda > 0$. Then $u := \mathcal{G}^0_\lambda f$ is well defined, $u \in \mathcal{C}^{2m+\alpha}_p$, and u satisfies* (1.1) *(in the sense of Definition* 2.4*) with L^0 in place of L. Furthermore,*

$$\|\,[u]_{2m+\alpha}\|_{L_p(\mathbb{R})} \le N \|\,[f]_\alpha\|_{L_p(\mathbb{R})}, \tag{2.7}$$

$$\lambda \|\,|u|_0\|_{L_p(\mathbb{R})} \le N \|\,|f|_0\,\|_{L_p(\mathbb{R})}, \tag{2.8}$$

where the constant N in (2.7) *depends only on α, p, d, m, δ, and n and the constant N in* (2.8) *depend only δ, n, m, d.*

REMARK 2.9. If we take $f(t,x) = 0$ for $t \le 0$, then $u(0,x) \equiv 0$ and we obtain a solution of the Cauchy problem with zero initial data. Other initial data can be considered as usual just by adding them to the solution of the Cauchy problem with zero initial data and appropriately modified f. By the way, even if the coefficients of L are independent of t, the above argument shows an advantage of considering f which are only measurable in t.

Theorem 2.8 is an existence theorem. The following result implies uniqueness. For $T \in \mathbb{R}$ and $S < T$, we introduce $\mathcal{C}^{2m+\alpha}_p(S,T]$ as the collection of the traces of functions in $\mathcal{C}^{2m+\alpha}_p$ on $(S,T] \times \mathbb{R}^d$. We write $\mathcal{C}^{2m+\alpha}_p(T) = \mathcal{C}^{2m+\alpha}_p(-\infty,T]$.

THEOREM 2.10. *Let $p \in (1,\infty]$, $\alpha \in (0,1)$, and $u \in \mathcal{C}^{2m+\alpha}_p(T)$. Then for any constant $\lambda \ge 0$,*

$$\|\,[u]_{2m+\alpha}\|_{L_p((-\infty,T])} \le N \|\,[\lambda u - L^0 u + \partial_t u]_\alpha\|_{L_p((-\infty,T])}, \tag{2.9}$$

$$\lambda \|\,|u|_0\|_{L_p((-\infty,T])} \le N \|\,|\lambda u - L^0 u + \partial_t u|_0\,\|_{L_p((-\infty,T])}; \tag{2.10}$$

and for any constants $\lambda, c \ge 0$ and any multi-index γ with $|\gamma| = 2m$, we have

$$c|\{t \le T : [u(t,\cdot)]_{2m+\alpha} > c\}| \le N \int_{-\infty}^{T} [(\lambda u - L^0 u + \partial_t u)(t,\cdot)]_\alpha dt. \tag{2.11}$$

Moreover, the constants N in (2.9) *and* (2.11) *depend only on α, p, d, m, δ, and n and the constant N in* (2.10) *depend only on δ, n, m, d.*

REMARK 2.11. It is known that for any $\mathbb{C}^n$-valued function $g(x)$, any integer $k \in [0, 2m]$ and any $\varepsilon > 0$, we have

$$[g]_{k+\alpha} \le N\varepsilon^{2m-k}[g]_{2m+\alpha} + N\varepsilon^{-k-\alpha}|g|_0,$$

$$[g]_k \le N\varepsilon^{2m+\alpha-k}[g]_{2m+\alpha} + N\varepsilon^{-k}|g|_0,$$

where N depends only on $d, n, 2m+\alpha$. By taking $\lambda > 0$ and $\varepsilon = \lambda^{-1/(2m+\alpha)}$, we obtain from (2.9) and (2.10) that

$$\sum_{k=0}^{2m} \big(\lambda^{(2m-k)(2m+\alpha)}\| \, [u]_{k+\alpha}\|_{L_p((-\infty,T])} + \lambda^{(2m+\alpha-k)(2m+\alpha)}\| \, [u]_k\|_{L_p((-\infty,T])}\big)$$

$$\le N\| \, |\lambda u - L^0 u + \partial_t u|_\alpha\|_{L_p((-\infty,T])}.$$

Since the powers of λ on the left are strictly greater than zero for $k < 2m$, this allows us to add lower-order terms to L^0 and to assert that there exists $N, \lambda_0 \ge 0$, depending only on α, p, d, m, δ, and n, such that under the assumptions of Theorem 2.10 if $\lambda \ge \lambda_0$, then

$$\| \, |u|_{2m+\alpha}\|_{L_p((-\infty,T])} \le N\| \, |\lambda u - Lu + \partial_t u|_\alpha\|_{L_p((-\infty,T])}.$$

After that one can use Theorem 2.8 and the method of continuity and prove the solvability of (1.1) for our general L.

REMARK 2.12. Obtaining a priori estimates for operators with coefficients only measurable in time, we have the following perhaps unexpected consequences. In the scalar case ($m = n = 1$) by Theorem 2.10 we have that

$$\sup_t [D^2 u(t,\cdot)]_\alpha \le N_1 \sup_t [f(t,\cdot)]_\alpha$$

with N_1 depending only on d, α if $u \in \mathcal{C}^{2+\alpha}_\infty$ is a solution of $\partial_t u = \Delta u + f$. It turns out that for any locally integrable $\mathbb{R}^d$-valued function $b = b(t)$ and solution $v \in \mathcal{C}^{2+\alpha}_\infty$ of the equation $\partial_t v(t,x) = \Delta v(t,x) + b_i(t)D_i v(t,x) + f(t,x)$, we have

$$\sup_t [D^2 v(t,\cdot)]_\alpha \le N_1 \sup_t [f(t,\cdot)]_\alpha$$

with the *same* constant N_1. This result is an immediate corollary of Theorem 2.10, and it suffices just to introduce $u(t,x) = v(t, x - x_t)$, where

$$x_t = \int_0^t b(s)\, ds,$$

and to use the fact that the C^α-norm is translation invariant.

In case u and v are independent of t, we get that

$$[D^2 v]_\alpha \le N_1[\Delta v + b_i D_i]_\alpha$$

with the same N_1 for any constant $b \in \mathbb{R}^d$.

These observations were crucial in [**6**] while considering second-order equations with growing coefficients in usual (not weighted) Hölder spaces.

Before stating one more result, we introduce a new object. If $f(t)$ is a locally integrable function on $\mathbb{R}$, we denote by $M_l(f)$ its left maximal function

$$M_l(f)(t) = \sup_{r>0} r^{-1} \int_{t-r}^t |f(s)| ds.$$

THEOREM 2.13. *Let* $\lambda \geq 0$, $\alpha \in (0,1)$, *and* $u \in C_1^{2m+\alpha}(S,T]$ *for any* $S \in (-\infty, T)$. *Define*

$$f = \lambda u - L^0 u + \partial_t u,$$

fix a $t \leq T$, *and assume that*

$$M_l([f]_\alpha)(t) < \infty, \quad \lim_{k\to\infty} \frac{1}{k} \int_{-k-2}^{-k} [u(\tau,\cdot)]_\alpha \, d\tau = 0.$$

Then $u(t,\cdot) \in C^{2m+\alpha}$ *and for any multi-index* γ *with* $|\gamma| = 2m$, *we have*

$$[D^\gamma u]_\alpha(t) \leq N M_l([f]_\alpha)(t), \tag{2.12}$$

where N *depends only on* α, d, m, δ, *and* n.

3. Fundamental solution and proof of Theorem 2.1

In the future we will use the summation convention over repeated indices, the range of which should be clear from the context. Let $\Im z$ denote the imaginary part of $z \in \mathbb{C}$.

LEMMA 3.1. *If* $\xi \in \mathbb{C}^d$ *and* $\theta \in \mathbb{C}^n$, *then*

$$\Re(i^{|\gamma|}\xi^\gamma A_t^\gamma \theta, \bar\theta) \leq -(\delta/2)|\Re\xi|^{2m}|\theta|^2 + N(|\Im\xi|^{2m} + N^*)|\theta|^2, \tag{3.1}$$

where the constants N *and* N^* *depend only on* δ, n, m, *and* d, *and* $N^* = 0$ *if* $A^\gamma \equiv 0$ *for* $|\gamma| \leq 2m-1$.

Proof. Denote by $f(\xi)$ the left-hand side of (3.1). Then $f(\xi) = f_1(\xi) + f_2(\xi)$, where

$$f_1(\xi) = \sum_{|\gamma|=2m} \Re((-1)^m \xi^\gamma A_t^\gamma \theta, \bar\theta), \qquad f_2(\xi) = \sum_{|\gamma|<2m} \Re(i^{|\gamma|}\xi^\gamma A_t^\gamma \theta, \bar\theta).$$

Since for any $\varepsilon > 0$ we have

$$a^k \leq \varepsilon a^r + N(k,r)\varepsilon^{-k/(r-k)}, \tag{3.2}$$

provided that $0 \leq k < r$ and $a \geq 0$, we conclude that

$$|f_2(\xi)| \leq \left(\varepsilon|\xi|^{2m} + N(\varepsilon)\right)|\theta|^2.$$

Furthermore, $|\xi|^{2m} \leq N|\Re\xi|^{2m} + N|\Im\xi|^{2m}$. Hence,

$$|f_2(\xi)| \leq \varepsilon|\Re\xi|^{2m}|\theta|^2 + N(\varepsilon)(|\Im\xi|^{2m} + 1)|\theta|^2,$$

where and below we denote by N various constants depending only on δ, n, m, and d, which may change from one appearance to another. If we write $N(\varepsilon)$, then this constant is allowed to depend on δ, n, d, m, and ε. Of course, $f_2 \equiv 0$ if $A^\gamma \equiv 0$ for $|\gamma| \leq 2m-1$.

Next, by evaluating $(\Re\xi + i\Im\xi)^\gamma$ we easily see that

$$f_1(\xi) = f_1(\Re\xi) + g(\xi),$$

where

$$|g(\xi)| \leq |\theta|^2 \sum_{k=0}^{2m-1} a_k |\Re\xi|^k |\Im\xi|^{2m-k}$$

and the constants a_k depend only on δ, n, and d. We now use (3.2) to get that

$$|g(\xi)| \leq \varepsilon|\Re\xi|^{2m}|\theta|^2 + N(\varepsilon)|\Im\xi|^{2m}|\theta|^2.$$

After that to finish proving the lemma it only remains to combine the above estimates with the inequality

$$f_1(\Re\xi) \le -\delta|\Re\xi|^{2m}|\theta|^2,$$

which is given by Assumption 2.1, and to choose the above ε's appropriately. The lemma is proved.

Next, for any matrix $C = (c_{ij})$, by C^* we denote its (usual) transpose, that is the matrix (c^*_{ij}) with $c^*_{ij} = c_{ji}$, and for two matrices $C_1 = (C_1^{jk})$ and $C_2 = (C_2^{jk})$ such that $C_1 C_2^*$ makes sense and is a square matrix, denote

$$\langle C_1, C_2\rangle = \operatorname{trace} C_1 \bar{C}_2^* = \sum_{j,k} C_1^{jk}\bar{C}_2^{jk},$$

so that

$$|C_1|^2 = \langle C_1, C_1\rangle = \sum_{j,k} |C_1^{jk}|^2.$$

Notice that if C_j^k denotes the k-th column of C_j, then

$$\langle C_1, C_2\rangle = \sum_k (C_1^k, \bar{C}_2^k).$$

If Φ is an $n\times n$-matrix and Φ^k denotes its k-th column, then the k-th column of $A^\gamma\Phi$ is $A^\gamma\Phi^k$, and hence

$$\langle A_t^\gamma\Phi, \Phi\rangle = \sum_{k=1}^n (A_t^\gamma\Phi^k, \bar{\Phi}^k).$$

This and Lemma 3.1 imply the following.

COROLLARY 3.2. *For any complex $n\times n$-matrix Φ and $\xi\in\mathbb{C}^d$,*

$$\Re\langle i^{|\gamma|}\xi^\gamma A_t^\gamma\Phi, \Phi\rangle \le -(\delta/2)|\Re\xi|^{2m}|\Phi|^2 + N(|\Im\xi|^{2m} + N^*)|\Phi|^2,$$

where N and N^* are the constants from Lemma 3.1.

For each $\xi\in\mathbb{C}^d$, define an $n\times n$ matrix-valued function $\Phi_{s,t}(\xi)$, $s,t\in\mathbb{R}$, as a unique solution (in the sense of Definition 2.3) of the linear system

$$\partial_t\Phi_{s,t}(\xi) = i^{|\gamma|}\xi^\gamma A_t^\gamma\Phi_{s,t}(\xi) \tag{3.3}$$

on $\mathbb{R}$ subject to $\Phi_{s,s}(\xi) = I_{n\times n}$. It follows from the theory of ODEs that $\Phi_{s,t}(\xi)$ is defined by the Peano–Baker series (which can be shown to converge for arbitrary t and s)

$$\Phi_{s,t}(\xi) = I_{n\times n} + i^{|\gamma|}\xi^\gamma\int_s^t A_r^\gamma\,dr + i^{|\gamma|}\xi^\gamma i^{|\sigma|}\xi^\sigma\int_s^t A_r^\gamma\int_s^r A_{r_1}^\sigma\,dr_1dr + \cdots,$$

and therefore it is an analytic function of ξ, which is also continuous with respect to (s,t) (see, for instance, [**2**]). As such, by Cauchy's theorem for any multi-index ν

$$|D_\xi^\nu\Phi_{s,t}(\xi)| \le N(n,|\nu|)\max_{\eta\in\mathbb{C}^d,|\eta|=1}|\Phi_{s,t}(\xi+\eta)|, \tag{3.4}$$

where D_ξ^ν is the sign of differentiation with respect to ξ.

Observe that $\langle C_1, C_2\rangle = \overline{\langle C_2, C_1\rangle}$ so that in this notation (3.3) implies that

$$\begin{aligned}\partial_t|\Phi_{s,t}(\xi)|^2 &= \langle \partial_t\Phi_{s,t}(\xi), \Phi_{s,t}(\xi)\rangle + \langle \Phi_{s,t}(\xi), \partial_t\Phi_{s,t}(\xi)\rangle \\ &= 2\Re\langle \partial_t\Phi_{s,t}(\xi), \Phi_{s,t}(\xi)\rangle = 2\Re(i^{|\gamma|}\xi^\gamma\langle A_t^\gamma\Phi_{s,t}(\xi), \Phi_{s,t}(\xi)\rangle).\end{aligned}$$

By Corollary 3.2 the last expression is dominated by

$$-\delta|\Re\xi|^{2m}|\Phi_{s,t}(\xi)|^2 + N(|\Im\xi|^{2m} + N^*)|\Phi_{s,t}(\xi)|^2.$$

By using this and the Gronwall inequality we come to a slightly stronger statement than the first part of the following lemma.

LEMMA 3.3. *For $t > s$ and $\xi \in \mathbb{C}^d$, we have*

$$|\Phi_{s,t}(\xi)| \le ne^{N^*(t-s)}\eta_{s,t}(\xi). \tag{3.5}$$

Furthermore, for any multi-index ν

$$|D_\xi^\nu\Phi_{s,t}(\xi)| \le N(t-s)^{|\nu|/(2m)}e^{N^*(t-s)}\eta_{s,t}(\xi), \tag{3.6}$$

where

$$\eta_{s,t}(\xi) = \exp\big([-(\delta/4)|\Re\xi|^{2m} + N|\Im\xi|^{2m}](t-s)\big).$$

In both estimates the constant N^* depends only on δ, n, m, and d, and the constant N depends only on the same parameters and $|\nu|$. Moreover, if $A_t^\gamma \equiv 0$ for $|\gamma| \le 2m-1$, then $N^* = 0$.

PROOF. We need only prove (3.6). We first prove it with a constant N^* depending only on δ, n, m, and d, and after that we show that one can take $N^* = 0$ if $A_t^\gamma \equiv 0$ for $|\gamma| \le 2m-1$. If $t - s \ge 1$, (3.6) is obtained from what is said before the lemma, (3.4), and the fact that for $|\eta| = 1$,

$$|\Re(\xi+\eta)| + 1 \ge |\Re(\xi+\eta) - \Re\eta| = |\Re\xi|, \qquad |\Im(\xi+\eta)| \le |\Im\xi| + 1,$$

which along with the inequality $(a+b)^r \le 2a^r + N(r)b^r$, $a, b \ge 0$, implies that

$$|\Re\xi|^{2m} \le 2|\Re(\xi+\eta)|^{2m} + N(m), \qquad |\Re(\xi+\eta)|^{2m} \ge (1/2)|\Re\xi|^{2m} - N(m),$$
$$|\Im(\xi+\eta)|^{2m} \le 2|\Im\xi|^{2m} + N(m).$$

To pass to the case $t - s \le 1$, fix $s_0 \in \mathbb{R}$ and $t_0 \in (s_0, s_0+1]$ and define

$$\check{A}_t^\gamma = (t_0 - s_0)^{1-|\gamma|/(2m)}A^\gamma_{s_0+t(t_0-s_0)},$$
$$\check{\Phi}_{0,t}(\xi) = \Phi_{s_0,s_0+t(t_0-s_0)}(\xi/(t_0-s_0)^{1/(2m)}).$$

Obviously, $\check{A}_t^\gamma$ satisfy Assumption 2.1 with the same constant δ. Furthermore, it is easy to check that $\check{\Phi}_{0,t}(\xi)$ satisfies

$$\partial_t\check{\Phi}_{0,t}(\xi) = i^{|\gamma|}\xi^\gamma\check{A}_t^\gamma\check{\Phi}_{0,t}(\xi),$$

and $\check{\Phi}_{0,0}(\xi) = I_{n\times n}$. By the above,

$$|D_\xi^\nu\check{\Phi}_{0,1}(\xi)| \le N\exp\big(-(\delta/4)|\Re\xi|^{2m} + N|\Im\xi|^{2m}\big),$$
$$|D_\xi^\nu[\Phi_{s_0,t_0}(\xi/(t_0-s_0)^{1/(2m)})]| \le N\exp\big(-(\delta/4)|\Re\xi|^{2m} + N|\Im\xi|^{2m}\big),$$
$$\begin{aligned}&|(D_\xi^\nu\Phi_{s_0,t_0})(\xi/(t_0-s_0)^{1/(2m)})| \\ &\qquad \le N(t_0-s_0)^{|\nu|/(2m)}\exp\big(-(\delta/4)|\Re\xi|^{2m} + N|\Im\xi|^{2m}\big).\end{aligned}$$

The arbitrariness of ξ allows us to conclude that (3.6) holds with $s = s_0$ and $t = t_0$, and this proves the lemma in the general case. In case $A_t^\gamma \equiv 0$ for $|\gamma| \le 2m-1$, the above self-similarity transformation reducing $t - s$ to one works equally well if

$t - s > 1$, which shows that indeed $N^* = 0$ in (3.6) in this case. The lemma is proved.

The next result is obtained by shifting the integral into a complex area. The idea of doing so in a similar situation of PDEs was applied in [**7**], where the shifting was justified by using the corresponding Cauchy theorem. In probability theory one can find an application of the same idea with the same justification earlier in the proof of Lemma 3 on page 12 of [**1**].

LEMMA 3.4. *Let P be a homogeneous polynomial of degree d_1 given on $\mathbb{C}^d$. Then for $t > s$ the function*

$$H_{s,t}[P](x) := \int_{\mathbb{R}^d} e^{i(x,\xi)} P(\xi)\Phi_{s,t}(\xi)\, d\xi$$

is well defined, infinitely differentiable with respect to x, and continuous with respect to (s,t), $s < t$. Furthermore, for any multi-index γ,

$$|D^\gamma H_{s,t}[P](x)| \leq N \frac{P_0 e^{N^*(t-s)}}{(t-s)^\mu} \exp[-\varepsilon\kappa(|x|^{2m}/(t-s))], \tag{3.7}$$

where $\mu = (|\gamma| + d + d_1)/(2m)$ and

$$\kappa(r) = r^{1/(2m-1)}.$$

Finally, if $A_t^\gamma \equiv 0$ for $|\gamma| \leq 2m-1$, then estimate (3.7) holds with $N^* = 0$. In both cases

$$P_0 := \max_{|\xi|=1} |P(\xi)|,$$

the constant N^* depends only on δ, n, m, and d, and the constants N and $\varepsilon > 0$ depend only on the same parameters and $d_1 + |\gamma|$.

PROOF. The first assertion follows from Lemma 3.3, the fact that $D^\gamma H_{s,t}[P] = H_{s,t}[Q]$, where $Q(\xi) = i^{|\gamma|}\xi^\gamma P(\xi)$, and the dominated convergence theorem allowing us to derive the continuity of integrals from the continuity of integrands given that they admit a uniform estimate. Here the issue of the possibility of differentiating under the integral sign can arise. We address such issues in a more complicated situation in the proof of Theorem 3.6 when we also differentiate with respect to t.

The relation $D^\gamma H_{s,t}[P] = H_{s,t}[Q]$ also shows that while proving the second assertion we may assume that $\gamma = 0$.

In that case without losing generality, we may assume that $s = 0$ and $t > 0$. Notice that Lemma 3.3 allows us to do the following manipulations:

$$\begin{aligned}\frac{\partial}{\partial \eta_k} &\int_{\mathbb{R}^d} e^{i(x,\xi+i\eta)} P(\xi+i\eta)\Phi_{0,t}(\xi+i\eta)\, d\xi \\ &= -x_k \int_{\mathbb{R}^d} e^{i(x,\xi+i\eta)} P(\xi+i\eta)\Phi_{0,t}(\xi+i\eta)\, d\xi \\ &\quad + \int_{\mathbb{R}^d} e^{i(x,\xi+i\eta)} i \frac{\partial}{\partial \xi_k}(P\Phi_{0,t})(\xi+i\eta)\, d\xi = 0,\end{aligned}$$

where the last equality is obtained by integrating by parts.

Hence, for any $\eta \in \mathbb{R}^d$, it follows from (3.5) that

$$|H_{0,t}[P](x)| = \Big| \int_{\mathbb{R}^d} e^{i(x,\xi+i\eta)} P(\xi + i\eta)\Phi_{0,t}(\xi + i\eta)\, d\xi \Big|$$
$$\leq Ne^{N^* t} P_0 e^{-(x,\eta)} \int_{\mathbb{R}^d} (|\xi|^{d_1} + |\eta|^{d_1}) e^{-(\delta/4)|\xi|^{2m}t + N|\eta|^{2m}t}\, d\xi.$$

We now use the fact that for any $r > 0$, there is a constant N depending only on d_1, m, r such that $a^{d_1} + b^{d_1} \leq N \exp(ra^{2m} + rb^{2m})$. By taking an appropriate r and $a = |\xi|t^{1/(2m)}$ and $b = |\eta|t^{1/(2m)}$, we see that

$$|H_{0,t}[P](x)| \leq Ne^{N^* t} t^{-d_1/(2m)} P_0 e^{-(x,\eta)+N_1|\eta|^{2m}t} \int_{\mathbb{R}^d} e^{-(\delta/8)|\xi|^{2m}t}\, d\xi$$
$$= Ne^{N^* t} t^{-(d+d_1)/(2m)} P_0 e^{-(x,\eta)+N_1|\eta|^{2m}t},$$

where the last equality is obtained by an obvious change of variables.

Now it only remains to take

$$\eta = 2\varepsilon |x/t|^{(2-2m)/(2m-1)} x/t \quad (0/0 := 0)$$

and choose $\varepsilon > 0$ such that,

$$-2\varepsilon + N_1 2^{2m} \varepsilon^{2m} = -\varepsilon.$$

The lemma is proved.

COROLLARY 3.5. *For any $t > s$ and multi-index γ*

$$\int_{\mathbb{R}^d} |D^\gamma H_{s,t}[P](x)|\, dx \leq Ne^{N^*(t-s)} P_0 (t-s)^{-(|\gamma|+d_1)/(2m)}.$$

PROOF. By Lemma 3.4

$$\int_{\mathbb{R}^d} |D^\gamma H_{s,t}[P](x)| dx$$
$$\leq Ne^{N^*(t-s)} P_0 (t-s)^{-(d+d_1+|\gamma|)/(2m)} \int_{\mathbb{R}^d} \exp[-\varepsilon\kappa(|x|^{2m}/(t-s))]\, dx$$

for any $t > s$. After making the change of variables

$$y = x(t-s)^{-1/(2m)},$$

we rewrite the right-hand side of the last inequality as

$$Ne^{N^*(t-s)} P_0 (t-s)^{-(d_1+|\gamma|)/(2m)} \int_{\mathbb{R}^d} \exp\left(-\varepsilon |y|^{2m/(2m-1)}\right) dy,$$

and the result follows.

Now we define the fundamental matrix $\Psi_{s,t}$ for system (1.1) with $\lambda = 0$ as the inverse Fourier transform of $(2\pi)^{-d/2}\Phi_{s,t}$:

$$\Psi_{s,t}(x) = I_{s<t}(2\pi)^{-d} \int_{\mathbb{R}^d} e^{i(x,\xi)} \Phi_{s,t}(\xi)\, d\xi. \tag{3.8}$$

Notice that, for any $t > s$, $\Psi_{s,t}(x) = (2\pi)^{-d/2} H_{s,t}[P](x)$, where $P \equiv 1$. By Lemma 3.4 for any $t > s$ and multi-index γ,

$$|D^\gamma \Psi_{s,t}(x)| \leq N \frac{e^{N^*(t-s)}}{(t-s)^{(|\gamma|+d)/(2m)}} \exp[-\varepsilon\kappa(|x|^{2m}/(t-s))], \tag{3.9}$$

and by Corollary 3.5,

$$\int_{\mathbb{R}^d} |D^\gamma \Psi_{s,t}(x)|\,dx \le N e^{N^*(t-s)}(t-s)^{-|\gamma|/(2m)}. \tag{3.10}$$

The argument used in Corollary 3.5 also implies the following fact. Namely, if $f \in C^\alpha$, $\alpha \in (0,1)$, and $\gamma \neq 0$, then

$$|\int_{\mathbb{R}^d} D^\gamma \Psi_{s,t}(y) f(x-y)\,dy| \le N e^{N^*(t-s)}(t-s)^{(\alpha-|\gamma|)/(2m)}[f]_\alpha. \tag{3.11}$$

Indeed,

$$\int_{\mathbb{R}^d} D^\gamma \Psi_{s,t}(y) f(x)\,dy = 0,$$

and the left-hand side of (3.11) is dominated by

$$[f]_\alpha \int_{\mathbb{R}^d} |D^\gamma \Psi_{s,t}(y)|\,|y|^\alpha\,dy.$$

Furthermore, since $(2\pi)^{-d/2}\Phi_{s,t}$ is the Fourier transform of $\Psi_{s,t}$,

$$(2\pi)^{-d/2}\Phi_{s,t}(\xi) = (2\pi)^{-d/2}\int_{\mathbb{R}^d} e^{-i(x,\xi)}\Psi_{s,t}(x)\,dx,$$

which for $\xi = 0$ yields

$$\int_{\mathbb{R}^d} \Psi_{s,t}(x)\,dx = 1. \tag{3.12}$$

Next, it is seen that Theorem 2.1 will be a direct consequence of Lemma 3.4 if we show that $\Psi_{s,t}(x)$ is indeed a fundamental solution, which means that we have to prove the following.

THEOREM 3.6. *Let ϕ be a $\mathbb{C}^n$-valued bounded function on $\mathbb{R}^d$. For fixed $s \in \mathbb{R}$ and $t > s$ define*

$$u(t,x) = \Psi_{s,t} * \phi(x) = \int_{\mathbb{R}^d} \Psi_{s,t}(x-y)\phi(y)\,dy.$$

Then u is infinitely differentiable with respect to x for $t > s$, and for any multi-index γ

$$D^\gamma u(t,x) = \int_{\mathbb{R}^d} \phi(y) D^\gamma \Psi_{s,t}(x-y)\,dy. \tag{3.13}$$

Moreover, the functions $\Psi_{s,t}(x)$ and $u(t,x)$ as functions of (t,x) satisfy the equation

$$\partial_t v(t,x) = Lv(t,x) \tag{3.14}$$

on $(s,\infty)\times\mathbb{R}^d$ (in the sense of Definition 2.4). Furthermore, if ϕ is continuous, then $u(t,x) \to \phi(x)$ at each point $x \in \mathbb{R}^d$ as $t \downarrow s$. Finally, if $\phi \in C^\alpha$ for an $\alpha \in (0,1)$, then, by extending the definition of $u(t,x)$ initially given only for $t > s$ as $u(s,x) = \phi(x)$, we obtain a function u which satisfies (3.14) *on $[s,\infty)\times\mathbb{R}^d$.*

PROOF. Unfortunately, we were unable to find a reference to the following fact, albeit a very well-known one. Suppose that we are given a measurable function $f(t,y)$ defined in a domain $G_1 \times G_2$, where G_1 is an interval in $\mathbb{R}$ and $G_2 \subset \mathbb{R}^d$. Then the integral

$$h(t) := \int_{G_2} f(t,y)\,dy$$

is absolutely continuous with respect to t in any subinterval of G_1 if

(i) the function $f(t, y)$ is absolutely continuous with respect to t for any $y \in G_2$ and for any $t \in G_1$ we have
$$\int_{G_2} |f(t, y)| \, dy < \infty;$$

(ii) we have
$$\int_{G_1} \int_{G_2} |D_t f(t, y)| \, dydt < \infty,$$
where, naturally, $D_t f(t, y)$ is the derivative of $f(t, y)$ with respect to t which exists for any y for almost any t since $f(t, y)$ is absolutely continuous.

In addition under the above conditions for almost any $t \in G_1$,

$$D_t \int_{G_2} f(t, y) \, dy = \int_{G_2} D_t f(t, y) \, dy. \tag{3.15}$$

These assertions easily follow from Fubini's theorem, which is applicable owing to (ii) and for any subinterval (a, t) of G_1 it yields

$$\begin{aligned} \int_a^t \Big(\int_{G_2} D_s f(s, y) \, dy \Big) \, ds &= \int_{G_2} \Big(\int_a^t D_s f(s, y) \, ds \Big) \, dy \\ &= \int_{G_2} [f(t, y) - f(a, y)] \, dy = h(t) - h(a). \end{aligned} \tag{3.16}$$

Here the first integral is absolutely continuous with respect to t as any integral of an integrable function and its derivative, which exists for almost all t, satisfies (3.15). Of course, we also used the fact that the increments of absolutely continuous functions equal integrals of their derivatives. These and many other properties of absolutely continuous functions can be found in [**10**]. Furthermore, if the right-hand side of (3.15) happens to be continuous on G_1, then (3.16) shows that $h(t) - h(a)$ equals the integral of a continuous function implying that $h(t)$ is continuously differentiable in t and (3.15) holds for all $t \in G_1$.

By applying this fact to $\Psi_{s,t}(x - y)\phi(y)$ as a function of $(x_1, y) \in G_1 \times \mathbb{R}^d$, where G_1 is any interval in $\mathbb{R}$ of unit length, with all other arguments being fixed and using Lemma 3.4, we see that $u(t, x)$ is absolutely continuous with respect to x_1 and (a.e.)
$$D_1 u(t, x) = \int_{\mathbb{R}^d} \phi(y)(D_1 \Psi_{s,t})(x - y) \, dy,$$
where the integral is a continuous function of x at each point of $\mathbb{R}^d$ in light of the dominated convergence theorem. Thus $D_1 u(t, x)$ is continuous in x. Proceeding further in an obvious way, we get that $u(t, x)$ is infinitely differentiable in x and equation (3.13) indeed holds for $t > s$ and for any multi-index γ.

We can also apply the above argument to formula (3.8) and see that, in light of equation (3.3) and estimate (3.6), the function $\Psi_{s,t}(x)$ as a function of (t, x) satisfies (3.14) on $(s, \infty) \times \mathbb{R}^d$ (in the sense of Definition 2.4). Furthermore,
$$|\partial_t \Psi_{s,t}(x)| \leq N \frac{e^{N^*(t-s)}}{(t - s)^{(2m+d)/(2m)}} \exp[-\varepsilon\kappa(|x|^{2m}/(t - s))],$$

owing to (3.3) and (3.9). The latter allows us to write that (in the sense of Definition 2.4) on $(s,\infty)\times\mathbb{R}^d$

$$\partial_t u(t,x)=\int_{\mathbb{R}^d}\phi(y)\partial_t\Psi_{s,t}(x-y)\,dy=\int_{\mathbb{R}^d}\phi(y)L_t\Psi_{s,t}(x-y)\,dy$$
$$=L_t\int_{\mathbb{R}^d}\phi(y)\Psi_{s,t}(x-y)\,dy=Lu(t,x),$$

where and below we use the notation L_t to show that the coefficients of L are taken at time t (and not s). This proves our assertion concerning (3.14).

To prove that $u(s+,x)=\phi(x)$ if ϕ is continuous, we use (3.9) for $\gamma=0$ and (3.12) to obtain that

$$|u(t,x)-\phi(x)|=\Big|\int_{\mathbb{R}^d}\Psi_{s,t}(y)[\phi(x-y)-\phi(x)]\,dy\Big|$$
$$\le N\frac{e^{N^*(t-s)}}{(t-s)^{d/(2m)}}\int_{\mathbb{R}^d}|\phi(x-y)-\phi(x)|\exp[-\varepsilon\kappa(|y|^{2m}/(t-s))]\,dy$$
$$=Ne^{N^*(t-s)}\int_{\mathbb{R}^d}|\phi(x-(t-s)^{1/(2m)}y)-\phi(x)|\exp(-\varepsilon|y|^{2m/(2m-1)})\,dy,$$

and our assertion follows by the dominated convergence theorem.

The fact that (3.14) holds for $t>s$ and the fact that $u(s+,x)=\phi(x)$ implies that for our extended function u and $t>s$, we have

$$u(t,x)-u(s,x)=\lim_{\varepsilon\downarrow 0}\int_{s+\varepsilon}^t Lu(r,x)\,dr.$$

The last limit is actually equal to

$$\int_s^t Lu(r,x)\,dr,$$

which converges absolutely owing to (3.11). The theorem is proved.

We have a similar result for adjoint systems.

THEOREM 3.7. *Let ϕ be a $\mathbb{C}^n$-valued bounded function on $\mathbb{R}^d$. For fixed $t\in\mathbb{R}$ and $s<t$ define*

$$u(s,x)=\Psi^*_{s,t}*\phi(x)=\int_{\mathbb{R}^d}\Psi^*_{s,t}(x-y)\phi(y)\,dy.$$

Then u is infinitely differentiable with respect to x for $t>s$ and satisfies

$$\partial_s u(s,x)=-(A^\gamma_s)^*D^\gamma u(s,x)\tag{3.17}$$

on $(-\infty,t)\times\mathbb{R}^d$ (always in the sense of Definition 2.4). Furthermore, if ϕ is continuous, then $u(s,x)\to\phi(x)$ at each point $x\in\mathbb{R}^d$ as $s\uparrow t$. Finally, if $\phi\in C^\alpha$ for an $\alpha\in(0,1)$, then, by extending the definition of $u(s,x)$ initially given only for $t>s$ as $u(t,x)=\phi(x)$, we obtain a function u which satisfies (3.17) on $(-\infty,t]\times\mathbb{R}^d$.

PROOF. From the theory of ODEs (see, for instance, Chapter 18 in [**2**]) it is known that

$$\partial_s\Phi^*_{s,t}(\xi)=-i^{|\gamma|}\xi^\gamma(A^\gamma_s)^*\Phi^*_{s,t}(\xi).$$

This equation allows us to repeat the proof of Theorem 3.6 almost word for word. The theorem is proved.

4. Resolvent operator and proof of Theorem 2.2

Introduce

$$(4.1)\qquad \begin{aligned}\mathcal{G}_\lambda f(t,x) &= \int_{-\infty}^{t} \left(\Psi_{s,t}(\cdot) * f(s,\cdot)\right)(x) e^{-\lambda(t-s)}\,ds\\ &= \int_{-\infty}^{t}\int_{\mathbb{R}^d} \Psi_{s,t}(x-y) f(s,y) e^{-\lambda(t-s)}\,dyds,\end{aligned}$$

whenever the right-most expression makes sense.

Observe that due to (3.10) we have

$$(4.2)\qquad \Big|\int_{\mathbb{R}^d} \Psi_{s,t}(x-y) f(s,y)\,dy\Big| I_{s<t} \le N e^{N^*(t-s)} |f(s,\cdot)|_0 I_{s<t},$$

and the right-hand side is locally integrable with respect to s if such is $|f(s,\cdot)|_0$. Add to this that the left-hand side of (4.2) is a continuous function of x for each $s<t$ such that $|f(s,\cdot)|_0<\infty$ (see Theorem 3.6), and combine this with the dominated convergence theorem. Then we see that if $|f(s,\cdot)|_0$ is summable with respect to $s\le t$ and vanishes for large negative s, then $\mathcal{G}_\lambda f(t,x)$ is well defined and is a bounded and continuous function of x. The same conclusion is also obviously true if $\lambda\ge 0$, $|f(s,\cdot)|_0$ is summable with respect to $s\le t$, and $L=L^0$ (so that $N^*=0$).

In any case $\Psi_{s,t}(x-y)f(s,y)e^{-\lambda(t-s)}I_{s<t}$ is a measurable function of its arguments (see Lemma 3.4), so that by a common addendum to Fubini's theorem, if in a domain of values of (t,x) we have

$$\int_{-\infty}^{t}\int_{\mathbb{R}^d} |\Psi_{s,t}(x-y) f(s,y)| e^{-\lambda(t-s)}\,dyds < \infty,$$

then $\mathcal{G}_\lambda f(t,x)$ is well defined in that domain and is a measurable function.

THEOREM 4.1. *Let $T\in\mathbb{R}$ and $u(t,x)$ be a $\mathbb{C}^n$-valued measurable function on*

$$\mathbb{R}^{d+1}_T := (-\infty,T]\times\mathbb{R}^d,$$

which vanishes for large values of $|t|$, $t<T$. Assume that, for almost any $t\le T$, $u(t,x)$ is $2m$ times continuously differentiable in x and $|u(t,\cdot)|_{2m}$ is summable over $(-\infty,T]$. Also assume that there exists a measurable function $g(t,x)$ on $\mathbb{R}^{d+1}_T$ such that $|g(t,\cdot)|_0$ is summable over $(-\infty,T]$ and, for any x, $\partial_t u(t,x)=g(t,x)$ for $t\le T$ (see Definition 2.3). Then, for any λ we have $u=\mathcal{G}_\lambda(\lambda u - Lu + \partial_t u)$ on $\mathbb{R}^{d+1}_T$.

PROOF. First assume that $u(t,x)$ vanishes for large values of $|x|+|t|$, $(t,x)\in\mathbb{R}^{d+1}_T$. Also for a moment assume $\lambda=0$ and set $f=\partial_t u - Lu$, that is

$$(4.3)\qquad u(t,x)-u(s,x) = \int_s^t [Lu(r,x)+f(r,x)]\,dr$$

for all $t,s\le T, x\in\mathbb{R}^d$. Since $u(t,x)$ vanishes for large values of $|x|$, we have

$$\int_{\mathbb{R}^d}\int_s^t (|Lu(r,x)|+|f(r,x)|)\,dr \le N\int_s^t (|u(r,\cdot)|_{2m}+|g(r,\cdot)|_0)\,dr<\infty.$$

This allows us to use Fubini's theorem while taking the spatial Fourier transform of both parts of (4.3) and yields

$$\partial_t\tilde{u}(t,\xi) = i^{|\gamma|}\xi^\gamma A^\gamma_t \tilde{u}(t,\xi) + \tilde{f}(t,\xi),$$

which is a system of ODEs (for each fixed ξ). From the theory of ODEs and the fact that $\tilde{u}(t,\xi)=0$ for large negative t, it follows that

$$\tilde{u}(t,\xi)=\int_{-\infty}^{t}\Phi_{s,t}(\xi)\tilde{f}(s,\xi)\,ds.$$

Now we take the spatial Fourier transform of $v:=\mathcal{G}_0 f$. Estimate (3.10) and the fact that f vanishes for large values of $|x|+|t|$, $(t,x)\in\mathbb{R}^{d+1}_T$ (which again allows us to restrict the integral with respect to $dyds$ to a bounded region) show that

$$\int_{\mathbb{R}^d}\int_{-\infty}^{t}\int_{\mathbb{R}^d}|\Psi_{s,t}(x-y)f(s,y)|\,dydsdx$$
$$\leq N\int_{|y|\leq N}\int_{-N}^{t}|f(s,y)|e^{N^*(t-s)}\,dsdy<\infty,$$

where N is an appropriate constant. Therefore, Fubini's theorem is applicable in the following computation, and we have

$$\tilde{v}(t,\xi)=(2\pi)^{-d/2}\int_{-\infty}^{t}\int_{\mathbb{R}^d}e^{-i(\xi,y)}\Psi_{s,t}(y)\int_{\mathbb{R}^d}e^{-i(\xi,x-y)}f(s,x-y)\,dxdyds$$
$$=\int_{-\infty}^{t}\Phi_{s,t}(\xi)\tilde{f}(s,\xi)\,ds.$$

Hence, $\tilde{u}(t,\xi)=\tilde{v}(t,\xi)$, and this along with the continuity of u and $\mathcal{G}_0 f$ in x proves our assertion when $\lambda=0$.

For arbitrary λ, by the definition of $\mathcal{G}_\lambda$ and by the previous step, it follows that

$$\mathcal{G}_\lambda(\lambda u-Lu+\partial_t u)(t,x)$$
$$=e^{-\lambda t}\int_{-\infty}^{t}\int_{\mathbb{R}^d}\Psi_{s,t}(x-y)\left(-L(e^{\lambda\cdot}u)+\partial_t(e^{\lambda\cdot}u)\right)(s,y)\,dyds$$
$$=e^{-\lambda t}e^{\lambda t}u(t,x)=u(t,x).$$

To pass to the case where $u(t,x)$ vanishes only for all large negative t, we take a $\zeta\in C_0^\infty(\mathbb{R}^d)$ which integrates to one, for a parameter $z\in\mathbb{R}^d$ denote $\zeta_z(x)=\zeta(x+z)$, and by the above write

$$u(t,x)\zeta_z(x)=\mathcal{G}_\lambda(\lambda u\zeta_z-L(u\zeta_z)+\zeta_z\partial_t u).$$

We integrate through this equality with respect to z over $\mathbb{R}^d$ and bring the integral inside $\mathcal{G}_\lambda$ which is possible by the same reasons as above involving Fubini's theorem. Then we easily get what we wanted. The theorem is proved.

PROOF OF THEOREM 2.2. Take a nonnegative $\zeta\in C_0^\infty(\mathbb{R}^d)$ with support in the unit ball centered at the origin and with unit integral, and for $\varepsilon>0$ set $\zeta^\varepsilon(x)=\varepsilon^{-d}\zeta(x/\varepsilon)$. Next, for any function $f(t,x)$ for which it makes sense define

$$f^{(\varepsilon)}(t,x)=\int_{\mathbb{R}^d}f(t,y)\zeta^\varepsilon(x-y)\,dy.$$

If we write (2.4) with $\zeta^\varepsilon(y-x)$ in place of $\zeta(x)$ with a fixed point $y\in\mathbb{R}^d$, then we see that

$$u^{(\varepsilon)}(t,y)-u^{(\varepsilon)}(s,y)=\int_s^t\sum_{|\gamma|\leq 2m}A_r^\gamma D^\gamma u^{(\varepsilon)}(r,y)\,dr.$$

Also $u^{(\varepsilon)}(0,x)\equiv 0$.

Moreover, $|x-y|^{2m/(2m-1)} \leq N(|x|^{2m/(2m-1)}+|y|^{2m/(2m-1)})$, which implies that, for $\varepsilon \leq 1$,

$$\begin{aligned}|u^{(\varepsilon)}(t,x)| &\leq N\int_{\mathbb{R}^d}\zeta^{\varepsilon}(y)\exp(N|x-y|^{2m/(2m-1)})\,dy\\ &\leq N\exp(N|x|^{2m/(2m-1)})\int_{\mathbb{R}^d}\zeta^{\varepsilon}(y)\,dy = N\exp(N|x|^{2m/(2m-1)}),\end{aligned}$$

where the constants N are independent of ε, t, and x. Therefore, $u^{(\varepsilon)}$ satisfies the conditions of Theorem 2.2, and if the theorem holds for $u^{(\varepsilon)}$, then $u^{(\varepsilon)} \equiv 0$. Since $u(t,x)$ is continuous with respect to x, by letting $\varepsilon \downarrow 0$, we then conclude that $u \equiv 0$. By adding to the above that $u^{(\varepsilon)}(t,x)$ is infinitely differentiable in x and for any multi-index γ and $\varepsilon \leq 1$

$$\begin{aligned}|D^{\gamma}u^{(\varepsilon)}(t,x)| &= \varepsilon^{-|\gamma|}\Big|\int_{\mathbb{R}^d}u(t,y)\varepsilon^{-d}(D^{\gamma}\zeta)((x-y)/\varepsilon)\,dy\Big|\\ &\leq N\varepsilon^{-|\gamma|}\exp(N|x|^{2m/(2m-1)}),\end{aligned}$$

we conclude that while proving the theorem we may additionally assume that $u(t,x)$ is infinitely differentiable with respect to x for each $t \in [0,T]$, and for any multi-index γ there is a constant N such that

$$\begin{aligned}&|D^{\gamma}u(t,x)| \leq N\exp(N|x|^{2m/(2m-1)}),\\ &|\partial_t u(t,x)| \leq N\exp(N|x|^{2m/(2m-1)}),\end{aligned} \tag{4.4}$$

where the last estimate comes from (2.3) and the first one.

Under these additional assumptions, extend $u(t,x)$ as zero for $t<0$ and apply Theorem 4.1 to the function $\eta(x/r)u(t,x)$, where $\eta \in C_0^{\infty}(\mathbb{R}^d)$ satisfies $\eta(x)=1$ for $|x| \leq 1$ and r is a parameter to be sent to infinity. Then we obtain that

$$u(t,0)=\eta(0)u(t,0)=\int_0^t\int_{\mathbb{R}^d}\Psi_{s,t}(y)[\partial_t(\eta u)(s,-y)-L(\eta u)(s,-y)]\,dyds. \tag{4.5}$$

Also observe that

$$\partial_t(\eta u)(t,x)-L(\eta u)(t,x)=\sum_{\substack{|\alpha|+|\beta|\leq 2m,\\ |\beta|\geq 1}}c^{\alpha\beta}(t)(D^{\alpha}u(t,x))r^{-|\beta|}(D^{\beta}\eta)(x/r),$$

where $c^{\alpha\beta}(t)$ are certain bounded functions, which implies that (for $r \geq 1$)

$$|\partial_t(\eta u)(t,x)-L(\eta u)(t,x)| \leq N_1 I_{|x|\geq r}\exp(N_1|x|^{2m/(2m-1)}),$$

where N_1 is independent of r, t, and x. This along with (4.5) and Theorem 2.1 yield that

$$|u(t,0)| \leq \int_0^t\frac{Ne^{N(t-s)}}{(t-s)^{d/(2m)}}\int_{|y|\geq r}\exp(\phi(t-s)|y|^{2m/(2m-1)})\,dyds,$$

where

$$\phi(t-s)=N_1-\varepsilon(t-s)^{-1/(2m-1)}.$$

Next, we restrict our attention to $t \leq t_0$, where t_0 is defined by

$$N_1-\varepsilon t_0^{-1/(2m-1)}=-(1/2)\varepsilon t_0^{-1/(2m-1)}.$$

Then $\phi(t-s) \leq -\varepsilon_1(t-s)^{-1/(2m-1)}$, where $\varepsilon_1=(1/2)\varepsilon$, and

$$|u(t,0)| \leq \int_0^t\int_{|y|\geq r}\frac{Ne^{N(t-s)}}{(t-s)^{d/(2m)}}\exp[-\varepsilon\kappa(|y|^{2m}/(t-s))]\,dyds. \tag{4.6}$$

We have seen before that the integrand in (4.6) has a finite integral over $[0,t]\times\mathbb{R}^d$. Furthermore, $I_{|y|\geq r}\to 0$ as $r\to\infty$ for any y. Therefore, by letting $r\to\infty$ in (4.6) and using the dominated convergence theorem, we conclude that $u(t,0)=0$ for all $t\leq t_0$. The origin can certainly be replaced with any other point, so that $u(t,x)=0$ for $t\leq t_0$ and all x.

We can now replace $t=0$ with $t=t_0$ in the above arguments and conclude that $u(t,x)=0$ for $t_0\leq t\leq 2t_0$ and so on. The theorem is proved.

5. Proof of Theorems 2.13, 2.10, and 2.8

Recall that in Theorems 2.13, 2.10, and 2.8 we deal with

$$L=L^0,$$

that is with an operator without lower-order terms. Before we prove Theorem 2.13, we point out that Theorem 4.1 makes it natural to estimate $\mathcal{G}_\lambda(\lambda u-Lu+\partial_t u)$ instead of u.

LEMMA 5.1. (i) *Let $\lambda\geq 0$, $\alpha\in(0,1)$, and let $f(s,x)$ be a $\mathbb{C}^n$-valued measurable function on $\mathbb{R}^{d+1}$ which is continuous with respect to x. Fix $t\in\mathbb{R}$ and assume that*

$$I_0(f,t):=\int_{-\infty}^t e^{-\lambda(t-s)}|f(s,\cdot)|_0\,ds<\infty.$$

Then $\mathcal{G}_\lambda f(t,x)$ is well defined, is a continuous function of x, and $|\mathcal{G}_\lambda f(t,x)|\leq NI_0(f,t)$, where N depends only δ,n,m, and d.

(ii) *If additionally, for an integer $k\geq 1$ and all $j=1,\dots,k$*

$$I_j(f,t):=\int_{-\infty}^t e^{-\lambda(t-s)}(t-s)^{(\alpha-j)/(2m)}[f(s,\cdot)]_\alpha\,ds<\infty,$$

then $\mathcal{G}_\lambda f(t,\cdot)\in C^k$ and for any multi-index γ with $|\gamma|\leq k$ and all x we have

$$|D^\gamma\mathcal{G}_\lambda f(t,x)|\leq NI_{|\gamma|}(f,t), \tag{5.1}$$

where N depends only on δ,n,m,d,α, and k.

This lemma follows immediately from (3.10), (3.11), and from what has been said in the beginning of the proof of Theorem 3.6.

LEMMA 5.2. *Suppose that the assumption of Lemma 5.1(i) is satisfied and also suppose that either $f(s,x)$ is vanishing for large negative s or $\lambda>0$. Let $f(s,\cdot)\in C^\alpha$ for almost all $s\leq t$ and*

$$M_l([f]_\alpha)(t)<\infty.$$

Then the following facts hold:

(i) *If $f(s,\cdot)=0$ for $s\leq t-2$, then $\mathcal{G}_\lambda f(t,\cdot)\in C^{2m}$ and, for any $x\in\mathbb{R}^d$ and any multi-index γ with $0<|\gamma|\leq 2m$,*

$$|D^\gamma\mathcal{G}_\lambda f(t,x)|\leq NM_l\left([f]_\alpha\right)(t). \tag{5.2}$$

(ii) *If $f(s,\cdot)=0$ for $s\geq t-1$, then $\mathcal{G}_\lambda f(t,\cdot)\in C^k$ for any integer $k\geq 1$, and, for any $x\in\mathbb{R}^d$ and any multi-index γ_1 with $|\gamma_1|=2m+1$,*

$$|D^{\gamma_1}\mathcal{G}_\lambda f(t,x)|\leq NM_l\left([f]_\alpha\right)(t). \tag{5.3}$$

(iii) $\mathcal{G}_\lambda f(t,\cdot) \in C^{2m+\alpha}$, *and for any* $x, y \in \mathbb{R}^d$ *and any multi-index* γ *with* $|\gamma| = 2m$, *we have*

$$|D^\gamma \mathcal{G}_\lambda f(t,x) - D^\gamma \mathcal{G}_\lambda f(t,y)| \le N M_l\left([f]_\alpha\right)(t)|x-y|^\alpha. \tag{5.4}$$

In all the above estimates the constants N *depend only on* n, m, d, δ, α.

PROOF. First observe the fact that $\mathcal{G}_\lambda f(t,x)$ is well defined and is a bounded and continuous function of x by Lemma 5.1.

(i) It turns out that under the assumption in Lemma 5.2(i), we have $I_k(f,t) \le N M_l\left([f]_\alpha\right)(t)$ for $k \le 2m$. Indeed, for $\beta = (\alpha - k)/(2m)$, we have $\beta > -1$ and

$$\begin{aligned}
\int_{t-2}^t [f(s,\cdot)]_\alpha (t-s)^\beta\, ds &= -\int_{t-2}^t (t-s)^\beta d_s \left(\int_s^t [f(r,\cdot)]_\alpha\, dr\right) \\
&\le N \int_{t-2}^t (t-s)^{-1} \left(\int_s^t [f(r,\cdot)]_\alpha\, dr\right)(t-s)^\beta ds + N\int_{t-2}^t [f(r,\cdot)]_\alpha\, dr \\
&\le N M_l([f]_\alpha)(t) \int_{t-2}^t (t-s)^\beta\, ds = N M_l([f]_\alpha)(t).
\end{aligned}$$

This and Lemma 5.1 finish the proof of assertion (i).

(ii) The proof of (ii) is basically the same. However, this time we do not need to worry about s close to t, because $f(s,x)$ vanishes for such s. Also $I_k(f,t) < \infty$ for all k. Indeed, if $\lambda = 0$, then this is true because $[f(s,\cdot)]_\alpha = 0$ for large negative s, and if $\lambda > 0$, then

$$\begin{aligned}
\int_{-\infty}^{t-1} e^{-\lambda(t-s)}[f(s,\cdot)]_\alpha (t-s)^\beta\, ds &\le \int_{-\infty}^t e^{-\lambda(t-s)}[f(s,\cdot)]_\alpha\, ds \\
&= \lambda \int_{-\infty}^t e^{-\lambda(t-s)}\Big(\int_s^t [f(r,\cdot)]_\alpha\, dr\Big)\, ds \\
&\le N\lambda M_l\left([f]_\alpha\right)(t) \int_{-\infty}^t (t-s)e^{-\lambda(t-s)}\, ds = N\lambda^{-1} M_l\left([f]_\alpha\right)(t).
\end{aligned}$$

This and Lemma 5.1 show that $\mathcal{G}_\lambda f(t,\cdot) \in C^k$ for any k. For $k = 2m+1$, we also have $\beta = (\alpha - k)/(2m) < -1$ and

$$\begin{aligned}
I_k(f,t) &\le \int_{-\infty}^{t-1} [f(s,\cdot)]_\alpha (t-s)^\beta\, ds \\
&= N \int_{-\infty}^{t-1} (t-s)^{-1}\left(\int_s^{t-1} [f(r,\cdot)]_\alpha\, dr\right)(t-s)^\beta\, ds \\
&\le N M_l\left([f]_\alpha\right)(t) \int_{-\infty}^{t-1} (t-s)^\beta\, ds = N M_l\left([f]_\alpha\right)(t).
\end{aligned}$$

Upon applying Lemma 5.1 one more time, we get (5.3).

(iii) Parabolic dilations (ρ times x and ρ^{2m} times t) easily show that it suffices to prove estimate (5.4) in the case where $|x-y| = 1$. Therefore, we assume that $|x-y| = 1$ and fix a function $\zeta \in C_0^\infty(\mathbb{R})$ such that $\zeta(s) = 1$ for $s \in [t-1, t]$ and $\zeta(s) = 0$ for $s \le t-2$. Denoting $\eta = 1 - \zeta$, we obtain

$$\mathcal{G}_\lambda f = \mathcal{G}_\lambda(\zeta f) + \mathcal{G}_\lambda(\eta f),$$

which along with assertions (i) and (ii) shows that $G_\lambda f(t,\cdot) \in C^{2m}$.

Then, for any multi-index γ of length $2m$,

$$|D^\gamma \mathcal{G}_\lambda f(t,x) - D^\gamma \mathcal{G}_\lambda f(t,y)| \le |D^\gamma \mathcal{G}_\lambda(\zeta f)(t,x)| + |D^\gamma \mathcal{G}_\lambda(\zeta f)(t,y)| + |D^\gamma \mathcal{G}_\lambda(\eta f)(t,x) - D^\gamma \mathcal{G}_\lambda(\eta f)(t,y)|.$$

Here by the mean value theorem the last term on the right does not exceed a constant times

$$\sup_{|\gamma_1|=2m+1} \sup_x |D^{\gamma_1}\mathcal{G}_\lambda(\eta f)(t,x)|.$$

Therefore, applying (i) and (ii), we get (5.4). The lemma is proved.

PROOF OF THEOREM 2.13. Let $\zeta(s)$ be a smooth function which equals one for $s \ge 0$ and zero for $s \le -2$. Assume that $|\zeta'| \le 1$. Introduce $\zeta_k(s) = \zeta(k+s)$, $u_k = \zeta_k u$, and

$$f_k := \lambda u_k - Lu_k + \partial_t u_k = \zeta_k f + \zeta_k' u.$$

By Theorem 4.1 we have $u_k(t,x) = \mathcal{G}_\lambda f_k(t,x)$.

Next we restrict our attention to k's such that $-k < t$. Then $u(t,x) = \mathcal{G}_\lambda f_k(t,x)$ and

$$M_l([f_k]_\alpha)(t) \le M_l([f]_\alpha)(t) + m_k,$$

where

$$m_k := \sup_{r>0} r^{-1} \int_{t-r}^t I_{k+\tau \in (-2,0)} [u(\tau,\cdot)]_\alpha \, d\tau.$$

Observe that the last integral is the integral of $[u(\tau,\cdot)]_\alpha$ over the intersection of the intervals $[-k-2,-k]$ and $[t-r,t]$, and for the intersection to be nonempty, we need at least $t-r \le -k$, that is $r \ge t+k$. It follows that

$$m_k \le \frac{1}{k+t} \int_{-2-k}^{k} [u(\tau,\cdot)]_\alpha \, d\tau \to 0$$

as $k \to \infty$. After that it only remains to apply Lemma 5.2 to f_k and send k to infinity. The theorem is proved.

Next, we prove Theorem 2.10, deriving it from Theorem 2.13.

PROOF OF THEOREM 2.10. Estimate (2.9) for $p = \infty$ is weaker than (2.12). For $p \in (1,\infty)$ this estimate follows immediately from Theorem 2.13 and the Hardy–Littlewood maximal function theorem. Estimate (2.11) also follows from Theorem 2.13 and the maximal function inequality estimating the weak $(1,1)$-norm of $M_l(f)$ by the L_p-norm of f.

Estimate (2.10) makes sense only when λ is strictly positive. In this case after setting $f = \lambda u - Lu + \partial_t u$, we observe that, thanks to (3.10), in the notation from the above proof we have

$$|u_k(t,x)| \le N \int_{-\infty}^t e^{-\lambda(t-s)} |f_k|_0(s) ds,$$

which after letting $k \to \infty$ implies that

$$|u(t,x)| \le N\xi(t) * |f|_0(t),$$

where $\xi(t) = I_{t>0}\exp(-\lambda t)$ with

$$\|\xi\|_{L^1(\mathbb{R})} = \lambda^{-1}.$$

Recalling that the L_p-norm of $\xi(t)*|f|_0(t)$ is dominated by the L_1-norm of $\xi(t)$ times the L_p-norm of $|f|_0$, we get (2.10). Thus, the proof of Theorem 2.10 is complete.

PROOF OF THEOREM 2.8. By Lemma 5.1(i) the function $u(t,x)$ is well defined and is a continuous function of x for each t. As noted before in Theorem 4.1, u is a measurable function of (t,x). Lemma 5.1(i) and Young's inequality also yield (2.8).

That $u(t,\cdot) \in C^{2m}$ for almost all t follows from Lemma 5.2 and the fact that the maximal function of an L_p-function with $p \in (1,\infty]$ is finite almost everywhere. It follows that $D^\gamma u$ are also measurable functions for $|\gamma| \le 2m$. Then estimate (5.1) and Young's inequality imply that $|D^\gamma u(t,\cdot)|_0$ are in $L_p(\mathbb{R})$.

Estimate (5.4) shows that the same holds for $[D^\gamma u(t,\cdot)]_\alpha$ if $|\gamma| = 2m$ and yields (2.7) as in the proof of Theorem 2.10. Lemma 5.1(ii), and (5.4), and Young's inequality also shows that

$$\|u\|_{\mathcal{C}_p^{2m+\alpha}} := \| \, |u|_{2m+\alpha} \|_{L_p(\mathbb{R})} \le N \|f\|_{\mathcal{C}_p^\alpha}, \tag{5.5}$$

where N depends only on λ, p, δ, α, m, n, d.

After that it remains to show that u is continuous, there exists a $g \in \mathcal{C}_p^\alpha$ such that $\partial_t u(t,x) = g(t,x)$ (a.e.) for any x, and that u satisfies (1.1).

We take $S, T \in \mathbb{R}$, $S < T$, and claim that for any $x \in \mathbb{R}^d$

$$\begin{aligned} u(T,x) - u(S,x) &= \int_S^T f(t,x)\,dt \\ &+ \int_S^T \int_{-\infty}^t \partial_t \int_{\mathbb{R}^d} f(s,y) e^{-\lambda(t-s)} \Psi_{s,t}(x-y)\,dy ds dt. \end{aligned} \tag{5.6}$$

Set

$$u(s,t,x) = \int_{\mathbb{R}^d} f(s,y)\Psi_{s,t}(x-y)\,dy, \qquad v(s,t,x) = e^{-\lambda(t-s)} u(s,t,x),$$

for $t > s$ and $u(s,s,x) = v(s,s,x) = f(s,x)$. Notice that

$$u(T,x) = \int_{-\infty}^T v(s,T,x)\,ds. \tag{5.7}$$

Then for any s such that $f(s,\cdot) \in C^\alpha$, which happens to be true for almost all s by assumption, by Theorem 3.6 on $[s,\infty) \times \mathbb{R}^d$ we have

$$\partial_t u(s,t,x) = L_t u(s,t,x)$$

(in the sense of Definition 2.4). In particular, $v(s,t,x)$ satisfies

$$\partial_t v(s,t,x) = L_t v(s,t,x) - \lambda v(s,t,x) \tag{5.8}$$

on $[s,\infty) \times \mathbb{R}^d$. Notice that for any x and $t \ge s$,

$$v(s,t,x) - f(s,x) = v(s,t,x) - v(s,s,x) = \int_s^t \partial_r v(s,r,x)\,dr. \tag{5.9}$$

Next, observe that the second term on the right in (5.6) equals

$$\int_S^T \int_{-\infty}^t \partial_t v(s,t,x)\,ds dt \tag{5.10}$$

and

$$
\begin{aligned}
(5.11)\qquad \int_S^T \int_{-\infty}^t |\partial_t v(s,t,x)|\,dsdt &= \int_S^T \int_{-\infty}^t (|L_t v(s,t,x) - \lambda v(s,t,x)|)\,dsdt \\
&= \int_S^T \int_{-\infty}^t e^{-\lambda(t-s)}(|L_t u(s,t,x) - \lambda u(s,t,x)|)\,dsdt,
\end{aligned}
$$

which in light of (3.10) (with $\gamma = 0$) and (3.11) is dominated by a constant times

$$
\begin{aligned}
&\int_S^T \int_{-\infty}^t e^{-\lambda(t-s)}([f(s,\cdot)]_\alpha (t-s)^{\alpha/(2m)-1} + |f(s,\cdot)|_0)\,dsdt \\
&= \int_0^\infty e^{-\lambda r}\Big(\int_S^T ([f(t-r,\cdot)]_\alpha\,dt\Big) r^{\alpha/(2m)-1}\,dr \\
&\quad + \int_0^\infty e^{-\lambda r}\Big(\int_S^T |f(t-r,\cdot)|_0\,dt\Big)\,dr.
\end{aligned}
$$

Here by Hölder's inequality

$$
\int_S^T [f(t-r,\cdot)]_\alpha\,dt + \int_S^T [f(t-r,\cdot)]_0\,dt \le N(p,S,T)\|f\|_{\mathcal{C}_p^\alpha},
$$

so that the expression (5.11) is finite. This allows us to use Fubini's theorem and, in light of (5.9), to transform (5.10) into

$$
\int_{-\infty}^T [v(s,T,x) - v(s,S\vee s,x)]\,ds,
$$

which according to (5.7) equals

$$
\begin{aligned}
&u(T,x) - \int_{-\infty}^S v(s,S,x)\,ds - \int_S^T f(s,x)\,ds \\
&\qquad = u(T,x) - u(S,x) - \int_S^T f(s,x)\,ds.
\end{aligned}
$$

Hence, the right-hand side of (5.6) equals $u(T,x) - u(S,x)$, and thus (5.6) is proved.

Equality (5.6) is rewritten (see (5.8)) as

$$
u(T,x) - u(S,x) = \int_S^T \Big[f(t,x) + \int_{-\infty}^t \big(L_t v(s,t,x) - \lambda v(s,t,x)\big)\,ds\Big]\,dt,
$$

and owing to the above estimate of (5.11) and our comments in the proof of Theorem 3.6 as

$$
\begin{aligned}
u(T,x) - u(S,x) &= \int_S^T \Big[f(t,x) + (L-\lambda)\int_{-\infty}^t v(s,t,x)\,ds\Big]\,dt \\
&= \int_S^T \big[f(t,x) + (L-\lambda)u(t,x)\big]\,dt.
\end{aligned}
$$

We draw two conclusions from this result. First is that $u(t,x)$ is a continuous function of t. Secondly, by definition, u satisfies (1.1).

We also observe that $u(t,x)$ is continuous in x uniformly with respect to t because

$$
\begin{aligned}
|u(t,x)-u(t,y)| &\le N\int_{-\infty}^{t} e^{-\lambda(t-s)}|f(s,x+\cdot)-f(s,y+\cdot)|_0\,ds \\
&\le N|x-y|^{\alpha}\int_{-\infty}^{t} e^{-\lambda(t-s)}[f(s,\cdot)]_{\alpha}\,ds \\
&\le N|x-y|^{\alpha}\|\,[f(s,\cdot)]_{\alpha}\|_{L_p(\mathbb{R})}.
\end{aligned}
$$

This and the continuity of $u(t,x)$ in t shows that u is a continuous function, which combined with all the above proves that $u \in C_p^{2m+\alpha}$. The theorem is proved.

References

[1] A. Ya. Khinchin, *Basic laws of probability theory*, GSSI, Moscow, Leningrad, 1932 (Russian)

[2] R. P. Agarwal and D. O'Regan, *An introduction to ordinary differential equations*, Universitext, Springer, New York, 2008. MR2439721 (2009h:34001)

[3] S. D. Eidelman, *Parabolic systems*, Nauka, Moscow, 1964 in Russian; English translation: North-Holland, Amsterdam, London, 1969. MR0167726 (29:4998)

[4] A. Friedman, *Partial differential equation of parabolic type*, Prentice Hall, Inc., Englewood Cliffs, NJ, 1964. MR0181836 (31:6062)

[5] N. V. Krylov, *Parabolic equations in L_p-spaces with mixed norms*, Algebra i Analiz, Vol. 14 (2002), No. 4, 91–106 Russian; English translation in St. Petersburg Math. J., Vol. 14 (2003), No. 4, 603–614. MR1935919 (2003j:35133)

[6] N. V. Krylov and E. Priola, *Elliptic and parabolic second-order PDEs with growing coefficients*, Comm. in PDEs, Vol. 35 (2010), No. 1, 1–22. MR2748616 (2011m:35142)

[7] O. A. Ladyžhenskaya, *On the uniqueness of the solution of Cauchy's problem for a linear parabolic quation.* (Russian) Mat. Sbornik N.S. 27(69) (1950), 175–184. MR0040542

[8] O. A. Ladyzhenskaya, V. A. Solonnikov and N. N. Ural'tseva, "Lineǐnye i Kvazilineǐnye Uravneniya Parabolicheskogo Tipa," (Russian) [Linear and Quasi-Linear Equations of Parabolic Type], "Nauka," Moscow, 1967; English translation, Amer. Math. Soc., Providence, RI, 1968. MR0241821 (39:3159a)

[9] Kijung Lee, *On a deterministic linear partial differential system*, J. Math. Anal. Appl., Vol. 353 (2009), 24–42. MR2504437 (2010h:35180)

[10] G. Leoni, *A first course in Sobolev spaces*, American Mathematical Society, Providence, RI, 2009. MR2527916 (2010m:46049)

[11] L. Lorenzi, *Optimal Schauder estimates for parabolic problems with data measurable with respect to time*, SIAM J. Math. Anal., Vol. 32, No. 3 (2000), 588-615. MR1786159 (2001h:35086)

[12] V. A. Solonnikov, *On boundary value problems for linear parabolic systems of differential equations of general form*, (Russian), Trudy Mat. Inst. Steklov., Vol. 83 (1965), 3–163; English translation: Proceedings of the Steklov Institute of Mathematics. no 83 (1965): Boundary value problems of mathematical physics. III. Edited by O. A. Ladyzhenskaya., American Mathematical Society, Providence, R.I. 1967. MR0211083 (35:1965)

Dipartimento di Matematica, Università degli Studi di Salerno, Via Ponte don Melillo, 84084, Fisciano (Sa), Italy

E-mail address: `seboccia@unisa.it`

127 Vincent Hall, University of Minnesota, Minneapolis, Minnesota 55455

E-mail address: `krylov@math.umn.edu`

Amer. Math. Soc. Transl.
Volume **232**, 2014

On variational ground of the m-Hessian operators

N. V. Filimonenkova and N. M. Ivochkina

To the memory of Ol'ga Aleksandrovna Ladyzhenskaya

ABSTRACT. In frames of the theory of m-Hessian equations there were introduced three types of functionals. Two of them are the basis of the theory of m-Hessian measures developed by N. S. Trudinger and X.-J. Wang in particular in order to obtain some analogs of classic embedding theorems. On the other hand, N. M. Ivochkina has introduced in the beginning of 1990s the m-Hessian analog of the Dirichlet integral in order to obtain variational description of the cone of m-admissible functions, i.e., in terms of local minimizers in C^2. In this paper we present a survey of some known properties of these functionals and establish some connection between them. It also contains new results.

1. Introduction

In mid 1970s the situation in the theory of elliptic second-order partial differential equations may be described as follows. On the one hand, the theory of linear and quasilinear equations had acquired the features of classic completeness [**10**]. In particular, the functional-analytic instruments developed to investigate properties of solutions to the classic boundary problems for such equations. On the other hand, it turned out that some methods from this theory might be applied to solutions of the Dirichlet problem for the elliptic Monge–Ampere equation (see for instance [**12**]), which was the classic example of the elliptic in the cone (of convex function) fully nonlinear second-order equation. In this course the problem of the description of the second-order differential operators contiguous to Monge–Ampere operator has sprung up. More precisely, the problem of the description of divergence-free second-order operators has to be considered. Its solution was presented in the paper [**1**]. In particular, it has been shown therein that such operators with necessity must be taken as a linear combination of minors of the Hesse matrix u_{xx}, $u \in C^2$.

This activity attracted attention of O. A. Ladyzhenskaya, and with many other deep remarks she advised the second author to take a look at the papers [**13**], [**25**], [**11**]. It turned out that in these papers there was in essence considered the same problem, although in other terminology and from the point of view of another kind of mathematics.

2010 *Mathematics Subject Classification.* Primary 35J20, 35J60.
Key words and phrases. m-Hessian operators, m-curvatures, functionals.
The paper supported by grants: RFFI No 12-01-31467, S.Pb.St.Univ. 6.38.670.2013 and Min. of Ed. and Sci. RF No 14.B37.21.1125.

Now we may say that this was a case of simultaneous and independent interest of different authors in the same problem. Namely, they discovered the algebraic description of the kernel of variational derivatives of functionals with Lagrangian depending on second derivatives of u in the scalar case either on the first derivatives of an arbitrary vector function in the general setting. It is remarkable that further plans of these authors were completely different. In our context it brought out some classes of fully nonlinear divergence-free operators, containing the Laplacian and $\det u_{xx}$ as the particular cases, with nonempty functional cones of ellipticity.

One may see the first steps for describing these cones in the paper [**2**], were has been shown, for instance, that if some second-order divergence-free operator is elliptic in C^2, then it is linear in second derivatives. Hence, there are no totally elliptic divergence-free fully nonlinear operators of second order. On the other hand, it is common knowledge that the set of ellipticity of the Monge–Ampere operator coincides with the cone of convex functions.

In this paper the operators $T_m[u]$, $u \in C^2(\Omega)$, $1 \leq m \leq n$, are introduced as the sum of all m-th order principal minors of the Hesse matrix u_{xx}, and are said to be of Monge–Ampere type. Similar to the latter all of them enjoy the property

$$(1.1) \qquad T_m[u] = \frac{1}{m}\sum_{i,j} \frac{\partial^2}{\partial x^i \partial x^j}\left(\frac{\partial T_m[u]}{\partial u_{ij}} u\right) = f_m, \quad m = 1, \dots, n.$$

Although the left-hand identity in (1.1) is well known by now, we derive it in Section 2 to demonstrate our approach.

It is also shown in [**2**] that equation (1.1) is elliptic in the cone of convex functions, and solutions of the Dirichlet problem for (1.1) may be treated by the methods developed in the book [**10**]. Namely, an a priori estimate of the gradient of solution at the boundary was constructed in [**3**] by the method of integral inequalities from [**10**].

To prove the classic solvability of the Dirichlet problem for equation (1.1) it is natural to apply the well-known continuity method (the method of extension by parameter). If so, the data of the problem must keep under control the ellipticity of the relevant equations on solutions for all values of the parameter. For the Monge–Ampere equation, $m = n$, such control is realized by the requirement $f_n(x;\tau) > 0$, $\tau \in [0;1]$ in (1.1), which guarantees that if we have started with a strictly convex solution u^0, then under this requirement all solutions u^τ will be strictly convex functions, i.e., equation (1.1), $m = n$, is elliptic on these solutions.

It does not work if $m < n$. For instance, the positiveness of the right-hand side of the Poisson equation, $m = 1$, by no means guarantees the convexity of its solutions. However, it is of no interest in this case, since we have the unique solution in C^2 of the Dirichlet problem for all $f_1 \in C$.

The crucial move to create the theory of classic solvability of the Dirichlet problem for equation (1.1) with arbitrary m was to associate with every $m = 1, \dots, n$ the functional cone $\mathbb{K}_m$:

$$(1.2) \qquad \mathbb{K}_m(\bar\Omega) := \{u \in C^2(\bar\Omega) : T_i[u] > 0, \quad i = 1, \dots, m\}.$$

These cones were introduced in the paper [**4**]. It has been proved therein that the sets (1.2) are the components of connectedness of positivity of the operators T_m in C^2 containing $u_0 = x^2$, and that equation (1.1) is elliptic onto $\mathbb{K}_m$. The convexity of these cones was established in [**4**] only for $m = 1, 2, n$, and in the paper [**5**] the classic solvability of the Dirichlet problem in convex domains and zero boundary data was proved for these m.

The most general approach to the problem of classic solvability of the Dirichlet problem for fully nonlinear elliptic in the functional cones equations $F(u_{xx}) = f$ has been presented in the paper of L. Caffarelli, N. Nirenberg and J. Spruck [**14**], with equation (1.1) as examples. What was of great importance there was also discovered in [**14**], the sufficient close to necessary conditions on the boundary for classic solvability of Hessian equations in the presence of arbitrary Dirichlet data. The authors of [**14**] assumed all the data as smooth as necessary and did not use in the argument the divergence speciality of operators. So, the theory that developed in [**14**] covers also nondivergence-free equations.

Functions from the relevant cones have been indicated in [**14**] as "admissible", and adapting this terminology to equation (1.1) we speak about m-admissible functions. Notice also that in order to separate curvature equations from $F(u_{xx}) = f$, Jh. Urbas has indicated the latter as Hessian equations, which is commonly accepted now. In this course we speak about m-Hessian operators and equations, keeping in mind equation (1.1).

It is remarkable that later on N. S. Trudinger has introduced a new class of Hessian equations $\{T_{m,l}[u] = f > 0\}_1^n$, $1 \leq l \leq m$ [**27**], and considered the problem of solvability in some closures of (1.2). In our paper the the following closure in $C^2(\bar{\Omega})$ will be involved

$$\bar{\mathbb{K}}_m(\bar{\Omega}) := \{u \in C^{1,1}(\bar{\Omega}) : \ T_i[u] \geq 0 \quad \text{a.e.}, \quad i = 1, \dots, m\}. \tag{1.3}$$

Below functions from (1.3) will be referred to as "m-admissible a.e."

As to the Dirichlet problem for the m-Hessian equations, the authors of the paper [**14**] consider them in the functional realization of algebraic cones discovered by L. Gårding in 1959 [**20**]. This outstanding paper actually contains the theory of a-hyperbolic polynomials, and one of its outcomes is convexity associated with these polynomials' algebraic cones. It follows that their functional analogs are also convex, which has been used in [**14**], to prove solvability of the Dirichlet problem for equation (1.1) by the continuity method.

The constructive definition (1.2) of the cones of m-admissible functions from [**4**] is equivalent to definitions generated by the L. Gårding algebraic cones associated with the polynomials $\{p_m(t; S) = P_m(S + t\,\mathrm{Id})\}_1^n$, $S \in \mathrm{Sym}(n)$. Hence the cones (1.2) are convex for all $\{m\}_1^n$.

On the other hand from [**21**] N. V. Krylov has started to develop his idea of uniformization of elliptic and parabolic second-order partial differential equations via reduction to Bellman equations, which are totally elliptic (parabolic), possibly with degeneration. Notice that all equations from [**14**] and [**27**] admit such uniformization. In this course N. V. Krylov has proved solvability of the Dirichlet problem for m-Hessian equations in [**22**]. The description of his approach to fully nonlinear equations may be found in [**23**].

The paper [**20**] had been delivered more than a half century ago and in the publications [**9**] and [**8**] one may see a renovated version or its principal part, which exposes the Gårding cones as the basis of the theory of m-Hessian equations and their application in differential geometry.

An updated survey of the situation in the theory of m-Hessian equations may be found in [**19**]. Here we extract from the results in [**19**] and [**18**] the following existence theorem.

THEOREM 1.1. *Let Ω be a bounded domain in R^n, let $\partial\Omega$ be $(m-1)$-convex hypersurface of C^4-smoothness, $\phi \in C^4(\partial\Omega)$, $f_m \geq 0$, $f_m \in C^4(\bar{\Omega})$.*

Then there exists the unique m-admissible a.e. solution $u \in C^{1,1}(\bar{\Omega})$ of equation (1.1) *equal to ϕ on $\partial\Omega$.*

Comparing with linear and quasilinear elliptic equations the smoothness requirements on the boundary and Dirichlet data looks excessive. That is one of the open problems in the theory of fully nonlinear second-order elliptic equations. Some steps to treat it are outlined in [**17**]. It also looks challenging that up to now the Neyman problem has been investigated only for the Monge–Ampere equation, $m = n$ [**24**] (see also [**33**]).

The goal of this paper is to present some properties of the principal functionals arising in this or that development involving m-Hessian operators T_m. Realization of our programme requires some purely algebraic relations and also a renovated point of view on classic geometric notions and functional formulae. It turns out, for instance, that it was more suitable in our context to imply the interior normal, when integrating by parts. Section 2 contains a description of algebraic and geometric equipment and derivations of some necessary relations for further proceeding.

In Section 3 we give a variational description of the cones (1.2), which is close to the relevant from the paper [**7**], where the Dirichlet integrals $J_p[v] = \int_\Omega v_x^2 T_{p-1}[v]dx$, $p = 1, \dots, n$ have been in command. We also show that the functionals $I_p[v] = -\int_\Omega vT_p[v]dx$, $p = 1, \dots, n$ from the papers [**15**], [**32**], [**29**], [**28**], [**30**], [**29**], [**16**], [**31**], [**26**] may be applied with the same outcome. Namely, we prove that the cone (1.2) contains all local in C^2-minimizers of $J_m[v]$ either $I_{m-1}[v]$ and only them. Notice that in Section 3 we apply only classic variation by functions with compact support in Ω.

In 1994 X.-J. Wang [**32**] showed that $I_{m-1}^{\frac{1}{m}}[v]$ may be considered as an integral norm of v in the cones $\mathbb{K}_m(\Omega)\hat{C}_0(\bar{\Omega})$, which estimates some Lebesgue norms of v. This activity was extended in the collection of papers by N. S. Trudinger and X.-J. Wang [**29**], [**30**], [**29**] and in [**16**]. See also the quite recent paper [**26**]. In Section 4 we present a brief survey of this trend, although it is far from complete.

It is known that in the theory of elliptic linear and quasilinear equations different types of integral monotonicity of operators are sometimes very efficient. In Sections 5 and 6 we apply nonclassic variational analysis to establish such properties of m-Hessian operators in (1.2). Some of them were derived by different methods in the papers [**28**], [**29**]. For instance, in Section 5 we prove the inequality

$$\int_\Omega (w-u)(T_m[w] - T_m[u])dx \leq 0,$$

which is true for all $u, w \in \mathbb{K}_m(\bar{\Omega})$ vanishing on the boundary $\partial\Omega$. As far as we know, this inequality is new.

In Section 6 we expose the connection of functionals I_m, J_m, and derive some consequences of it. It also contains some results of variational analysis of the functionals J_m, $m = 1, \dots, n$.

2. The principal notions and auxiliary propositions

Denote by $\mathrm{Sym}(n)$ the space of $n \times n$ symmetric matrices, and fix up some integer m, $1 \leqslant m \leqslant n$. We introduce the m-th trace of a matrix $S = (s_{ij}) \in \mathrm{Sym}(n)$ as the sum of all its principal m-th-order minors, and denote it by $T_m(S)$. For instance, $T_1(S) = \operatorname{tr} S$, $T_n(S) = \det S$. It is clear that a matrix-function T_m is orthogonal invariant in the sense that

$$T_m(S) \equiv T_m(BSB^T), \quad S \in \mathrm{Sym}(n), \quad BB^T = I. \tag{2.1}$$

In the sequel we will use the following notations:

$$T_m^{ij}(S) = \frac{\partial T_m(S)}{\partial s_{ij}}, \quad T_m^{ij,kl}(S) = \frac{\partial^2 T_m(S)}{\partial s_{ij}\partial s_{kl}}, \quad i,j,k,l = 1,\dots,n. \tag{2.2}$$

It is also obvious that T_m is m-homogeneous:

$$T_m^{ij}(S)s_{ij} = mT_m(S). \tag{2.3}$$

Fine properties (2.1) and (2.3) give rise to many not obvious algebraic relations for m-traces, which are necessary for successful development of the theory of fully nonlinear second-order partial differential equations. In the context of this paper the following identity is of importance:

$$T_1(S)T_{m-1}(S) - \sum_{k=1}^{n} s_{ik}s_{jk}T_{m-1}^{ij}(S) = mT_m(S). \tag{2.4}$$

Indeed, due to (2.1) it is sufficient to consider the left-hand side of relation (2.4) onto diagonal matrices $\Lambda = \mathrm{diag}\{\lambda_1,\dots,\lambda_n\}$. We compute

$$T_1(\Lambda)T_{m-1}(\Lambda) = \sum_{1}^{n}(\lambda_i^2 T_{m-2,i}(\Lambda) + \lambda_i T_{m-1,i}(\Lambda)), \quad T_{m-1,i}(\Lambda) = \frac{\partial T_m(\Lambda)}{\partial \lambda_i},$$

which in combination with identity (2.3) rewritten as $\sum_1^n \lambda_i T_{m-1,i}(\Lambda) = mT_m(\Lambda)$ carries out the line (2.4).

We formulate one more algebraic identity leaving its proof to the reader. Namely, for arbitrary $S \in \mathrm{Sym}(n)$, $a = (a_1,\dots,a_n) \in R^n$, the following equality is valid:

$$a_n T_{m-1} = a_k T_{m-1}^{ni}(S)s_{ki} + a_i T_m^{ni}(S). \tag{2.5}$$

Let $\Omega \subset R^n$ be a bounded domain, and let $u \in C^2(\Omega)$, u_{xx} be the Hesse matrix of u. The operators generated by m-traces as

$$T_m[u] = T_m(u_{xx}), \quad u \in C^2(\Omega),$$

are said to be m-Hessian operators. Notice that $T_1[u]$ is the Laplacian and $T_n[u]$ is the Monge–Ampere operator.

We now prove the first auxiliary proposition, which justifies representation (1.1) for m-Hessian operators. It confirms a long-since known fact that minors of gradient matrices belong to the kernel of variational derivatives of relevant functionals, i.e., divergence of the gradient of such minors is identical to zero.

LEMMA 2.1. *Let $v = (v^1,\dots,v^n) \in C^2(\Omega)$, $\Omega \subset R^n$, $m_p[v]$ be some minor of p-th order of the matrix $(\bigtriangledown v)$. The identity*

$$\frac{d}{dx^i}\frac{\partial m_p[v]}{\partial v_i^k} = 0, \quad k = 1,\dots,n. \tag{2.6}$$

is valid.

PROOF. We represent the left-hand side of (2.6) as

$$\frac{d}{dx^i}\frac{\partial m_p[v]}{\partial v_i^k} = \sum_{i<j}\left(\frac{\partial^2 m_p[v]}{\partial v_i^k \partial v_j^l} + \frac{\partial^2 m_p[v]}{\partial v_j^k \partial v_i^l}\right) v_{ij}^l. \tag{2.7}$$

Here and further on the lower indices are differential, i.e., $v_i^k := \partial v^k / \partial x^i$. Since i, j are the numbers of rows of a chosen minor $m_p[v]$, every summand in brackets in (2.7), i.e., with fixed indices i, j, k, l, is identical 0. Hence the identity (2.6) holds true. □

Consider now the scalar case, $v = \nabla u, u \in C^2(\Omega)$. Then the representation

$$T_m[u] = \frac{1}{m}\sum_{i,j}\frac{\partial^2}{\partial x^i \partial x^j}\left(T_m^{ij}[u]u\right)$$

is a simple consequence of (2.3) and Lemma 2.1, as well as the useful identities

$$\sum_{\{jkl\}} T_m^{ij,kl}[u] = 0, \quad i,j,k,l = 1,\dots,n, \tag{2.8}$$

where $\{jkl\}$ is the set of permutations,

$$\frac{d}{dx^i}T_m^{ij}[u] = 0, \qquad j = 1,\dots,n. \tag{2.9}$$

In order to describe the geometric speciality of a closed hypersurface $\partial\Omega \subset R^n$, we denote by $X = X(M)$ its position vector and consider some local parametrization, i.e.,

$$\partial\Omega = \{X = (X^1, X^2, \dots, X^n)(\theta),\ \theta \in R^{n-1}\}.$$

Denote also

$$X_i = \frac{\partial X}{\partial \theta^i}, \quad X_{ij} = \frac{\partial^2 X}{\partial \theta^i \partial \theta^j}, \quad X_{\theta\theta} = (X_{ij}).$$

Then the metric tensor $g[\partial\Omega]$ locally looks as $g = (g_{ij}) = (X_i, X_j)$. Eventually, we introduce tensors $\eta = (\eta_i^j(\theta))$, $\tau = \eta^{-1}$, by the requirement $g = \eta^T\eta$.

Let $x \in \partial\Omega$ and $\mathbf{n} = \mathbf{n}(x)$ be the interior unit normal to the hypersurface $\partial\Omega$ at x. Denote

$$X_{(k)} = X_i\tau_k^i, \quad X_{(\theta\theta)} = \tau^T X_{\theta\theta}\tau.$$

Notice that $(X_{(i)}, X_{(j)})\delta_{ij}$ and the systems

$$X_{(k)} = X_i\tau_k^i, \quad X_{(\theta\theta)} = \tau^T X_{\theta\theta}\tau,$$

represent the orthogonal moving frame of the hypersurface $\partial\Omega$.

We are now ready to introduce the curvature matrix $\mathcal{K}[\partial\Omega]$ by

$$\mathcal{K}[\partial\Omega](x) = (X_{(\theta\theta)}, \mathbf{n})(x). \tag{2.10}$$

The curvature matrix is symmetric and geometric invariant in the sense that its eigenvalues are the principal curvatures of the hypersurface $\partial\Omega$. Notice that in contrast to g the matrix (2.10) is not a tensor.

The first steps to construction (2.10) were made in the paper [**6**]. The precise description of the curvature matrix was given in the papers [**8**] and [**9**].

In frames of the theory of m-Hessian operators, it is more credible to operate with other than principal curvatures geometric invariants of hypersurfaces. In this course we introduce a notion of p-curvature $\mathbf{k}_p[\partial\Omega]$ by the line

$$\mathbf{k}_p[\partial\Omega](x) = T_p(\mathcal{K}[\partial\Omega])(x), \quad p = 1,\dots,n. \tag{2.11}$$

Notice that in computing p-curvatures, it is sufficient to work with the set of cartesian local parametrizations, which is natural for m-Hessian equations.

LEMMA 2.2. *Let v, Φ be arbitrary functions from $C^2(\bar{\Omega})$. Assume that $v - \Phi =$ const on $\partial\Omega$. Then in notation (2.2) and (2.11) for all $x \in \partial\Omega$, the identity*

$$T_p^{ij}[v](v-\Phi)_i(\mathbf{n}, e_j) = T_p(\Phi_{(\theta\theta)} - v_n\mathcal{K}), \quad p = 1, \dots, n-1, \tag{2.12}$$

is true. In the case $\Phi = 0, x \in \partial\Omega$, the relation

$$-T_p^{ij}[v]v_i(\mathbf{n}, e_j) = (-v_n)^{p+1}\mathbf{k}_p[\partial\Omega], \quad p = 1, \dots, n-1, \tag{2.13}$$

holds.

PROOF. We associate with $x_0 \in \partial\Omega$ a cartesian basis $\{e_i(x)\}_1^n$, $e_n = \mathbf{n}(x_0)$. Then $(v-\Phi)_i(x_0) = 0$, $(e_i, \mathbf{n}(x_0)) = 0$, $i = 1, \dots, n-1$, and the left-hand sides of (2.12) and (2.13) read as $T_p^{nn}[v](v-\Phi)_n$, $T_p^{nn}[v](v)_n$, respectively.

Now we choose a special parametrization of $\partial\Omega$ in some neighborhood of x_0. Namely, let $\theta^i = x^i, i = 1, \dots, n$. Then the hypersurface $\partial\Omega$ may be considered as the graph of function ω if our neighborhood is sufficiently small, i.e.,

$$\partial\Omega \cap B_r(x_0) = \{x : x^n = \omega(\theta), \omega_\theta(0) = 0, x^i = \theta^i\}, \quad 0 < r \ll 1.$$

It is obvious that $g(\theta)(x_0) = \mathrm{Id}$, and then $\mathcal{K}[\partial\Omega](x_0) = \omega_{\theta\theta}(x_0)$,

$$v_{(ij)}(x_0) = v_{ij}(x_0) = (\Phi_{(ij)}(x_0) - v_n\mathcal{K}_{ij}[\partial\Omega](x_0)), \quad i, j = 1, \dots, n-1.$$

Since $T_p^{nn}[v](x_0) = T_{p-1}(v_{(\theta\theta)}(x_0))$, we arrive at (2.12) and (2.13). □

We conclude this section with a collection of precise definitions arising in the theory of solvability of the Dirichlet problem for m-Hessian equations (1.1).

DEFINITION 2.3. *A matrix $S \in \mathrm{Sym}(n)$ is said to be m-positive if $T_i(S) > 0$, $i = 1, 2, \dots, m$. The set of m-positive matrices forms the following cone in the space $\mathrm{Sym}(n)$:*

$$K_m = \{S \in \mathrm{Sym}(n) : \quad T_i(S) > 0, \quad i = 1, 2, \dots, m\}.$$

Definition 2.3 brings out the sequence of inclusions

$$K_1 \supset K_2 \supset K_3 \supset \dots \supset K_n,$$

where K_n is the cone of positive definite matrices.

DEFINITION 2.4. *Function $u \in C^2(\bar{\Omega})$ is said to be m-admissible in $\bar{\Omega}$ if its Hesse matrix u_{xx} is m-positive in all points $x \in \bar{\Omega}$. The set of m-admissible functions forms the following cone in the space $C^2(\bar{\Omega})$:*

$$\mathbb{K}_m(\bar{\Omega}) := \{u \in C^2(\bar{\Omega}) : \ T_i[u] > 0, \quad i = 1, \dots, m\}. \tag{2.14}$$

In particular, $\mathbb{K}_n(\bar{\Omega})$ is the cone of C^2-smooth strictly convex functions. In the papers [**9**] and [**17**], the idea of Gårding from [**20**] has been adapted to show that the cones $\mathbb{K}_m(\bar{\Omega})$ are convex and belong to the set of ellipticity of the operators T_m, respectively.

DEFINITION 2.5. *A hypersurface $\partial\Omega \in C^2$ is said to be $(m-1)$-convex if $\mathcal{K}[\partial\Omega](x) \in K_{m-1}$ for all $x \in \partial\Omega$, in other words, $\mathbf{k}_p[\partial\Omega](x) > 0$ for all $x \in \partial\Omega$, $p = 1, 2, \dots, m-1$.*

The notion of an $(n-1)$-convex C^2-hypersurface coincides with the classic notion of a strictly convex hypersurface.

LEMMA 2.6. *Let $u \in \mathbb{K}_m(\bar{\Omega})$. Then the level surfaces of u are $(m-1)$-convex.*

This property of m-admissible functions has long since been established in other terminology (see for instance [**27**]). Our formulation Lemma 2.6 was proved in [**17**].

3. The classic approach to variational problems

Consider in $C^2(\bar{\Omega})$, $\Omega \subset R^n$, the functionals

$$\mu_{p-1}[v] = \int_\Omega T_{p-1}[v]dx, \quad J_p[v] = \int_\Omega v_x^2 T_{p-1}[v]dx, \quad I_p[v] = -\int_\Omega vT_p[v]dx, \tag{3.1}$$

where $p = 1, \ldots, n+1$. Since by definition $T_0[v] = 1$, the functional $\mu_0[v]$ is of no interest as well as $I_{n+1}[v]$ in view of identity $T_{n+1}[v] = 0$, $v \in C^2$. The functional $J_1[v]$ is the classic Dirichlet integral, and in [**7**] this name was expanded on the whole set $\{J_p[v]\}_0^{n+1}$.

In this section we vary the functionals (3.1) in classic style. Following this pattern, we use common notation for the first and second variations. For instance, it looks as follows for μ_{p-1}:

$$\delta\mu_{p-1}[v] = \frac{d\mu_{p-1}(u+t\eta)}{dt}\Big|_{t=0}, \quad \delta^2\mu_{p-1}[v] = \frac{d^2\mu_{p-1}(u+t\eta)}{dt^2}\Big|_{t=0}.$$

LEMMA 3.1. *Let $v = u + t\eta$, $u \in C^2(\bar{\Omega})$, $t \in R$, $\eta \in C^\infty$ be with compact support in Ω. Then the identities*

$$\delta\mu_{p-1}[v] = \delta^2\mu_{p-1}[v] = 0, \tag{3.2}$$

$$\delta J_p[v] = \frac{2p}{p+1}\delta I_p[v] = -2p\int_\Omega \eta T_p[u]dx, \tag{3.3}$$

$$\delta^2 J_p[v] = \frac{2p}{p+1}\delta^2 I_p[v] = -2p\int_\Omega T_p^{ij}[u]\eta_i\eta_j dx, \quad p = 1, \ldots, n+1, \tag{3.4}$$

are true.

PROOF. Indeed, the identity (3.2) is an obvious consequence of (2.9). It is not too hard to derive (3.3) and (3.4). For instance, the (2.4) and integration by parts carry out the first equality from (3.3) as follows

$$\begin{aligned}\frac{d}{dt}\int_\Omega v_x^2 T_{p-1}[v]dx &= \int_\Omega (2(v_x, \eta_x)T_{p-1}[v] + v_x^2 T_{p-1}^{ij}[v]\eta_{ij})dx \\ &= \int_\Omega (-2T_1[v]T_{p-1}[v] + 2(v_{xi}, v_{xj})T_{p-1}^{ij}[v])\eta dx = -2p\int_\Omega T_p[v]\eta dx,\end{aligned}$$

and letting $t = 0$ we arrive at a desirable relation. □

Similar to quadratic functionals considered in the theory of linear second-order elliptic equations, we will explore the extremal properties of the functionals

$$\begin{aligned}J_p[u; f_p] &:= J_p[u] + 2p\int_\Omega uf_p dx, \\ I_p[u; f_p] &:= I_p[u] + (p+1)\int_\Omega uf_p dx, \quad p = 1, \ldots, n,\end{aligned} \tag{3.5}$$

where f_p is some known function. For fixed p the Euler–Lagrange equation for these functionals is the same and coincides with (1.1), $p = m$. Here we are interested in the local minimums in $C^2(\bar{\Omega})$ of the functionals (3.5) in the following sense.

DEFINITION 3.2. *A function $w \in C^2(\bar{\Omega})$ is said to be the local in $C^2(\Omega)$-minimizer of some functional $J[u]$ if there is a constant $c(w)$ such that $J[w] \leq J[w+\eta]$ for all $\eta \in C^\infty$ with finite support in Ω, satisfying the inequality $||\eta||_{C^2(\Omega)} \leq c(w)$.*

In order to precisely describe the set of all local minimizers for the functionals (3.5) with $f_p \in C(\bar{\Omega})$, $f > 0$ we utilize the idea from the paper [**7**].

LEMMA 3.3. *Let $w \in C^2(\Omega)$. If $w \in C^2(\Omega)$ is a local minimizer for the functional $J_m[u; f_m]$ either $I_m[u; f_m]$ with some $f_m \in C(\Omega)$, then the operator T_m is elliptic on w, not strictly in general.*

On the other hand, if T_m is strictly elliptic on w, i.e., the inequality

$$(\nabla T_m[w]\xi, \xi) = T_m^{ij}[w]\xi^i\xi^j > 0, \quad x \in \Omega, \quad \xi \in R^n, \quad |\xi| = 1, \tag{3.6}$$

is satisfied, then w is a local minimizer for the functionals (3.5), $p = m$ with $f_m = T_m[w]$.

PROOF. If the function $w \in C^2(\bar{\Omega})$ provides a local minimum to the functionals (3.5), $p = m$ with some $f_m \in C(\Omega)$, then the first variation of the latter is vanishing at w, while the second is not negative. That means that a nonstrong version of inequality (3.6), which is equivalent to not strict ellipticity of the operator T_m on w. □

The identity (2.8) brings out the following version of the inequality (3.6):

$$\mathrm{tr}_m(w_{xx} + \xi \times \xi) - \mathrm{tr}_m\, w_{xx} > 0, \quad \xi \in R^n, \quad |\xi| = 1, \tag{3.7}$$

which may be qualified as ξ-monotonicity of the operator T_m on w. As was shown in [**9**], the set of functions satisfying (3.7) may be not connected in $C^2(\Omega)$ for $1 < m < n$. In the following proposition we describe the set of another type of monotonicity for the operator T_m, which is a convex part of the set of its ξ-monotonicity. In other words we reasonably narrow down the set of local minimizers of T_m.

LEMMA 3.4. *Let w be an arbitrary function from $C^2(\Omega)$. The inequality*

$$\mathrm{tr}_m(w_{xx} + t\,\mathrm{Id}) > \mathrm{tr}_m\, w_{xx} > 0, \quad x \in \Omega, \quad t \in R^+, \tag{3.8}$$

suffices to prove the existence of function $f_m \in C(\bar{\Omega})$ such that w delivers a local in a $C^2(\Omega)$-minimum to functionals (3.5) with $p = m$.

Actually, Lemma 3.3 also brings out the basis of the variational description of the cone of m-admissible functions (1.2) and (2.14).

THEOREM 3.5. *Assume $T_m[w] > 0$. Then w delivers a local in $C^2(\Omega)$-minimum to all functionals (3.5) with $f_p = T_p[w]$, $p = 1, \ldots, m$, if and only if $w \in \mathbb{K}_m(\Omega)$.*

PROOF. Indeed, if w provides a local minimum to all functionals (3.5) with $p \leq m$, then Lemma 3.3 guarantees correctness of the inequalities

$$(\nabla T_p[w]\xi, \xi) \geq 0, \quad x \in \Omega, \quad \xi \in R^n, \quad |\xi| = 1, \quad p = 1, \ldots, m, \tag{3.9}$$

and hence $T_p[w] \geq 0$, $1 \leq p \leq m - 1$. However the requirement $T_m[w] > 0$ is compatible with only strong versions of inequalities (3.9), which follows from the Gårding theory of a-hyperbolic polynomials (see, for instance, [**9**]). Then $T_p[w] > 0$, $1 \leq p \leq m$, and we have arrived at a constructive definition of $\mathbb{K}_m(\Omega)$.

Now let $w \in \mathbb{K}_m(\Omega)$. It is well known that all operators T_p, $1 \leq p \leq m$, are strictly elliptic in the cone of m-admissible functions, hence on w. To finish our proof, we notice that w satisfies equations $T_p[w] = f_p$ by the choice of f_p and it follows from the second part of Lemma 3.3 that it delivers local minimums to the whole set of functionals (3.5). □

It is rather remarkable that in the theory of m-Hessian equations there are many equivalent definitions of the cone of m-admissible functions, and the line (3.8) is one of them. In this course Lemma 3.4 is an obvious consequence of Theorem 3.5.

Lemma 3.4 and Theorem 3.5 make credible the idea that the cone of p-admissible functions should be a natural set of global minimization of functionals (3.5). Then it is reasonable to vary a function $u \in \mathbb{K}_p(\Omega)$ not leaving this cone. However, the classic approach presented above cannot be applied in this context because functions with compact support in Ω are not p-admissible. Proceeding further, we suggest a different approach to the problem of minimization of functionals (3.5).

4. On some special norms in $\mathbb{K}_m$

The functionals $I_m[u]$, $m = 1, \dots, n$, first appeared in the paper of X.-J. Wang [**32**]. In particular, one remarkable property of these functionals was discovered therein. Namely, a theorem was proved in [**32**], which in our notations reads as follows.

THEOREM 4.1. *Let $u, w \in \bar{\mathbb{K}}^0_m(\bar{\Omega})$, where*

$$\bar{\mathbb{K}}^0_m(\bar{\Omega}) = \{u \in \bar{\mathbb{K}}_m(\bar{\Omega}), \quad u|_{\partial\Omega} = 0\}. \tag{4.1}$$

Then for all $m = 1, \dots, n$ the following assertions are true:

(*i*) *if $I_m^{\frac{1}{m+1}}[u] = 0$, then $u \equiv 0$;*

(*ii*) *$I_m^{\frac{1}{m+1}}[tu] = tI_m^{\frac{1}{m+1}}[u]$ for all $t > 0$;*

(*iii*) *$I_m^{\frac{1}{m+1}}[u + w] \le I_m^{\frac{1}{m+1}}[u] + I_m^{\frac{1}{m+1}}[w]$.*

But the collection (*i*)–(*iii*) means that $I_m^{\frac{1}{m+1}}[u]$ may be considered as a norm in $\mathbb{K}^0_m$ and the author of [**32**] denotes it by $\mathbb{K}^0_m$. We also keep to this notation.

There are some discrepancies in the proof of Theorem 4.1, and we suggest its slightly reconstructed version in order to demonstrate our "nonclassic" approach to variational problems.

PROOF. It is well known that m-admissible a.e. functions vanishing on $\partial\Omega$ are nonpositive and the uniqueness of solutions from Theorem 1.1 guarantees the validity of (*i*). The inequality (*ii*) is an obvious consequence of the $(m+1)$-homogeneity of $T_m[u]$.

In order to prove (*iii*), we introduce the following function in one variable,

$$h(t) = I_m^{\frac{1}{m+1}}[u + t\eta], \quad t \in [0; 1], \quad \eta = w - u,$$

and show that it is convex if $u, w \in \mathbb{K}^0_m$.

Indeed, integrating by parts and implying the identities (2.9) and (2.3), we arrive at

$$\frac{d}{dt} I_m[v] = -\int_\Omega \left(\frac{1}{m} T_m^{ij}[v] v_{ij} \eta + v T_m^{ij}[v] \eta_{ij} \right) dx = \frac{m+1}{m} \int_\Omega T_m^{ij}[v] v_i \eta_j dx. \tag{4.2}$$

Also

$$\begin{aligned}\frac{d^2}{dt^2}I_m[v] &= \frac{m+1}{m}\int_\Omega \left(T_m^{ij,kl}[v]v_i\eta_{kl}\eta_j + T_m^{ij}[v]\eta_i\eta_j\right)dx \\ &= (m+1)\int_\Omega T_m^{ij}[v]\eta_i\eta_j dx.\end{aligned} \tag{4.3}$$

Here we are making use of (2.8).

Notice that under conditions of Theorem 4.1 and due to the convexity of $\mathbb{K}$, we have $v = (1-t)u + tw \in \bar{\mathbb{K}}_m$ for $t \in [0;1]$. Hence, v satisfies inequality (3.9), which leads to

$$\left(\int_\Omega T_m^{ij}[v]v_i\eta_j dx\right)^2 \leq \int_\Omega T_m^{ij}[v]v_iv_j dx \int_\Omega T_m^{ij}[v]\eta_i\eta_j dx. \tag{4.4}$$

It is easy to see that equalities (4.2) and (4.3) together with (4.4) carry out the inequality $h''(t) \geq 0$, $t \in [0;1]$, and (iii) is proved. □

Theorem 4.1 suggests that there are analogs of imbedding theorems in the cones $\mathbb{K}_m^0$, $1 < m \leq n$. Exactly this idea was developed in the papers [**32**], [**30**], [**16**]. It is well known that the theory of imbedding theorems implies relevant integral inequalities and most of the cited papers contains derivations of this. Consider two examples.

The first is a particular case of the inequality from Theorem 5.2 proved in [**32**]:

$$||u||_{L^{m+1}} \leq c_0||u||_{\mathbb{K}_m^0}, \quad m = 1, \dots, n. \tag{4.5}$$

In order to prove this theorem, X.-J. Wang was constructing a nonlinear analog of the theory of critical power for semilinear equations. It is remarkable that in [**32**] the precise value or the constant c_0 in (4.5) has been found. Namely, $c_0 = \lambda_1^{-1}$, where λ_1 is the unique positive "eigenvalue" of the problem

$$T_m[u] + \lambda u = 0, \quad u \in \mathbb{K}_m^0(\bar{\Omega}).$$

Notice that earlier inequality (4.5) for $m = n$ has been proved in [**15**].

Our second example is taken from the paper [**30**]:

$$||u||_{\mathbb{K}_l^0} \leq ||u||_{\mathbb{K}_m^0}, \quad 0 \leq l < m, \quad m = 1, \dots, n, \tag{4.6}$$

and it is labeled therein as "Poincare inequality".

To produce some imbedding theorems on the basis of (4.5) and (4.6), it is necessary first to find the correct supplement of the cones $\mathbb{K}_m^0$ in the class of functions with weaker differential properties. This difficult problem is the subject of a group of papers, [**28**], [**29**], [**31**].

Now we expose an analog of Theorem 4.1 for the functionals $\mu_m[u]$, $m = 1, \dots, n$.

THEOREM 4.2. *Let u, w be arbitrary functions from* (4.1). *Then for all $m = 1, \dots, n$ the following assertions are true:*

(i) if $\mu_m^{\frac{1}{m}}[u] = 0$, then $u \equiv 0$;

(ii) $\mu_m^{\frac{1}{m}}[tu] = t\mu_m^{\frac{1}{m}}[u]$ for $t > 0$;

(iii) $\mu_m^{\frac{1}{m}}[u+w] \leq \mu_m^{\frac{1}{m}}[u] + \mu_m^{\frac{1}{m}}[w]$.

PROOF. To begin with, notice that relations (i) and (ii) are obvious. In order to prove (iii) we remind the reader first that according to Lemma 2.6 the level surfaces of m-admissible functions are $(m-1)$-convex, and under the conditions of the theorem we have $|v_x(M)| = -v_n(M)$, where $M \in \partial\Omega$, $v_n(M)$ is the derivative in the direction of the interior to $\partial\Omega$ normal $\mathbf{n}$. Also consider $v = u + t\eta$, $\eta = w - u$, $0 \leq t \leq 1$.

Keeping in mind identity (2.13) from Lemma 2.2, we represent $\mu_m[v]$ in the form

$$(4.7)\quad \mu_m[v] = \int_\Omega T_m[v]dx = -\frac{1}{m}\int_{\partial\Omega} T_m^{ij}[v]v_i(\mathbf{n}, e_j)ds = -\frac{1}{m}\int_{\partial\Omega} |v_x|^m \mathbf{k}_{m-1}ds.$$

The following lines are similar to (4.2) and (4.3):

$$(4.8)\quad \begin{aligned} \frac{d}{dt}\mu_m[v] &= -\int_{\partial\Omega} |v_x|^{m-1}\eta_n \mathbf{k}_{m-1}ds, \\ \frac{d^2}{dt^2}\mu_m[v] &= (m-1)\int_{\partial\Omega} |v_x|^{m-2}\eta_n^2 \mathbf{k}_{m-1}ds. \end{aligned}$$

Denote $\hat{h}(t) = \mu^{\frac{1}{m}}[v]$. Straightforward computation on the basis of (4.7), (4.8), and the relevant analog of (4.4) carries out the inequality $\hat{h}''(t) \geq 0$; hence, assertion (iii) is indeed true. □

The following simple consequence of identity (2.13) and Theorem 4.2 justifies in a way our reluctance to involve into reasoning functions with compact support.

COROLLARY 4.3. *Assume that u is m-admissible a.e. in Ω (see (1.3)), and in addition $u = 0$, $u_n = 0$ on $\partial\Omega$. Then $u \equiv 0$.*

5. Some properties of the functionals $I_m[u]$, $\mu_m[u]$ in $\mathbb{K}_m$

Here and further on we vary functionals not leaving the relevant cone of admissible functions. Namely, the functionals I_m, μ_m are considered on the set $\mathbf{V} = \{v = tw + (1-t)u, u, w \in \mathbb{K}_m, t \in [0;1]\}$. Since $\mathbb{K}_m$ is convex, we have the inclusion $\mathbf{V} \subset \mathbb{K}_m$. In proceeding further it is often convenient to work with functions from $\mathbf{V}$ represented as $v = u + t\eta$, $\eta = w - v$. As to η we may only guarantee it vanishes at the boundary of Ω, when $u = w, x \in \partial\Omega$. It is exactly the case considered below, where we associate with given function $\Phi \in C^4(\bar{\Omega})$ the cones

$$(5.1)\quad \mathbb{K}^\Phi(\bar{\Omega}) := \{u \in \mathbb{K}_m(\bar{\Omega}), \quad u|_{\partial\Omega} = \Phi\}.$$

Denote also $\Omega_r = \{x \in \Omega : \mathrm{dist}\{x; \partial\Omega\} > r, 0 < r \ll 1\}$.

THEOREM 5.1. *Let $u, w \in \mathbb{K}_m^\Phi(\bar{\Omega})$. Assume there exists $r > 0$ such that $w \geq u$ for all $x \in \Omega\backslash\Omega_r$. Then the inequalities*

$$(5.2)\quad \mu_p[w] \leq \mu_p[u], \quad p = 1, \dots, m$$

are valid. If in addition $u \leq w$ in Ω and $\Phi \leq 0$ on $\partial\Omega$, together with (5.2) the inequalities

$$(5.3)\quad I_p[w] \leq I_p[u], \quad p = 1, \dots, m,$$

are satisfied.

PROOF. To prove (5.2), it is sufficient to show that under conditions of our theorem the inequality $d\mu_p[v]/dt \leq 0$ with $v = u + t\eta$, $\eta = w - u$ is true. Indeed,

$$\frac{d}{dt}\mu_p[v] = -\int_\Omega T_p^{ij}[v]\eta_i\eta_j dx - \int_{\partial\Omega} T_p^{nn}[v]\eta_n ds \leq 0. \tag{5.4}$$

Here we made use of ellipticity (3.6) and representation (2.12) from Lemma 2.2.

To prove the inequalities (5.3) we make use of relations (3.6) and (2.12) to derive the lines

$$\frac{d}{dt}I_p[v] = -\int_\Omega (\eta T_p[v] + vT_p^{ij}[v]\eta_{ij})dx = -(p+1)\int_\Omega \eta T_p[v]dx + \int_{\partial\Omega} \Phi T_p^{nn}[v]\eta_n ds \leq 0,$$

which concludes the proof. □

One may find different approaches to the proof of inequalities (5.2) in [**29**, Lemma 2.1]. Concerning inequalities (5.3), they are claimed to be valid for arbitrary boundary functions $\{\Phi\}$ in the paper [**29**, Lemma 5.2], what can not be true in such generality.

It looks probable that

$$\min_{\bar{\mathbb{K}}_m^\Phi(\Omega)} \mu_m[u] = \min_{\bar{\mathbb{K}}_m^\Phi(\Omega)} I_m[u] = 0, \tag{5.5}$$

and Theorems 1.1 and 5.1 with $f_m \equiv 0$ contain sufficient conditions of existence and uniqueness of the relevant minimizer from (5.1). In particular, $(m-1)$-convexity of the boundary of Ω (see Definition 2.5) is assumed in Theorem 1.1. On the other hand, Lemma 2.6 confirms that level surfaces of m-admissible functions are $(m-1)$-convex, and hence this requirement turns into a necessary one if $\Phi = \text{const}$.

Consider now the problem of minimization of the functional $I_m[u; f]$ in $\mathbb{K}_m^0$ (see (3.5)).

THEOREM 5.2. *Let $f \in C^4(\bar{\Omega})$, $f \geq \delta > 0$, $\partial\Omega$ be a C^4-smooth $(m-1)$-convex surface. Then there exists the unique function $u \in \mathbb{K}_m^0(\bar{\Omega})$, such that for all $w \in \mathbb{K}_m^0(\bar{\Omega})$ the following inequality is valid*

$$I_m[w; f] > m\int_\Omega uT_m[u]dx = -mI_m[u] = \inf_{\mathbb{K}_m^0(\bar{\Omega})} I_m[w; f], \quad w \neq u, \tag{5.6}$$

the equality holds only for $w = u$.

PROOF. As in the proof of Theorem 4.2, we associate with u a function $v = u + t\eta$, $t \in [0; 1]$, where w is a fixed function from $\mathbb{K}_m^0(\Omega)$. This time the following presentation of equality (4.2) will be applied

$$\frac{d}{dt}I_m[v; f](t) = -(m+1)\int_\Omega \eta(T_m[v] - f)dx. \tag{5.7}$$

In the capacity of u we choose the unique in $\mathbb{K}_m^0$ solution of equation (1.1) with $f_m = f$. Notice that the function $I_m[v; f](t)$ satisfies equality (4.3), which together with (5.7) carry out the following

$$I_m[v; f](t) = I_m[u; f] + \frac{m+1}{2}t^2\int_\Omega T_m^{ij}[\hat{v}]\eta_i\eta_j dx, \quad \hat{v} = u + \hat{t}\eta, \quad \hat{t} \in (0; 1). \tag{5.8}$$

Notice also that $I_m[u; f] = -mI_m[u; 0] = -mI_m[u]$. We see that the ellipticity of the operator T_m in the cone $\mathbb{K}_m$, the convexity of the latter, and (2.14) together with (5.8) bring out the desirable inequality (5.6). □

The classic calculus of variations construction of minimizing sequences is a powerful tool of investigation. However, in Theorem 5.2 minimums of functionals under consideration are negative, and for the aims of proceeding further we prefer to associate sequences of the following type with the functional $\{I_m\}_1^n$:

$$(5.9) \qquad \{u^k; f^k\}_1^\infty = \{u^k \in \mathbb{K}_m^0, \quad |I_m[u^k; f_k]| \geq |I_m[u^{k+1}; f_{k+1}]|, \quad k \in N\}.$$

The source of sequences in (5.9) for $I_m[u; f]$ is the set of solutions to equation (1.1) vanishing at the boundary. Namely, let $f \geq 0$ be some function, and let $f^k = f + \delta^k$, where $\delta^k > 0$, $\lim_{k\to\infty} \delta^k = 0$. The relevant sequence of $\{\}_1^\infty$ is taken as the sequence of solutions of equations (1.1) with right-hand sides f^k, $k \in N$. Then, by limit passage and Theorem 1.1, we expand Theorem 5.2, perhaps loosing strong inequality, for $f \geq 0$.

Notice eventually that the operator T_m enjoys integral monotonicity in the cone (4.1) in the following sense.

COROLLARY 5.3. *For arbitrary $w, u \in \mathbb{K}^0(\bar{\Omega})$, the following inequality is true:*

$$(5.10) \qquad \int_\Omega (w-u)(T_m[w] - T_m[u])dx \leq 0.$$

Moreover, equality is attained only on $w = u$.

Indeed, for $f = T_m[u]$ relation (5.6) is equivalent to

$$(5.11) \qquad -\int_\Omega w(T_m[w] - (m+1)T_m[u])dx > m\int_\Omega uT_m[u]dx.$$

Since the setting is symmetric with respect to u, w we may exchange them in (5.11). Let this version of (5.11) be labeled as "inversion". Summing up (5.11) with its inversion, we arrive at (5.10).

6. Some properties of the functional $J_m[u]$ in the cone $\mathbb{K}_m(\Omega)$

The functionals $\{J_p\}_1^n$, (3.1) were introduced in the paper [**7**] as straightforward analogs of the classic Dirichlet integral, $m = 1$, in the cones $\mathbb{K}_p$. In the following lemma a natural connection of these was established with $\{I_m\}_1^n$ from above section in C^2.

LEMMA 6.1. *Any $v \in C^2(\bar{\Omega})$ satisfies the equality*

$$(6.1) \qquad J_m[v] = \frac{2m}{m+1}I_m[v] - \frac{1}{m+1}\int_{\partial\Omega}(2vT_m^{ni}[v] + v_x^2 T_{m-1}^{ni}[v])v_i ds.$$

To deduce (6.1) it is sufficient to make use of identities (2.4), (2.5), and Lemma 2.2 after appropriate integration by parts in $J_m[v]$. Consider some particular cases of (6.1). To begin with let $m = n+1$. Since $T_{n+1}[v] \equiv 0$, our identity looks like

$$(6.2) \qquad J_{n+1}[v] = \int_\Omega v_x^2 \det v_{xx} dx = -\frac{1}{n+2}\int_{\partial\Omega} v_x^2 T_n^{ni}[v]v_i ds.$$

We remind the reader that at the boundary by $()_n$ we always bear in mind differentiation along the interior to $\partial\Omega$ normal.

Now let $v|_{\partial\Omega} = 0$. Then it follows from Lemma 2.2 and equality (6.1) that

(6.3)
$$J_m[v] = \frac{2m}{m+1}I_m[v] + \frac{1}{m+1}\int_{\partial\Omega} v_x^2(-v_n)^{m-1}\mathbf{k}_{m-2}[\partial\Omega]ds, \quad m = 1, \ldots, n+1.$$

In view of (6.2) the line (6.3) reads particularly simply, when $m = n+1$. Namely, an arbitrary function $u \in C_0^2(\bar\Omega)$ satisfies

$$\int_\Omega u_x^2(1+u_x^2)^{-\frac{n+2}{2}}\mathbf{G}[\Gamma]dx = \frac{1}{n+2}\int_{\partial\Omega}|u_x|^2(-u_n)^n\mathbf{G}[\partial\Omega]ds. \tag{6.4}$$

Here $\Gamma = \Gamma[u]$ is the graph of function u, and by $\mathbf{G}$ we denote Gauss curvature of the relevant hypersurface.

Now we restrict the situation to functions from $\mathbb{K}_m^0(\bar\Omega)$.

LEMMA 6.2. *If $v \in \mathbb{K}_m^0(\bar\Omega)$, then*

$$J_m[v] = \frac{2m}{m+1}I_m[v] + \frac{1}{m+1}\int_{\partial\Omega}|v_x|^{m+1}\mathbf{k}_{m-2}[\partial\Omega]ds, \quad m = 1,\dots,n+1. \tag{6.5}$$

Indeed, since vanishing at $\partial\Omega$ m-admissible functions are negative in Ω, on $\partial\Omega$ we have the identity $-v_n = |v_x|$, and (6.3) transforms into (6.5) (see Lemma 2.2).

It follows from Lemma 6.2 that values of I_m, J_m may be compared in the cone (4.1). Indeed, by the setting given hypersurface $\partial\Omega$ is the level surface of m-admissible function v. Hence, it is $(m-1)$-convex and $\mathbf{k}_{m-2}[\partial\Omega] > 0$ (see Definition 2.5, Lemma 2.6). But then the identities (6.5) considered in $\mathbb{K}_m^0(\bar\Omega)$ make obvious the inequalities $(m+1)J_m[v] \geq 2mI_m[v]$, $m = 1,\dots,n$, which are strong if $v \neq 0$.

As some kind of application we expose one simple consequence of the X.-J. Wang inequality (4.5) combined with identity (6.5).

THEOREM 6.3. *Functions $v \in \mathbb{K}_m^0(\bar\Omega)$, $m = 1,\dots,n$, satisfy the inequality*

$$\frac{2m}{C}||v||_{L^{m+1}(\Omega)}^{m+1} + \int_{\partial\Omega}|v_x|^{m+1}\mathbf{k}_{m-2}[\partial\Omega]ds \leq (m+1)\int_\Omega |v_x|^2 tr_{m-1}v_{xx}dx. \tag{6.6}$$

Note that we cannot increase inequality (6.6) in accordance with Theorem 5.2 of [**32**]. For example, if $2m > n$, then we can take $\sup_\Omega |v|^{m+1}$ with a large constant $C = C(n,m,\Omega) > 0$ instead of $||v||_{L^{m+1}(\Omega)}^{m+1}$ on the left-hand side of (6.6).

We also note that inequality (4.5) does not work for $m = n+1$. So instead of (6.6) we get an appropriate version of the identity (6.4) for functions vanishing on the $\partial\Omega$. Using a variational approach, we show that the functionals J_m are monotone in $\mathbb{K}_m$ in the sense of the sentence.

THEOREM 6.4. *Let $u, w \in \mathbb{K}_m(\bar\Omega)$. Assume that the following inequality is true:*

$$w|_{\partial\Omega} = u|_{\partial\Omega}, \quad w(x) \leq u(x), \quad x \in \Omega. \tag{6.7}$$

Then

$$J_p[w] \geq J_p[u], \quad p = 1,\dots,m. \tag{6.8}$$

PROOF. Assume that $v = u + t\eta$, $t \in [0;1]$, $\eta = w - u$. The first condition (6.7) gives $\eta = 0$ on $\partial\Omega$, and the identities (2.4), (2.5), and Lemma 2.2 lead to the equality

$$\frac{d}{dt}J_m[v](t) = -2m\int_\Omega \eta T_m[v]dx - \int_{\partial\Omega}\eta_n v_x^2 T_{m-1}^{nn}[v]ds. \tag{6.9}$$

Since conditions (6.7) imply $\eta \leq 0$ in Ω and $\eta_n \leq 0$ on $\partial\Omega$, then the derivative (6.9) is nonnegative for all $t \in [0;1]$, and therefore inequality (6.8) is true. □

Unlike the trivial equality (5.5), $\inf_{\mathbb{K}_m^\phi} J_m[v] > 0$ for $\phi \neq 0$. Moreover, Theorem 1.1 gives sufficient conditions for the existence of an absolute minimizer of the functional J_m in cone (5.1). Namely, the following theorem holds:

THEOREM 6.5. *Let $\partial\Omega$ be a C^4-smooth $(m-1)$-convex surface, $\phi \in C^4(\partial\Omega)$, $\phi \not\equiv 0$. Then there exists an unique function $u^m \in \bar{\mathbb{K}}^{\phi}_m(\bar{\Omega})$ such that*

$$\inf_{\mathbb{K}^{\phi}_m(\bar{\Omega})} J_m[v] = J_m[u^m] > 0, \quad m = 1, \dots, n. \tag{6.10}$$

In addition $u^m \in C^{1,1}(\bar{\Omega})$.

Indeed, in this case the conditions of Theorem 1.1 with $f_m \equiv 0$ are satisfied. On the other hand, the comparison theorem for m-admissible functions implies the following conclusion: $v - u^m \leq 0$ in Ω for all $v \in \mathbb{K}^{\phi}_m(\bar{\Omega})$. So we are in the conditions of Theorem 6.4. Note that we cannot extend the range of m in (6.10) because obviously $J_{n+1}[u^n] \equiv 0$.

References

[1] Ivochkina N. M., About possibility integral formulas in Rn. Zap. Nauchn. Sem. LOMI, 52 (1975), pp. 3551 in Russian. MR0410055 (53:13805)

[2] Ivochkina N. M., Second order equations with d-elliptic operators. Trudy Mat. Inst. Steklov, 147(1980), pp. 40–56 in Russian; English transl. in Proc. Steclov Inst. Math., 2(1981). MR573899 (81h:35018)

[3] Ivochkina N. M., The integral method of barrier functions and the Dirichlet problem for equations with operators of MongeAmpere type. Mathematics of the USSR-Sbornik(1981),40:2, pp. 179–192 in Russian; English transl. in Mat. Sb. (N.S.), 112(156) 1980, pp. 193206. MR585774 (82d:35049)

[4] Ivochkina N. M., A description of the stability cones generated by differential operators of Monge - Ampere type. Mat. Sb., 122(1983), pp. 265–275 in Russian; English transl. in Math. USSR Sb., 50(1985). MR717679 (85g:35043)

[5] N. M. Ivochkina, Solution of the Dirichlet problem for some equations of Monge - Ampere type. Mat. Sb., 128(1985), pp. 403–415 in Russian; English transl. in Math. USSR Sb., 56(1987). MR815272 (88a:35092)

[6] Ivochkina N. M., Solution of the Dirichlet problem for the curvature equation order m. Algebra i Analiz, 2(1990), pp. 192–217 in Russian; English transl. in Leningrad Math. J., 2(1991). MR1073214 (92e:35072)

[7] Ivochkina N. M., The Dirichlet principle in the theory of equations of Monge - Ampere type. Algebra i Analiz, 4(1993) in Russian; English transl. in St. Petersburg Math. J., 4(1993). MR1199638 (94d:35057)

[8] Ivochkina N. M., From Gårding cones to p-convex hipersurfaces. Proc. of the Sixth Int. Conf. on Diff. and Funct.-Diff. Equations (Moscow, August 1421, 2011). Part 1, CMFD, 45, PFUR, M., 2012, pp. 94104. MR3087053

[9] Ivochkina N. M., Prokof'eva S. I., Yakunina G. V., he Garding cones in the modern theory of fully nonlinear second order differential equations. Journal of Mathematical Sciences, July 2012, Volume 184, Issue 3, pp. 295-315. MR2962808

[10] Ladyzhenskaya O. A., Uralt'seva N. N., Linear and Quasilinear Elliptic Equations. M.: Nauka, 1973, 576p. in Russian. MR0509265 (58:23009)

[11] Lychagin V.V., Contact geometry and non-linear second-order differential equations. Russian Mathematical Surveys 34 (1), 1979, 149-180. MR525652 (80h:58057)

[12] Pogorelov A. V., The Minkowsky multidimensional problem. Washington: Scripta, 1979, 97 p. in Russian; English transl. in M.: Nauka, 1975, 96p.

[13] Ball J.M., Convexity conditions and existence theorems in nonlinear elasticity. Arch. Rat. Mech. Analyse, 63 (1997), no. 4, pp. 339-403. MR0475169 (57:14788)

[14] Caffarelli L., Nirenberg L., Spruck J., The Dirichlet problem for nonlinear second order elliptic equations III. Functions of the Hessian. Acta Math. 155, 1985, pp. 261-301. MR806416 (87f:35098)

[15] Chou K.S. (Tso K.), On a real Monge - Ampere functional. Invent. Math. 101 (1990), pp. 425-448. MR1062970 (91i:35082)

[16] Chou Kai-Seng, Wang Xu-Jia, A variational theory of the Hessian equations. Comm. Pure Appl. Math., 54 (2001), pp. 1029-1064. MR1835381 (2002e:35072)

[17] Ivochkina, N. M., Filimonenkova N.V., On the bakgrounds of the theory of m-Hessian equations. Comm. Pure Appl. Anal., 12 (2013), no.4, pp. 1687-1703. MR2997536
[18] Ivochkina N. M., Trudinger N., Wang X.-J., The Dirichlet problem for degenerate Hessian equations. Comm.Partial Differ. Equations., 29 (2004), pp. 219–235. MR2038151 (2005d:35076)
[19] Filimonenkova N.V., On the classical solvability of the Dirichlet problem for nondegenerate m-Hessian equations. J. of Math. Sci., 178 (2011), pp. 666–694. MR3014090
[20] Gårding L., An inequality for hyperbolic polynomials. J. Math. Mech., 8, (1959), pp. 957–965. MR0113978 (22:4809)
[21] Krylov N.V., Sequences of convex functions and estimates of the maximum of the solution of a parabolic equation. Sib. Math. Journal., vol.17, no. 2 (1976), pp. 290–303 in Russian. MR0420016 (54:8033)
[22] Krylov N.V., On the first boundary value problem for nonlinear degenerate elliptic equations. Izv. Akad. Nauk SSSR, ser. math., 51 (1987), no. 2, pp. 242–269 in Russian; English transl. Math. USSR-Izv., 30 (1988), no. 2, MR896996 (88h:35040)
[23] Krylov N.V., On general notion of fully nonlinear second order elliptic equation. Trans. Amer. Math. Soc., 347, (1995), no. 3, pp. 857-895. MR1284912 (95f:35075)
[24] Lions P.L., Trudinger N.S., Urbas Jh.I., The Neumann problem for equations of Monge - Ampere type. Comm. Pure Appl. Math., 39 (1986), pp. 539–563. MR840340 (87j:35114)
[25] Rund H., Integral formulae associated with Euler - Lagrange operators of multiple integral problems in the calculus of variations. Aequations Math., 11 (1974), no. 23, pp. 212–229 MR0361971 (50:14413)
[26] Tian G.-J., Wang X.-J., Moser - Trudinger type inequality for the Hessian equation. J. Funct. Anal. 259 (2010), no.8, pp. 1974–2002. MR2671118 (2011f:35098)
[27] Trudinger N.S., On the Dirichlet problem for Hessian equations. Acta. Math. 175 (1995), pp. 151–164. MR1368245 (96m:35113)
[28] Trudinger N.S., Weak solutions of Hessian equations. Comm. Partial Differential Equation. 22, 1997, pp. 1251–1261. MR1466315 (99a:35077)
[29] Trudinger N.S., Wang X.-J., Hessian measures I. Topol. Methods in Nonlinear Anal., 10 (1997), pp. 225–239. MR1634570 (2000a:35061)
[30] Trudinger N.S., Wang X.-J., A Poincare type inequality for Hessian integrals. Calc. Var. and PDF, 6 (1998), pp. 315–328. MR1624292 (99d:58163)
[31] Trudinger N.S., Wang X.-J., Hessian measures II. Ann. of Math., 150 (1999), pp. 579–604. MR1726702 (2001f:35141)
[32] Wang X.-J., A class of fully nonlinear elliptic equations and related functionals. Indiana Univ. Math. J., 43 (1994), pp. 25–54. MR1275451 (95f:35089)
[33] Urbas Jh.I, Nonlinear oblique boundary value problems for Hessian equations in two dimensions. Ann. Inst. Henri Poincare, 12 (1995), pp. 507–575. MR1353259 (96h:35071)

St. Petersburg State University of Architecture and Civil Engineering, Vtoraya Krasnoarmeiskaya ul. 4, St. Petersburg, 190005, Russia

E-mail address: NF33@yandex.ua

St. Petersburg State University, Universitetskii pr. 28, St. Petersburg, 198504, Russia

E-mail address: ivochkina@gmail.com

Amer. Math. Soc. Transl.
Volume **232**, 2014

The magneto-geostrophic equations: a survey

S. Friedlander, W. Rusin, and V. Vicol

Abstract. We discuss recent results obtained by the authors, regarding the analysis of the magneto-geostrophic equation: a model proposed by Moffatt and Loper to study the geodynamo and turbulence in the Earth's fluid core. We conclude this review by indicating some open problems around the MG equation that remain to be addressed in the future.

1. Introduction

Physicists have long realized the importance of the Earth's magnetic field and that this field originates in the Earth's fluid core. The *geodynamo* is the process by which the rotating, convecting, electrically conducting molten iron in the Earth's fluid core maintains the geomagnetic field against ohmic decay. The convective processes in the core that produce the velocity fields required for dynamo action are a combination of thermal and compositional convection. The full dynamo problem requires the examination of the full three-dimensional partial differential equations governing convective, incompressible magneto-hydrodyamics (MHD). In the past decades computer models have been used to simulate the actual geodynamo. See, for example, the review articles by Roberts et al. on modeling the geodynamo [**BR95, GR97, RG00, RG01**] and references therein. However, as Glatzmaier et al. [**GOC04**] remark, no three-dimensional dynamo model has been run at the spatial resolution required to simulate the broad spectrum of turbulence which surely exists in the Earth's fluid core.

Current computers and numerical methods require the imposition of diffusivities that are several orders of magnitude larger than those which are realistic. It is therefore reasonable to attempt to gain some insight into the geodynamo by considering a reduction of the full MHD equations to a system that is more tractable, but one that retains many of the essential features of the problem. The magneto-geostrophic equation proposed by Moffatt and Loper [**ML94, Mof08**] is one such model. The physical postulates of this model are the following: slow cooling of the Earth leads to slow solidification of the liquid metal core onto the solid inner core and releases latent heat of solidification that drives compositional convection in the

2010 *Mathematics Subject Classification.* Primary 76D03, 35Q35, 76W05.

Key words and phrases. Magneto-geostrophic models, active scalar equations, De Giorgi iteration, nonlinear instability, Hadamard ill-posedness, thin Fourier support.

This survey article was invited for the special volume of "Proceedings of the St. Petersburg Mathematical Society" dedicated to O.A. Ladyzhenskaya.

fluid core. The arguments for the appropriate ranges of the characteristic length, velocity, and perturbation density are based on these physical postulates.

We first present the full coupled three-dimensional MHD equations for the evolution of the velocity vector $u(x,t)$, the magnetic field vector $B(x,t)$, and the buoyancy field $\theta(x,t)$ in the Boussinesq approximation. We write these equations in the frame of reference rotating with angular velocity Ω. The physical forces governing this system are Coriolis forces, Lorentz forces, and gravity. Following the notation of Moffatt and Loper [**ML94**], we write the equations in terms of dimensionless variables. The orders of magnitude of the resulting nondimensional parameters are motivated by the physical postulates of the Moffatt and Loper model:

$$N^2\left[R_o\left(\partial_t u + u\cdot\nabla u\right) + e_3 \times u\right] = -\nabla P + e_2\cdot\nabla b + R_m b\cdot\nabla b + N^2\theta e_3 + \varepsilon\Delta u, \tag{1.1}$$

$$R_m\left[\partial_t b + u\cdot\nabla b - b\cdot\nabla u\right] = e_2\cdot\nabla u + \Delta b, \tag{1.2}$$

$$\partial_t\theta + u\cdot\nabla\theta = \varepsilon_\kappa\Delta\theta + S, \tag{1.3}$$

$$\nabla\cdot u = 0, \nabla\cdot b = 0. \tag{1.4}$$

Here $S(x,t)$ is a given smooth function that represents the forcing of the system.

The mathematical statement of the geodynamo problem asks whether there are initial data for the MHD system for which the evolution of the perturbation of the magnetic field $b(x,t)$ grows for a sufficiently long time. This can be interpreted as a question of the existence of instabilities of (1.1)–(1.4).

The notation in (1.1)–(1.4) is the following. The Cartesian unit vectors are denoted by e_1, e_2, and e_3. For simplicity, we have assumed that the axis of rotation and the gravity g are aligned in the direction of e_3. We have assumed that the magnetic field $B(x,t)$ consists of the sum of an underlying constant field $B_0 e_2$ and a perturbation $b(x,t)$. Our choice of $B_0 e_2$ as the underlying magnetic field is consistent with the models where the magnetic field is believed to be predominantly toroidal (cf. [**ML94**]). The dimensionless parameters are

$$N^2 = \frac{2\Omega\mu_0\eta\rho}{B_0^2}, \quad R_o = \frac{V}{2L\Omega}, \quad R_m = \frac{VL}{\eta}, \quad \varepsilon = \frac{\nu\eta\mu_0\rho}{B_0^2L^2}, \quad \varepsilon_\kappa = \frac{\kappa}{LV}, \tag{1.5}$$

where the typical velocity scale V is chosen to be

$$V = \frac{\theta_0 g}{2\Omega}. \tag{1.6}$$

Here θ_0 is a typical amplitude of the buoyancy θ and L is a characteristic length scale. P is the sum of the fluid and magnetic pressures, ν is the kinematic viscosity, η is the magnetic diffusivity, and κ is the molecular diffusivity of the compositional variation that creates an ambient density ρ. The usual estimate for η in the fluid core is $3\text{m}^2\ \text{cm}^{-1}$. The values of ν and κ are speculative, but likely to be extremely small. For a more detailed discussion of plausible ranges of the physical parameters that are appropriate for the geodynamo, we refer the reader to the book of Ghil and Childress [**GC87**].

For the regions in the Earth's fluid core modeled by Moffatt and Loper, it is argued that the dimensionless parameters are the following orders of magnitude. The inverse Elsasser number N^2 will be of order unity, or somewhat less in the condition of the fluid core, while the Rossby number R_o is of order 10^{-3}. The value of ε is somewhat uncertain since the viscosity ν is a matter of guess work, but plausibly $\varepsilon \leq 10^{-8}$ in the core conditions. The inverse Peclet number ε_κ is also likely to be extremely small, perhaps of the order of 10^{-8}. According to Moffatt and Loper [**ML94**], the magnetic Reynolds number R_m is relatively small in the range of L considered. With these estimates of the sizes of the parameters, the dominant balance of the leading order terms in equations (1.1) and (1.2) give the following reduced system:

$$N^2 e_3 \times u = -\nabla P + e_2 \cdot \nabla b + N^2 \theta e_3, \tag{1.7}$$

$$0 = e_2 \cdot \nabla u + \Delta b \tag{1.8}$$

along with the incompressibility condition

$$\nabla \cdot u = 0, \quad \nabla \cdot b = 0. \tag{1.9}$$

The *linear* system of equations (1.7)–(1.9) determine the differential operators that relate the vector fields $u(x,t)$ and $b(x,t)$ with the scalar buoyancy $\theta(x,t)$. These operators encode the vestiges of the physics in the problem, namely Coriolis force, Lorentz force, and gravity. Vector manipulations of (1.7)–(1.9) give the expression

$$\begin{aligned} &\left[N^4 (e_3 \cdot \nabla)^2 \Delta + (e_2 \cdot \nabla)^4\right] u \\ &\quad = N^4 (e_3 \cdot \nabla) \Delta (e_3 \times \nabla \theta) + N^2 (e_2 \cdot \nabla)^2 \nabla \times (e_3 \times \nabla \theta). \end{aligned} \tag{1.10}$$

The sole remaining nonlinearity in the system comes from the coupling of (1.10) with the equation (1.4) for the time evolution of the $\theta(x,t)$. We call this nonlinear active scalar equation the *magneto-geostrophic* (MG) equation. It is the subject of this article.

We study the properties and behavior of the active scalar equation in $\mathbb{T}^3$

$$\partial_t \theta + u \cdot \nabla \theta = \varepsilon_\kappa \Delta \theta + S, \tag{1.11}$$

$$u = M[\theta] \tag{1.12}$$

via an examination of the Fourier multiplier operator M obtained from (1.10), which relates u and θ. More precisely,

$$u_j = M_j[\theta] = (\widehat{M}_j \widehat{\theta})^\vee$$

for $j \in \{1,2,3\}$, and the explicit expressions for the components of $\widehat{M}$ as functions of the Fourier variable $k = (k_1, k_2, k_3) \in \mathbb{Z}^3_* := \mathbb{Z}^3 \setminus \{k_3 = 0\}$ are

$$\widehat{M}_1(k) = \left(N^4 k_2 k_3 |k|^2 - N^2 k_1 k_2^2 k_3\right) D(k)^{-1}, \tag{1.13}$$

$$\widehat{M}_2(k) = \left(-N^4 k_1 k_3 |k|^2 - N^2 k_2^3 k_3\right) D(k)^{-1}, \tag{1.14}$$

$$\widehat{M}_3(k) = \left(N^2 k_1^2 k_2^2 + N^2 k_2^4\right) D(k)^{-1}, \tag{1.15}$$

where

$$D(k) = N^4 |k|^2 k_3^2 + k_2^4. \tag{1.16}$$

On $\{k_3 = 0\}$ we define $\widehat{M}_j(k) = 0$, for all $j \in \{1,2,3\}$, since for self-consistency of the model we assume that θ and u have zero vertical mean, i.e., $\int \theta dx_3 = \int u dx_3 = 0$. Several important properties of the $\widehat{M}_j$'s are immediately obvious. First, the functions are strongly anisotropic with respect to the dependence on the integers k_1, k_2, and k_3. This is a consequence of the interplay of the three above-mentioned physical forces. Second, the function $\widehat{M}_3$ is an even function of k_1, k_2, and k_3.

Third, we have $\widehat{M_j}(k)k_j = 0$ for any $k \in \mathbb{Z}^3$ (we use the standard summation convention on repeated indices), which is built into the system since the velocity field is divergence-free

$$\nabla \cdot u = 0.$$

Last, although the symbols $\widehat{M_j}$ are bounded (by a constant that depends only on N) in regions of Fourier space where $k_2 \leq \max\{k_2, k_3\}$; this is not the case in the "curved" regions where $k_3 = \mathcal{O}(1)$, and $k_2 = \mathcal{O}(|k_1|^r)$ with $0 < r \leq 1/2$. In such regions the symbols are unbounded as $|k_1| \to \infty$ with

$$|\widehat{M}(k)| \leq C|k|, \tag{1.17}$$

for some suitable chosen constant C that *is independent of* N, and this bound is sharp. Hence the MG equation falls in the class of abstract active scalar equations (see (2.1)–(2.2) below) where the relation between the divergence-free velocity field u and the scalar θ is given as a singular operator of order 1. These properties of $\widehat{M}$ make the MG equation interesting and challenging mathematically, as well as having a clear physical basis in its derivation from the MHD equations.

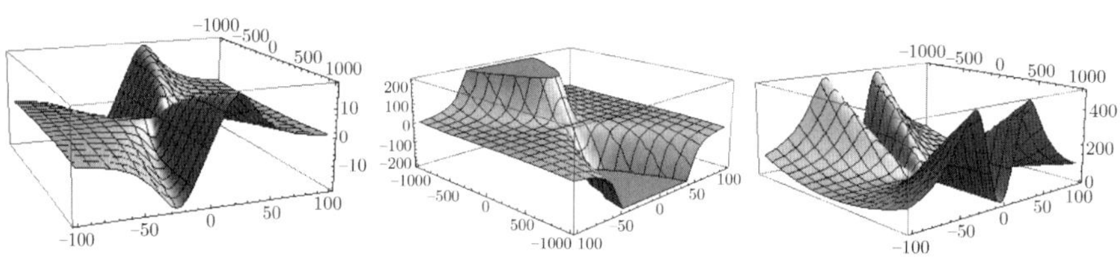

FIGURE 1. From left to right: plot of $\widehat{M_1}(k_1, k_2, 1)$, $\widehat{M_2}(k_1, k_2, 1)$, and $\widehat{M_3}(k_1, k_2, 1)$, with $N = 1$. The range of the frequencies is $k_1 \in [-10^3, 10^3]$ and $k_2 \in [-10^2, 10^2]$.

In Section 2 we state results proved in [**FV11a, FV11b, FV12**] for the diffusive MG equation, i.e., the case when $\varepsilon_\kappa > 0$. We recall that in the fluid core ε_κ is positive but probably extremely small. In this case the MG equation is globally well posed. To prove this fact, we use the control on $\|\theta(t)\|_{L^\infty}$, a quantity that is nonincreasing along solutions, and which gives a bound on the drift velocity u in the space $L_t^\infty BMO_x^{-1}$. The latter information on the drift is "critical" for the natural scaling of the *linear* drift-diffusion equation

$$\partial_t \theta + v \cdot \nabla \theta = \Delta \theta. \tag{1.18}$$

Indeed, for $v \in L_t^\infty BMO_x^{-1} \cap L_t^2 L_x^2$ divergence-free, we proved in [**FV11a**] that the solution θ of (1.18) becomes Hölder continuous in finite time, by appealing to the classical ideas of De Giorgi [**DG57**], in the spirit of [**CV10**]. This result was also established independently in [**SSŠZ12**]. See also [**Nas58, Osa87, Zha04, MV06**], the more recent [**NU12, SV12, Fil12, IKR12**], and references therein, for regularity issues of linear parabolic equations with singular coefficients and divergence-free drift. The Hölder continuity of solutions to the linear equation (1.18) may then be bootstrapped to show that the solutions of the dissipative MG equations become smooth in arbitrarily short finite time [**FV11b**].

Having established the global well-posedness of (1.11)–(1.12) for any $\varepsilon_\kappa > 0$, we can proceed to examine the stability of equilibria. We show that the existence of an unstable eigenvalue for the linearized equations implies nonlinear (Lyapunov) instability of the nonlinear MG equation (see also [**FSV97, BGS02, VF03, FPS06, FPV09**]). We then exhibit the existence of an unstable eigenvalue in a particular example, using the method of continued fractions; cf. [**MS61, FSV97**]. The magnitude of this eigenvalue is $\mathcal{O}(\varepsilon_\kappa^{-1})$ which, in the context of the fluid core, is extremely large. Thus the MG equation can be used to model very rapid exponential growth in time of θ, and hence, through (1.11)–(1.12), of the perturbation velocity and magnetic fields, i.e.; a dynamo type instability. We emphasize that this "MG-dynamo" instability requires the presence of all three forces—Coriolis, Lorentz, and gravity—because the dimensionless parameter N^2 (defined above in (1.5)) must be strictly in the range $(0, \infty)$.

In Section 3 we examine the nondiffusive MG equation and set $\varepsilon_\kappa = 0$ (cf. [**FV11b**]). The situation is then dramatically different from the case $\varepsilon_\kappa > 0$. Without the Laplacian to control the unbounded operator M, the MG equation is *ill-posed* in the sense of Hadamard in Sobolev spaces. Such a result utilizes in an essential way both the unboundedness of $\widehat{M}_3$ and the fact that it is an even function of k. The proof again uses the existence of an unstable eigenvalue of the linearized equation; however, when $\varepsilon_\kappa = 0$ it is possible to construct such an eigenvalue that is arbitrarily large, and it then follows that there is no Lipschitz solution map at time $t = 0$ (see also [**Tao06, Ren09, GVD10, GVN10, GN10**]).

In Section 4 we describe results for the fractionally diffusive MG equation (cf. [**FRV12**]), i.e., replace the Laplacian by $-(-\Delta)^\gamma$, for $\gamma \in (0,1)$. This situation is nonphysical but mathematically interesting in the context of the tools required to treat the fractional Laplacian. We now have a dichotomy across the value $\gamma = 1/2$. In the range $1/2 < \gamma < 1$ the equation is *locally* well-posed, while it is Hadamard *Lipschitz ill-posed* for $0 < \gamma < 1/2$. At the critical value $\gamma = 1/2$ the problem is globally well-posed for suitably small initial data (and source term), but is ill-posed for sufficiently large initial data. In terms of the nondimensional parameters introduced above in (1.5), this dichotomy can be seen as global well-posedness for $\varepsilon_\kappa \gg 1$, and Lipschitz ill-posedness for $\varepsilon_\kappa \ll 1$. A further feature of interest is that the anisotropy of the symbol can be explored to obtain an improvement in the regularity of the solutions when the initial data and source are *supported on a plane in Fourier space*. For such well-prepared data the local existence and uniqueness of solutions can be obtained for all values $\gamma \in (0,1)$, and the global existence holds for all initial data when $\gamma \in (1/2, 1)$.

We note that some of the theorems discussed in this article are valid for certain classes of partial differential equations (cf. (2.1)–(2.2) below) and in any spatial dimension $d \geq 2$. For such theorems, the three-dimensional MG equation is a particular example of an active scalar equation for which these results hold.

We conclude the manuscript with Section 5, in which we present a few remaining open problems regarding the mathematical analysis of the MG equations.

2. Diffusive MG equation

In this section we discuss the global regularity of solutions to (1.11)–(1.12) and the Lyapunov stability of steady states, in the case that $\varepsilon_\kappa > 0$.

2.1. Global regularity of singular drift-diffusion equations. In order to show that there is no finite time loss of regularity of solutions to (1.11)–(1.12) with $\varepsilon_\kappa > 0$, let us first perform a rough derivative count. The dissipation is smoothing by two derivatives, while the nonlinearity is also losing two derivatives: the first coming from $\nabla\theta$, while the second one derivative loss is coming from the constitutive law $u = M[\theta]$, in view of (1.17). In this sense we may call the problem "critical" (we make this more precise below).

Since the operator M relating u and θ is an order 1 pseudo-differential operator with $\widehat{M}(k) \cdot k = 0$ for any wave-number k, one may view the MG equation as an example of a more general class of of active scalar equations:

$$\partial_t \theta + (u \cdot \nabla)\theta = \Delta\theta, \tag{2.1}$$

$$u_j = \partial_i T_{ij}\theta, \quad \partial_j u_j = 0, \tag{2.2}$$

with T_{ij} a matrix of zero-order pseudo-differential operators, which is skew-symmetric, i.e., $T_{ij} = -T_{ji}$. Here and throughout the paper we use the summation convention on repeated indices. We have dropped the smooth source term S from (1.11) since it is not an obstacle to the well-posedness theory, and we have rescaled the time variable so that the diffusivity coefficient is 1 instead of ε_κ (this is not an issue since we will construct global in time solutions). The matrix $\{T_{ij}\}_{i,j=1}^3$ corresponding to the MG equations (1.12) is given by

$$T = \begin{pmatrix} 0 & N_3 & -N_2 \\ -N_3 & 0 & N_1 \\ N_2 & -N_1 & 0 \end{pmatrix}; \quad \text{where} \quad N = (-\Delta)^{-1}\nabla \times M. \tag{2.3}$$

More explicitly, we have that N is given as the Fourier multiplier operator with symbol

$$\widehat{N}_1(k) = \left(N^4 k_1 k_3^2 + N^2 k_2^3\right) D(k)^{-1}, \tag{2.4}$$

$$\widehat{N}_2(k) = \left(N^4 k_2 k_3^2 - N^2 k_1 k_2^2\right) D(k)^{-1}, \tag{2.5}$$

$$\widehat{N}_3(k) = \left(-N^4 k_3 (k_1^2 + k_2^2)\right) D(k)^{-1}, \tag{2.6}$$

and $D(k)$ is defined in (1.16). In particular, one may check from (2.4)–(2.6) that there exits $C > 0$ such that $|\widehat{N}_j(k)| \le C$ for any $k \in \mathbb{Z}^3$ with $k_3 \neq 0$. Thus the matrix T given in (2.3), corresponding to the MG equations, indeed has entries which are bounded Fourier multipliers, and hence zero-order operators.

It is clear that any M such that $\partial_j M_j = 0$ may we written as $M_j = \partial_i T_{ij}$ for a skew-symmetric matrix T (see also [**SSŠZ12**]). We call the class of active scalar equations (2.1)–(2.2) "singular" since the drift velocity is given in terms of the advected scalar by a constitutive law which is losing derivatives. We refer the interested reader in to the work of Chae, Constantin, Cordoba, Gancedo, and Wu [**CCC$^+$12**] and also [**FGSV12, MX12**] for further discussions about the well-posedness theory of singular active scalar equations.

To make the idea of criticality more precise, we note that the active scalar equation (2.1)–(2.2) has a natural scaling invariance built into it: if $\theta(t, x)$ is a solution, then $\theta_\lambda(t, x) = \theta(\lambda^2 t, \lambda x)$ is *also* a solution, with corresponding drift given by $u_\lambda(t, x) = \lambda u(\lambda^2 t, \lambda x) = \partial_i T_{ij}\theta_\lambda(t, x)$. Hence $L^\infty(\mathbb{R}^d)$ is the critical Lebesgue space with respect to the natural scaling of the equation. Additionally (2.1) is a scalar equation, so that L^p-norms are nonincreasing in time for $1 \le p \le \infty$, and hence the L^∞-norm is the "strongest" a priori controlled norm along solutions. In this sense we say that (2.1)–(2.2) is a "critical" equation.

Another active scalar equation derived from the Navier–Stokes equations for a rapidly rotating fluid in a thin shell is the surface quasi-geostrophic (SQG) equation, introduced by Constantin, Majda, and Tabak [**CMT94**] (see also [**CCW01, CC04, Wu05, Kis10**] and references therein), which in the critically dissipative case reads

$$\partial_t \theta + (u \cdot \nabla)\theta = -(-\Delta)^{1/2}\theta, \tag{2.7}$$

$$u = \nabla^{\perp}(-\Delta)^{-1/2}\theta \equiv (R_2\theta, -R_1\theta), \tag{2.8}$$

and R_i represents the ith Riesz transform. There are strong analogies between the systems (2.7)–(2.8) and (2.1)–(2.2). The velocity field is divergence-free in both cases, and the L^∞-norm is critical with respect to the natural scaling of the equation. The main difference is that for SQG the dissipative operator is nonlocal, but the drift velocity is not more singular than the advected scalar.

Inspired by the De Giorgi-type proof of global regularity for the critically dissipative SQG equation of Caffarelli and Vasseur [**CV10**] (see also [**KNV07, KN09, CV12**] for different proofs of global well-posedness of the critical SQG), to show that solutions of (2.1)–(2.2) do not blow up in finite time, we first look at the associated linear problem, where the drift velocity is *given* and obeys the bounds that are a priori available for weak solutions of the nonlinear problem: $u \in L^\infty(0,\infty; BMO^{-1}) \cap L^2(0,\infty;L^2)$. In [**FV11a**] we use a De Giorgi iteration technique to prove that weak solutions of the *linear* drift-diffusion equation

$$\partial_t \theta + (v \cdot \nabla)\theta = \Delta\theta, \tag{2.9}$$

$$\nabla \cdot v = 0, \quad v_j = \partial_i V_{ij}, \tag{2.10}$$

$$V_{ij} \in L^\infty(0,\infty; BMO) \cap L^2(0,\infty; H^1), \quad V_{ij} = -V_{ji}, \tag{2.11}$$

instantly become Hölder continuous. More precisely, the result is:

THEOREM 2.1 ([**FV11a**]). *Let the initial data $\theta_0 \in L^2$ and $\theta \in L^\infty(0,\infty;L^2) \cap L^2(0,\infty;H^1)$ be a weak solution of* (2.9), *where the drift velocity v is given by* (2.10)–(2.11) *for a suitable matrix V_{ij}. Then, for any t_0, there exists $\alpha > 0$ such that $\theta \in C^{\alpha/2}(t_0,\infty;C^\alpha)$.*

The above result was obtained independently by Sergin, Silvestre, Sverak, and Zlatos in [**SSŠZ12**]. See also [**NU12**] for a closely related result. In view of the constitutive law (2.2) given explicitly by (2.3), a direct corollary of Theorem 2.1 is

COROLLARY 2.2. *Weak solutions θ of* (2.1)–(2.2) *are Hölder continuous in positive time; i.e., they obey $\theta \in L^\infty(t_0,\infty;C^\alpha)$ for any $t_0 > 0$, and a suitable $\alpha > 0$.*

In view of the above discussion about scaling, Hölder regularity is subcritical information for (2.1)–(2.2), and one may bootstrap to prove that the solution is classical:

THEOREM 2.3 ([**FV12**]). *Let $\theta \in L^\infty(t_0,\infty;C^\alpha)$ be a weak solution of the initial value problem associated to* (2.1)–(2.2), *with $\alpha \in (0,1)$. Then $\theta \in L^\infty(t_1,\infty;C^{1+\delta})$, for any $t_1 > t_0$, for some $\delta \in (0,1)$.*

The proof of Theorem 2.3 relies on a suitable bootstrap procedure in Besov spaces, in the spirit of [**CW08**]. Repeating the argument for the derivative of θ, one may further obtain that the solution of (2.1)–(2.2) is in fact C^∞-smooth in finite time.

We conclude this subsection with a very brief discussion of the proof of Theorem 2.1. We refer the reader to [**FV11a**] for all details. Along the lines of Caffarelli and Vasseur [**CV10**, Theorem 3] for the critical SQG equation, the first step consists of showing that a weak solution is bounded in the spatial variable (see [**FV11a**, Appendix A] for a construction of weak solutions which obey the energy inequalities). This is achieved by proving that the function $(\theta - h)_+ = \max\{\theta - h, 0\}$ has zero energy if h is chosen large enough. The key ingredient is the inequality

$$\partial_t(\theta - h)_+ - \Delta(\theta - h)_+ + (v \cdot \nabla)(\theta - h)_+ \leq 0, \tag{2.12}$$

which can be shown to hold for weak solutions, in the sense of distributions. Indeed, given a weak solution $\theta \in L_t^\infty L_x^2 \cap L_t^2 H_x^1$, we have that for any ϕ which is Lipschitz continuous, that $\nabla\phi(\theta) \in L_{t,x}^2$, and since $v \in L_{t,x}^2$ is divergence-free, one may give meaning in the sense of distributions to the equality $\phi'(\theta)v \cdot \nabla\theta = v \cdot \nabla\phi(\theta) = \nabla \cdot (v\phi(\theta))$. To obtain (2.12) one then approximates $(\theta - h)_+$ by $\phi(\theta)$, for a suitable sequence of C^2 smooth, nondecreasing, convex functions ϕ. See e.g., [**CV10, NU12**] for details.

Using that v is divergence-free and that $v = \nabla \cdot V$ for a skew-symmetric matrix $V = \{V_{ij}\} \in L_t^\infty BMO_x$, or equivalently that $(-\Delta)^{-1}\nabla \times v \in L_t^\infty BMO_x$, one may use the argument in [**MV06**] to prove from (2.12) that the energy inequality

$$\int_{\mathbb{R}^d} |(\theta(t_2, \cdot) - h)_+|^2 dx + 2\int_{t_1}^{t_2} \int_{\mathbb{R}^d} |\nabla(\theta - h)_+|^2 dx dt \leq \int_{\mathbb{R}^d} |(\theta(t_1, \cdot) - h)_+|^2 dx \tag{2.13}$$

holds for all $h > 0$ and $0 < t_1 < t_2 < \infty$. This was already observed in [**NU12**, p. 100], where the localized version of (2.13) is also discussed (see also [**SSŠZ12**]). The argument in [**MV06**] is quite ingenious and uses a hidden cancellation property which arises due to the skew-symmetry of V. Indeed, one may show that the form

$$B(\phi, \eta) = \int_{t_1}^{t_2} \int_{\mathbb{R}^d} \partial_i V_{ij}\, \phi\, \partial_j \eta\, dx dt \tag{2.14}$$

extends continuously to $L_t^2 H_x^1 \times L_t^2 H_x^1$ by factoring the form into determinants which have better than expected regularity: using compensated-compactness, the determinants may be shown to lie not just in $L_{t,x}^1$ but in fact in $L_t^1 \mathcal{H}_x^1$ (here $\mathcal{H}^1$ the Hardy space, precisely the dual of BMO). Since $\nabla \cdot v = 0$, we have $B(\phi, \eta) = -B(\eta, \phi)$ for $\eta, \phi \in L_t^2 H_x^1$, and in view of the above discussion we may therefore infer $B((\theta - h)_+, (\theta - h)_+) = 0$ to obtain (2.13). We also refer the reader to [**Kuk06**] for the validity of the energy inequalities for weak solutions of the Navier–Stokes equations, given a priori assumptions on only the pressure. We note that the system under consideration in our present paper, being scalar, does not contain a pressure term.

For $t_0 > 0$, and $H > 0$ to be chosen sufficiently large, letting $t_n = t_0 - t_0/2^n$, $h_n = H - H/2^n$, and

$$c_n = \sup_{t \geq t_n} \int_{\mathbb{R}^d} |(\theta(t, \cdot) - h_n)_+|^2 dx + 2\int_{t_n}^{\infty} \int_{\mathbb{R}^d} |\nabla(\theta - h_n)|^2 dx dt,$$

it follows from (2.13) that

$$c_{n+1} \leq \frac{C}{t_0 H^{4/d}} 2^{n(1+4/d)} c_n^{1+2/d}$$

for some constant $C > 0$. Letting $H = H(\|\theta_0\|_{L^2(\mathbb{R}^d)}, t_0^{d/4})$ be sufficiently large, we thus obtain that $c_n \to 0$ exponentially as $n \to \infty$, and hence $\theta(t_0, \cdot) \leq H$.

Having established the boundedness of the weak solutions, the next step is to prove Hölder regularity, which requires a more delicate argument. The main obstruction to applying the classical parabolic De Giorgi estimates via an L^p-based Caccioppoli inequality ($1 < p < \infty$) is that $v_j = \partial_i V_{ij}$ is the derivative of a BMO function. By essentially using the divergence-free nature of v and appealing to the John–Nirenberg inequality, we may however prove a suitable local energy inequality: for any $0 < r < R$ and $h \in \mathbb{R}$, we have

$$\begin{aligned} &\|(\theta - h)_+\|^2_{L_t^\infty L_x^2(Q_r)} + \|\nabla(\theta - h)_+\|^2_{L^2_{t,x}(Q_r)} \\ &\leq \frac{CR}{(R-r)^2} \|(\theta - h)_+\|^{2-\frac{2}{d+2}}_{L^2_{t,x}(Q_R)} \|(\theta - h)_+\|^{\frac{2}{d+2}}_{L^\infty_{t,x}(Q_R)}, \end{aligned} \tag{2.15}$$

where $C = C(d, \|V_{ij}\|_{L_t^\infty BMO_x})$ is a fixed positive constant. Here we have denoted by Q_ρ the parabolic cylinder $[t_0 - \rho^2, t_0] \times B_\rho(x_0)$ for $\rho > 0$ and $(t_0, x_0) \in (0, \infty) \times \mathbb{R}^d$. Moreover, estimate (2.15) also holds with θ replaced by $-\theta$. The bound (2.15) may be in turn used, using a suitable iteration argument to obtain an improvement of oscillation result. Assume that $h_0 \leq \sup_{Q_{r_0}} \theta$, where $r_0 > 0$ is arbitrary. Then we have

$$\sup_{Q_{r_0/2}} \theta \leq h_0 + C \left(\frac{|\{\theta > h_0\} \cap Q_{r_0}|^{1/(d+2)}}{r_0} \right)^{1/2} \left(\sup_{Q_{r_0}} \theta - h_0 \right) \tag{2.16}$$

for some positive constant $C = C(d, \|V_{ij}\|_{L_t^\infty BMO_x})$. We note that the prefactor of $\sup_{Q_{r_0}} \theta - h_0$ of the right-hand side of (2.16) *does not scale with* r_0. Besides estimate (2.16), the main ingredient in the proof of Hölder continuity is the following estimate on the growth of level sets of the solution. There exist $\kappa_0 \in (0,1)$, $\delta_0 = \delta_0(\kappa_0, \|V_{ij}\|_{L_t^\infty BMO_x})$, and $n_0 \geq 2$ such that for $t_1, R > 0$ if we know that

$$|\{\theta(t_1, \cdot) \geq h\} \cap B_r| \leq \frac{1}{2} |B_r|,$$

then for all $t_2 \in [t_1, t_1 + \delta_0 r^2]$ we have

$$|\{\theta(t_2, \cdot) \geq H\} \cap B_R| \leq \frac{7}{8} |B_R|, \tag{2.17}$$

where $r = \kappa_0 R$, $M = \sup_{(t_1, t_1+\delta_0 R^2) \times B_R} \theta$, $m = \inf_{(t_1, t_1+\delta_0 R^2) \times B_R} \theta$, $h = (M + m)/2$, and $H = M - (M - m)/2^{n_0}$. To prove the above estimate, one uses a suitable version of the local parabolic energy inequality.

Bounds (2.16) and (2.17) are the main ingredients in the De Giorgi iteration procedure, which at the end of the day yields the improvement in oscillation

$$\left(\sup_{Q_1} \theta - \inf_{Q_1} \theta \right) \leq \gamma \left(\sup_{Q_2} \theta - \inf_{Q_2} \theta \right) \tag{2.18}$$

for some $\gamma \in (0,1)$, where $Q_1 = [t_1, t_1 + \delta_0 r^2] \times B_4$, $Q_2 = [t_1, t_1 + \delta_0 R^2] \times B_R$, $t_1, R > 0$ are arbitrary, $r = \kappa_0 R$, and $\delta_0, \kappa_0 \in (0,1)$ is as above. Since $\gamma < 1$, it is then classical to show that (2.18) implies the solution θ is Hölder continuous, with exponent that may be computed explicitly.

2.2. Lyapunov stability of steady states. In this subsection we consider the stability of steady states to (1.11)–(1.12). First we prove for steady states of the more general class of equations (1.11), with drift velocity given by (2.2), that linear implies nonlinear instability. We then turn to the specific MG equations (1.11)–(1.12), for which we construct an explicit unstable eigenvalue of the linear operator.

2.2.1. *Linear implies nonlinear instability.* Let $\Theta_0 \in C^\infty$ be a steady solution of (1.11)–(2.2), i.e.,

$$U_0 \cdot \nabla \Theta_0 = \varepsilon_\kappa \Delta \Theta_0 + S, \quad U_{0j} = \partial_i T_{ij} \Theta_0, \quad \nabla \cdot U_0 = 0, \tag{2.19}$$

where $S \in C^\infty$ is a time-independent force, $\varepsilon_\kappa > 0$, and T_{ij} are zero-order pseudo-differential operators. Writing the solution of (1.11)–(2.2) as $\theta + \Theta_0$, we obtain that the perturbation temperature θ obeys

$$\partial_t \theta = L\theta + N\theta, \tag{2.20}$$

where the *dissipative linear* operator L is defined as

$$L\theta = -U_0 \cdot \nabla \theta - u \cdot \nabla \Theta_0 + \varepsilon_\kappa \Delta \theta, \tag{2.21}$$

with $u_j = \partial_i T_{ij} \theta$, and the *nonlinear* operator N is given by

$$N\theta = -u \cdot \nabla \theta = -\nabla \cdot (u\theta). \tag{2.22}$$

Recall the following notion of Lyapunov stability:

DEFINITION 2.4 ([**FSV97**]). *Let (X, Z) be a pair of Banach spaces. A steady state Θ_0 is called (X, Z) nonlinearly stable if for any $\rho > 0$, there exists $\widetilde{\rho} > 0$ so that if $\theta(\cdot, 0) \in X$ and $\|\theta(\cdot, 0)\|_Z < \widetilde{\rho}$, then*

(i) *there exists a global in time solution θ to the initial value problem* (2.20)–(2.22) *with $\theta \in C([0, \infty); X)$;*
(ii) *and we have the bound $\|\theta(\cdot, t)\|_Z < \rho$ for a.e. $t \in [0, \infty)$.*

An equilibrium Θ_0 that is not stable in the above sense is called Lyapunov unstable.

In the above definition, the Banach space X is the space where a local existence theorem for the nonlinear equations is available (e.g., $X = H^s$), while Z is the space where the spectrum of the linear operator is analyzed and where the instability is measured (e.g., $Z = L^2$). The following result states that the linear Lyapunov instability of a steady state Θ_0 implies the nonlinear one.

THEOREM 2.5 ([**FV11b**]). *Let Θ_0 be a smooth, mean zero solution of* (2.19). *If the associated linear operator L, as defined in* (2.21), *has spectrum in the unstable region, then the steady state is (H^s, L^2) Lyapunov nonlinearly unstable, for arbitrary $s > d/2$.*

The proof relies on a bootstrap argument, that is by now well-known for fluids equations [**FSV97, BGS02, VF03, FPS06, FPV09**], and an a priori time-independent bound on the H^2-norm of the solution, which we obtain from the De Giorgi C^α estimate.

We only briefly sketch the argument given in full detail in [**FV11b**], to which refer the interested reader. First, we fix a linearly unstable smooth eigenfunction

ϕ, with eigenvalue of *maximal* real part λ. In order to prove Theorem 2.5, for $\varepsilon \in (0, \varepsilon_*]$ we consider solutions θ^ε of

$$\partial_t \theta^\varepsilon = L\theta^\varepsilon + N(\theta^\varepsilon), \quad \theta^\varepsilon|_{t=0} = \varepsilon\phi, \tag{2.23}$$

where L, N, and ϕ are as above, and we show that there exits a constant $C_* > 0$ such that for any $\varepsilon \in (0, \varepsilon_*]$, we may find a time $T_\varepsilon > 0$ with

$$\|\theta^\varepsilon(T_\varepsilon)\|_{L^2} \geq C_*. \tag{2.24}$$

By Definition 2.4, the bound (2.24) shows that the trivial solution $\theta = 0$ of (2.20)–(2.22) is (H^s, L^2) Lyapunov unstable: indeed, an $\mathcal{O}(\varepsilon)$ perturbation leads in finite time to an $\mathcal{O}(1)$-size solution, in finite time.

In order to prove (2.24) we use the following classical idea: the initial data, being an eigenfunction of L, will grow exponentially in time, while the nonlinear term may be suitably controlled so that it cannot cancel this exponential growth. To make this idea concrete, it is convenient to define $L_\delta = L - (\lambda + \delta)I$, where $0 < \delta \leq \lambda/8$. Due to the shift, the resolvent of L_δ contains the full right-half of the complex plane, and being a bounded perturbation of the Laplacian, we also have that L_δ generates an analytic semigroup over L^2, and L_δ^{-1} is smoothing by two derivatives. Then from the Duhamel formula and the interpolation bound $\|L_\delta^{-7/8} N(\theta^\varepsilon)\|_{L^2} \leq C\|\theta^\varepsilon\|_{L^2}^{9/8}\|\Delta\theta^\varepsilon\|_{L^2}^{7/8}$, we obtain the estimate

(2.25)
$$\begin{aligned}\|\theta^\varepsilon(t)\|_{L^2} &\leq \|\varepsilon e^{Lt}\phi\|_{L^2} + \int_0^t e^{(\lambda+\delta)(t-s)}\|L_\delta^{7/8} e^{L_\delta(t-s)}\|_{\mathcal{L}(L^2)}\|L_\delta^{-7/8} N(\theta^\varepsilon(s))\|_{L^2} ds \\ &\leq C_\phi \varepsilon e^{\lambda t} + C\int_0^t e^{(\lambda+\delta)(t-s)}\frac{1}{(t-s)^{7/8}}\|\theta^\varepsilon(s)\|_{L^2}^{9/8}\|\Delta\theta^\varepsilon(s)\|_{L^2}^{7/8} ds.\end{aligned}$$

The key step which allows us to close the estimates is proving the existence of a constant $C^* = C^*(\Theta_0, \phi, S)$ such that

$$\|\Delta\theta^\varepsilon\|_{L^2} \leq C^* < \infty \tag{2.26}$$

for all sufficiently small ε. The proof of estimate (2.26) relies on the C^α-bound which may be obtained from Theorem 2.1 and a suitable bootstrap procedure to reach the H^2 regularity level (see [**FV11b**, Appendix A]). Lastly, let $R > C_\phi := \|\phi\|_{L^2}$, and let $T = T(R, \varepsilon)$ be the maximal time such that

$$\|\theta^\varepsilon(t)\|_{L^2} \leq \varepsilon R e^{\lambda t} \quad \text{for all} \quad 0 \leq t \leq T.$$

From (2.25) and (2.26) one may then obtain

$$\|\theta^\varepsilon(t)\|_{L^2} \leq C_\phi \varepsilon e^{\lambda t} + C_1\left(\varepsilon R e^{\lambda t}\right)^{9/8} \tag{2.27}$$

and then in turn that

$$T \geq T_\varepsilon = \lambda^{-1}\log(C_2/\varepsilon).$$

A lower bound on $\|\theta^\varepsilon(t)\|_{L^2}$, similar to (2.27) but with the linear term minus the 9/8 power term, may be obtained from the Duhamel formula. In turn, this implies that $\|\theta^\varepsilon(T_\varepsilon)\|_{L^2} \geq C_2(2C_\phi - R) =: C_* > 0$ and concludes the proof of Theorem 2.5.

2.2.2. *The existence of unstable eigenvalues for the linearized MG operator.* In order to apply Theorem 2.5 to the (nonlinear) MG equations (1.11)–(1.12), we need to find a steady state Θ_0 for which the corresponding linear operator L defined in (2.21) has unstable spectrum.

To achieve this we use an idea that dates back at least to [**MS61**] for Navier–Stokes and [**FSV97**] for Euler. First, note that $\Theta_0 = F(x_3)$, with F smooth, is automatically a steady state, with associated velocity $U_0 = 0$, if we assume that the source term is given by $S = -\varepsilon_\kappa F''(x_3)$. We thus consider the source term to equal $S = \varepsilon_\kappa m^2 \sin(mx_3)$, for which we have a steady state solution of (1.11) given by

$$\Theta_0 = \sin(mx_3), \quad U_0 = 0. \tag{2.28}$$

For the above choice of the steady states, the equation $\partial_t \theta = L\theta$ becomes

$$\partial_t \theta(x,t) = L\theta(x,t) = -mu_3(x,t)\cos(mx_3) + \varepsilon_\kappa \theta(x,t), \quad u_3 = M_3[\theta], \tag{2.29}$$

where M_3 is the Fourier multiplier defined earlier in (1.15)–(1.16). Next, we make the *ansatz* that the temperature θ which solves (2.29) has the form

$$\theta(x,t) = e^{\sigma t} \sin(k_1 x_1)\sin(k_2 x_2) \sum_{p\geq 1} c_p \sin(mpx_3) \tag{2.30}$$

for some fixed k_1, k_2 in $\mathbb{Z}_*$, to be chosen precisely later. Using (1.15)–(1.16) we may compute from the third component of the velocity field u associated with θ given by (2.30), insert this into (2.29), and obtain that σ and $\{c_p\}_{p\geq 1}$ must solve

$$\sigma_p \sum_{p\geq 1} c_p \sin(mpx_3) + \sum_{p\geq 1} \frac{c_p}{\alpha_p} \left(\sin(m(p+1)x_3) + \sin(m(p-1)x_3)\right) = 0, \tag{2.31}$$

where we have denoted

$$\sigma_p = \sigma + \varepsilon_\kappa \left(k_1^2 + k_2^2 + (mp)^2\right) \tag{2.32}$$

and

$$\frac{2}{\alpha_p} = mN^2 \frac{k_2^2(k_1^2 + k_2^2)}{N^4 (mp)^2 (k_1^2 + k_2^2 + (mp)^2) + k_2^4} \tag{2.33}$$

with N being the nondimensional constant in (1.5). An essential feature of the coefficients α_p is that they form an increasing sequence which grows rapidly, as p^4 when $p \to \infty$. Equation (2.31) gives a recurrence relation for the sequence $\{c_p\}_{p\geq 1}$ (see e.g. [**FV11b, FRV12**]):

$$\sigma_p c_p + \frac{c_{p+1}}{\alpha_{p+1}} + \frac{c_{p-1}}{\alpha_{p-1}} = 0, \quad \text{for } p \geq 2, \tag{2.34}$$

$$\sigma_1 c_1 + \frac{c_2}{\alpha_2} = 0, \text{for } p = 1. \tag{2.35}$$

We emphasize that given any $\sigma > 0$ (which then uniquely defines σ_p for all $p \geq 1$) and given any $c_1 > 0$, one can solve for all $\{c_p\}_{p\geq 2}$ using the recursion relations (2.34)–(2.35). The key is, however, that only for suitable values of σ do the c_p vanish sufficiently fast so that θ defined by (2.30) is sufficiently smooth. Solving for such σ amounts to solving the infinite continued fraction equation

$$\sigma_1 \alpha_1 = \cfrac{1}{\sigma_2 \alpha_2 - \cfrac{1}{\sigma_3 \alpha_3 - \cfrac{1}{\sigma_4 \alpha_4 - \cdots}}}. \tag{2.36}$$

Solving (2.36) is the main difficulty in this argument, since we need not just show there is a solution, but also give bounds on it in terms of the α_p. We omit these details and summarize the results obtained in [**FV11b**, Lemma 2.6], [**FRV12**, Lemma 4.2], and [**FGSV12**, Theorem 2.2] as follows.

LEMMA 2.6 (**Bound of solution to continued fraction equation**). *Let α_p, σ_p be defined by (2.32) and (2.33), respectively, and let $c_1 = \alpha_1$. There exits a real eigenvalue $\sigma_* = \sigma_*(k_1, k_2, m, N, \varepsilon_\kappa) > 0$ which solves (2.36), and we have the estimate*

$$\frac{1}{\alpha_1\alpha_2} \leq \sigma_1\sigma_2 \leq \frac{2}{\alpha_1\alpha_2}. \tag{2.37}$$

Moreover, the corresponding sequence $\{c_p\}_{p\geq 1}$ decays exponentially in p.

A direct consequence of estimate (2.37) is that the eigenvalue σ_* obeys the upper and lower bounds

$$\sigma_* \geq \frac{mN^2}{2}\frac{k_2^2(k_1^2+k_2^2)}{4N^4m^2(k_1^2+k_2^2+4m^2)+k_2^4} - \varepsilon_\kappa(k_1^2+k_2^2+4m^2), \tag{2.38}$$

$$\sigma_* \leq mN^2\frac{k_2^2(k_1^2+k_2^2)}{N^4m^2(k_1^2+k_2^2+m^2)+k_2^4} - \varepsilon_\kappa(k_1^2+k_2^2+m^2). \tag{2.39}$$

For N, ε_κ, m fixed, we hence obtain from (2.39) that there exists at most finitely many frequency pairs $(k_1, k_2) \in \mathbb{Z}_*^2$ for which $\sigma_* > 0$. Moreover, one may show that in the parameter range of the geodynamo, i.e., $0 < \varepsilon_\kappa \ll 1$, for suitable (k_1, k_2) the left-hand side of (2.38) is positive (see (2.40) below), thereby proving

THEOREM 2.7 ([**FV11b**]). *Associated to the steady state Θ_0 defined in (2.28), the linearized MG operator L defined in (2.29), has unstable spectrum.*

Lastly, note that one may optimize in (k_1, k_2), and compute a lower bound for the largest eigenvalue of L in terms of the physical parameters N, ε_κ, and of m. More precisely, we have that

$$\begin{aligned}\sup_{k_1,k_2\in\mathbb{Z}_*}&\left(\frac{mN^2}{2}\frac{k_2^2(k_1^2+k_2^2)}{4N^4m^2(k_1^2+k_2^2+4m^2)+k_2^4} - \varepsilon_\kappa(k_1^2+k_2^2+4m^2)\right)\\ &\geq \frac{1}{2^8\varepsilon_\kappa} - 8m^2\varepsilon_\kappa,\end{aligned} \tag{2.40}$$

with the lower bound being attained, e.g., for (of course, we need to take integer part)

$$k_1 = \frac{1}{16\varepsilon_\kappa} \quad \text{and} \quad k_2 = \frac{N\sqrt{m}}{\sqrt{8\varepsilon_\kappa}}, \quad \text{for } 0 < \varepsilon_\kappa \ll 1.$$

In particular, we note that as $\varepsilon_\kappa \to 0$, the lower bound given by the right-hand side of (2.40) blows up (this will be important in Sections 3 and 4). Moreover, for the above choice of k_1 and k_2, we have $k_1k_2^{-2} \to 1/(2N^2m)$ as $\varepsilon_\kappa \to 0$. This is in precise accordance to the fact that the absolute value of the Fourier multiplier associated to M grows at the fastest rate when $|k_2| = \mathcal{O}(|k_1|^{1/2})$ as $|k_1| \to \infty$; see (1.17).

Combining the results in this subsection with Theorem 2.5, we have thus constructed an example of solutions to the MG equations which displays strong (exponential) growth in θ and hence in the velocity field u by (1.12) and also in the magnetic field $b(x, t)$ through (1.8). The rate of this growth can be shown to be proportional to $\exp(t/2^8\varepsilon_\kappa)$ which is consistent with the dynamo instability scenario.

3. Nondiffusive MG equation

As discussed in the introduction, in the fluid core the value of ε_κ, given by the quotient of κ the molecular diffusivity coefficient, and LV, the product of the typical length scale and typical velocity in the fluid, is very small.

Therefore, from the physical point of view it is natural to study the nondiffusive (unforced) MG equation

$$\partial_t\theta + (u\cdot\nabla)\theta = 0, \tag{3.1}$$

$$u = M[\theta], \quad \nabla\cdot u = 0, \tag{3.2}$$

or the corresponding nondiffusive singular active scalar equation when (3.2) is replaced by (2.2), i.e., $u_j = \partial_i T_{ij}\theta$. As we shall see below, the system (3.1)–(3.2) is *ill-posed* in Sobolev spaces, in the sense that there is no operator which is Lipschitz in time at $t = 0$.

3.1. On the (lack of) well-posedness for the MG equation in Sobolev spaces. Before we discuss the issue of ill-posedness, let us briefly discuss why the *evenness* of the operator M breaks the classical proof of local existence in Sobolev spaces for the class of singular active scalar equations (3.1)–(3.2). The standard energy-approach to local existence in H^s would entail applying the $(-\Delta)^{s/2}$ to (3.1), and then take L^2 inner products with $(-\Delta)^{s/2}\theta$. Morally speaking (this is rigorous when $s/2$ is an integer in view of the Leibniz rule), we have only one term which seems to prevent closing the estimates at the H^s level:

$$\mathcal{T} = \int (-\Delta)^{s/2}u\cdot\nabla\theta\,(-\Delta)^{s/2}\theta\,dx = \int M_j(-\Delta)^{s/2}\theta\,\partial_j\theta\,(-\Delta)^{s/2}\theta\,dx. \tag{3.3}$$

Indeed, the other term where more than s derivatives lie on a single function is $\int u\cdot\nabla(-\Delta)^{s/2}\theta(-\Delta)^{s/2}\theta dx$, but this term vanishes upon integrating by parts since $\nabla\cdot u = 0$. The only hope to treat the term $\mathcal{T}$ would be to discover a commutator structure. However, since M is *not anti-symmetric*, i.e., *even* in Fourier space, we *cannot* write $\mathcal{T} = -\mathcal{S}$, where

$$\begin{aligned}\mathcal{S} &= \int M_j\left((-\Delta)^{s/2}\theta\,\partial_j\theta\right)(-\Delta)^{s/2}\theta\,dx\\ &= \mathcal{T} + \int [M_j,\partial_j\theta](-\Delta)^{s/2}\theta\,(-\Delta)^{s/2}\theta dx = \mathcal{T} + \mathcal{C}\end{aligned} \tag{3.4}$$

and $[\cdot,\cdot]$ is the usual commutator bracket. Instead for the MG, we have $\mathcal{T} = \mathcal{S}$, and we *cannot* explore the fact that $\mathcal{C}$ may be bounded in terms of $\|\theta\|_{H^s}^3$ (if $s > 2$). This is the main reason why we do not seem to be able to close estimates at the Sobolev level.

On the other hand, it was observed in [**CCC+12**] (see also [**FV11b**]) that if we replace $u = M[\theta]$ (cf. (3.2)) by a certain class of singular constitutive laws $u_j = \partial_i T_{ij}\theta$ (cf. (2.2)), and the matrix T_{ij} is such that

$$\langle T_{ij}f, g\rangle = \langle f, T_{ij}g\rangle$$

for any f, g and i, j (here $\langle\cdot,\cdot\rangle$ is the usual L^2 inner product), then the corresponding singular active scalar equation (3.1)–(2.2) is locally well-posed in H^s for $s > 2$. The main reason is that under these assumptions the operator $\partial_i T_{ij}$ is anti-symmetric, and we are able to write $2\mathcal{T} = \mathcal{C}$, using the notation of (3.3) and (3.4). The commutator term $\mathcal{C}$ would enable one estimate to close at the H^s level.

Lastly, we mention that for analytic initial data, the nondiffusive MG equation is indeed locally well-posed in the class of real-analytic functions, since each term in the equation loses at most one derivative. This was proven in [**FV11b**, Theorem 2.1] by using a Cauchy–Kowalewski-type argument, in the spirit of the Gevrey-class regularity proof of [**FT89**] (see also [**LO97, OT00, KV09**]). In particular, this result justifies the short-time numerical simulations of the MG equation, as they only keep track of finitely many Fourier modes (hence the associated functions are real analytic).

3.2. Hadamard ill-posedness for the nondiffusive MG equation. A Cauchy problem for a certain partial differential equation is called *well-posed in X in the sense of Hadamard*, if for any initial data in X, the problem has a unique solution in $L^\infty(0,T;X)$, with T depending only on the initial data in the X-norm, and the solution map $Y \mapsto L^\infty(0,T;X)$ satisfies some continuity properties, for a suitable space $Y \subset X$. If one of these properties fail, the Cauchy problem is called *ill-posed*. In this section we discuss the failure of the solution map for the nondiffusive MG equation to be *Lipschitz continuous* with respect to perturbations in the initial data (in the topology of a Sobolev space X) around a specific steady profile Θ_0. Such results have been recently obtained for several models in fluid dynamics; see, e.g., [**Gre00, BP08, Ren09, CS10, GVD10, GVN10, GN10, DQS11, FGSV12, FR12**] and references therein.

The proof consists of a linear and a nonlinear step. We construct a steady state and a sequence of eigenfunctions, with arbitrarily large eigenvalues, for the linearized MG equation around this steady state. Once these eigenvalues are exhibited, one may use a fairly robust argument (see, e.g., [**Ren09**]) to show that this severe linear ill-posedness implies the Lipschitz ill-posedness for the nonlinear problem. We emphasize that the key feature of the MG constitutive law (1.12) which gives this ill-posedness is the evenness and unboundedness of the symbol associated to M. In fact, the evenness also gives nonuniqueness for bounded weak solutions of the nondiffusive MG equation, as was recently established by Shvydkoy [**Shv11**].

3.2.1. *Linear ill-posedness.* Let $m \geq 1$ be a positive integer. We use the steady state

$$\Theta_0 = \sin(mx_3), \tag{3.5}$$

with associated velocity field $U_0 = 0$. Similarly to (2.29), we consider the associated linearized (nondiffusive) MG operator defined by

$$L\theta = -M_3[\theta]\partial_3\Theta_0, \tag{3.6}$$

where the M_3 operator is defined in terms of its Fourier symbol (1.15)–(1.16). With this notation, we have

THEOREM 3.1 ([**FV11b**]). *The Cauchy problem associated to the linearized MG equation*

$$\partial_t\theta = L\theta \tag{3.7}$$

is ill-posed in the sense of Hadamard over L^2. That is, for any $T, K > 0$, there exits a real-analytic initial datum $\theta(0,\cdot)$, such that the IVP associated to (3.7) does *not have a solution $\theta \in L^\infty(0,T;L^2)$, which obeys*

$$\sup_{t\in[0,T]} \|\theta(t,\cdot)\|_{L^2} \leq K\|\theta(0,\cdot)\|_Y, \tag{3.8}$$

where Y is any Sobolev space that embeds compactly in L^2.

The proof of the above theorem builds on the construction of unstable eigenvalues in the critically diffusive case, more specifically on Lemma 2.6.

As in Section 2.2.2 (cf. (2.30)) we make the ansatz that θ is given in terms of a Fourier sine series

$$\theta(t,x) = e^{\sigma t}\sin(k_1x_1)\sin(k_2x_2)\sum_{p\geq 1} c_p \sin(mpx_3),$$

with k_1 and k_2 to be chosen precisely later. Note that since we are in the non-diffusive case, we need to set $\varepsilon_\kappa = 0$, and hence (cf. (2.32)) we have $\sigma_p = \sigma$ for any $p \geq 1$. The value of α_p remains however unchanged. Using Lemma 2.6, we obtain an eigenvalue $\sigma_* = \sigma_*(k_1,k_2)$, and an exponentially decaying sequence $\{c_p\}_{p\geq 1}$, such that (cf. (2.38)) the bound

$$\sigma_* = \sigma_*(k_1,k_2) \geq \frac{mN^2}{2}\frac{k_2^2(k_1^2+k_2^2)}{4N^4m^2(k_1^2+k_2^2+4m^2)+k_2^4} \tag{3.9}$$

holds. We set $k_1 = j^2$ and $k_2 = j$ in (3.9), where $j \geq 2m$ is a positive integer, and obtain the lower bound

$$\sigma_{*,j} := \sigma_*(j^2,j) \geq j\cdot\frac{mN^2}{24N^4m^2+2} = \frac{j}{C_1} \tag{3.10}$$

on the eigenvalue, associated to this choice of (k_1,k_2). Here $C_1 = C_1(m,N)$. We also denote by $\theta^{(j)}$ the eigenfunction constructed in this way. Using an additional argument, which requires suitably bounding the exponentially decaying sequence $\{c_p\}_{p\geq 1}$ which defines our eigenfunction, one may in fact additionally normalize θ and prove (cf. [**FV11b**, Lemma 2.7], [**FRV12**, Lemma 4.3]) that

$$\|\theta^{(j)}(0,\cdot)\|_Y = 1 \quad \text{and} \quad \|\theta^{(j)}(0,\cdot)\|_{L^2} \geq \frac{1}{C_2 j^2}, \tag{3.11}$$

where $C_2 = C_2(m,N,Y)$ is a sufficiently large constant. For example, if $Y = H^s$ with $s > 0$, we may explicitly compute a suitable such constant.

We conclude the proof of Theorem 3.1 by using a contradiction argument. If such a solution would exist, for a fixed T and K, we pick $j = j(T,K,m,N,Y)$ large enough so that

$$\frac{\exp(Tj/(2C_1))}{C_2 j^2} \geq 2K$$

holds. The linear problem (3.7) can be shown to have a unique L^2 solution, and hence by construction we have

$$\theta^{(j)}(t,x) = \theta^{(j)}(0,x)\exp(\sigma_{*,j}t),$$

the assumed estimate (3.8) is contradicted by our choice of j and the lower bound (3.10). This concludes the proof of Theorem 3.1.

3.2.2. *Nonlinear ill-posedness.* Before stating our main ill-posedness result, let us recall the definition of Lipschitz well-posedness.

DEFINITION 3.2 ([**GN10, FV11b**]). *Let $Y \subset X \subset W^{1,4}$ be Sobolev spaces. The Cauchy problem for the nondiffusive MG equation is locally Lipschitz* (X, Y) well-posed, *if there exist continuous functions $T, K : [0,\infty)^2 \to (0,\infty)$, so that for every pair of initial data $\theta^{(1)}(0,\cdot), \theta^{(2)}(0,\cdot) \in Y$ there exist unique solutions $\theta^{(1)}, \theta^{(2)} \in L^\infty(0,T;X)$ of the initial value problem associated to* (3.1)–(3.2)*, that additionally satisfy*

$$\|\theta^{(1)}(t,\cdot) - \theta^{(2)}(t,\cdot)\|_X \leq K\|\theta^{(1)}(0,\cdot) - \theta^{(2)}(0,\cdot)\|_Y \tag{3.12}$$

for every $t \in [0,T]$, where

$$T = T(\|\theta^{(1)}(0,\cdot)\|_Y, \|\theta^{(2)}(0,\cdot)\|_Y) \quad \textit{and} \quad K = K(\|\theta^{(1)}(0,\cdot)\|_Y, \|\theta^{(2)}(0,\cdot)\|_Y).$$

We note that Definition 3.2 allows the solution map to lose regularity. Since the MG equation is first order, the typical pairs of spaces (X, Y) that we have in mind here are $X = H^s$ and $Y = H^{s+1}$, with $s > 1 + 3/4$.

The main result of this section is then

THEOREM 3.3 ([**FV11b**]). *The nondiffusive MG equations* (3.1)–(3.2) *are locally Lipschitz* (X, Y) ill-posed *in Sobolev spaces $Y \subset X$ embedded in $W^{1,4}$, in the sense of Definition* 3.2.

The proof of the above result is based on the severe linear ill-posedness obtained in Theorem 3.1, and a standard perturbative argument [**Ren09**, Theorem 2] or [**Tao06**, pp. 183], which we briefly sketch below.

Let $\Theta_0 = \sin(mx_3)$ be the steady state defined above in (3.5), let $L\theta = -M_3[\theta]\partial_r\Theta_0$ be its associated linearized MG operator, and denote the MG nonlinearity by $N(\theta) = M_j[\theta]\partial_j\theta$. By our assumptions on X and Y, we have

$$\|L\theta\|_{L^2} \leq C_1\|\theta\|_X \quad \text{and} \quad \|N(\theta)\|_{L^2} \leq C_1\|\theta\|_X^2 \tag{3.13}$$

for some constant $C_1 = C_1(\Theta_0) > 0$. Also, let $C_0 = \|\Theta_0\|_Y$.

In definition (3.2), we chose $\theta^{(2)}(0) = \Theta_0 = \theta^{(2)}(t)$ to be the steady state, and let $\theta^{(1)}(0) = \theta_\delta^{(1)}(0) = \Theta_0 + \delta\psi_0$, for some smooth function ψ_0 to be chosen later. Assume by contradiction that the problem is Lipschitz locally well-posed in (X, Y). One may then show that there exist $T = T(C_0)$ and $K = K(C_0)$ such that

$$\sup_{t\in[0,T]} \|\theta_\delta^{(1)}(t) - \Theta_0\|_X \leq K\delta$$

for *any* $\delta \in (0, \delta_0(C_0)]$. Writing the solution $\theta_\delta^{(1)}(t)$ as an order δ perturbation around the steady state $\theta^{(1)}(t) = \Theta_0 + \delta\psi_\delta(t)$, we thus have

$$\sup_{t\in[0,T]} \|\psi_\delta(t)\|_X \leq K, \tag{3.14}$$

and ψ_δ solves the Cauchy problem

$$\partial_t\psi_\delta = L\psi_\delta + \delta N(\psi_\delta), \quad \psi_\delta(0) = \psi_0. \tag{3.15}$$

The key observation is that when combined with (3.13), estimate (3.14) yields

$$\delta\|N(\psi_\delta)\|_{L^2} \leq C_1\delta K^2$$

for any $t \in [0,T]$. Therefore, when passing $\delta \to 0$, the nonlinearity in (3.15) disappears, and we should obtain the linear limiting equation. Indeed, from (3.14),

(3.15), and the bounds (3.13), we obtain the necessary estimates to apply the Aubin–Lions compactness argument and to show that $\psi_\delta \to \psi$ strongly in $L^\infty(0,T;L^2)$, and on this time interval the limiting function obeys

$$\partial_t \psi = L\psi, \quad \psi(0) = \psi_0. \tag{3.16}$$

But ψ also inherits the bound (3.14) from the sequence ψ_δ, which contradicts Thereom 3.1—choose ψ_0 an eigenfunction of L, as constructed in Section 3.2.1, so that in time $T/2$ the exponential growth of the L^2-norm caused the solution to exceed the level $2K$. This concludes our sketch of the proof of Theorem 3.3.

4. Fractionally diffusive MG equation and well-prepared initial data

The fractionally diffusive MG equation, although nonphysical, presents a very challenging and interesting mathematical problem. Replacing the Laplacian in (1.11) by $-(-\Delta)^\gamma$, for $\gamma \in (0,1)$, we obtain the active scalar equation

$$\partial_t \theta + u \cdot \nabla \theta = \varepsilon_\kappa (-\Delta)^\gamma \theta + S, \tag{4.1}$$

$$u = M[\theta], \tag{4.2}$$

$$\theta(0,\cdot) = \theta_0, \tag{4.3}$$

where the symbol of the Fourier multiplier M is given by (1.13)–(1.16). Note that here we work with nondimensional variables, so that the initial data and the source term are *normalized to be of size unity.*

The main feature of this situation is that the structure of $\widehat{M_j}$ (anisotropy, evenness, and unboundedness) and a relatively weak smoothing effect of the nonlocal operator $(-\Delta)^\gamma$ produce a sharp dichotomy across the value $\gamma = 1/2$, as follows. If $1/2 < \gamma < 1$, the equation is *locally* well-posed, while it is Hadamard *Lipschitz ill-posed* for $0 < \gamma < 1/2$. At the threshold value $\gamma = 1/2$, the problem is globally well-posed for $\varepsilon_\kappa \gg 1$, respectively ill-posed for $\varepsilon_\kappa \ll 1$. Furthermore, the anisotropy of the symbols $\widehat{M_j}$ can be used to improve the regularity of the solutions when the initial data and source have a suitably chosen Fourier support. For such *well-prepared* data, the local existence and uniqueness of solutions can be obtained for all values $\gamma \in (0,1)$, and the global existence holds for all initial data when $\gamma \in (1/2, 1)$. We remark that in all above-mentioned regimes, the problem is *super-critical* in the sense described in Section 2.

4.1. Regime of well-posedness in Sobolev spaces. In Section 2 we stated the well-posedness result when the diffusion in the equation is given by the full Laplacian. If the diffusion in the equation is introduced by the fractional operator $(-\Delta)^\gamma$, $\gamma \in (1/2,1)$, then we have the following result.

THEOREM 4.1 ([**FRV12**]). *Let $\gamma \in (1/2,1)$, and fix $s > 5/2 + (1-2\gamma)$. Assume that $\theta_0 \in H^s(\mathbb{T}^3)$ and $S \in L^\infty(0,\infty; H^{s-\gamma}(\mathbb{T}^3))$ have zero-mean on $\mathbb{T}^3$. Then there exists a time $T > 0$ and a unique smooth solution $\theta \in L^\infty(0,T;H^s(\mathbb{T}^3)) \cap L^2(0,T;H^{s+\gamma}(\mathbb{T}^3))$ of the Cauchy problem (4.1)–(4.2).*

Under an additional assumption on the size of ε_κ (equivalently, the size of the initial data in the unscaled equations), we obtain a global existence result.

THEOREM 4.2 ([**FRV12**]). *Let γ and S be as in the statement of Theorem 4.1, and let $\theta_0 \in H^s(\mathbb{T}^3)$ have zero-mean on $\mathbb{T}^3$, where $s > 5/2 + (1-2\gamma)$. There exists* *a large enough number $\bar{C}$, such that if $\varepsilon_\kappa \geq \bar{C}$, then the unique smooth solution θ of the Cauchy problem (4.1)–(4.2) is global in time, i.e., $\theta \in L^\infty(0,\infty; H^s(\mathbb{T}^3))$.*

REMARK 4.3. Note that in [**FRV12**] we did not use the nondimensional variables with which we are working here. Therefore, the global well-posedness condition stated in [**FRV12**] is that the initial data and source have to be *sufficiently small*, compared to the diffusivity coefficient κ. In this paper, however, the data and source have been normalized to be of order unity, and the only coefficient remaining is ε_κ. It is hence natural that the small data condition turns to a *large enough* ε_κ condition. This remains valid throughout this section.

The detailed proofs of Theorems 4.1 and 4.2 are given in [**FRV12**], and rely mainly on energy methods and the use of suitable commutator estimates. We omit these details here.

4.2. Regime of ill-posedness in Sobolev spaces. In this subsection we consider the stability of steady states to (4.1)–(4.2). The construction follows the outline described in Section 2.2.2. In this case, we choose $\Theta_0 = \sin(mx_3)$ and $S = \varepsilon_\kappa m^{2\gamma}\sin(mx_3)$, for some integer $m \geq 1$. As before, the first step in proving that the Cauchy problem (4.1)–(4.2) is Lipschitz ill-posed is to consider the linearized problem around the given steady state and exhibit the existence of unstable eigenvalues with arbitrarily large real part. Thus, we consider

$$\partial_t\theta = L\theta, \tag{4.4}$$

where the linear operator L is given by

$$L\theta(x,t) = -M_3\theta(x,t)\partial_3\theta_0(x_3) - \varepsilon_\kappa(-\Delta)^\gamma\theta.$$

We seek the temperature θ which solves (4.4) and has the form

$$\theta(t,x) = e^{\sigma t}\sin(k_1x_1)\sin(k_2x_2)\sum_{p\geq 1} c_p\sin(mpx_3) \tag{4.5}$$

for some fixed $k_1, k_2 \in \mathbb{Z}_*$ to be chosen later. Inserting (4.5) into (4.4) and using the definition of M_3, we obtain that (2.31) holds, but with σ_p denoted by

$$\sigma_p = \sigma + \varepsilon_\kappa(k_1^2 + k_2^2 + (mp)^2)^\gamma \tag{4.6}$$

and with α_p given by (2.33) above. As before, these assumptions lead to solving the continued fraction equation (2.36) for some real $\sigma > 0$. This is achieved by suitably modifying Lemma 2.6 to obtain

LEMMA 4.4. *Let α_p, σ_p be defined by (2.33) and (4.6), respectively, and let $c_1 = \alpha_1$. There exists a real eigenvalue $\sigma_* > 0$ and a rapidly decaying sequence $\{c_p\}_{p\geq 2}$ so that the function θ defined by (4.5) solves (4.4). Furthermore, we have the lower bound*

$$\sigma_* \geq \frac{mN^2}{2}\frac{k_2^2(k_1^2+k_2^2)}{4N^4m^2(k_1^2+k_2^2+4m^2)+k_2^4} - \varepsilon_\kappa(k_1^2+k_2^2+4m^2)^\gamma. \tag{4.7}$$

In particular, setting for instance $k_1 = j^2$ and $k_2 = j$ in (4.7), for each $j \geq 2m$, we obtain a real eigenvalue $\sigma_{*,j}$ of L, which is bounded from below as

$$\sigma_{*,j} \geq j^{2\gamma}\left(j^{1-2\gamma}\frac{mN^2}{24N^4m^2+2} - 3\varepsilon_\kappa\right). \tag{4.8}$$

Moreover, given a Sobolev space Y that is compactly embedded in L^2, we may ensure the existence of a constant $C_2 = C_2(m, N, Y, \gamma)$ such that the smooth eigenfunction $\theta^{(j)}$ corresponding to $\sigma_{*,j}$ obeys (3.11). The immediate consequence of estimate (4.8) is that due to $1 - 2\gamma > 0$, we have

$$\sigma_{*,j} \to \infty, \quad \text{as} \quad j \to \infty.$$

The above construction proves that the linearized problem (4.4) is ill-posed in L^2, as we have exhibited eigenvalues of arbitrarily large real part. Following the argument outlined in Section 3.2.2, this type of strong ill-posedness for the linearized problem automatically implies that the nonlinear problem is Lipschitz ill-posed in Sobolev spaces (recall Definition 3.2 of Lipschitz well-posedness). Thus we obtain

THEOREM 4.5 ([**FRV12**]). *The Cauchy problem* (4.1)–(4.3), *with* $\gamma \in (0, 1/2)$, *is locally Lipschitz* (X, Y) ill-posed *in Sobolev spaces* $Y \subset X$ *embedded in* $W^{1,4}(\mathbb{T}^3)$.

4.3. The threshold case $\gamma = 1/2$. The value $\gamma = 1/2$ marks the threshold for proving the local well-posedness in Sobolev spaces for (4.1)–(4.3). If the initial datum is small with respect to κ or, equivalently, if ε_κ is sufficiently large (see Remark 4.3 above), then one may use energy estimates to show that there exists a unique global smooth solution evolving from this data. In this direction we have

THEOREM 4.6 ([**FRV12**]). *Let* $s > 5/2$, *and assume that* $\theta_0 \in H^s(\mathbb{T}^3)$ *and* $S \in L^\infty(0, \infty; H^{s-\gamma}(\mathbb{T}^3))$ *have zero-mean on* $\mathbb{T}^3$. *There exists a sufficiently large number* $\bar{C}$, *such that if* $\varepsilon_\kappa \geq \bar{C}$, *then the unique smooth solution* $\theta \in L^\infty(0, T; H^s(\mathbb{T}^3))$ *of the Cauchy problem* (4.1)–(4.3) *is global in time.*

As in the case of $\gamma \in (1/2, 1)$, the proof is based on the classical Kato–Ponce commutator estimate. The proof however does not apply for the case of "small" ε_κ (large data), not even to prove the local existence of solutions.

In the case where ε_κ is suitably small, we may construct a steady state such that solutions evolving from initial data which are arbitrarily close to this steady state diverge from it at an arbitrarily large rate, i.e., the problem is Hadamard ill-posed. More precisely, we have

THEOREM 4.7 ([**FRV12**]). *Let* $Y \subset X \subset W^{1,4}(\mathbb{T}^3)$ *be Sobolev spaces, let* $\gamma = 1/2$, *and fix the integer* $m = 1$. *Consider the steady state solution of* (4.1) *given by* $\theta_0^{(1)} = \sin(mx_3) \in Y$, *with source term* $S = \varepsilon_\kappa m \sin(mx_3)$, *and assume that*

$$\varepsilon_\kappa < \frac{1}{26\sqrt{3}}. \tag{4.9}$$

Then, given any $T > 0$ *and any* $K > 0$, *there exists an initial condition* $\theta_0^{(2)} \in Y$ *and a corresponding solution* $\theta^{(2)} \in L^\infty(0, T; X)$ *of the Cauchy problem* (4.1)–(4.2), *such that*

$$\sup_{t \in [0,T]} \|\theta^{(2)}(t, \cdot) - \theta^{(1)}(t, \cdot)\|_X \geq 2K \|\theta_0^{(2)} - \theta_0^{(1)}\|_Y,$$

where $\theta^{(1)}(t, \cdot) = \theta_0^{(1)}$. *That is, the problem* (4.1)–(4.2) *is locally Lipschitz* (X, Y) *ill-posed when* $\varepsilon_\kappa \ll 1$.

To prove the above result, we follow the ill-posedness result in the case $\gamma \in (0, 1/2)$. The main difference is that while for $\gamma \in (0, 1/2)$ by (4.8) we have real eigenvalues of arbitrarily large real part, when $\gamma = 1/2$, the lower bound (4.8) only blows up for ε_κ sufficiently small, depending on N^2 and our choice of m. This condition is rather unsatisfactory, since it implies $\varepsilon_\kappa \to 0$ as $N^2 \to 0$. In fact, we can establish a more robust smallness condition on ε_κ, in the sense that it is independent of N^2 (and m). To see this, insert $k_1 = j^2 N^{-2}$ and $k_2 = jm^{1/2}$ in the lower bound (4.7), where $j \geq \max\{N^4 m, 4m\}$. We obtain the lower bound

(4.10)
$$\sigma_{*,j} \geq \frac{mN^2}{2} \frac{j^6 m N^{-4}}{12N^4 m^2 j^4 N^{-4} + j^4 m^2} - \sqrt{3}\varepsilon_\kappa j^2 N^{-2} = j^2 N^{-2} \left(\frac{1}{26} - \sqrt{3}\varepsilon_\kappa \right).$$

Therefore, under assumption (4.9), which is independent of m or N, the right-hand side of (4.10) diverges to ∞ as $j \to \infty$ and hence we again have a sequence of unstable eigenvalues of the linear problem, with arbitrarily large real part. Combined with the nonlinear argument presented in Section 3.2.2 this concludes the sketch of the proof of Theorem 4.7.

4.4. Well-prepared initial data. In this subsection we present results which take advantage of the curved nature (paraboloid-type) of the regions in frequency space where the symbol $\widehat{M}(k)$ grows like $|k|$. We observe that if the support of $\widehat{\theta}$ lies on a plane, then the operator M behaves like an operator of order zero acting on this θ, and hence the velocity u is as smooth as θ. This allows us to improve the well-posedness results in the range $\varepsilon_\kappa > 0$ and $0 < \gamma < 1$, over the generic case when no conditions on the Fourier transforms of the initial data (and source term) are imposed.

To be more precise, let us introduce some notation. For a rational number $q \in \mathbb{Q}$ with $q \neq 0$, define frequency plane $\mathcal{P}_q$ as

$$\mathcal{P}_q = \{k = (k_1, k_2, k_3) \in \mathbb{Z}^3 : k_2 = qk_1\}.$$

One may show that whenever $q \neq 0$,

$$|\widehat{M}_j(k)| \leq C_q < \infty, \quad \text{for all } k \in \mathcal{P}_q \text{ and } j \in \{1, 2, 3\}. \tag{4.11}$$

This observation, combined with the fact that when the data θ_0 and source S have frequency support on $\mathcal{P}_q$, then so does the solution $\theta(t, \cdot)$ of the Cauchy problem (4.1)–(4.3) at all later times $t > 0$, means that the smoothing properties of the fractional diffusion term are stronger than they are for the generic data situation. Hence it is possible to prove stronger regularity results, and in particular the *local existence* and uniqueness of Sobolev solutions holds *for all* values of $\gamma \in (0, 1)$.

THEOREM 4.8 ([**FRV12**]). *Let $\gamma \in (0, 1)$, and fix $s > 5/2 - \gamma$. Assume that $\theta_0 \in H^s(\mathbb{T}^3)$ and $S \in L^\infty(0, \infty; H^{s-\gamma}(\mathbb{T}^3))$ have zero-mean on $\mathbb{T}^3$ and are such that $\operatorname{supp}(\widehat{\theta}_0) \subset \mathcal{P}_q$ and $\operatorname{supp}(\widehat{S}(t, \cdot)) \subset \mathcal{P}_q$ for some fixed $q \in \mathbb{Q} \setminus \{0\}$ and all $t \geq 0$. Then there exists a $T > 0$ and a unique smooth solution $\theta \in L^\infty(0, T; H^s(\mathbb{T}^3)) \cap L^2(0, T; H^{s+\gamma}(\mathbb{T}^3))$ of the Cauchy problem* (4.1)–(4.3), *such that* $\operatorname{supp}(\widehat{\theta}(t, \cdot)) \subset \mathcal{P}_q$ *for all $t \in [0, T)$.*

We note that the above theorem holds even if $\gamma \in (0, 1/2)$, a setting in which we know that for *generic* initial data the problem is ill-posed. Moreover, one may also prove that if the initial data is small with respect to κ, or equivalently, $\varepsilon_\kappa \gg 1$ in the nondimensional setting of this manuscript, the solution obtained in Theorem 4.8 is global in time.

The main difficulty in the proof of Theorem 4.8 is the construction and uniqueness of solutions, which live on the frequency plane $\mathcal{P}_q$. This relies on the construction of a suitable sequence of approximations supported on the Fourier space on $\mathcal{P}_q$ and showing its convergence using energy methods. The main obstacle one has to overcome is designing an iteration scheme which is both suitable for energy estimates and is such that in each iteration step the frequency support of the approximation lies on $\mathcal{P}_q$. More explicitly, a scheme in which $\partial_t\theta_{n+1}$ is given in terms of $v\cdot\nabla\theta_n$ preserves the frequency support on $\mathcal{P}_q$; however, it is not good for closing the estimates at the level of Sobolev spaces. On the other hand if we define $\partial_t\theta_{n+1}$ in terms of $v\cdot\nabla\theta_{n+1}$, the energy estimates are now clear but it seems difficult to inductively prove that $\widehat{\theta}_{n+1}$ is supported on $\mathcal{P}_q$. Therefore, we work with a regularized problem, where we introduce an artificial term $-\epsilon\Delta\theta$. The advantage of considering a stronger dissipative problem is that we can *construct* smooth solutions which have frequency support lying on $\mathcal{P}_q$. For details of the construction we refer the interested reader to [**FRV12**]. Lastly, we pass to the limit $\epsilon\to 0$ and obtain a solution of (4.1)–(4.3) with the desired properties.

If the initial data θ_0 and the source term S have Fourier support in $\mathcal{P}_q$, we show that for $\gamma\in[1/2,1)$ the local in time solution may be continued for all time.

THEOREM 4.9 ([**FRV12**]). *Let s,θ_0, and S be as in the statement of Theorem* 4.8. *If $\gamma\geq 1/2$, the unique smooth solution θ of the Cauchy problem* (4.1)–(4.3) *is global in time.*

The proof hinges on the observation that the solution constructed in Theorem 4.8 satisfies $\mathrm{supp}(\widehat{\theta})\subset\mathcal{P}_q$. Then, in view of (4.11) and the Hörmander–Mikhlin theorem, we have $\|u\|_{W^{s,p}}\leq C\|\theta\|_{W^{s,p}}$ for all $2\leq p<\infty, s\geq 0$, and so the velocity u is as smooth as the scalar θ. Hence, the case $\gamma>1/2$ is "subcritical" for the a priori controlled quantities, and energy estimates in the spirit of those used to prove the global regularity of the subcritical SQG equation [**CW99, CC04, Wu05**] may be used to prove Theorem 4.9. The case $\gamma=1/2$ in Theorem 4.9 is more subtle, and is analogous with the problem of global regularity for the critical SQG equation [**CV10, KNV07, KN09, CV12**]. The anisotropic nature of the symbol M makes it more convenient to apply the De Giorgi–inspired proof of [**CV10**] in the context of the fractional MG equation with $\gamma=1/2$. The main ingredient is that if $\theta\in L^\infty$ and $\mathrm{supp}(\widehat{\theta})\in\mathcal{P}_q$, then $u\in BMO$. Hence, considering first the linear problem with divergence-free BMO drift, we obtain Hölder continuity of the solution. The later norm is subcritical, and one may bootstrap to higher regularity. We refer the interested reader to [**FRV12**] for details.

5. Conclusions

We have proven that the magneto-geostrophic model proposed by Moffatt and Loper is capable of sustaining dynamo action. The MG-dynamo instability exhibited here relies crucially on the fact that the dimensionless parameter N^2 (1.5) is neither 0 nor ∞. The exponential growth is of order $\mathcal{O}(\exp(t/\varepsilon_\kappa))$, for all $N^2\in(0,\infty)$, and $0<\varepsilon_\kappa\ll 1$. In terms of the physics in the model, this condition on N^2 requires that *all three forces*, Coriolis, Lorentz, and gravity, *are nonzero*, making the MG instability somewhat different in nature from the usual Rayleigh–Taylor instability.

At the level of mathematical analysis several problems remain to be addressed, including some well-/ill-posedness issues. For the nondiffusive equation, *is there a "stronger" form of ill-posedness, such as norm inflation or even finite time blowup?* Can we refine the well-posedness results for the fractionally diffusive problem in the range $1/2 < \gamma < 1$? That is, *can one establish the global existence of smooth solutions or, on the contrary, prove finite time blowup?*

Several, interesting, mathematical issues arise when returning to the derivation of the MG equation from the full three dimensional MHD equations in the spirit of [**ML94, Mof08**]. In particular, we note that Moffatt and Loper's initial discussion includes the kinematic viscosity ν, which is directly proportional to ε in (1.1). The values of ε and ε_κ are highly conjectural, but both are believed to be *extremely* small. Therefore, as *opposed* to the MG system, if we keep $\varepsilon > 0$, we derive from (1.1)–(1.4) the system

$$N^2 e_3 \times u = -\nabla P + e_2 \cdot \nabla b + N^2 \theta e_3 + \varepsilon \Delta u, \tag{5.1}$$

$$0 = e_2 \cdot \nabla u + \Delta b, \tag{5.2}$$

$$\partial_t \theta + u \cdot \nabla \theta = \varepsilon_\kappa \Delta \theta + S, \tag{5.3}$$

$$\nabla \cdot u = 0, \nabla \cdot b = 0. \tag{5.4}$$

The presence of the Laplacian term in (5.1) makes the constitutive law relating u and θ "better", in the sense that the M operator corresponding to (5.1)–(5.4), now depends both on N and ε, and is in fact smoothing of degree 2 (see equations (29)–(30) in [**ML94**]). Hence, in this case the constitutive law is even better than that for the two-dimensional Euler equation in vorticity form, implying that *in the range* $\varepsilon > 0$, *the system* (5.1)–(5.4) *is globally well-posed, even if* $\varepsilon_\kappa = 0$.

This observation is interesting in view of the much more challenging problem of *establishing properties of the solutions to* (5.1)–(5.4) *in the vanishing parameters limits* $\varepsilon \to 0$ *and* $\varepsilon_\kappa \to 0$, as proposed by Moffatt and Loper. The later question is directly related to the turbulent scaling regime for θ conjectured by Moffatt in [**Mof08**]. "Turbulence" in the framework of vanishing diffusivity ε_κ for the forced system (5.1)–(5.4), and the presence of anomalous dissipation of energy for θ, or "thetargy", may be plausible for very weak solutions (below Onsager critical) of the nondiffusive equation.

Acknowledgments. We are very grateful to the anonymous referee for helpful remarks and suggestions. The work of the first author is supported in part by the NSF grants DMS-0803268 and DMS-1207780. The work of the third author is supported in part by the NSF grant DMS-1211828.

References

[BGS02] C. Bardos, Y. Guo, and W. Strauss. Stable and unstable ideal plane flows. *Chinese Ann. Math. Ser. B*, 23(2):149–164, 2002. Dedicated to the memory of Jacques-Louis Lions. MR1924132 (2003k:35192)

[BP08] J. Bourgain and N. Pavlović. Ill-posedness of the Navier-Stokes equations in a critical space in 3D. *J. Funct. Anal.*, 255(9):2233–2247, 2008. MR2473255 (2010g:35222)

[BR95] S.I. Braginsky and P.H. Roberts. Equations governing convection in earth's core and the geodynamo. *Geophysical & Astrophysical Fluid Dynamics*, 79(1-4):1–97, 1995.

[CC04] A. Córdoba and D. Córdoba. A maximum principle applied to quasi-geostrophic equations. *Comm. Math. Phys.*, 249(3):511–528, 2004. MR2084005 (2005f:76011)

[CCC+12] D. Chae, P. Constantin, D. Córdoba, F. Gancedo, and J. Wu. Generalized surface quasi-geostrophic equations with singular velocities. *Communications on Pure and Applied Mathematics*, 65(8):1037—1066, 2012. MR2928091

[CCW01] P. Constantin, D. Cordoba, and J. Wu. On the critical dissipative quasi-geostrophic equation. *Indiana Univ. Math. J.*, 50(Special Issue):97–107, 2001. Dedicated to Professors Ciprian Foias and Roger Temam (Bloomington, IN, 2000). MR1855665 (2002h:35246)

[CMT94] P. Constantin, A.J. Majda, and E. Tabak. Formation of strong fronts in the 2-D quasi-geostrophic thermal active scalar. *Nonlinearity*, 7(6):1495–1533, 1994. MR1304437 (95i:76107)

[CS10] A. Cheskidov and R. Shvydkoy. Ill-posedness of the basic equations of fluid dynamics in besov spaces. *Proc. Amer. Math. Soc*, 138(3):1059–1067, 2010. MR2566571 (2011b:35364)

[CV10] L.A. Caffarelli and A. Vasseur. Drift diffusion equations with fractional diffusion and the quasi-geostrophic equation. *Ann. of Math. (2)*, 171(3):1903–1930, 2010. MR2680400 (2011m:35281)

[CV12] P. Constantin and V. Vicol. Nonlinear maximum principles for dissipative linear nonlocal operators and applications. *Geometric And Functional Analysis*, 22(5):1289–1321, 2012. MR2989434

[CW99] P. Constantin and J. Wu. Behavior of solutions of 2D quasi-geostrophic equations. *SIAM J. Math. Anal.*, 30(5):937–948, 1999. MR1709781 (2000j:76019)

[CW08] P. Constantin and J. Wu. Regularity of Hölder continuous solutions of the supercritical quasi-geostrophic equation. *Ann. Inst. H. Poincaré Anal. Non Linéaire*, 25(6):1103–1110, 2008. MR2466323 (2009h:35328)

[DG57] E. De Giorgi. Sulla differenziabilità e l'analiticità delle estremali degli integrali multipli regolari. *Mem. Accad. Sci. Torino. Cl. Sci. Fis. Mat. Nat. (3)*, 3:25–43, 1957. MR0093649 (20:172)

[DQS11] M. Dai, J. Qing, and M.E. Schonbek. Norm inflation for incompressible magneto-hydrodynamic system. *arXiv preprint arXiv:1110.2723*, 2011. MR2829502 (2012e:76146)

[FGSV12] S. Friedlander, F. Gancedo, W. Sun, and V. Vicol. On a singular incompressible porous media equation. *Journal of Mathematical Physics*, 53(11):1–20, 2012. MR3026547

[Fil12] N. Filonov. On the regularity of solutions to the equation $-\Delta u + b\nabla u = 0$. *arXiv preprint arXiv:1209.1243*, 2012. MR3048265

[FPS06] S. Friedlander, N. Pavlović, and R. Shvydkoy. Nonlinear instability for the Navier-Stokes equations. *Comm. Math. Phys.*, 264(2):335–347, 2006. MR2215608 (2007g:35171)

[FPV09] S. Friedlander, N. Pavlović, and V. Vicol. Nonlinear instability for the critically dissipative quasi-geostrophic equation. *Comm. Math. Phys.*, 292(3):797–810, 2009. MR2551794 (2010j:86014)

[FR12] S. Friedlander and W. Rusin. On the second iterate for active scalar equations. *arXiv preprint arXiv:1206.3349*, 06 2012.

[FRV12] S. Friedlander, W. Rusin, and V. Vicol. On the supercritically diffusive magneto-geostrophic equations. *Nonlinearity*, 25(11):3071–3097, 2012. MR2991429

[FSV97] S. Friedlander, W. Strauss, and M. Vishik. Nonlinear instability in an ideal fluid. *Ann. Inst. H. Poincaré Anal. Non Linéaire*, 14(2):187–209, 1997. MR1441392 (99a:76057)

[FT89] C. Foias and R. Temam. Gevrey class regularity for the solutions of the Navier-Stokes equations. *J. Funct. Anal.*, 87(2):359–369, 1989. MR1026858 (91a:35135)

[FV11a] S. Friedlander and V. Vicol. Global well-posedness for an advection-diffusion equation arising in magneto-geostrophic dynamics. *Ann. Inst. H. Poincaré Anal. Non Linéaire*, 28(2):283–301, 2011. MR2784072 (2012g:35257)

[FV11b] S. Friedlander and V. Vicol. On the ill/well-posedness and nonlinear instability of the magneto-geostrophic equations. *Nonlinearity*, 24(11):3019–3042, 2011. MR2844825 (2012k:86010)

[FV12] S. Friedlander and V. Vicol. Higher regularity of Hölder continuous solutions of parabolic equations with singular drift velocities. *J. Math. Fluid Mech.*, 14(2):255–266, 2012. MR2925107

[GC87] M. Ghil and S. Childress. *Topics in geophysical fluid dynamics: atmospheric dynamics, dynamo theory, and climate dynamics.* Springer-Verlag New York Inc., New York, NY, 1987. MR878761 (88h:86001)

[GN10] Y. Guo and T. Nguyen. A note on the prandtl boundary layers. *arXiv:1011.0130v3 [math.AP]*, 2010. MR2849481 (2012m:35239)

[GOC04] G.A. Glatzmaier, D.E. Ogden, and T.L. Clune. Modeling the earth's dynamo. *Geophysical Monograph Series*, 150:13–24, 2004.

[GR97] G.A. Glatzmaier and P.H. Roberts. Simulating the geodynamo. *Contemporary physics*, 38(4):269–288, 1997.

[Gre00] E. Grenier. On the nonlinear instability of Euler and Prandtl equations. *Comm. Pure Appl. Math.*, 53(9):1067–1091, 2000. MR1761409 (2001i:76056)

[GVD10] D. Gérard-Varet and E. Dormy. On the ill-posedness of the Prandtl equation. *J. Amer. Math. Soc.*, 23(2):591–609, 2010. MR2601044 (2011f:35259)

[GVN10] D. Gérard-Varet and T. Nguyen. Remarks on the ill-posedness of the prandtl equation. *arXiv:1008.0532v1 [math.AP]*, 2010. MR2601044 (2011f:35259)

[IKR12] M. Ignatova, I. Kukavica, and L. Ryzhik. The harnack inequality for second-order elliptic equations with divergence-free drifts. *Preprint*, 2012. MR2871790 (2012m:35044)

[Kis10] A. Kiselev. Regularity and blow up for active scalars. *Math. Model. Nat. Phenom.*, 5(4):225–255, 2010. MR2662457 (2011e:35303)

[KN09] A. Kiselev and F. Nazarov. A variation on a theme of Caffarelli and Vasseur. *Zap. Nauchn. Sem. S.-Peterburg. Otdel. Mat. Inst. Steklov. (POMI)*, 370(Kraevye Zadachi Matematicheskoi Fiziki i Smezhnye Voprosy Teorii Funktsii. 40):58–72, 220, 2009. MR2749211 (2012c:35350)

[KNV07] A. Kiselev, F. Nazarov, and A. Volberg. Global well-posedness for the critical 2D dissipative quasi-geostrophic equation. *Invent. Math.*, 167(3):445–453, 2007. MR2276260 (2008f:35308)

[Kuk06] I. Kukavica. Role of the pressure for validity of the energy equality for solutions of the Navier-Stokes equation. *J. Dynam. Differential Equations*, 18(2):461–482, 2006. MR2229985 (2007c:35131)

[KV09] I. Kukavica and V. Vicol. On the radius of analyticity of solutions to the three-dimensional Euler equations. *Proc. Amer. Math. Soc.*, 137(2):669–677, 2009. MR2448589 (2009j:35272)

[LO97] C.D. Levermore and M. Oliver. Analyticity of solutions for a generalized Euler equation. *J. Differential Equations*, 133(2):321–339, 1997. MR1427856 (97k:35198)

[ML94] H.K. Moffatt and D.E. Loper. The magnetostrophic rise of a buoyant parcel in the earth's core. *Geophysical Journal International*, 117(2):394–402, 1994.

[Mof08] H.K. Moffatt. Magnetostrophic turbulence and the geodynamo. In Y. Kaneda, editor, *IUTAM Symposium on Computational Physics and New Perspectives in Turbulence, Nagoya, Japan, September, 11–14, 2006*, volume 4 of *IUTAM Bookser.*, pages 339–346. Springer, Dordrecht, 2008. MR2432632 (2009j:76227)

[MS61] L.D. Mešalkin and J.G. Sinaĭ. Investigation of the stability of a stationary solution of a system of equations for the plane movement of an incompressible viscous liquid. *J. Appl. Math. Mech.*, 25:1700–1705, 1961. MR0137419 (25:871)

[MV06] V.G. Maz'ya and I.E. Verbitsky. Form boundedness of the general second-order differential operator. *Comm. Pure Appl. Math.*, 59(9):1286–1329, 2006. MR2237288 (2008d:47089)

[MX12] C. Miao and L. Xue. Global well-posedness for a modified critical dissipative quasi-geostrophic equation. *Journal of Differential Equations*, 252(1):792–818, 2012. MR2852227

[Nas58] J. Nash. Continuity of solutions of parabolic and elliptic equations. *Amer. J. Math.*, 80:931–954, 1958. MR0100158 (20:6592)

[NU12] A.I. Nazarov and N.N. Ural'tseva. The harnack inequality and related properties for solutions of elliptic and parabolic equations with divergence-free lower-order coefficients. *St. Petersburg Math. J*, 23:93–115, 2012. MR2760150 (2012d:35026)

[Osa87] H. Osada. Diffusion processes with generators of generalized divergence form. *J. Math. Kyoto Univ.*, 27(4):597–619, 1987. MR916761 (89b:60188)

[OT00] M. Oliver and E.S. Titi. Remark on the rate of decay of higher order derivatives for solutions to the Navier-Stokes equations in $\mathbf{R}^n$. *J. Funct. Anal.*, 172(1):1–18, 2000. MR1749867 (2000m:35147)

[Ren09] M. Renardy. Ill-posedness of the hydrostatic Euler and Navier-Stokes equations. *Arch. Ration. Mech. Anal.*, 194(3):877–886, 2009. MR2563627 (2011e:35313)

[RG00] P.H. Roberts and G.A. Glatzmaier. Geodynamo theory and simulations. *Reviews of modern physics*, 72(4):1081, 2000.

[RG01] P.H. Roberts and G.A. Glatzmaier. The geodynamo, past, present and future. *Geophysical & Astrophysical Fluid Dynamics*, 94(1-2):47–84, 2001.

[Shv11] R. Shvydkoy. Convex integration for a class of active scalar equations. *J. Amer. Math. Soc.*, 24(4):1159–1174, 2011. MR2813340 (2012d:35295)

[SSŠZ12] G. Seregin, L. Silvestre, V. Šverák, and A. Zlatoš. On divergence-free drifts. *Journal of Differential Equations*, 252(1):505–540, 2012. MR2852216 (2012k:35434)

[SV12] L. Silvestre and V. Vicol. Hölder continuity for a drift-diffusion equation with pressure. *Annales de l'Institut Henri Poincare (C) Non Linear Analysis*, 20(4):637–652, 2012. MR2948291

[Tao06] T. Tao. *Nonlinear dispersive equations*, volume 106 of *CBMS Regional Conference Series in Mathematics*. Published for the Conference Board of the Mathematical Sciences, Washington, DC, 2006. Local and global analysis. MR2233925 (2008i:35211)

[VF03] M. Vishik and S. Friedlander. Nonlinear instability in two dimensional ideal fluids: the case of a dominant eigenvalue. *Comm. Math. Phys.*, 243(2):261–273, 2003. MR2021907 (2004k:76055)

[Wu05] J. Wu. Global solutions of the 2D dissipative quasi-geostrophic equation in Besov spaces. *SIAM J. Math. Anal.*, 36(3):1014–1030 (electronic), 2004/05. MR2111923 (2005i:76014)

[Zha04] Q.S. Zhang. A strong regularity result for parabolic equations. *Comm. Math. Phys.*, 244(2):245–260, 2004. MR2031029 (2005b:35116)

Department of Mathematics, University of Southern California, Los Angeles, California 90089
E-mail address: susanfri@usc.edu

Department of Mathematics, University of Southern California, Los Angeles, California 90089
E-mail address: wrusin@usc.edu

Department of Mathematics, Princeton University, Princeton, New Jersey 08544
E-mail address: vvicol@math.princeton.edu

Amer. Math. Soc. Transl.
Volume **232**, 2014

Variations on Liouville's theorem in the setting of stationary flows of generalized Newtonian fluids in the plane

M. Fuchs

In memory of Olga A. Ladyzhenskaya (1922–2004)

ABSTRACT. We consider entire solutions $u : \mathbb{R}^2 \to \mathbb{R}^2$ of the equations describing the stationary flow of a generalized Newtonian fluid in $2D$ and prove the constancy of the velocity field u if either u is bounded (or even not exceeding a certain growth rate at infinity) or if u has finite energy.

1. Introduction

A large part of the work of Olga A. Ladyzhenskaya is devoted to the deep analysis of mathematical problems arising in fluid mechanics, especially to the study of the Navier–Stokes equations (NSE) including the stationary variant as well as the time dependent situation in dimensions 2 and 3. Some of her fundamental results are highlighted in the monograph [**La1**] and in the papers [**La2**], [**La3**], [**La4**], [**La5**], [**LK**]. In our note we limit ourselves to the stationary case in $2D$ and try to exhibit conditions which imply the constancy of entire solutions, i.e., of solutions defined on the whole plane. Roughly speaking, our goal is to obtain Liouville-type results not only for NSE but also for the case of fluids with shear dependent viscosity. We first look at entire solutions of the homogeneous NSE

$$-\mu\Delta u + u^k \partial_k u + \nabla \pi = 0 \tag{1.1}$$

together with the incompressibility condition

$$\operatorname{div} u = 0\,. \tag{1.2}$$

Here $u : \mathbb{R}^2 \to \mathbb{R}^2$ denotes the velocity field, $\pi : \mathbb{R}^2 \to \mathbb{R}$ is the pressure and μ stands for the constant viscosity. Without loss of generality we can assume that $\mu = 1$; moreover, throughout this paper we adopt the convention of summation with respect to indices repeated twice. In 1978 the following result was obtained by Gilbarg and Weinberger [**GW**]:

2010 *Mathematics Subject Classification.* 76D05, 76D07, 76M30, 35Q30, 35Q35.

Key words and phrases. Generalized Newtonian fluids, stationary flows in $2D$, equations of Navier–Stokes type, Liouville theorems.

The author thanks the organizers of the workshop "Advances in mathematical analysis of partial differential equations" at Institut Mittag-Leffler, 9-13 July 2012.

THEOREM 1.1. *Suppose that* (1.1) *and* (1.2) *hold on the entire plane and that the (weak and thereby classical) solution* $u : \mathbb{R}^2 \to \mathbb{R}^2$ *is of finite energy in the sense that*

$$\int_{\mathbb{R}^2} |\nabla u|^2 \, dx < \infty . \tag{1.3}$$

Then the constancy of u *follows.*

One major ingredient of their proof is to introduce the vorticity $\omega = -\partial_1 u^2 + \partial_2 u^1$ satisfying the linear elliptic equation

$$\Delta \omega - u \cdot \nabla \omega = 0 , \tag{1.4}$$

and to give precise estimates for the growth of Dirichlet's energy of ω on annuli in terms of the radius with the help of the maximum principle valid for solutions of equation (1.4). Approximately 30 years later Koch, Nadirashvili, Seregin, Sverǎk [**KNSS**] studied entire solutions $u(t,x)$ of the much more delicate instationary variant of (1.1) and, as a byproduct of their investigations, they could replace (1.3) by the requirement that u is a bounded function, more precisely:

THEOREM 1.2. *If* (1.1) *and* (1.2) *hold on* $\mathbb{R}^2$ *and if*

$$\sup_{\mathbb{R}^2} |u| < \infty , \tag{1.5}$$

then $u = u_0$ *for some vector* $u_0 \in \mathbb{R}^2$.

We just wish to remark that the proof of Theorem 1.2 uses the linearity of the leading part of NSE in an essential way. An interesting question is whether we can weaken condition (1.5). In [**BFZ**] we could show:

THEOREM 1.3. *Suppose that* $u : \mathbb{R}^2 \to \mathbb{R}^2$ *is a solution of* (1.1) *and* (1.2) *on* $\mathbb{R}^2$ *such that*

$$\limsup_{|x| \to \infty} |u(x)| \, |x|^{-\alpha} < \infty \tag{1.6}$$

for some $\alpha < 1/3$. *Then the constancy of* u *follows.*

Since the proof of Theorem 1.3 is rather self-contained, we will present it in Section 2. It would be very interesting to know whether $\alpha < 1$ is actually admissible in (1.6).

In her book [**La1**] Ladyzhenskaya proposes also to study fluids with variable viscosity. One approach is to replace (1.1) through the equation

$$-\operatorname{div} \left[T^D \left(\varepsilon(u) \right) \right] + u^k \partial_k u + \nabla \pi = 0 , \tag{1.7}$$

where $\varepsilon(u)$ is the symmetric gradient of the velocity field and where T^D is the deviatoric part of the stress tensor. For further mathematical and also physical explanations, we refer to the monographs of Ladyzhenskaya [**La1**], Galdi [**Ga1**], [**Ga2**], and Málek, Necǎs, Rokyta, Růžička [**MNRR**] as well as to the book [**FS**]. We assume that

$$T^D = \nabla H \tag{1.8}$$

for a potential H being of the form

$$H(\varepsilon) = h(|\varepsilon|) \tag{1.9}$$

with a given density $h : [0, \infty) \to [0, \infty)$ of class C^2. Since

$$T^D(\varepsilon) = \frac{h'(|\varepsilon|)}{|\varepsilon|}\,\varepsilon\,,$$

we see that the variable viscosity is given by

$$\mu = \mu\,(|\varepsilon|) = \frac{h'(|\varepsilon|)}{|\varepsilon|}\,.$$

Following standard terminology we say that the flow is shear-thickening (-thinning), if μ is an increasing (decreasing) function of $|\varepsilon|$.

Next we formulate our hypotheses imposed on the density h occurring in the structural condition (1.9). We suppose that h satisfies the following:

(A1) h is strictly increasing and convex; we have $h''(0) > 0$ and $\lim_{t\to 0} \dfrac{h(t)}{t} = 0\,.$

(A2) There is a constant $a > 0$ such that $h(2t) \le ah(t)$ for all $t \ge 0$ (doubling property).

($\mathrm{A3_I}$) In the shear-thickening case we have $\dfrac{h'(t)}{t} \le h''(t)$ for all $t > 0\,.$

($\mathrm{A3_{II}}$) In the shear-thinning case we have $h''(t) \le \dfrac{h'(t)}{t}$ for all $t > 0\,.$

REMARK 1.1. i) *From* (A1) *it directly follows that* $h(0) = h'(0) = 0$ *and* $h'(t) > 0$ *for any* $t > 0$.

ii) *By considering* $\frac{d}{dt}\frac{h'(t)}{t}$, *it is immediate that* ($\mathrm{A3_I}$) *and* ($\mathrm{A3_{II}}$) *express the fact that the fluid is shear-thickening and shear-thinning, respectively.*

iii) (A1) *together with* (A2) *implies the balancing condition*

$$\text{(1.10)} \qquad c\,th'(t) \le h(t) \le th'(t) \quad \textit{for all } t \ge 0$$

and for a suitable positive constant c. *In fact,* $0 = h(0) \ge h(t) - th'(t)$ *holds by convexity, whereas by* (A2) *and the monotonicity of* h'

$$h(t) \ge \frac{1}{a}h(2t) = \frac{1}{a}\int_0^{2t} h'(s)\,ds \ge \frac{1}{a}\int_t^{2t} h'(s)\,ds \ge \frac{1}{a}th'(t)\,.$$

iv) *It is easy to see that from* (A2) *it follows*

$$h(t) \le h(1)t^a \quad \textit{for all} \quad t \ge 1\,,$$

thus

$$\text{(1.11)} \qquad h(t) \le c\,[t^a + 1] \quad \textit{for all } t \ge 0\,.$$

v) *If we are in the shear-thickening case* ($\mathrm{A3_I}$), *then* $\frac{h'(t)}{t} \ge \lim_{s\to 0}\frac{h'(s)}{s} = h''(0)$ *gives*

$$\text{(1.12)} \qquad h(t) \ge \frac{1}{2}h''(0)t^2 \quad \textit{for all } t \ge 0\,,$$

and (A1) *implies on account of* $h''(0) > 0$ *that our energy is of at least quadratic growth.*

vi) *In the shear-thinning case we have*

$$h(t) \le \frac{1}{2}h''(0)t^2 \tag{1.13}$$

and

$$h'(t)^2 \le ch(t) \tag{1.14}$$

for any $t \ge 0$. *For* (1.14) *we observe* $h'(t) \le th''(0)$, *which is an immediate consequence of* $h'(t)/t \le \lim_{s\to 0} h'(s)/s$, *thus*

$$h'(t)^2 \le th''(0)h'(t) \overset{(1.10)}{\le} ch''(0)h(t)\,.$$

Note that according to (1.13) *the condition* ($\mathrm{A3_{II}}$) *implies that the energy has subquadratic growth.*

Actually, even the case of linear growth is covered, which means that we can easily give examples of densities h *satisfying* (A1)–($\mathrm{A3_{II}}$) *for which* $\lim_{t\to\infty} h(t)/t \in (0,\infty)$.

vii) *It is not hard to show that* (A1) *and* ($\mathrm{A3_{II}}$) *already imply* (A2); *we refer to the Appendix of* [**BF1**].

REMARK 1.2. *Standard examples are non-degenerate power law fluids for which* $h(t) = \left(1+t^2\right)^{p/2}$, $t \ge 0$, *with* $p \in (1,\infty)$ *being of shear-thickening type, if* $p \ge 2$, *and of shear-thinning type otherwise. Another model we can include is the Prandtl–Eyring fluid described by* $h(t) = t\ln(1+t)$, $t \ge 0$.

Summarizing the results obtained in [**Fu**], [**FZ**] and [**Zh**], we have

THEOREM 1.4. *Suppose that* $u \in C^1(\mathbb{R}^2,\mathbb{R}^2)$ *is a (weak) solution of* (1.7) *and* (1.2) *on* $\mathbb{R}^2$ *under the hypothesis* (1.8) *and* (1.9). *Then* u *is constant, if either*

a) h *satisfies* (A1), (A2), ($\mathrm{A3_I}$) *and it holds that*

$$\int_{\mathbb{R}^2} h\left(|\nabla u|\right)\,dx < \infty \quad \text{or} \quad \sup_{\mathbb{R}^2}|u| < \infty\,,$$

or

b) h *satisfies* (A1), (A2), ($\mathrm{A3_{II}}$) *and in addition*

$$h''(t) \ge c/(1+t),\ t \ge 0\,,$$

and u *is assumed to be bounded.*

Note that the lower bound imposed on h'' in part b) of Theorem 1.4 together with $h(0) = h'(0) = 0$ implies that $h(t) \ge ct\ln(1+t), t \ge 0$. Since the case of bounded solutions for shear-thinning flows is treated in the cited references only for the Prandtl–Eyring model, a proof of part b) of Theorem 1.4 will be provided in Section 3. Let us clarify our terminology: a function $u \in C^1(\mathbb{R}^2,\mathbb{R}^2)$ is called a weak solution of (1.7) and (1.2), if (1.2) holds in the pointwise sense and if

$$\int_{\mathbb{R}^2} DH\left(\varepsilon(u)\right) : \varepsilon(\varphi)\,dx + \int_{\mathbb{R}^2} u^k\partial_k u \cdot \varphi\,dx = 0 \tag{1.15}$$

holds all $\varphi \in C_0^1(\mathbb{R}^2,\mathbb{R}^2)$, $\operatorname{div}\varphi = 0$. A discussion of the hypothesis $u \in C^1$ can be found in, e.g., [**Fu**], [**FZ**], [**Zh**].

We emphasize that Theorem 1.4 requires the existence as well as the positivity of $D^2H(0)$ so that we can cover the non-degenerate p-case and also the Prandtl–Eyring fluid. Concerning the degenerate p-case, we proved in [**BFZ**]:

THEOREM 1.5. *Let $u \in C^1(\mathbb{R}^2, \mathbb{R}^2)$ denote a weak solution of* (1.7) *and* (1.2) (*cf. equation* (1.15)) *with $H(\varepsilon) = |\varepsilon|^p$, $1 < p < \infty$.*

i) a) *Suppose that $1 < p \le 2$. If u is bounded, then the constancy of u follows.*

b) *Let $1 < p < 2$. Then*

$$\limsup_{|x|\to\infty} |x|^{-\alpha}\,|u(x)| < \infty$$

for some $\alpha \in \left[0, \frac{2-p}{p+6}\right)$ implies the constancy of u.

ii) *If $\frac{6}{5} < p \le 3$ together with $\int_{\mathbb{R}^2} |\nabla u|^p\,dx < \infty$, then again $u \equiv const$ follows.*

iii) *Let $p > 2$, and let $u_\infty \in \mathbb{R}^2$ denote a vector such that*

a) *in case $2 < p < 6$,*

$$\sup_{|x|\ge R} |u(x) - u_\infty|\,|x|^{\frac{p-2}{p+6}} \to 0 \text{ as } R \to \infty\,,$$

b) *in case $p = 6$,*

$$\limsup_{|x|\to\infty} |x|^{1/3}\,|u(x) - u_\infty| < \infty\,;$$

c) *in case $p > 6$,*

$$\sup_{|x|\ge R} |u(x) - u_\infty|\,|x|^{1/3} \longrightarrow 0 \text{ as } R \to \infty\,.$$

Then we obtain $u = u_\infty$.

REMARK 1.3. *Letting $p = 2$ in* i) *and* ii), *we recover the statements of Theorems 1.1 and 1.2. For $p < 2$, even a certain growth is allowed depending on the number $(2-p)/(p+6)$. Clearly, one should try to optimize this a priori growth condition, and what happens in the finite energy case should also be investigated, if $p > 3$ or $p < 6/5$. For $p > 2$ actually a certain decay of the solution is required being stated in* iii) a)–c).

REMARK 1.4. *The C^1-regularity of $W^1_{p,\mathrm{loc}}$-solutions of* (1.7) *and* (1.2) *for the degenerate p-case has been investigated by Naumann* [**Na**] *and Wolf* [**Wo**] *for $\frac{3}{2} < p < 2$. For $p > 2$, the C^1-regularity of weak local solutions seems to be open.*

2. Proof of Theorem 1.3

Let u denote an entire solution of (1.1) and (1.2) satisfying (1.6). Introducing the vorticity $\omega := \partial_2 u^1 - \partial_1 u^2$, we have for $q, \ell \in \mathbb{N}$ sufficiently large with $\eta \in C_0^\infty(\mathbb{R}^2)$

$$(2.1)\quad \int_{\mathbb{R}^2} \omega^{2q}\eta^{2\ell}\,dx = \int_{\mathbb{R}^2} \left(\partial_2 u^1 - \partial_1 u^2\right)\omega^{2q-1}\eta^{2\ell}\,dx$$

$$= \int_{\mathbb{R}^2} \operatorname{div}\left(-u^2, u^1\right)\omega^{2q-1}\eta^{2\ell}\,dx = -\int_{\mathbb{R}^2} \left(-u^2, u^1\right)\cdot\nabla\left[\omega^{2q-1}\eta^{2\ell}\right]dx$$

$$= (2q-1)\int_{\mathbb{R}^2} \nabla\omega\cdot\left(u^2, -u^1\right)\omega^{2q-2}\eta^{2\ell}\,dx$$

$$+2\ell\int_{\mathbb{R}^2} \left(u^2, -u^1\right)\cdot\nabla\eta\,\omega^{2q-1}\eta^{2\ell-1}\,dx\,,$$

and from $\operatorname{div} u = 0$ we infer

$$(2.2)\qquad \begin{aligned}\int_{\mathbb{R}^2} u \cdot \nabla\omega\,\omega^{2q-3}\eta^{2\ell}\,dx &= \frac{1}{2q-2}\int_{\mathbb{R}^2} u\cdot\nabla\omega^{2q-2}\eta^{2\ell}\,dx\\ &= -\frac{1}{2q-2}\int_{\mathbb{R}^2} u\cdot\nabla\eta^{2\ell}\omega^{2q-2}\,dx\,.\end{aligned}$$

Recall (cf. (1.4)) that $\Delta\omega - u\cdot\nabla\omega = 0$ on $\mathbb{R}^2$, hence

$$\int_{\mathbb{R}^2}\nabla\omega\cdot\nabla\varphi\,dx + \int_{\mathbb{R}^2} u\cdot\nabla\omega\varphi\,dx = 0$$

for $\varphi\in C_0^1(\mathbb{R}^2)$. We specify $\varphi = \eta^{2\ell}\omega^{2q-3}$ and get

$$(2.3)\qquad \begin{aligned}&\int_{\mathbb{R}^2}\eta^{2\ell}\,(2q-3)\,|\nabla\omega|^2\,\omega^{2q-4}\,dx\\ &= -\int_{\mathbb{R}^2}\nabla\omega\cdot\nabla\eta^{2\ell}\omega^{2q-3}\,dx - \int_{\mathbb{R}^2} u\cdot\nabla\omega\omega^{2q-3}\eta^{2\ell}\,dx\,.\end{aligned}$$

By Young's inequality the first term on the right-hand side of (2.3) is estimated through

$$\delta\int_{\mathbb{R}^2}|\nabla\omega|^2\,\omega^{2q-4}\eta^{2\ell}\,dx + c\,(\delta,\ell)\int_{\mathbb{R}^2}|\nabla\eta|^2\,\eta^{2\ell-2}\omega^{2q-2}\,dx\,;$$

to the second term on the right-hand side of (2.3) we apply (2.2). After appropriate choice of δ, this yields

$$(2.4)\qquad \begin{aligned}&\int_{\mathbb{R}^2}|\nabla\omega|^2\,\omega^{2q-4}\eta^{2\ell}\,dx\\ &\le c\,(\ell,q)\left[\int_{\mathbb{R}^2}\omega^{2q-2}\eta^{2\ell-2}\,|\nabla\eta|^2\,dx + \int_{\mathbb{R}^2}|u|\,\left|\nabla\eta^{2\ell}\right|\omega^{2q-2}\,dx\right].\end{aligned}$$

Now we return to (2.1) and apply Young's inequality one more time with the result

$$\begin{aligned}\int_{\mathbb{R}^2}\omega^{2q}\eta^{2\ell}\,dx &\le (2q-1)\int_{\mathbb{R}^2}|\nabla\omega|\,|u|\omega^{2q-2}\eta^{2\ell}\,dx\\ &\quad +2\ell\int_{\mathbb{R}^2}|u|\,|\nabla\eta|\,\omega^{2q-1}\eta^{2\ell-1}\,dx\\ &\le \delta\int_{\mathbb{R}^2}\omega^{2q}\eta^{2\ell}\,dx + c(\delta,q)\int_{\mathbb{R}^2}|\nabla\omega|^2\,|u|^2\omega^{2q-4}\eta^{2\ell}\,dx\\ &\quad +2\ell\int_{\mathbb{R}^2}|u|\,|\nabla\eta|\,\omega^{2q-1}\eta^{2\ell-1}\,dx\,,\end{aligned}$$

hence for δ sufficiently small
(2.5)

$$\int_{\mathbb{R}^2}\eta^{2\ell}\omega^{2q}\,dx \le c(\ell,q)\left[\int_{\mathbb{R}^2}|\nabla\omega|^2\,|u|^2\omega^{2q-4}\eta^{2\ell}\,dx + \int_{\mathbb{R}^2}|u|\,|\nabla\eta|\,\omega^{2q-1}\eta^{2\ell-1}\,dx\right].$$

Next we specify η. Let $R\ge 1$ and choose $\eta = 1$ on $B_R(0)$, $0\le\eta\le 1$, $\operatorname{spt}\eta\subset B_{2R}(0)$, $|\nabla\eta|\le c/R$. From (1.6) we obtain (without loss of generality we assume $\alpha > 0$)

$$(2.6)\qquad |u(x)|\le cR^{\alpha},\ x\in B_R(0)\,.$$

We use (2.6) on the right-hand side of (2.5) and get

$$\int_{B_{2R}(0)} \eta^{2\ell}\omega^{2q}\,dx$$
$$\leq c(\ell,q)\left[R^{2\alpha}\int_{B_{2R}(0)} |\nabla\omega|^2\,\omega^{2q-4}\eta^{2\ell}\,dx + R^{\alpha}\int_{B_{2R}(0)} |\nabla\eta|\,\omega^{2q-1}\eta^{2\ell-1}\,dx\right],$$

and if we apply (2.4) on the right-hand side quoting (2.6) one more time it follows that

$$\begin{aligned}\text{(2.7)}\quad \int_{B_{2R}(0)} \eta^{2\ell}\omega^{2q}\,dx \;\leq\; & c(\ell,q)\left[R^{2\alpha}\int_{B_{2R}(0)} \omega^{2q-2}\eta^{2\ell-2}|\nabla\eta|^2\,dx\right. \\ & + R^{3\alpha}\int_{B_{2R}(0)} \left|\nabla\eta^{2\ell}\right|\omega^{2q-2}\,dx \\ & \left. + R^{\alpha}\int_{B_{2R}(0)} \omega^{2q-1}|\nabla\eta|\,\eta^{2\ell-1}\,dx\right] \\ =:\; & c(\ell,q)\left[T_1+T_2+T_3\right].\end{aligned}$$

Young's inequality yields

$$\begin{aligned} T_1 &\leq \int_{B_{2R}(0)} \omega^{2q-2}\eta^{2\ell-2}R^{2\alpha-2}\,dx \\ &\leq \delta\int_{B_{2R}(0)} \omega^{2q}\eta^{(2\ell-2)2q/(2q-2)}\,dx + c(\delta)R^{2+q(2\alpha-2)}, \\ T_2 &\leq \int_{B_{2R}(0)} \omega^{2q-2}\eta^{2\ell-1}R^{3\alpha-1}\,dx \\ &\leq \delta\int_{B_{2R}(0)} \omega^{2q}\eta^{(2\ell-1)2q/(2q-2)}\,dx + c(\delta)R^{2+q(3\alpha-1)}, \\ T_3 &\leq \int_{B_{2R}(0)} \omega^{2q-1}\eta^{2\ell-1}R^{\alpha-1}\,dx \\ &\leq \delta\int_{B_{2R}(0)} \omega^{2q}\eta^{(2\ell-1)2q/(2q-1)}\,dx + c(\delta)R^{2+2q(\alpha-1)}, \end{aligned}$$

and for $\ell \gg 1$ we have

$$2\ell \leq (2\ell-2)2q/(2q-2)$$

and

$$2\ell \leq (2\ell-1)2q/(2q-1),$$

hence, for δ small enough, we obtain from (2.7)

$$\text{(2.8)}\qquad \int_{B_{2R}(0)} \eta^{2\ell}\omega^{2q}\,dx \leq c(\ell,q)\left[R^{2+q(2\alpha-2)} + R^{2+(3\alpha-1)q} + R^{2+2q(\alpha-1)}\right].$$

Recall that $\alpha < 1/3$. Therefore, we can fix a sufficiently large exponent q with the property that

$$2+(3\alpha-1)q < 0,$$

and (2.8) shows

$$\int_{B_R(0)} \omega^{2q}\,dx \le c(\ell,q)R^{2+(3\alpha-1)q} \longrightarrow 0$$

as $R\to\infty$, hence $\omega\equiv 0$ on $\mathbb{R}^2$. This together with $\operatorname{div} u = 0$ shows that u is harmonic, and the constancy of u then follows from (1.6) and results concerning entire harmonic functions. □

3. Bounded solutions in the shear-thinning case

Throughout this section we assume that h satisfies (A1), (A2), and (A3$_{\rm II}$) together with $h''(t)\ge c/(1+t)$ and that $u\in C^1(\mathbb{R}^2,\mathbb{R}^2)$ denotes a bounded solution of (1.7) and (1.2). Then it is an easy exercise to show that u is in the space $W^2_{2,\text{loc}}(\mathbb{R}^2,\mathbb{R}^2)$, which follows from the non-degeneracy of D^2H and the local boundedness of ∇u. For proving the constancy of u we first claim:

LEMMA 3.1. *There is a constant* $c=c(\|u\|_{L^\infty(\mathbb{R}^2)})$ *such that*

$$\int_{Q_R(x_0)} h\left(|\varepsilon(u)|\right)dx \le c(R+1) \tag{3.1}$$

for any square $Q_R(x_0) := \left\{x\in\mathbb{R}^2 : \left|x^i - x_0^i\right| < R,\ i=1,2\right\}$.

Proof of Lemma 3.1. Recalling (1.15), we have

$$0 = \int_{Q_{2R}(x_0)} DH\left(\varepsilon(u)\right):\varepsilon\left(\eta^2 u - w\right)dx + \int_{Q_{2R}(x_0)} u^k\partial_k u^i\left(\eta^2u^i - w^i\right)dx\,. \tag{3.2}$$

Here $\eta\in C^1_0\left(Q_{2R}(x_0)\right)$ denotes a cut-off function such that $\eta=1$ on $Q_R(x_0)$, $0\le\eta\le 1$, $|\nabla\eta|\le c/R$, and $w\in \overset{\circ}{W}{}^1_2\left(Q_{2R}(x_0),\mathbb{R}^2\right)$ is chosen to satisfy (cf. [**Ga1**])

$$\operatorname{div} w = \operatorname{div}(\eta^2u) = \nabla\eta^2\cdot u \quad \text{on } Q_{2R}(x_0)$$

together with

$$\|\nabla w\|_{L^2(Q_{2R}(x_0))} \le c\left\|\nabla\eta^2\cdot u\right\|_{L^2(Q_{2R}(x_0))}\,.$$

From (3.2) we obtain

$$\begin{aligned}
&\int_{Q_{2R}(x_0)} DH\left(\varepsilon(u)\right):\varepsilon(u)\eta^2\,dx\\
&\quad+\int_{Q_{2R}(x_0)} 2\frac{\partial H}{\partial\varepsilon_{i\alpha}}\left(\varepsilon(u)\right)\partial_\alpha\eta u^i\eta\,dx\\
&\quad-\int_{Q_{2R}(x_0)} DH\left(\varepsilon(u)\right):\varepsilon(w)\,dx\\
&\quad+\int_{Q_{2R}(x_0)} u^k\partial_k u^i u^i\eta^2\,dx\\
&\quad-\int_{Q_{2R}(x_0)} u^k\partial_k u^i w^i\,dx = 0\,.
\end{aligned}$$

We have

$$\int_{Q_{2R}(x_0)} DH(\varepsilon(u)) : \varepsilon(u)\eta^2\, dx \geq \int_{Q_{2R}(x_0)} \eta^2 H(\varepsilon(u))\, dx\,,$$

$$\begin{aligned}
&\left| 2\int_{Q_{2R}(x_0)} \frac{\partial H}{\partial \varepsilon_{i\alpha}}(\varepsilon(u))\, \partial_\alpha \eta u^i \eta\, dx \right| \\
&\quad < c\int_{Q_{2R}(x_0)} h'(|\varepsilon(u)|)\, |\nabla\eta||u|\eta\, dx \\
&\quad \leq \delta\int_{Q_{2R}(x_0)} h'(|\varepsilon(u)|)\, |\varepsilon(u)|\eta^2\, dx + c(\delta)\int_{Q_{2R}(x_0)} \frac{h'(|\varepsilon(u)|)}{|\varepsilon(u)|}|u|^2|\nabla\eta|^2\, dx\,,
\end{aligned}$$

and if δ is chosen small enough and if we take into account the inequality $\frac{h'(t)}{t} \leq$ const (compare the remarks after (1.14)) in combination with (1.10) it follows

$$\begin{aligned}
(3.3)\qquad &\int_{Q_{2R}(x_0)} \eta^2 H(\varepsilon(u))\, dx \\
&\leq c\left[\int_{Q_{2R}(x_0)} |u|^2|\nabla\eta|^2\, dx + \left|\int_{Q_{2R}(x_0)} DH(\varepsilon(u)) : \varepsilon(w)\, dx\right| \right.\\
&\qquad \left. + \left|\int_{Q_{2R}(x_0)} u^k\partial_k u^i u^i \eta^2\, dx\right| + \left|\int_{Q_{2R}(x_0)} u^k\partial_k u^i w^i\, dx\right|\right] \\
&=: c\left[T_1 + T_2 + T_3 + T_4\right]\,.
\end{aligned}$$

The quantities T_i are estimated as follows. It clearly holds

$$T_1 \leq cR^{-2}\int_{Q_{2R}(x_0)} |u|^2\, dx\,. \tag{3.4}$$

We have for any $\delta > 0$ by Young's inequality and the properties of w,

$$\begin{aligned}
T_2 &\leq \delta\int_{Q_{2R}(x_0)} h'(|\varepsilon(u)|)^2\, dx + c(\delta)\int_{Q_{2R}(x_0)} |\nabla w|^2\, dx \\
&\leq \delta\int_{Q_{2R}(x_0)} h'(|\varepsilon(u)|)^2\, dx + c(\delta)\int_{Q_{2R}(x_0)} |\operatorname{div}(\eta^2 u)|^2\, dx \\
&\leq \delta\int_{Q_{2R}(x_0)} h'(|\varepsilon(u)|)^2\, dx + c(\delta)R^{-2}\int_{Q_{2R}(x_0)} |u|^2\, dx\,,
\end{aligned}$$

hence (replacing δ by $\delta/4$ and quoting (1.14))

$$T_2 \leq \delta\int_{Q_{2R}(x_0)} H(\varepsilon(u))\, dx + c(\delta)R^{-2}\int_{Q_{2R}(x_0)} |u|^2\, dx\,. \tag{3.5}$$

Next we observe

$$\begin{aligned}
(3.6)\qquad T_3 &= \frac{1}{2}\left|\int_{Q_{2R}(x_0)} u^k\partial_k|u|^2\eta^2\, dx\right| \\
&= \frac{1}{2}\left|\int_{Q_{2R}(x_0)} u^k|u|^2\partial_k\eta^2\, dx\right| \leq cR^{-1}\int_{Q_{2R}(x_0)} |u|^3\, dx\,,
\end{aligned}$$

and from

$$\int_{Q_{2R}(x_0)} u^k \partial_k u^i w^i\, dx = -\int_{Q_{2R}(x_0)} u^k u^i \partial_k w^i\, dx$$

it follows using Hölder's inequality

$$(3.7)\qquad \begin{aligned} T_4 &\leq \left(\int_{Q_{2R}(x_0)} |u|^4\, dx\right)^{1/2} \left(\int_{Q_{2R}(x_0)} |\nabla w|^2\, dx\right)^{1/2} \\ &\leq c\left(\int_{Q_{2R}(x_0)} |u|^4\, dx\right)^{1/2} \left(R^{-2}\int_{Q_{2R}(x_0)} |u|^2\, dx\right)^{1/2} \\ &= cR^{-1}\left[\int_{Q_{2R}(x_0)} |u|^4\, dx \int_{Q_{2R}(x_0)} |u|^2\, dx\right]^{1/2} \\ &\leq cR^{-1}\left[\int_{Q_{2R}(x_0)} |u|^4\, dx + \int_{Q_{2R}(x_0)} |u|^2\, dx\right]. \end{aligned}$$

Inserting (3.4)–(3.7) into (3.3), we get

$$\begin{aligned} \int_{Q_R(x_0)} H(\varepsilon(u))\, dx &\leq \delta \int_{Q_{2R}(x_0)} H(\varepsilon(u))\, dx \\ &+c(\delta)\left[R^{-1}\int_{Q_{2R}(x_0)} \left(|u|^2 + |u|^3 + |u|^4\right) dx + R^{-2}\int_{Q_{2R}(x_0)} |u|^2\, dx\right] \end{aligned}$$

for any $\delta > 0$ and all $Q_R(x_0)$. The δ-lemma of Giaquinta and Modica (cf. Lemma 0.5 in [**GM**]) and its slight extension stated in Lemma 3.1 of [**FZ**] then yields

$$\begin{aligned} &\int_{Q_R(x_0)} H(\varepsilon(u))\, dx \\ &\quad\leq c\left[R^{-1}\int_{Q_{2R}(x_0)} \left(|u|^2 + |u|^3 + |u|^4\right) dx + R^{-2}\int_{Q_{2R}(x_0)} |u|^2\, dx\right], \end{aligned}$$

and since u is bounded we have established (3.1). □

Next we like to prove the validity of

$$(3.8)\qquad \int_{\mathbb{R}^2} D^2H(\varepsilon(u))(\partial_k\varepsilon(u), \partial_k\varepsilon(u))\, dx < \infty.$$

Note that from (3.8) we immediately get (recalling our hypotheses imposed on h)

$$(3.9)\qquad \int_{\mathbb{R}^2} \frac{1}{1+|\varepsilon(u)|} |\nabla\varepsilon(u)|^2\, dx < \infty.$$

For the discussion of (3.8) we return to equation (1.15). Replacing φ by $\partial_\alpha\varphi$ for $\varphi \in C_0^\infty(Q_{\frac{3}{2}R}(x_0), \mathbb{R}^2)$ with $\operatorname{div}\varphi = 0$, we obtain by partial integration

$$(3.10)\qquad \begin{aligned} 0 &= \int_{Q_{\frac{3}{2}R}(x_0)} D^2H(\varepsilon(u))(\partial_\alpha\varepsilon(u), \varepsilon(\varphi))\, dx \\ &\quad - \int_{Q_{\frac{3}{2}R}(x_0)} u^k \partial_k u^i \partial_\alpha \varphi^i\, dx, \qquad \alpha = 1, 2. \end{aligned}$$

Let $\eta \in C_0^\infty(Q_{\frac{3}{2}R}(x_0))$ such that $\eta = 1$ on $Q_R(x_0)$, $0 \le \eta \le 1$, and $|\nabla \eta| \le c/R$. Let $f_\alpha := \operatorname{div}(\partial_\alpha u \eta^2) = \partial_\alpha u \cdot \nabla \eta^2$ and select w_α according to [**Ga1**] from the space $\overset{\circ}{W}{}_2^1\left(Q_{\frac{3}{2}R}(x_0), \mathbb{R}^2\right)$ such that

$$
\begin{aligned}
&\operatorname{div} w_\alpha = f_\alpha \text{ on } Q_{\frac{3}{2}R}(x_0)\,,\\
&\int_{Q_{\frac{3}{2}R}(x_0)} |\nabla w_\alpha|^2\,dx \le c \int_{Q_{\frac{3}{2}R}(x_0)} |\partial_\alpha u \cdot \nabla \eta|^2\,dx\,.
\end{aligned}
\tag{3.11}
$$

Finally we choose $\varphi := \eta^2 \partial_\alpha u - w_\alpha$ in (3.10). Equation (3.10) then yields

$$
\begin{aligned}
\int_{Q_{\frac{3}{2}R}(x_0)} & D^2H(\varepsilon(u))\left(\partial_\alpha \varepsilon(u), \partial_\alpha \varepsilon(u)\right)\eta^2\,dx \\
= & -\int_{Q_{\frac{3}{2}R}(x_0)} D^2H(\varepsilon(u))\left(\partial_\alpha \varepsilon(u), \nabla\eta^2 \odot \partial_\alpha u\right)dx \\
& + \int_{Q_{\frac{3}{2}R}(x_0)} D^2H(\varepsilon(u))\left(\partial_\alpha \varepsilon(u), \varepsilon(w_\alpha)\right)dx \\
& + \int_{Q_{\frac{3}{2}R}(x_0)} u^k \partial_k u^i \partial_\alpha\left(\eta^2 \partial_\alpha u^i\right)dx - \int_{Q_{\frac{3}{2}R}(x_0)} u^k \partial_k u^i \partial_\alpha w_\alpha^i\,dx \\
=: & \; -S_1 + S_2 + S_3 - S_4\,,
\end{aligned}
\tag{3.12}
$$

where "$\odot$" is the symmetric product of vectors. Using the Cauchy–Schwarz inequality for the bilinear form $D^2H(\varepsilon(u))$ in combination with Young's inequality, we obtain for any $\delta > 0$ (observe that $D^2H(\varepsilon)(\sigma,\sigma) \le c|\sigma|^2$ on account of (A3$_{\mathrm{II}}$) and the remarks after (1.14))

$$
\begin{aligned}
|S_2| \le\; & \delta \int_{Q_{\frac{3}{2}R}(x_0)} D^2H(\varepsilon(u))\left(\partial_\alpha \varepsilon(u), \partial_\alpha \varepsilon(u)\right)dx \\
& + \frac{1}{\delta}\int_{Q_{\frac{3}{2}R}(x_0)} D^2H(\varepsilon(u))\left(\varepsilon(w_\alpha), \varepsilon(w_\alpha)\right)dx \\
\le\; & \delta \int_{Q_{\frac{3}{2}R}(x_0)} D^2H(\varepsilon(u))\left(\partial_\alpha \varepsilon(u), \partial_\alpha \varepsilon(u)\right)dx \\
& + c(\delta)\int_{Q_{\frac{3}{2}R}(x_0)} \nabla w_\alpha : \nabla w_\alpha\,dx \\
\overset{(3.11)}{\le}\; & \delta \int_{Q_{\frac{3}{2}R}(x_0)} D^2H(\varepsilon(u))\left(\partial_\alpha \varepsilon(u), \partial_\alpha \varepsilon(u)\right)dx + c(\delta)\int_{Q_{\frac{3}{2}R}(x_0)} |\nabla u|^2 |\nabla \eta|^2\,dx\,,
\end{aligned}
$$

hence

$$
|S_2| \le \delta \int_{Q_{\frac{3}{2}R}(x_0)} b\,dx + c(\delta)\int_{Q_{\frac{3}{2}R}(x_0)} |\nabla u|^2 |\nabla \eta|^2\,dx\,. \tag{3.13}
$$

Here we have abbreviated

$$b := D^2 H(\varepsilon(u))\,(\partial_\alpha \varepsilon(u), \partial_\alpha \varepsilon(u))\,. \tag{3.14}$$

By applying exactly the same arguments to S_1 we see

$$|S_1| \le \delta \int_{Q_{\frac{3}{2}R}(x_0)} b\,dx + c(\delta) \int_{Q_{\frac{3}{2}R}(x_0)} |\nabla \eta|^2 |\nabla u|^2\,dx\,, \tag{3.15}$$

and (3.15) is valid for any choice of $\delta > 0$.

Next we look at S_3. It holds

$$\begin{aligned} S_3 &= \int_{Q_{\frac{3}{2}R}(x_0)} u^k \partial_k u^i \partial_\alpha \left(\eta^2 \partial_\alpha u^i\right)\,dx = -\int_{Q_{\frac{3}{2}R}(x_0)} \partial_\alpha (u^k \partial_k u^i) \eta^2 \partial_\alpha u^i\,dx \\ &= -\int_{Q_{\frac{3}{2}R}(x_0)} \partial_\alpha u^k \partial_k u^i \partial_\alpha u^i \eta^2\,dx - \int_{Q_{\frac{3}{2}R}(x_0)} u^k \partial_\alpha \partial_k u^i \partial_\alpha u^i \eta^2\,dx\,, \end{aligned}$$

and since we are in the 2D case, the first integral on the right-hand side vanishes. This shows

$$|S_3| = \frac{1}{2}\left|\int_{Q_{\frac{3}{2}R}(x_0)} u^k \partial_k |\nabla u|^2 \eta^2\,dx\right| = \frac{1}{2}\left|\int_{Q_{\frac{3}{2}R}(x_0)} u \cdot \nabla \eta^2 |\nabla u|^2\,dx\right|,$$

and we obtain

$$|S_3| \le cR^{-1} \int_{Q_{\frac{3}{2}R}(x_0)} |\nabla u|^2\,dx \tag{3.16}$$

for a constant c depending on $\|u\|_{L^\infty(\mathbb{R}^2)}$. Finally, we discuss S_4 again using the boundedness of the velocity field:

$$\begin{aligned} |S_4| &\le c \int_{Q_{\frac{3}{2}R}(x_0)} |\nabla u| |\partial_\alpha w_\alpha|\,dx \\ &\overset{(3.11)}{\le} c \left(\int_{Q_{\frac{3}{2}R}(x_0)} |\nabla u|^2\,dx\right)^{1/2} \left(\int_{Q_{\frac{3}{2}R}(x_0)} |\nabla \eta|^2 |\nabla u|^2\,dx\right)^{1/2}, \end{aligned}$$

thus

$$|S_4| \le cR^{-1} \int_{Q_{\frac{3}{2}R}(x_0)} |\nabla u|^2\,dx\,. \tag{3.17}$$

Putting together our estimates (3.13)–(3.17) and returning to (3.12), we have shown for any $\delta > 0$ the validity of the inequality

$$\begin{aligned}(3.18)\qquad \int_{Q_{\frac{3}{2}R}(x_0)} \eta^2 b\,dx \le \delta \int_{Q_{\frac{3}{2}R}(x_0)} b\,dx \\ +c(\delta)\left[R^{-2}\int_{Q_{\frac{3}{2}R}(x_0)} |\nabla u|^2\,dx + R^{-1}\int_{Q_{\frac{3}{2}R}(x_0)} |\nabla u|^2\,dx\right]\end{aligned}$$

with $c(\delta)$ also depending on $\|u\|_{L^\infty(\mathbb{R}^2)}$. In order to control Dirichlet's integral on the right-hand side of (3.18) in an appropriate way, let us select $\varphi \in C_0^\infty(Q_{2R}(x_0))$, $0 \le \varphi \le 1$, $\varphi = 1$ on $Q_{\frac{3}{2}R}(x_0)$, $|\nabla\varphi| \le c/R$. We have by Korn's inequality

$$\begin{aligned}\int_{Q_{\frac{3}{2}R}(x_0)} |\nabla u|^2\,dx &\le \int_{Q_{2R}(x_0)} \varphi^2|\nabla u|^2\,dx \\ &\le c\left[\int_{Q_{2R}(x_0)} |\nabla(\varphi u)|^2\,dx + \int_{Q_{2R}(x_0)} |\nabla\varphi|^2|u|^2\,dx\right] \\ &\le c\left[\int_{Q_{2R}(x_0)} |\varepsilon(\varphi u)|^2\,dx + \int_{Q_{2R}(x_0)} |\nabla\varphi|^2|u|^2\,dx\right] \\ &\le c\left[\int_{Q_{2R}(x_0)} \varphi^2|\varepsilon(u)|^2\,dx + R^{-2}\int_{Q_{2R}(x_0)} |u|^2\,dx\right],\end{aligned}$$

and if we recall the support property of η, inequality (3.18) in combination with the above estimates implies

$$\begin{aligned}(3.19)\qquad \int_{Q_R(x_0)} b\,dx \le \delta \int_{Q_{2R}(x_0)} b\,dx \\ +c(\delta)\Bigg[R^{-4}\int_{Q_{2R}(x_0)} |u|^2\,dx + R^{-3}\int_{Q_{2R}(x_0)} |u|^2\,dx \\ +R^{-2}\int_{Q_{2R}(x_0)} \varphi^2|\varepsilon(u)|^2\,dx + R^{-1}\int_{Q_{2R}(x_0)} \varphi^2|\varepsilon(u)|^2\,dx\Bigg].\end{aligned}$$

We have by Hölder's and Young's inequality

$$\begin{aligned}
\int_{Q_{2R}(x_0)} \varphi^2|\varepsilon(u)|^2\,dx &= \int_{Q_{2R}(x_0)} \varepsilon_{ij}(u)\varepsilon_{ij}(u)\varphi^2\,dx \\
&= -\int_{Q_{2R}(x_0)} u^i\partial_j\left(\varepsilon_{ij}(u)\varphi^2\right)\,dx \\
&= -\int_{Q_{2R}(x_0)} u^i\partial_j\varepsilon_{ij}(u)\varphi^2\,dx - \int_{Q_{2R}(x_0)} u^i\varepsilon_{ij}(u)\partial_j\varphi^2\,dx \\
&\le c\left[\int_{Q_{2R}(x_0)} |\nabla\varepsilon(u)|\,dx + R^{-1}\int_{Q_{2R}(x_0)} |\varepsilon(u)|\,dx\right] \\
&= c\left[\int_{Q_{2R}(x_0)} (1+|\varepsilon(u)|)^{-1/2}\,|\nabla\varepsilon(u)|\,(1+|\varepsilon(u)|)^{1/2}\,dx\right. \\
&\qquad \left. + R^{-1}\int_{Q_{2R}(x_0)} |\varepsilon(u)|\,dx\right] \\
&\le \left[\left(\int_{Q_{2R}(x_0)} \frac{|\nabla\varepsilon(u)|^2}{1+|\varepsilon(u)|}\,dx\right)^{1/2}\left(\int_{Q_{2R}(x_0)} (1+|\varepsilon(u)|)\,dx\right)^{1/2}\right. \\
&\qquad \left. + R^{-1}\int_{Q_{2R}(x_0)} |\varepsilon(u)|\,dx\right] \\
&\le \tau\int_{Q_{2R}(x_0)} b\,dx + c\tau^{-1}\int_{Q_{2R}(x_0)} (1+|\varepsilon(u)|)\,dx \\
&\qquad | + cR^{-1}\int_{Q_{2R}(x_0)} |\varepsilon(u)|\,dx\,,
\end{aligned}$$

where τ is any positive number. During our calculation, we have used that

$$D^2H(\varepsilon)(\sigma,\sigma) \ge \min\left\{\frac{h'(|\varepsilon|)}{|\varepsilon|}, h''(|\varepsilon|)\right\}|\sigma|^2\,,$$

hence

$$D^2H(\varepsilon)(\sigma,\sigma) \ge h''(|\varepsilon|)\,|\sigma|^2$$

on account of $(A3_{II})$. Recalling our assumption

$$h''(t) \ge c(1+t)^{-1}$$

concerning the shear-thinning case in Theorem 1.4, the above chain of inequalities is established. Choosing

$$\tau = \delta c(\delta)^{-1}R^2,\ c(\delta) \text{ from } (3.19)\,,$$

we get with a new constant $\widetilde{c}(\delta)$ recalling also (3.14)

$$\begin{aligned}
(3.20)\quad c(\delta)R^{-2}\int_{Q_{2R}(x_0)} \varphi^2\,|\varepsilon(u)|^2\,dx &\le \delta\int_{Q_{2R}(x_0)} b\,dx \\
+\widetilde{c}(\delta)\left[R^{-4}\int_{Q_{2R}(x_0)} (1+|\varepsilon(u)|)\,dx + R^{-3}\int_{Q_{2R}(x_0)} |\varepsilon(u)|\,dx\right]&,
\end{aligned}$$

whereas the choice

$$\tau := \delta c(\delta)^{-1}R$$

leads to

$$(3.21)\quad c(\delta)R^{-1}\int_{Q_{2R}(x_0)}\varphi^2\left|\varepsilon(u)\right|^2 dx \le \delta\int_{Q_{2R}(x_0)} b\,dx$$
$$+\widetilde{c}(\delta)\left[R^{-2}\int_{Q_{2R}(x_0)}(1+|\varepsilon(u)|)\,dx + R^{-2}\int_{Q_{2R}(x_0)}|\varepsilon(u)|\,dx\right].$$

With (3.20) and (3.21) we return to (3.19) writing again $c(\delta)$ for constants depending on δ (and $\|u\|_{L^\infty(\mathbb{R}^2)}$) and replacing the parameter δ by $\delta/3$. We obtain:

$$(3.22)\quad \int_{Q_R(x_0)} b\,dx \le \delta\int_{Q_{2R}(x_0)} b\,dx + c(\delta)\left[R^{-4}\int_{Q_{2R}(x_0)}|u|^2\,dx\right.$$
$$+R^{-3}\int_{Q_{2R}(x_0)}|u|^2\,dx + R^{-4}\int_{Q_{2R}(x_0)}(1+|\varepsilon(u)|)\,dx$$
$$\left.+R^{-3}\int_{Q_{2R}(x_0)}(1+|\varepsilon(u)|)\,dx + R^{-2}\int_{Q_{2R}(x_0)}(1+|\varepsilon(u)|)\,dx\right].$$

To estimate (3.22) we can apply Lemma 3.1 from [**FZ**] and get for all squares $Q_R(x_0)$ with $c = c(\|u\|_{L^\infty(\mathbb{R}^2)})$

$$(3.23)\quad \int_{Q_R(x_0)} b\,dx \le c\left[R^{-4}\int_{Q_{2R}(x_0)}|u|^2\,dx\right.$$
$$+R^{-3}\int_{Q_{2R}(x_0)}|u|^2\,dx + R^{-4}\int_{Q_{2R}(x_0)}(1+|\varepsilon(u)|)\,dx$$
$$\left.+R^{-3}\int_{Q_{2R}(x_0)}(1+|\varepsilon(u)|)\,dx + R^{-2}\int_{Q_{2R}(x_0)}(1+|\varepsilon(u)|)\,dx\right].$$

Now, if the case $R \ge 1$ is considered, inequality (3.23) implies the bound

$$(3.24)\quad \int_{Q_R(x_0)} b\,dx \le c\left[1 + R^{-2}\int_{Q_{2R}(x_0)}|\varepsilon(u)|\,dx\right].$$

Clearly the assumptions imposed on h yield ($Q^\pm := Q_{2R}(x_0)\cap\left[|\varepsilon(u)|\gtrless 1\right]$)

$$\int_{Q_{2R}(x_0)}|\varepsilon(u)|\,dx = \int_{Q^-}|\varepsilon(u)|\,dx + \int_{Q^+}|\varepsilon(u)|\,dx$$
$$\le \left(\int_{Q^-}1\,dx\right)^{1/2}\left(\int_{Q^-}|\varepsilon(u)|^2\,dx\right)^{1/2} + c\int_{Q^+}H(\varepsilon(u))\,dx$$
$$\le cR\left(\int_{Q_{2R}(x_0)}H(\varepsilon(u))\,dx\right)^{1/2} + c\int_{Q_{2R}(x_0)}H(\varepsilon(u))\,dx,$$

and since we still assume that $R \ge 1$, we get from (3.1) the bound

$$(3.25)\quad \int_{Q_{2R}(x_0)}|\varepsilon(u)|\,dx \le cR^{3/2}.$$

Now, if we insert (3.25) into (3.24), our claims (3.8) and (3.9) are clearly established.

In a final step we show

$$\int_{\mathbb{R}^2} b\,dx = 0\,. \tag{3.26}$$

Obviously (recall (3.14)) equation (3.26) gives $\nabla\varepsilon(u) = 0$, hence $\nabla^2 u = 0$ so that u must be affine. But since we assume $u \in L^\infty(\mathbb{R}^2, \mathbb{R}^2)$, the constancy of u clearly follows.

It remains to prove (3.26). Let

$$b_\infty := \int_{\mathbb{R}^2} b\,dx\,.$$

Going through the calculations leading to (3.18) with the choice $x_0 = 0$, a closer look at the quantities $S_i, i = 1, \ldots, 4$, implies the inequality

$$\begin{aligned}\int_{Q_R} b\,dx \le \delta \int_{Q_{\frac{3}{2}R}} b\,dx + c(\delta)\Bigg[R^{-2}\int_{T_{\frac{3}{2}R}} |\nabla u|^2\,dx + R^{-1}\int_{T_{\frac{3}{2}R}} |\nabla u|^2\,dx \\ + R^{-1}\left(\int_{Q_{\frac{3}{2}R}} |\nabla u|^2\,dx\right)^{1/2}\left(\int_{T_{\frac{3}{2}R}} |\nabla u|^2\,dx\right)^{1/2}\Bigg],\end{aligned} \tag{3.27}$$

where we have abbreviated $T_{\frac{3}{2}R} := Q_{\frac{3}{2}R} - \overline{Q}_R$ and where on the right-hand side of (3.27) the integration over $T_{\frac{3}{2}R}$ has to be performed in appropriate places due to the support properties of $\nabla\eta$. In the calculations after (3.18) we estimated $\int_{Q_{\frac{3}{2}R}} |\nabla u|^2\,dx$, but of course we can bound $\int_{T_{\frac{3}{2}R}} |\nabla u|^2\,dx$ in the same way by choosing $\varphi \equiv 1$ on $T_{\frac{3}{2}R}$, $0 \le \varphi \le 1$, $|\nabla\varphi| \le c/R$ and $\operatorname{spt}\varphi \subset Q_{2R} - \overline{Q}_{R/2} =: T_{2R}$. This yields

$$\int_{T_{\frac{3}{2}R}} |\nabla u|^2\,dx \le c\left[\int_{T_{2R}} \varphi^2|\varepsilon(u)|^2\,dx + R^{-2}\int_{T_{2R}} |u|^2\,dx\right], \tag{3.28}$$

and from the arguments used after (3.19) we deduce

$$\begin{aligned}\int_{T_{2R}} \varphi^2|\varepsilon(u)|^2\,dx \;\le\; & c\Bigg[\left(\int_{T_{2R}} b\,dx\right)^{1/2}\left(\int_{T_{2R}} (1+|\varepsilon(u)|)\,dx\right)^{1/2} \\ & + R^{-1}\int_{T_{2R}} |\varepsilon(u)|\,dx\Bigg] =: \Phi(R)\,.\end{aligned} \tag{3.29}$$

Putting together (3.28) and (3.29) and going back to (3.27), we obtain choosing $\delta = 1/2$

$$\begin{aligned}\int_{Q_R} b\,dx \;\le\; & \frac{1}{2}b_\infty + c\Bigg\{R^{-4}\int_{Q_{2R}} |u|^2\,dx + R^{-3}\int_{Q_{2R}} |u|^2\,dx \\ & + R^{-2}\Phi(R) + R^{-1}\Phi(R) \\ & + R^{-1}\left(\int_{Q_{\frac{3}{2}R}} |\nabla u|^2\,dx\right)^{1/2}\left(\int_{T_{\frac{3}{2}R}} |\nabla u|^2\,dx\right)^{1/2}\Bigg\}.\end{aligned} \tag{3.30}$$

Clearly $R^{-4}\int_{Q_{2R}}|u|^2\,dx+R^{-3}\int_{Q_{2R}}|u|^2\,dx\longrightarrow 0$ as $R\to\infty$, and from (3.25) we obtain ($R\geq 1$)

$$\Phi(R)\leq c\left[R\left(\int_{T_{2R}}b\,dx\right)^{1/2}+R^{1/2}\right],$$

hence $R^{-2}\Phi(R)+R^{-1}\Phi(R)\longrightarrow 0$ as $R\to\infty$ on account of (3.8). At this stage we like to remark that here it is essential to integrate b just over the set T_{2R}. Let us finally look at the quantity

$$\Psi(R):=R^{-1}\left(\int_{Q_{\frac{3}{2}R}}|\nabla u|^2\,dx\right)^{1/2}\left(\int_{T_{\frac{3}{2}R}}|\nabla u|^2\,dx\right)^{1/2}.$$

By (3.28) and (3.29) we have

$$\int_{T_{\frac{3}{2}R}}|\nabla u|^2\,dx\leq c\left[R^{-2}\int_{T_{2R}}|u|^2\,dx+\Phi(R)\right],$$

thus

$$\Psi(R)\leq c\left(R^{-1}\int_{Q_{\frac{3}{2}R}}|\nabla u|^2\,dx\right)^{1/2}\left(R^{-3}\int_{T_{2R}}|u|^2\,dx+R^{-1}\Phi(R)\right)^{1/2},$$

and the second factor on the right-hand side goes to zero as $R\to\infty$, as observed earlier. Returning to our previous bound

$$\begin{aligned}\int_{Q_{\frac{3}{2}R}}|\nabla u|^2\,dx\ \leq\ & c\left[R^{-2}\int_{Q_{2R}}|u|^2\,dx\right.\\ &+\left(\int_{Q_{2R}}b\,dx\right)^{1/2}\left(\int_{Q_{2R}}(1+|\varepsilon(u)|)\,dx\right)^{1/2}\\ &\left.+R^{-1}\int_{Q_{2R}}|\varepsilon(u)|\,dx\right]\end{aligned}$$

we see in combination with (3.25) and (3.8) that $R^{-1}\int_{Q_{\frac{3}{2}R}}|\nabla u|^2\,dx$ stays bounded, which means that also $\Psi(R)\longrightarrow 0$ as $R\to\infty$. Therefore the passage to the limit in (3.30) finally yields our claim $b_\infty=0$. □

4. Some related problems

Let $\Omega\subset\mathbb{R}^d$, $d=2,3$, denote a bounded domain having a smooth boundary $\partial\Omega$. In 1933 J. Leray [**Le**] investigated the solvability of an exterior problem for NSE which reads as follows (without loss of generality $\mu=1$)

$$\left\{\begin{array}{l}-\Delta u+u^k\partial_k u+\nabla\pi=0\,,\\ \operatorname{div}u=0\quad\text{in }\mathbb{R}^d-\overline{\Omega}\,,\\ u=0\quad\text{on }\partial\Omega\end{array}\right.\tag{4.1}$$

together with the asymptotic condition

$$\lim_{|x|\to\infty}u(x)=u_\infty\,.\tag{4.2}$$

Here u_∞ is a prescribed vector from $\mathbb{R}^d$, and (4.2) has to be interpreted as uniform convergence. The $3D$ case of (4.1) along with (4.2) is well understood and different from the $2D$ situation. A nice explanation together with the historical background is provided in Galdi's paper [**Ga3**]. There it is also outlined how in the $2D$ case the solvability of (4.1) and (4.2) with arbitrary vector $u_\infty \in \mathbb{R}^2$ is related to the until now open question of whether for $u_\infty = 0$ the equations (4.1) and (4.2) admit only the trivial solution $u \equiv 0$. Of course we could not prove or disprove the conjecture that u must be identically zero, however it holds (see [**BF2**]).

THEOREM 4.1. *Let $d = 2$, and consider a weak solution $u \in C^1(\mathbb{R}^2 - \Omega, \mathbb{R}^2)$ (cf. equation (1.15)) of (1.7) and (1.2) on $\mathbb{R}^2 - \overline{\Omega}$ with potential H given by (1.8) and (1.9). Let h satisfy (A1), (A2), (A3$_\mathrm{I}$) or (A1), (A2), (A3$_\mathrm{II}$). If u vanishes on $\partial\Omega$, then u is identically zero in each of the following cases:*

(i) *$u(x) \to 0$ uniformly as $|x| \to \infty$ and $u^k \partial_k u$ is neglected,*

(ii) *$|x|^{1/3}\,|u(x)| \to 0$ uniformly as $|x| \to \infty$.*

These results apply to the non-degenerate p-case, $1 < p < \infty$, as well as to the Prandtl–Eyring fluid. The proof of Theorem 4.1 is based on energy estimates in the spirit of Lemma 3.1, where now one has to work on the exterior domain $\mathbb{R}^2 - \overline{\Omega}$. Note that in (ii) actually a rather strong decay of u at infinity is required.

Next recall that in Theorem 1.4 the Prandtl–Eyring fluid model is included with potential H given by $H(\varepsilon) = |\varepsilon| \ln(1 + |\varepsilon|)$. However, the applicability of Theorem 1.4 is limited to solutions u of class C^1, and in order to motivate Theorem 1.4 for this particular fluid model, one should firstly discuss the (weak) solvability of the, e.g., homogeneous boundary value problem on some bounded smooth domain $\Omega \subset \mathbb{R}^2$ with given volume forces $f : \Omega \to \mathbb{R}^2$, and secondly investigate the differentiability properties of such weak solutions. The first question was recently answered in the paper [**BDF**]. Letting $h(t) = t\ln(1 + t)$, we introduce the space

$$V^{1,h}_{0,\mathrm{div}} := \left\{ v \in L^1(\Omega, \mathbb{R}^2) : \int_\Omega h\left(|\varepsilon(v)|\right) dx < \infty,\ \operatorname{div} v = 0, v|_{\partial\Omega} = 0 \right\}$$

and consider the problem (cf. (1.15)) to find $u \in V^{1,h}_{0,\mathrm{div}}$ such that

$$\int_\Omega DH\left(\varepsilon(u)\right) : \varepsilon(\varphi)\, dx = \int_\Omega f \cdot \varphi\, dx + \int_\Omega u \otimes u : \varepsilon(\varphi)\, dx \tag{4.3}$$

holds for all $\varphi \in C_0^\infty(\Omega, \mathbb{R}^2)$, $\operatorname{div} \varphi = 0$. Note that $V^{1,h}_{0,\mathrm{div}}$ is a subspace of the class $BD(\Omega)$ (= fields of bounded deformation) so that

$$V^{1,h}_{0,\mathrm{div}} \hookrightarrow L^2(\Omega, \mathbb{R}^2) \text{ continuously},$$

which means that the second integral on the right-hand side is well defined. In their paper [**FMS**] Frehse, Málek, and Steinhauer investigated the existence problem for the model case $H_p(\varepsilon) = |\varepsilon|^p$ and proved existence in $\overset{\circ}{W}{}^1_p(\Omega, \mathbb{R}^2)$ of weak solutions for any $p > 1$. Since they work in the Sobolev space $\overset{\circ}{W}{}^1_p(\Omega, \mathbb{R}^2)$, it is possible to adjust the Lipschitz truncation technique, which roughly speaking replaces a function u through a Lipschitz function $\widetilde{u}$ different from u only on a small set. In [**BDF**] we provide a similar procedure where now a field w from $V^{1,h}_{0,\mathrm{div}}$ is replaced through a solenoidal one having bounded symmetric gradient and being different from w again on a set with small measure. This yields

THEOREM 4.2. *Let $\Omega \subset \mathbb{R}^2$ denote a bounded smooth domain, and let $f \in L^{p_o}(\Omega, \mathbb{R}^2)$ for some $p_o > 1$. Then (4.3) has at least one solution $u \in V^{1,h}_{0,\mathrm{div}}$.*

However, we know nothing about the regularity properties of this solution u, and up to now there are no existence results in the $3D$ case.

References

[BDF] Breit, D., Diening, L., Fuchs, M., Solenoidal Lipschitz truncation and applications in fluid mechanics. J. Diff. Equ. 253, 1910–1942 (2012). MR2943947

[BF1] Bildhauer, M., Fuchs, M., Variational integrals of splitting type: higher integrability under general growth conditions. Annali Mat. Pura Appl. 188, 467–496 (2009). MR2512159 (2010j:49057)

[BF2] Bildhauer, M., Fuchs, M., On the exterior problem in $2D$ for stationary flows of fluids with shear dependent viscosity. Comm. M. U. C. 53,2, 221–236 (2012) MR3017256

[BFZ] Bildhauer, M., Fuchs, M., Zhang, G., Liouville type theorems for steady flows of degenerate power law fluids in the plane. J. Math. Fluid Mech. 15, 583–616 (2013). MR3084326

[FMS] Frehse, J., Málek, J., Steinhauer, M., On analysis of steady flows of fluids with shear dependent viscosity based on the Lipschitz truncation method. SIAM J. Math. Anal. 34(5), 1064–1083 (2003). MR2001659 (2005c:76007)

[Fu] Fuchs, M., Liouville theorems for stationary flows of shear thickening fluids in the plane. J. Math. Fluid Mech. 14, 421–444 (2012). MR2964742

[FS] Fuchs, M., Seregin, G.A., Variational methods for problems from plasticity theory and for generalized Newtonian fluids. Lect. Notes Math. 1749, Springer-Verlag, Berlin-Heidelberg-New York 2000. MR1810507 (2001k:74043)

[FZ] Fuchs, M., Zhang, G., Liouville theorems for entire local minimizers of energies defined on the class $L\log L$ and for entire solutions of the stationary Prandtl-Eyring fluid model. Calc. Var. 44(1-2), 271–295 (2012). MR2898779

[Ga1] Galdi, G., An introduction to the mathematical theory of the Navier-Stokes equations Vol.I, Springer Tracts in Natural Philosophy Vol.38, Springer, Berlin-Heidelberg-New York 1994. MR1284205 (95i:35216a)

[Ga2] Galdi, G., An introduction to the mathematical theory of the Navier-Stokes equations Vol.II, Springer Tracts in Natural Philosophy Vol.39, Springer, Berlin-Heidelberg-New York 1994. MR1284206 (95i:35216b)

[Ga3] Galdi, G., On the existence of symmetric steady-state solutions to the plane exterior Navier–Stokes problem for arbitrary large Reynolds number. Advances in fluid dynamics, Quad. Mat.,4, Aracne, Rome, 1–25 (1999). MR1770187 (2001h:35143)

[GM] Giaquinta, M., Modica, G., Nonlinear systems of the type of stationary Navier-Stokes system. J. Reine Angew. Math. 330, 173–214 (1982). MR641818 (83h:35041)

[GW] Gilbarg, D., Weinberger, H.F., Asymptotic properties of steady plane solutions of the Navier-Stokes equations with bounded Dirichlet integral. Ann. S.N.S. Pisa (4), 5, 381–404 (1978). MR501907 (80c:35079)

[KNSS] Koch, G., Nadirashvili, N., Seregin, G., Sverăk, V., Liouville theorems for the Navier-Stokes equations and applications. Acta Math.203, 83–105 (2009). MR2545826 (2010i:35281)

[La1] Ladyzhenskaya, O. A., The mathematical theory of viscous incompressible flow. Gordon and Breach, 1969. MR0254401 (40:7610)

[La2] Ladyzhenskaya, O. A., Solution "in the large" of boundary value problems for the Navier-Stokes equations in two space variables. Dokl. Akad. Nauk SSSR 123 (1958), 427–429; English transl., Soviet Phys. Dokl. 3 (1959), 1128–1131 and Comm. Pure App. Math. 12 (1959), 427–433. MR0108963 (21:7675)

[La3] Ladyzhenskaya, O. A., Uniqueness and smoothness of generalized solutions of the Navier-Stokes equations. Zap. Nauchn. Sem. Leningrad. Otdel. Mat. Inst. Steklov (LOMI) 5 (1967), 169-185; English transl., Sem. Math. V. A. Steklov Math. Inst. Leningrad 5 (1969), 60-66. MR0236541 (38:4836)

[La4] Ladyzhenskaya, O. A., Unique global solvability of the three-dimensional Cauchy problem for the Navier-Stokes equations in the presence of axial symmetry. Zap. Nauchn. Sem. Leningrad. Otdel. Mat. Inst. Steklov (LOMI) 7 (1968), 155-177; English transl., Sem. Math. V. A. Steklov Math. Inst. Leningrad 7 (1970), 70-79. MR0241833 (39:3170)

[La5] Ladyzhenskaya, O. A., A dynamical system generated by the Navier- Stokes equations. Zap. Nauchn. Sem. Leningrad. Otdel. Mat. Inst. Steklov (LOMI) 27 (1972), 91–115; English transl., J. Soviet. Math. 3 (1975), 458–479. MR0328378 (48:6720)

[LK] Ladyzhenskaya, O. A., Kiselev, A. A., On the existence and uniqueness of the solution of the non-stationary problem for a viscous incompressible fluid. Izv. Akad. Nauk SSSR Ser. Mat. 21 (1957), 665-680; English transl., Amer. Math. Soc. Transl. (2) 24 (1963), 79–106. MR0100448 (20:6881)

[Le] Leray, J., Etude de diverses équations intégrals non linéaires et de quelques problémes que pose l'hydrodynamique. J. Math. Purcs Appl. 12, 1–82 (1933).

[MNRR] Málek, J., Necǎs, J., Rokyta, M., Růžička, M., Weak and measure valued solutions to evolutionary PDEs. Chapman & Hall, London–Weinheim–New York, 1996. MR1409366 (97g:35002)

[Na] Naumann, J., On the differentiability of weak solutions of a degenerate system of PDE's in fluid mechanics. Annali Mat. Pura Appl. (IV)151, 225–238 (1988). MR964511 (90g:35027)

[Wo] Wolf, J., Interior $C^{1,\alpha}$-regularity of weak solutions to the equations of stationary motions of certain non-Newtonian fluids in two dimensions. Boll. U.M.I. (8)10-B, 317–340 (2007). MR2339444 (2008f:35320)

[Zh] Zhang, G., Liouville theorems for stationary flows of generalized Newtonian fluids. Ph.D. thesis. University of Jyväskylä 2012.

Fachbereich 6.1 Mathematik, Universität des Saarlandes, Postfach 15 11 50, D-66041 Saarbrücken, Germany

E-mail address: `fuchs@math.uni-sb.de`

Amer. Math. Soc. Transl.
Volume **232**, 2014

On the normal-type parabolic system corresponding to the three-dimensional Helmholtz system

A. V. Fursikov

Dedicated to the memory of Olga Alexandrovna Ladyzhenskaya

ABSTRACT. A semilinear normal parabolic system is derived corresponding to the three-dimensional Helmholtz equations for the curl of the velocity vector field of viscous incompressible fluid. An explicit formula for the solution to the normal parabolic system is applied to examine the structure of the corresponding dynamic flow. The phase space of this dynamical system is shown to consist of the set of stability (on which the solution tends to zero with unboundedly increasing time $t \to \infty$), the set of explosions (on which the solutions explode at a finite time), and the set of increase (on which the solutions unboundedly increase as $t \to \infty$). A geometrical description of these sets is given.

1. Introduction

It is well known (see, for example, Hopf [**1**], Ladyzhenskaya [**2**], Temam [**3**]) that the existence of a weak solution of a three-dimensional Navier–Stokes system is based on the energy estimate, which, in turn, holds because the image $B(v)$ under the nonlinear operator generated by the nonlinear terms of the Navier–Stokes equations is L_2-orthogonal to the velocity vector v; that is, the vector $B(v)$ is tangent to the L_2-sphere of radius $\|v\|_{L_2}$ and center at the origin. However, insufficient smoothness does not allow one to prove the uniqueness of a weak solution. If the vectors $B(v)$ were tangent to the analogues sphere in the Sobolev space H^1, then similar methods would be capable of proving the existence of the strong solution of the Navier–Stokes system—in this case the solution is sufficiently smooth to guarantee its uniqueness. It appears that $B(v)$ fails to satisfy this property in the space H^1: in this case $B(v) = B_\tau(v) + B_n(v)$, where $B_\tau(v)$ is the component of

2010 *Mathematics Subject Classification.* Primary 76D03, 35K58, 35A20.
Key words and phrases. Normal parabolic system, Helmholtz system, structure of the dynamic flow.
This research was carried out with the financial support of the Russian Foundation for Basic Research (grant no. 12-01-00262a) and the programme "Theoretical problems of contemporary mathematics" of RAS (the project "Optimization of numerical algorithm of solution of problems of mathematical physics").

the nonlinear operator which is tangent to the H^1-sphere, and $B_n(v)$ is the component normal to the sphere (collinear to the velocity vector v). The presence of the component $B_n(v)$ prevents the application of the methods based on the energy inequality to prove the existence of a strong solution.

Our aim in this paper is to implement the following plan. First, replacing the nonlinear operator $B(v)$ in the system under study by the normal component $B_n(v)$, we shall study the resulting system, to be referred to as the Normal Parabolic System (NPS). Then, using the information obtained, we shall try to better understand the problem of existence of a strong solution to the original system with the nonlinear operator $B(v) = B_\tau(v) + B_n(v)$.

This paper is devoted to realizing the first part of this plan: we examine the NPS constructed from the three-dimensional Helmholtz system describing the curl of the vector field of viscous incompressible fluid with periodic boundary conditions. Note that the NPS constructed from the Burgers equation was studied earlier in **[4]**, **[5]**. The study of the NPS corresponding to a three-dimensional Helmholtz system was initiated in **[6]**, in which the main results were only stated.

In the present paper, we give complete proofs of all these results. Moreover, we continue the study of the structure of the dynamic flow that was initiated in **[4]**, **[5]**, **[6]**. In particular, we introduce the parameter-independent set of stability, emphasizing that the parameter-dependent set of stability of **[4]**, **[5]**, **[6]** is employed only for technical purposes. Thanks to this, the use of the intermediate set from previous papers has proved unnecessary—instead there appears the set of increase.

We show that the phase space of the dynamical system corresponding to the NPS is the union of the set of stability, the set of increase, and the set of explosions—a fairly detailed geometrical description of these sets is given at the end of the paper. In Section 2, we derive the NPS corresponding to the Helmholtz system. The explicit formula for the solution of the NPS is obtained in Section 3. This is a key result providing quite sufficient insight into the NPS. The proofs of the unique solvability of the boundary-value problem for the NPS in corresponding function spaces and of the continuous dependence of the solution on the initial data are given in Section 4. Section 5 is concerned with the study of the functional entering the explicit formula for the solution of the NPS and governing many properties of the solutions. Finally, in Section 6 we examine the structure of the dynamic flow generated by the NPS.

2. Semilinear normal parabolic system

2.1. The Navier–Stokes system. Consider the three-dimensional system of Navier–Stokes equations with periodic boundary conditions:

$$\partial_t v(t,x) - \Delta v + (v,\nabla)v + \nabla p(t,x) = 0, \quad \operatorname{div} v = 0, \tag{2.1}$$

$$v(t,\dots,x_i,\dots) = v(t,\dots,x_i+2\pi,\dots), \qquad i = 1,2,3, \tag{2.2}$$

$$v(t,x)|_{t=0} = v_0(x); \tag{2.3}$$

here $t \in \mathbb{R}_+, x = (x_1,x_2,x_3) \in \mathbb{R}^3$, $v(t,x) = (v_1,v_2,v_3)$ is the fluid velocity vector field, ∇p is the pressure gradient, Δ is the Laplacian, and $(v,\nabla)v = \sum_{j=1}^3 v_j \partial_{x_j} v$. The periodic boundary conditions (2.2) mean in particular that system (2.1) and the initial conditions (2.3) are defined on the torus $\mathbb{T}^3 = (\mathbb{R}/2\pi\mathbb{Z})^3$.

For each $m \in \mathbb{Z}_+ = \{j \in \mathbb{Z} : j \geqslant 0\}$, we define the space

$$V^m = V^m(\mathbb{T}^3) = \Big\{v(x) \in (H^m(\mathbb{T}^3))^3 : \operatorname{div} v = 0,\ \int_{\mathbb{T}^3} v(x)\,dx = 0\Big\}, \tag{2.4}$$

where $H^m(\mathbb{T}^3)$ is the Sobolev space.

It is well known that the nonlinear term $(v, \nabla)v$ of problem (2.1)–(2.3) satisfies the relation

$$\int_{\mathbb{T}^3} (v(t,x), \nabla)v(t,x) \cdot v(t,x)\,dx = 0.$$

Hence, multiplying scalarly the Navier–Stokes system (2.1) by v in $L_2(\mathbb{T}^3)$, integrating by parts (in x), and then integrating in t, we arrive at the well-known energy inequality

$$\int_{\mathbb{T}^3} |v(t,x)|^2 dx + 2\int_0^t \int_{\mathbb{T}^3} |\nabla_x v(\tau,x)|^2 dx\,d\tau \leqslant \int_{\mathbb{T}^3} |v_0(x)|^2 dx, \tag{2.5}$$

which enables one to prove the existence of a weak solution of problem (2.1)–(2.3). Note that in the case of a two-dimensional Navier–Stokes system (that is, for $x = (x_1, x_2) \in \mathbb{R}^2$, $v(t,x) = (v_1, v_2)$), Ladyzhenskaya **[7]**, **[2]** employed the energy inequality (2.5) to prove the uniqueness and smoothness of a weak solution. However, in the three-dimensional setting, the smoothness of a weak solution is not enough to guarantee its uniqueness.

Unfortunately, the method of energy estimates is incapable of proving the existence of smooth solutions of system (2.1). Indeed, multiplying (2.1) scalarly by v in $V^1(\mathbb{T}^3)$, we will not obtain an analogue of the energy inequality in the phase space V^1 because the image of the quadratic operator $(v, \nabla)v$ from (2.1) is not V^1-orthogonal to v. Nevertheless, we shall try to better understand this situation by passing from the Navier–Stokes system to the Helmholtz system.

2.2. The Helmholtz system. We transform the boundary-value problem (2.1)–(2.3) with respect to the fluid velocity to the one with unknown curl of the velocity field:

$$y(t,x) = \operatorname{curl} v(t,x) = (\partial_{x_2} v_3 - \partial_{x_3} v_2,\ \partial_{x_3} v_1 - \partial_{x_1} v_3,\ \partial_{x_1} v_2 - \partial_{x_2} v_1). \tag{2.6}$$

Recall that

$$(v, \nabla)v = y \times v + \nabla \frac{|v|^2}{2}, \tag{2.7}$$

$$\operatorname{curl}(y \times v) = (v, \nabla)y - (y, \nabla)v, \quad \text{provided that} \quad \operatorname{div} v = 0, \quad \operatorname{div} y = 0, \tag{2.8}$$

where $y \times v = (y_2 v_3 - y_3 v_2,\ y_3 v_1 - y_1 v_3,\ y_1 v_2 - y_2 v_3)$ is the vector product of vector y, v, and $|v|^2 = v_1^2 + v_2^2 + v_3^2$. We substitute (2.7) into (2.1) and apply the curl operator to both sides of the resulting equality. Then, using (2.6), (2.8), and the formula $\operatorname{curl} \nabla F = 0$, we obtain the system of Helmholtz equations:

$$\partial_t y(t,x) - \Delta y + (v, \nabla)y - (y, \nabla)v = 0, \quad \operatorname{div} y = 0. \tag{2.9}$$

We augment these equations with the initial conditions

$$y(t,x)|_{t=0} = y_0(x), \tag{2.10}$$

where $y_0 = \operatorname{curl} v_0$.

REMARK 2.1. Note that in the two-dimensional case $y(t,x) = \operatorname{curl} v(t,x) = \partial_{x_1} v_2 - \partial_{x_2} v_1$ is a scalar function, and hence the first of equations (2.9) is a scalar equation without the term $(y,\nabla)v$. Hence, the solution of this equation satisfies the analogue of the energy inequality. Starting out from this, Yudovich [**8**] constructed the nonlocal existence and uniqueness theory for smooth solutions of the Euler equations governing the flow of an ideal incompressible fluid.

Taking into account that there is no analogue of the energy inequality in the three-dimensional setting for solving problem (2.9), (2.10), we shall proceed as follows.

2.3. The normal parabolic system and its derivation. Using the Fourier series expansion

$$v(x) = \sum_{k\in\mathbb{Z}^3} \widehat{v}(k)e^{ix\cdot k}, \quad \text{where} \quad \widehat{v}(k) = (2\pi)^{-3}\int_{\mathbb{T}^3} v(x)e^{-ix\cdot k}\,dx,$$

$x\cdot k = \sum_{j=1}^3 x_j k_j$, $k = (k_1,k_2,k_3)$, and applying the well-known formula

$$\operatorname{curl}\operatorname{curl} v = -\Delta v \quad \text{for} \quad \operatorname{div} v = 0,$$

one easily verifies that the inverse to the curl operator is well defined on V^m and is given by

$$\operatorname{curl}^{-1} y(x) = i \sum_{k\in\mathbb{Z}^3\setminus\{0\}} \frac{k\times\widehat{y}(k)}{|k|^2} e^{ix\cdot k}. \tag{2.11}$$

Hence, the operator

$$\operatorname{curl}:\ V^1 \longrightarrow V^0$$

is an isomorphism of the spaces, and so the V^1-sphere for problems (2.1), (2.3) is equivalent to the V^0-sphere for problems (2.9), (2.10).

We let B denote the nonlinear term in the Helmholtz system,

$$B(y) = (v,\nabla)y - (y,\nabla)v \tag{2.12}$$

(the dependence of B on v is not indicated, since v is given in terms of y by (2.11)).

Multiplying scalarly equality (2.12) by $y = (y_1,y_2,y_3)$ in V^0 and integrating by parts, this gives

$$(B(y),y)_{V^0} = -\int_{\mathbb{T}^3} \sum_{j,k=1}^3 y_j\partial_j v_k y_k\,dx, \tag{2.13}$$

which, in general, is not zero. By this reason, the three-dimensional Helmholtz system does not satisfy the energy inequality. In other words, the operator B can be expanded as

$$B(y) = B_n(y) + B_\tau(y), \tag{2.14}$$

where the vector $B_n(y)$ is orthogonal to the sphere

$$\Sigma(\|y\|_{V^0}) = \{u\in V^0 : \|u\|_{V^0} = \|y\|_{V^0}\}$$

at the point y (that is, the vectors $B_n(y)$ and y are collinear), and the vector $B_\tau(y)$ touches the sphere $\Sigma(\|y\|_{V^0})$ at y. Note that, in general, both the operators B_n, B_τ in (2.14) are nonzero. Given the derivation of the energy inequality being hindered by the presence of the operator B_n, it is very likely that it is B_n that is responsible for possible singularities of the solution. It would therefore seem advisable to drop the component B_τ in the Helmholtz system and be concerned, in the early stage, with the analogue of system (2.9) in which the nonlinear operator $B(y)$ is replaced by its normal component $B_n(y)$. The so-obtained system will be referred to as a *normal-type system* or a *normal parabolic system* (*NPS*).[1]

Let us build the NPS for problems (2.9), (2.10) with respect to the V^0-sphere.

The term $(v, \nabla)y$ in (2.9) is a tangential operator; that is,

$$\int_{\mathbb{T}^3} (v, \nabla)y \cdot y\, dx = 0.$$

Consequently, the normal part of the nonlinear operator in (2.9) is governed by the nonlinear term $(y, \nabla)v$. We shall seek it in the form $\Phi(y)y$, where Φ is an unknown functional determined from the equation

$$\int_{\mathbb{T}^3} \Phi(y)y(x) \cdot y(x)\, dx = \int_{\mathbb{T}^3} (y(x), \nabla)v(x) \cdot y(x)\, dx. \tag{2.15}$$

In view of (2.15),

$$\Phi(y) = \begin{cases} \int_{\mathbb{T}^3} (y(x), \nabla) \operatorname{curl}^{-1} y(x) \cdot y(x)\, dx \,/ \int_{\mathbb{T}^3} |y(x)|^2 dx, & y \neq 0, \\ 0, & y \equiv 0, \end{cases} \tag{2.16}$$

where $\operatorname{curl}^{-1} y(x)$ is defined in (2.11).

Thus, the normal parabolic system corresponding to the Helmholtz equations (2.9) is defined as follows:

$$\partial_t y(t,x) - \Delta y - \Phi(y)y = 0, \qquad \operatorname{div} y = 0; \tag{2.17}$$

here Φ is the functional from (2.16). This system, as equipped with the initial condition (2.10) and periodic boundary data, is the main object of study of the present paper.

3. Explicit formula for the solutions of NPS

Our aim in this section is to derive an explicit formula for the solutions of the NPS. This is a key result underlying many important properties thereof, some of which will be put forward in subsequent sections.

Let $S(t, x; z_0)$ be the solution operator of the following Stokes system with periodic boundary conditions:

$$\partial_t z(t,x) - \Delta z(t,x) = 0, \quad \operatorname{div} z = 0, \quad z(t,x)|_{t=0} = z_0(x); \tag{3.1}$$

that is, $S(t, x; z_0) = z(t,x)$ (of course, it is assumed that $\operatorname{div} z_0 = 0$).

---

[1]It has already been observed in the Introduction that one needs to return in the sequel to the original Helmholtz system and then, using the available information about the NPS, examine the interaction of the operators $B_n(y)$ and $B_\tau(y)$. This question will be addressed in papers to follow.

The following result holds.

THEOREM 3.1. *The solution $y(t,x) := y(t,x;y_0)$ of problems* (2.17), (2.10) *is given by*

$$y(t,x;y_0) = \frac{S(t,x;y_0)}{1-\int_0^t \Phi(S(\tau,\,\cdot\,;y_0))\,d\tau}. \tag{3.2}$$

PROOF. The solution of problems (2.17), (2.10) will be sought in the form $y(t,x) = c(t)S(t,x;y_0)$, where $S(t,x;y_0)$ is the solution of problem (3.1) with $z_0 = y_0$, and $c(t)$ is an unknown scalar function. Substituting this function $y(t,x)$ into (2.17), (2.10) and taking into account that the functional $\Phi(v)$ is homogeneous in v, we see that

$$\begin{aligned}\partial_t y(t,x) &- \Delta y - \Phi(y)y \\ &= c(t)(\partial_t S(t,x;y_0) - \Delta S(t,x;y_0)) \\ &\quad + (S(t,x;y_0)(dc(t)/dt - c^2(t)\Phi(S(t,\,\cdot\,;y_0)) = 0,\end{aligned}$$

and now, in view of (3.1) and the straightforward relation $c(0) = 1$,

$$1/c(t) = 1 - \int_0^t \Phi(S(\tau,\,\cdot\,;y_0))\,d\tau,$$

which proves (3.2). □

For brevity, we shall often conventionally write:

$$y(t;y_0) = y(t,\,\cdot\,;y_0), \qquad S(t;y_0) = S(t,\,\cdot\,;y_0). \tag{3.3}$$

4. Fundamental properties of the boundary-value problem for NPS

We start with examining the most elementary properties of the solution of problems (2.17), (2.10). We shall first investigate for which classes of initial data the solution is well defined by formula (3.2). Further, we prove the unique solvability of the problem in corresponding function spaces and the continuous dependence of the solution on the initial data.

4.1. Properties of the functional $\Phi(u)$. Let $s \in \mathbb{R}$. By definition, the Sobolev space $H^s(\mathbb{T}^3)$ is the space of periodic real distributions with finite norm

$$\|z\|^2_{H^s(\mathbb{T}^3)} \equiv \|z\|_s^2 = \sum_{k\in\mathbb{Z}^3\setminus\{0\}} |k|^{2s}|\widehat{z}(k)|^2 < \infty, \tag{4.1}$$

where $\widehat{z}(k)$ are the Fourier coefficients[2] of the function z.

We shall use the following generalization of the spaces (2.4) of solenoidal vector fields:

$$\begin{aligned}V^s &\equiv V^s(\mathbb{T}^3) \\ &= \Big\{v(x) \in (H^s(\mathbb{T}^3))^3 :\ \operatorname{div} v(x) = 0,\ \int_{\mathbb{T}^3} v(x)\,dx = 0\Big\}, \quad s \in \mathbb{R}.\end{aligned} \tag{4.2}$$

LEMMA 4.1. *Let $\Phi(u)$ be the functional* (2.16). *Then there exists a constant $c > 0$ such that, for every $u \in V^{3/2}$,*

$$|\Phi(u)| \leqslant c\|u\|_{3/2}. \tag{4.3}$$

[2]Strictly speaking, one needs to add $|\widehat{z}(0)|^2$ to the right-hand side of (4.1). However, we did not do this, since starting from (4.2) it will be assumed that this Fourier coefficient is zero.

PROOF. The estimate

$$
\begin{aligned}
(4.4)\qquad |\Phi(u)| &\leqslant \frac{\|u\|^2_{L_3(\mathbb{T}^3)}\|\nabla \operatorname{curl}^{-1} u\|_{L_3(\mathbb{T}^3)}}{\|u\|_0^2} \leqslant c\frac{\|u\|^2_{1/2}\|\nabla \operatorname{curl}^{-1} u\|_{1/2}}{\|u\|_0^2} \\
&\leqslant c\frac{\|u\|^3_{1/2}}{\|u\|_0^2} \leqslant c\frac{\|u\|_0^2\|u\|_{3/2}}{\|u\|_0^2} = c\|u\|_{3/2}
\end{aligned}
$$

follows from definitions (2.16), (2.11), Sobolev's embedding theorem (according to which $H^{1/2}(\mathbb{T}^3) \subset L_3(\mathbb{T}^3)$), and the interpolation inequality $\|v\|^3_{1/2} \leqslant c\|v\|_0^2\|v\|_{3/2}$. □

LEMMA 4.2. *Let Φ be the functional* (2.16). *For any $\beta < 1/2$, there exists a constant $c_1 > 0$ such that*

$$
(4.5)\qquad \int_0^t |\Phi(S(\tau; y_0))|\, d\tau \leqslant c_1\|y_0\|_{-\beta}
$$

for any $y_0 \in V^{-\beta}(\mathbb{T}^3)$ and $t > 0$. Here, $S(t; y_0)$ is the solution operator of problem (3.1).

PROOF. Using (4.3) and the representation of the solution of problem (3.1) in terms of the Fourier series, we see that

$$
\begin{aligned}
(4.6)\qquad &\int_0^t \big|\Phi(S(\tau; y_0))\big|\, d\tau \\
&\leqslant c\int_0^t e^{-\tau/2}\left(\sum_{k\neq 0}(|\widehat{y}_0(k)|^2|k|^{-2\beta})|k|^{3+2\beta}e^{-(k^2-1)\tau}\right)^{1/2} d\tau,
\end{aligned}
$$

where $\widehat{y}_0(k)$ are the Fourier coefficients of the function y_0. The solution $\widehat{\rho} = \widehat{\rho}(t)$ of the extremal problem

$$
f(t,\rho) = \rho^{3+2\beta}e^{-(\rho^2-1)t} \to \max, \quad \rho \geqslant 1,
$$

is given by $\widehat{\rho}(t) = \sqrt{\frac{3+2\beta}{2t}}$. We also have

$$
(4.7)\qquad f(t,\widehat{\rho}(t)) = \begin{cases} \left(\frac{3+2\beta}{2t}\right)^{\frac{3+2\beta}{2}} e^{-(3+2\beta-2t)/2}, & t \leqslant \frac{3+2\beta}{2}, \\ 1, & t \geqslant \frac{3+2\beta}{2}. \end{cases}
$$

Substituting (4.7) into (4.6), we arrive at (4.5), the result required. □

REMARK 4.1. In view of Lemma 4.2, the functional on the left-hand side of (4.5) is well defined for $y_0 \in V^{-\beta}(\mathbb{T}^3)$ with $\beta < 1/2$. In particular, Lemma 4.2 and (3.2) show that the solution of problems (2.17), (2.10) is well defined for any initial data $y_0 \in V^0$ and is infinitely differentiable in these variables for each $t > 0$ and $x \in \mathbb{T}^3$.

In the following two subsections we will justify the choice of the space V^0 as the phase space of the corresponding dynamical system.

4.2. Unique solvability of problems (2.17), (2.10). We claim that problems (2.17), (2.10) have no other solutions than (3.2). Consider the space of solutions to this problem:

$$V^{1,2(-1)}(Q_T) = \Big\{ y(t,x) \in L_2(0,T;V^1(\mathbb{T}^3)) : \\ \frac{\partial y(t,x)}{\partial t} \in L_2(0,T;V^{-1}(\mathbb{T}^3)) \Big\}; \tag{4.8}$$

here $Q_T = (0,T)\times\mathbb{T}^3$, $T > 0$ is some number or $T = \infty$. Assume that $y_0 \in V^0(\mathbb{T}^3)$, and impose on solution the following

CONDITION 4.1. *Solution $y(t,x;y_0) \in V^{1,2(-1)}(Q_T)$ of problems* (2.17), (2.10) *with initial condition $y_0 \in V^0 \setminus \{0\}$ satisfies relation $\|y(t,\cdot,y_0)\|_0 \neq 0$ for any $t \in [0,T]$.*

THEOREM 4.1. *For any $y_0 \in V^0$ there exist $T > 0$ such that problems* (2.17), (2.10), *has a unique solution $y(t,x;y_0)$ in the space $V^{1,2(-1)}(Q_T)$ satisfying Condition* 4.1.

PROOF. By Lemma 4.2 the functional $\int_0^t \Phi(S(\tau;y_0))\,d\tau$ is well defined for any $t > 0$ and is continuous in t. Hence, the inequality

$$1 - \int_0^t \Phi(S(\tau;y_0))\,d\tau \geqslant \varepsilon \quad \forall t \in [0,T] \tag{4.9}$$

holds in view of (4.5) for some $T > 0$ and sufficiently small $\varepsilon > 0$.

By Theorem 3.1, (4.9), and the well-known properties of the solution operator $S(t,x;y_0)$ (which in turn follow by expanding S into the Fourier series), the function $y(t,x;y_0)$, as defined by (3.2), lies in the space $V^{1,2(-1)}(Q_T)$. Moreover,

$$\|y(t,\cdot;y_0)\|_0 \geqslant \frac{1}{\varepsilon}\|S(t;y_0)\|_0 \geqslant \frac{1}{\varepsilon}\|S(T;y_0)\|_0 \quad \forall t \in [0,T]; \tag{4.10}$$

that is, Condition 4.1 is satisfied. This proves the solvability of problems (2.17), (2.10).

We next claim that this problem has a unique solution. Assume that $y(t,x) \in V^{1,2(-1)}(Q_T)$ is a solution of this problem satisfying Condition 4.1. It will be shown that there exists a function $\gamma(t) \in C[0,T]$ such that $\gamma(t)y(t,x)$ is the solution of the problem

$$\partial_t z(t,x) - \Delta z(t,x) = 0, \quad z|_{t=0} = y_0(x). \tag{4.11}$$

Assuming that there exists a function $\gamma(t)$ and substituting $\gamma(t)y(t,x)$ into equation (4.11), we obtain

$$y\partial_t\gamma + \gamma(\partial_t y - \Delta y) = 0.$$

Since $y(t,x)$ is the solution of equation (2.17), it follows from the previous equality that

$$y[\partial_t\gamma + \gamma\Phi(y)] = 0,$$

and hence, since y satisfies Condition 4.1 and since $y(t,x)$ and $\gamma(t)y(t,x)$ agree for $t = 0$,

$$\gamma(t) = e^{-\int_0^t \Phi(y(\tau,\cdot))\,d\tau}.$$

Thus, the function $y(t,x)e^{-\int_0^t \Phi(y(\tau,\,\cdot\,))\,d\tau}$ is the solution of problem (4.11), and hence, by the definition of the function $S(t,x;y_0)$ (see (3.1) and below),

$$S(t,x;y_0) = y(t,x)e^{-\int_0^t \Phi(y(\tau,\,\cdot\,))\,d\tau}. \tag{4.12}$$

Solving (4.12) for $y(t,x)$ and substituting the result into (2.17) gives

$$e^{\int_0^t \Phi(y(\tau,\,\cdot\,))\,d\tau}\Big(\Phi(y(t,\,\cdot\,))S(t;y_0) + \partial_t S - \Delta S - e^{\int_0^t \Phi(y)\,d\tau}\Phi(S)S\Big) = 0,$$

whence it follows that

$$\Phi(S(t;y_0)) = \Phi(y(t,\,\cdot\,))e^{-\int_0^t \Phi(y)\,d\tau} = -\partial_t e^{-\int_0^t \Phi(y)\,d\tau}.$$

Consequently,

$$e^{-\int_0^t \Phi(y)\,d\tau} = 1 - \int_0^t \Phi(S(\tau;y_0))\,d\tau.$$

Substituting this into (4.12), we see that $y(t,x)$ is given by (3.2). This proves the uniqueness of the solution of problems (2.17), (2.10). □

4.3. Continuous dependence of the solution on the initial data.

LEMMA 4.3. *Assume that $y_m \to y_0$ as $m \to \infty$ in the space $V^0(\mathbb{T}^3)$. Then*

$$\int_0^t \Phi(S(\tau;y_m))\,d\tau \longrightarrow \int_0^t \Phi(S(\tau;y_0))\,d\tau \quad \text{as} \quad m\to\infty \quad \forall t \in (0,T]. \tag{4.13}$$

PROOF. Since for $y_0 = 0$ the proof of this lemma is evident, we assume that $y_0 \neq 0$. We let

$$A(y(x)) = (y(x),\nabla)\operatorname{curl}^{-1} y(x)\cdot y(x)$$

denote the integrand in the numerator of the fraction in (2.16). Formula (2.16) and the following well-known estimates for $S(t,x;y_0)$,

$$\gamma(T)\|y\|_0 \leqslant \|S(t;y)\|_0 \leqslant C\|y\|_0 \quad \forall t\in(0,T], \quad \|S(\,\cdot\,;y)\|_{L_2(0,T;V^1(\mathbb{T}^3))} \leqslant C\|y\|_0,$$

with some constants $\gamma(T) > 0, C > 0$ imply that

$$\begin{aligned} &\left|\int_0^t \Phi(S(\tau;y_0)) - \Phi(S(\tau;y_m))\,d\tau\right| \\ &\quad = \left|\int_0^t \frac{(A(S(\tau;y_0)) - A(S(\tau;y_m)))}{\|S(\tau;y_0)\|_0^2}\,d\tau\right. \\ &\qquad \left. + \int_0^t \frac{(\|S(\tau;y_m)\|_0^2 - \|S(\tau;y_0)\|_0^2)A(S(\tau;y_m))}{\|S(\tau;y_0)\|_0^2\|S(\tau;y_m)\|_0^2}\,d\tau\right| \\ &\quad \leqslant c\int_0^t |A(S(\tau;y_0)) - A(S(\tau;y_m))| \\ &\qquad + \|S(\tau;y_0-y_m)\|_0 \frac{|A(S(\tau;y_m))|}{\|S(\tau;y_m)\|_0^2}\,d\tau. \end{aligned} \tag{4.14}$$

By Lemma 4.2,

$$\begin{aligned} \int_0^t \|S(\tau;y_0-y_m)\|_0 \frac{|A(S(\tau;y_m))|}{\|S(\tau;y_m)\|_0^2}\,d\tau &\leqslant c\|y_0-y_m\|_0\int_0^t |\Phi(S(\tau;y_m))|\,d\tau \\ &\leqslant c_1\|y_0-y_m\|_0\|y_m\|_0. \end{aligned} \tag{4.15}$$

Besides,

$$
\begin{aligned}
&\int_0^t |A(S(\tau;y_0)) - A(S(\tau;y_m))|\,d\tau \\
&\quad = \int_0^t |(S(\tau;y_0-y_m),\nabla)\operatorname{curl}^{-1}S(\tau;y_0)\cdot S(\tau;y_0) \\
&\qquad + (S(\tau;y_m),\nabla)\operatorname{curl}^{-1}S(\tau;y_0-y_m)\cdot S(\tau;y_0) \\
&\qquad + (S(\tau;y_m),\nabla)\operatorname{curl}^{-1}S(\tau;y_m)\cdot S(\tau;y_0-y_m)|\,d\tau \\
&\quad \leqslant c\|y_0-y_m\|_0 \int_0^t (\|S(\tau;y_0)\|_{3/4}^2 \\
&\qquad + \|S(\tau;y_m)\|_{3/4}\|S(\tau;y_0)\|_{3/4} + \|S(\tau;y_m)\|_{3/4}^2)\,d\tau \\
&\quad \leqslant c_1\|y_0-y_m\|_0(\|y_0\|_0^2+\|y_m\|_0^2).
\end{aligned}
\tag{4.16}
$$

Finally, (4.13) follows from (4.14), (4.15), (4.16). □

Let us prove the theorem on the continuous dependence of the solution to problems (2.17), (2.10) on the initial data. This result will be used for later purposes.

THEOREM 4.2. *The solution* $y(t,x;y_0) \in V^{1,2(-1)}(Q_T)$ *of problems* (2.17), (2.10) *depends continuously on the initial data* $y_0 \in V^0(\mathbb{T}^3)$; *that is, for any* $t \in (0,T]$, *the mapping*

$$V^0(\mathbb{T}^3) \ni y_0 \longrightarrow y(t,\,\cdot\,;y_0) \in V^0(\mathbb{T}^3)$$

is continuous.

PROOF. Assume that $y_m \to y_0$ as $m \to \infty$ in the space $V^0(\mathbb{T}^3)$. Then, using (3.2), employing the well-known properties of the operator function $S(t,x;y_0)$ (which readily follow on expanding into the Fourier series) and invoking (4.13), we obtain the assertion of the theorem:

$$
\begin{aligned}
&\|y(t;y_0)-y(t;y_m)\|_0 \leqslant \frac{\|S(t;y_0-y_m)\|_0}{1-\int_0^t \Phi(S(\tau;y_0))\,d\tau} \\
&\qquad + \frac{\left|\int_0^t \Phi(S(\tau;y_m))\,d\tau - \int_0^t \Phi(S(\tau;y_0))\,d\tau\right| \|S(t;y_m)\|_0}{\left(1-\int_0^t \Phi(S(\tau;y_0))\,d\tau\right)\left(1-\int_0^t \Phi(S(\tau;y_m))\,d\tau\right)} \\
&\quad \leqslant c_1\|y_0-y_m\|_0 \\
&\qquad + c_2\left|\int_0^t \Phi(S(\tau;y_m))\,d\tau - \int_0^t \Phi(S(\tau;y_0))\,d\tau\right| \to 0 \\
&\qquad\qquad \text{as} \quad m\to\infty.
\end{aligned}
\tag{4.17}
$$

□

5. Properties of the functional $\Phi(S(t;y_0))$

The explicit formula (3.2) implies that in order to study the dynamics generated by problems (2.17), (2.10), one first needs to examine in detail the properties of the functional $\Phi(S(t;y_0))$.

5.1. On the kernel of the functional $\Phi(S(t;u))$. In view of the results obtained in the previous section, it is natural to take the space V^0 as the phase space of the dynamical system corresponding to problems (2.17), (2.10). The kernel of the functional $\Phi(S(t;u))$ is defined by

$$K\Phi = \{u \in V^0 : \Phi(S(t;u)) \equiv 0 \quad \forall t \in \mathbb{R}_+\}. \tag{5.1}$$

The functional $u \to \Phi(S(t;u))$ is first-order homogeneous (that is, $\Phi(S(t;\lambda u)) = \lambda\Phi(S(t;u))$ for all $\lambda \in \mathbb{R}$), and hence $K\Phi$ is a double-sided cone; that is, if $u \in K\Phi$, then $\lambda u \in K\Phi$ for all $\lambda \in \mathbb{R}$.

We claim that the cone $K\Phi$ is nonempty. To this end, we introduce the linear subspace

$$L = \{z \in V^0 : z(x) = \sum_{k \in \mathcal{U}_L} \hat{z}(k)e^{ik\cdot x},\ \hat{z}(-k) = \overline{\hat{z}(k)}\}, \tag{5.2}$$

where

$$\mathcal{U}_L = \{k = (k_1, k_2, k_3) \in \mathbb{Z}^3 \setminus \{0\} : k_1 + k_2 + k_3 - \text{is an odd number}\}. \tag{5.3}$$

The space L consists of vector fields $z(x) \in V^0$ whose Fourier coefficients $\hat{z}(k)$ are nonzero for $k \notin \mathcal{U}_L$. Clearly, L is an infinite-dimensional subspace of the space V^0.

LEMMA 5.1. *The following inclusion holds:*

$$L \subset K\Phi; \tag{5.4}$$

here $L, K\Phi$ *are defined in* (5.2), (5.3), (5.1).

PROOF. In view of (2.11),

$$\int_{\mathbb{T}^3} (z(x), \nabla) \operatorname{curl}^{-1} z(x)\cdot z(x)\, dx = -\sum_{k,m \in \mathcal{U}_L} \hat{z}(k)\cdot m\frac{m \times \hat{z}(m)}{|m|^2}\cdot\overline{\hat{z}(k+m)} = 0 \tag{5.5}$$

for any $z \in L$.

Indeed, $\overline{\hat{z}(k+m)} = 0$, because $(k_1+m_1)+(k_2+m_2)+(k_3+m_3)$ is an even number for $k, m \in \mathcal{U}$. Therefore, by (5.5) and the definition of the operator $S(t,x;z)$,

$$\begin{aligned}\int_{\mathbb{T}^3} &(S(t,x;z), \nabla) \operatorname{curl}^{-1} S(t,x;z)\cdot S(t,x;z)\, dx \\ &= -\sum_{k,m \in \mathcal{U}_L} e^{-2(k^2+m^2+k\cdot m)t}\hat{z}(k)\cdot m\frac{m \times \hat{z}(m)}{|m|^2}\cdot\overline{\hat{z}(k+m)} = 0,\end{aligned} \tag{5.6}$$

for any $t \in \mathbb{R}_+$.

Finally, $\Phi(S(t;z)) = 0$ for any $t \in \mathbb{R}_+$ with $z \in L$ by (5.6) and the definition (2.16) of the functional Φ. □

5.2. Properties of the function $t \to \Phi(S(t;u))$.

LEMMA 5.2. *For any* $z \in V^0 \setminus K\Phi$, *there exists* $t_0 > 0$ *such that for* $t > t_0$ *the function* $t \to \Phi(S(t;z))$ *is sign-definite.*

PROOF. From definition (2.16) of the functional Φ we have, similarly to (5.6),

$$\Phi(S(t,z)) = \frac{-\sum_{k,m\in\mathbb{Z}^3\setminus\{0\}} e^{-2(k^2+m^2+k\cdot m)t}\hat{z}(k)\cdot m\frac{m\times\hat{z}(m)}{|m|^2}\cdot\overline{\hat{z}(k+m)}}{\sum_{k\neq 0}|\hat{z}(k)|^2 e^{-2k^2 t}}. \tag{5.7}$$

Assume that $z(x)$ is such that $\hat{z}(k) = 0$ for $|k| < \gamma$, where $\gamma \geqslant 1$ is some number, the first nonzero Fourier coefficients appearing for $|k| = \gamma$. The numerator in (5.7) can be written down as follows

$$-\sum_{k,m\in\mathbb{Z}^3\setminus\{0\}} e^{-2(k^2+m^2+k\cdot m)t}\hat{z}(k)\cdot m\frac{m\times\hat{z}(m)}{|m|^2}\cdot\overline{\hat{z}(k+m)}$$
$$= e^{-(2\gamma^2+\alpha)t}\left(a_0 + \sum_{m=1}^{\infty} a_m e^{-mt}\right),$$

where $\alpha \geqslant 2$ is an even number, $a_0 \neq 0$, $a_m \in \mathbb{R}$. For the denominator, we have

$$\sum_{k\neq 0}|\hat{z}(k)|^2 e^{-2k^2 t} = e^{-2\gamma^2 t}\left(b_0 + \sum_{m=1}^{\infty} b_m e^{-mt}\right),$$

where $b_0 > 0, b_m \geqslant 0$. Hence, in view of (5.7),

$$\begin{aligned}
\Phi(S(t,z)) &= e^{-\alpha t}\frac{a_0 + \sum_{m=1}^{\infty} a_m e^{-mt}}{b_0 + \sum_{m=1}^{\infty} b_m e^{-mt}} \\
&= e^{-\alpha t}\left(\frac{a_0}{b_0} + \sum_{m=1}^{\infty}\frac{a_m}{b_0}e^{-mt}\right)\sum_{j=0}^{\infty}(-1)^j\left(\sum_{m=1}^{\infty}\frac{b_m}{b_0}e^{-mt}\right)^j \qquad (5.8)\\
&= e^{-\alpha t}\left(\frac{a_0}{b_0} + \sum_{m=1}^{\infty} c_m e^{-mt}\right),
\end{aligned}$$

where $c_m \in \mathbb{R}$ are some constants, the series on the right-hand side of (5.8) converging absolutely for all sufficiently large t. Clearly, for large t, the sign of the bracketed expression in the last line of (5.8) agrees that of a_0/b_0. □

LEMMA 5.3. *For any $z \notin K\Phi$ and any $\alpha \in \mathbb{R}$, the equation*

$$\int_0^t \Phi(S(\tau; z))\,d\tau = \alpha \qquad (5.9)$$

has at most a finite number of solutions $0 \leqslant t_1 < t_2 < \cdots < t_N$, where either $t_N < \infty$ or $t_N = \infty$.

PROOF. By Lemma 5.2 the function $t \to \int_0^t \Phi(S(\tau; z))\,d\tau$ is analytic for all $t \in \mathbb{R}_+$ and is strictly monotone for $t > t_0$ for some sufficiently large t_0. Hence, all the solutions of (5.9), with at most one exception, lie on the interval $[0, t_0]$. If the interval $[0, t_0]$ contains an infinite number of solutions, then one may choose a subsequence converging to $\hat{t} \in [0, t_0]$. But then all the derivatives of the function $\int_0^t \Phi(S(\tau; z))\,d\tau$ vanish at $t = \hat{t}$, and hence, being analytic, this function is identically constant. But this contradicts its strict monotonicity for $t \in [t_0, \infty]$. Hence, the number of solutions of (5.9) is at most finite. □

5.3. Some auxiliary results. In this section, we perform some auxiliary calculations to be used later for the study of the behaviour of the function $t \to \int_0^t \Phi(S(\tau; z))\,d\tau$.

Consider the vectors

$$e_1 = (1, 0, 0), \qquad e_2 = (0, 1, 0),$$

and specify the Fourier coefficients

$$\begin{aligned} &\hat{z}(\pm ne_1) = (0, \alpha_2, \alpha_3), && \hat{z}(\pm ne_2) = (\beta_1, 0, \beta_3), \\ &\hat{z}(\pm n(e_1 + e_2)) = (\gamma, -\gamma, 0), && \hat{z}(\pm n(e_1 - e_2)) = (\delta, \delta, 0), \end{aligned} \tag{5.10}$$

where n is some fixed odd number and $\alpha_2, \alpha_3, \beta_1, \beta_3, \gamma, \delta \in \mathbb{R}$ are some real numbers.

LEMMA 5.4. *Assume that all the Fourier coefficients of the vector field $z(x)$ are zero, except for those indicated in* (5.10). *Then*

$$\begin{aligned} &\int_{\mathbb{T}^3} (S(t, x; z), \nabla) \operatorname{curl}^{-1} S(t, x; z) \cdot S(t, x; z)\, dx \\ &\qquad = - \sum_{k, m \in \mathbb{Z}^3 \setminus \{0\}} e^{-2(k^2 + m^2 + k \cdot m)t} \hat{z}(k) \cdot m \frac{m \times \hat{z}(m)}{|m|^2} \cdot \overline{\hat{z}(k + m)} \\ &\qquad = (\alpha_2 \beta_3 + \alpha_3 \beta_1)(\gamma + \delta) e^{-4n^2 t}. \end{aligned} \tag{5.11}$$

PROOF. We have

$$I(k, m) := \hat{z}(k) \cdot m \frac{m \times \hat{z}(m)}{|m|^2} \cdot \overline{\hat{z}(k + m)} = \frac{\hat{z}(k) \cdot m}{|m|^2} \det(m, \hat{z}(m), \overline{\hat{z}(k + m)}),$$

and hence, using (5.10) it follows by simple calculations that

$$\begin{aligned} I(\pm n(e_1, e_2)) &= -I(\pm n(e_1, -e_1 - e_2)) = 2I(\pm n(e_1 + e_2, -e_1)) = \alpha_2 \beta_3 \gamma, \\ I(\pm n(e_1, -e_2)) &= -I(\pm n(e_1, -e_1 + e_2)) = 2I(\pm n(e_1 - e_2, -e_1)) = \alpha_2 \beta_3 \delta, \\ I(\pm n(e_2, e_1)) &= -I(\pm n(e_2, -e_2 - e_1)) = 2I(\pm n(e_2 + e_1, -e_2)) = \alpha_3 \beta_1 \gamma, \\ I(\pm n(e_2, e_1 - e_2)) &= -I(\pm n(e_2, -e_1)) = 2I(\pm n(e_2 - e_1, -e_2)) = \alpha_3 \beta_1 \delta. \end{aligned}$$

Substituting these formulae on the right-hand side of the first equality in (5.11), we obtain the right-hand side of the second equality. □

5.4. Concrete examples of the function $t \to \Phi(S(t; z))$. Below we will employ Lemma 5.4 to construct a series of concrete examples of the function $t \to \Phi(S(t; z))$ useful for later purposes.

PROPOSITION 5.1. *We have*

$$K\Phi \setminus L \neq \varnothing, \tag{5.12}$$

where the sets $K\Phi$ and L are defined in (5.1) *and* (5.2), *respectively.*

PROOF. As an element of the nonempty set (5.12) one may take the vector field $z(x)$. For this field only the Fourier coefficients indicated in (5.10) are nonzero with $n = 1$, for which $\gamma + \delta = 0$. Now the inclusion $z \in K\Phi \setminus L$ readily follows from (5.11). □

REMARK 5.1. In fact, the set $K\Phi \setminus L$ is much broader than that indicated in the proof of Proposition 5.1. Indeed, one easily sees that as $z \in K\Phi \setminus L$, one may also take the vector field $z(x)$ whose nonzero Fourier coefficients are only those indicated in (5.10) with any odd $n > 0$, and for which $\gamma_n + \delta_n = 0$.

PROPOSITION 5.2. *There exists $z \in V^0 \setminus K\Phi$ such that*

$$\int_0^t \Phi(S(\tau;z))\,d\tau \leqslant 1 \quad \forall t>0 \quad \text{and} \quad \max_{t>0} \int_0^t \Phi(S(\tau;z))\,d\tau = 1, \tag{5.13}$$

where the maximum is attained only at $t=\infty$. Moreover, for such z there exist constants $0<c_1<c_2$ such that

$$c_1 e^{-2t} \leqslant 1 - \int_0^t \Phi(S(\tau;z))\,d\tau \leqslant c_2 e^{-2t} \tag{5.14}$$

for sufficiently large t.

PROOF. Consider $z(x)$ whose nonzero Fourier coefficients are only those indicated in (5.10) with $n=1$, where α_i, β_j, γ, and δ are such that

$$(\alpha_2\beta_3+\alpha_3\beta_1)(\gamma+\delta)=A, \quad \alpha_2^2+\alpha_3^2+\beta_1^2+\beta_3^2=a_1, \quad 2\gamma^2+2\delta^2=a_2, \tag{5.15}$$

with some $A>0$, $a_1>0$, $a_2>0$. Hence, in view of (5.11),

$$\int_0^t \Phi(S(\tau;z))\,d\tau = \int_0^t \frac{Ae^{-4\tau}}{a_1e^{-2\tau}+a_2e^{-4\tau}}\,d\tau = \int_0^t \frac{Ae^{-2\tau}}{a_1+a_2e^{-2\tau}}\,d\tau.$$

The integrand on the right-hand side is positive, and hence the corresponding integral increases to some limit as $t\to\infty$. Further, replacing in this equality z by λz with suitable $\lambda>0$, we can achieve that this limit is 1. Finally, (5.14) follows, since, clearly,

$$\int_t^\infty \frac{Ae^{-2\tau}}{a_1+a_2e^{-2\tau}}\,d\tau \leqslant \frac{A}{a_1}\int_t^\infty e^{-2\tau}\,d\tau = \frac{A}{2a_1}e^{-2\tau},$$

and since, for sufficiently large t,

$$\int_t^\infty \frac{Ae^{-2\tau}}{a_1+a_2e^{-2\tau}}\,d\tau \geqslant \frac{A}{a_1(1+\varepsilon)}\int_t^\infty e^{-2\tau}\,d\tau = \frac{A}{2a_1(1+\varepsilon)}e^{-2\tau}. \qquad \square$$

PROPOSITION 5.3. *There exists $z \in V^0 \setminus K\Phi$ such that*

$$\max_{t>0}\int_0^t \Phi(S(\tau;z))\,d\tau = 1 = \int_0^{t_0} \Phi(S(\tau;z))\,d\tau, \tag{5.16}$$

that is, this maximum is attained at finite $t=t_0$.

PROOF. Consider $z(x)$ whose nonzero Fourier coefficients are only those indicated in (5.10) with $n=1$ and $n=3$, where α_i, β_j, γ, and δ corresponding to $n=1$ satisfy (5.15) with $A<0$, $a_1>0$, $a_2>0$, while α_i, β_j, γ, and δ corresponding to $n=3$ are such that

$$(\alpha_2\beta_3+\alpha_3\beta_1)(\gamma+\delta)=B, \quad \alpha_2^2+\alpha_3^2+\beta_1^2+\beta_3^2=b_1, \quad 2\gamma^2+2\delta^2=b_2,$$

with some $B>0$, $b_1>0$, $b_2>0$. Hence, by (5.11),

$$\begin{aligned}\int_0^t \Phi(S(\tau;z))\,d\tau &= \int_0^t \frac{Ae^{-4\tau}+Be^{-36\tau}}{a_1e^{-2\tau}+a_2e^{-4\tau}+b_1e^{-18\tau}+b_2e^{-36\tau}}\,d\tau\\ &= \int_0^t \frac{e^{-2\tau}(A+Be^{-32\tau})}{a_1+a_2e^{-2\tau}+b_1e^{-16\tau}+b_2e^{-34\tau}}\,d\tau.\end{aligned}$$

Let $A+B>0$. Then, clearly, the integrand on the right-hand side of this equality is positive for small τ, it becomes negative for large τ, and it monotonically increases to zero as $\tau\to\infty$. It readily follows that this integrand assumes the maximum positive value at some finite $t = t_0$. In order that this maximum value be 1, it suffices to multiply z by a suitably chosen $\lambda > 0$. $\square$

6. The structure of the dynamics generated by the NPS

The purpose of this section is to examine the structure of the dynamic flow generated by the NPS. To this end we partition the phase space of the dynamical system into three sets on each of which the behaviour of the dynamic flow is substantially different.

6.1. Characteristic sets of the phase space of the NPS. We first give the definition of three component subsets of the phase space, the NPS-generated dynamics being substantially different on these sets. Of course, these sets are invariant relative to the dynamical system under study. Recall that the space $V^0(\mathbb{T}^3) \equiv V^0$ (see (2.4)) was taken as the phase space of problems (2.17), (2.10).

DEFINITION 6.1. A subset $M_- \subset V^0$ of initial data y_0, for which the solution $y(t, x; y_0)$ of problems (2.17), (2.10) exists, is unique, and satisfies the inequality

$$\|y(t, \cdot; y_0)\|_0 \leqslant \alpha \|y_0\|_0 e^{-t} \qquad \forall t > 0, \tag{6.1}$$

and is called the *set of stability*. Here, $\alpha \geqslant 1$ is some fixed number depending, in general, on $\|y_0\|_0$.

Such a dependence of α on $\|y_0\|_0$ in (6.1) is not felt if y_0 lies in the subset

$$M_-(\alpha) = \{y_0 \in M_- : \ y(t, x; y_0) \quad \text{satisfies (6.1)}\} \tag{6.2}$$

of the set of stability. Here, $\alpha \geqslant 1$ is a fixed number and $y(t, x; y_0)$ is the solution of problems (2.17), (2.10). Clearly,

$$\bigcup_{\alpha \geqslant 1} M_-(\alpha) = M_- . \tag{6.3}$$

The sets $M_-(\alpha)$ prove useful in examining the set of stability M_-. The following result holds.

LEMMA 6.1. *Suppose that $y_0 \in V^0$ satisfies the estimate*

$$\sup_{t \in \mathbb{R}_+} \int_0^t \Phi(S(\tau; y_0))\, d\tau \leqslant \frac{\alpha - 1}{\alpha} \tag{6.4}$$

with $\alpha \geqslant 1$. Then $y_0 \in M_-(\alpha)$.

PROOF. Indeed, in view of (3.2) and (6.4),

$$\|y(t, \cdot; y_0)\|_0 \leqslant \frac{\|S(t; y_0)\|_0}{1 - \sup_{t \in \mathbb{R}_+} \int_0^t \Phi(S(\tau; y_0))\, d\tau} \leqslant \alpha \|y_0\|_0 e^{-t}.$$ $\square$

DEFINITION 6.2. A subset $M_+ \subset V^0$ of initial data y_0 of problems (2.17), (2.10) for which the solution $y(t, x; y_0)$ exists only on a finite time interval $t \in (0, t_0)$ with $t_0 > 0$ depending on y_0 and is exploded at $t = t_0$,[3] is called the *set of explosions*.

[3] i.e., $\|y(t; y_0)\| \to \infty$ as $t \nearrow t_0$

In view of (3.2),

$$M_+ = \left\{ y_0 \in V^0 : \exists\, 0 < t_0 < \infty,\ \int_0^{t_0} \Phi(S(\tau; y_0))\, d\tau = 1 \right\} \tag{6.5}$$

for the solution $y(t, x; y_0)$.

The smallest t_0 at which the equality in (6.5) is attained is called the *explosion moment* of the solution $y(t, x; y_0)$.

DEFINITION 6.3. A subset $M_g \subset V^0$ of initial data y_0 of problems (2.17), (2.10) for which the solution $y(t, x; y_0)$ exists on an infinite time interval $t \in \mathbb{R}_+$ and $\|y(t, x; y_0)\|_0 \to \infty$ as $t \to \infty$ is called the *set of increase*.

LEMMA 6.2. *Any of the sets* M_-, M_+, M_g *is nonempty.*

PROOF. We have $K\Phi \subset M_-$ in view of (3.2) and (5.1), and moreover, $K\Phi \neq \varnothing$, by Lemma 5.1 and Proposition 5.1. Hence, $M_- \neq \varnothing$. Further, $M_+ \neq \varnothing$ in view of (3.2) and Proposition 5.3, and finally, $M_g \neq \varnothing$ by (3.2) and Proposition 5.2. □

LEMMA 6.3. *The sets* M_-, M_+, M_g *are invariant under the operator* $y_0 \to y(t, \cdot\,; y_0)$ *of translation along the solution of problems* (2.17), (2.10); *that is,*

$$y(t; M_-) \subset M_-, \quad y(t; M_g) \subset M_g \quad \forall t > 0; \quad y(t; y_0) \in M_+, \quad \forall t \in (0, t_0), \tag{6.6}$$

where t_0 *is the explosion moment of the initial data* $y_0 \in M_+$.

PROOF. Let $y_0 \in M_-$. Then estimate (6.4) with some $\alpha \geqslant 1$ is satisfied. We set $y_1 = y(t_1, \cdot\,; y_0)$. Hence, in view of (3.2), since the operator $S(t; y_0)$ is linear in y_0, further since the functional $\Phi(v)$ is homogeneous in v, and by (6.4),

$$\begin{aligned}
\|y(t, \cdot\,; y_1)\|_0 &= \frac{\|S(t; y_1\|_0}{1 - \int_0^t \Phi\big(S\big(\tau; \frac{S(t_1; y_0)}{1 - \int_0^{t_1} \Phi(S(\tau_1; y_0)\, d\tau_1}\big)\big)\, d\tau} \\
&= \frac{(1 - \int_0^{t_1} \Phi(S(\tau_1; y_0))\, d\tau_1)\|S(t; y_1\|_0}{1 - \int_0^{t_1} \Phi(S(\tau_1; y_0))\, d\tau_1 - \int_{t_1}^{t+t_1} \Phi(S(\tau; y_0))\, d\tau} \leqslant \alpha_1 e^{-t}\|y_1\|_0,
\end{aligned}$$

where $\alpha_1 = \alpha(1 - \int_0^{t_1} \Phi(S(\tau_1; y_0))\, d\tau_1) < \infty$, proving the first inclusion in (6.6).

Assume that $t_1 > 0$, $t_2 > 0$ and that the solution $y(t; y_0)$ is defined for $t = t_1$ and $t = t_1 + t_2$. If $y_1 = y(t_1; y_0)$, then, in view of (3.2), we have as before,

$$\begin{aligned}
y(t_2; y_1) &= \frac{S\big(t_2; \frac{S(t_1; y_0)}{1 - \int_0^{t_1} \Phi(S(\tau; y_0))\, d\tau}\big)}{1 - \int_0^{t_2} \Phi\big(S\big(\tau_1; \frac{S(t_1; y_0)}{1 - \int_0^{t_1} \Phi(S(\tau; y_0))\, d\tau}\big)\big)\, d\tau_1} \\
&= \frac{S(t_1 + t_2; y_0)}{1 - \int_0^{t_1+t_2} \Phi(S(\tau; y_0))\, d\tau} = y(t_1 + t_2, \cdot\,; y_0).
\end{aligned} \tag{6.7}$$

Now if $y_0 \in M_+$, t_0 is an explosion moment, $y_1 = y(t_1; y_0)$, $t_1 > 0$, $t_2 > 0$, $t_1 + t_2 < t_0$, then, in view of (6.7),

$$\|y(t_2; y_1)\|_0 = \|y(t_1 + t_2; y_0)\|_0 \to \infty \quad \text{for} \quad t_2 \to t_0 - t_1,$$

and hence, $y_1 \in M_+$, proving the third inclusion in (6.6).

Finally, if $y_0 \in M_g$, $y_1 = y(t_1; y_0)$, $t_1 > 0$, $t_2 > 0$, then a similar analysis shows that

$$\|y(t_2; y_1)\|_0 = \|y(t_1 + t_2; y_0)\|_0 \to \infty \quad \text{as} \quad t_2 \to \infty,$$

proving the second inclusion in (6.6). □

Below we shall show that $M_- \cup M_+ \cup M_g = V^0$, and we shall study geometrical properties of these sets in V^0.

6.2. Some subsets of the unit sphere of the space V^0. We let

$$\Sigma = \{v \in V^0 : \quad \|v\|_0 = 1\} \tag{6.8}$$

denote the unit sphere of the phase space V^0.

In order to describe the structure of the phase flow corresponding to problems (2.17), (2.10), we need to define several subsets of the sphere Σ. Namely, we set

$$A_-(t) = \{v \in \Sigma : \int_0^t \Phi(S(\tau; v))\, d\tau \leqslant 0\},$$

$$A_0(t) = \{v \in \Sigma : \int_0^t \Phi(S(\tau; v))\, d\tau = 0\},$$

and define

$$A_- = \bigcap_{t \geqslant 0} A_-(t), \quad A_0 = \bigcap_{t \geqslant 0} A_0(t). \tag{6.9}$$

All these sets are nonempty and closed. We have $A_0 = K\Phi \cap \Sigma$ by (5.1), (6.8), and hence $A_0 \neq \varnothing$ by Lemma 5.1 and Proposition 5.1. It is also clear that $A_0 \subset A_-$. The sets $A_-(t)$ and A_- have nonempty relative interior with respect to the induced topology on the sphere Σ from V^0. This follows from Lemma 4.3.

Consider also the set

$$B_+ = \Sigma \setminus A_- \equiv \Big\{v \in \Sigma : \ \exists t_0 > 0 \ \int_0^{t_0} \Phi(S(\tau; v))\, d\tau > 0\Big\}. \tag{6.10}$$

Clearly, the set B_+ is open with respect to the induced topology on Σ. Moreover, the boundary ∂B_+ of B_+ is as follows:

$$\partial B_+ = \Big\{v \in \Sigma : \ \forall t > 0 \int_0^t \Phi(S(\tau; v))\, d\tau \leqslant 0,$$
$$\exists t_0 \in \mathbb{R}_+ \cup \{\infty\} : \ \int_0^{t_0} \Phi(S(\tau; v))\, d\tau = 0\Big\}.$$

Also, it is an easy matter to check that $A_0 \subset \partial B_+$ and $\partial B_+ \setminus A_0 \neq \varnothing$.

6.3. On the structure of the phase space V^0. We introduce the function

$$B_+ \ni v \to b(v) = \max_{t \geqslant 0} \int_0^t \Phi(S(\tau; v))\, d\tau, \tag{6.11}$$

which is defined on the subset B_+ of the sphere Σ. Clearly, $b(v) > 0$ and $b(v) \to 0$ as $v \to \partial B_+$.

Next we define the mapping $\Gamma(v)$, which plays an important role in describing the phase flow generated by the boundary-value problems (2.17), (2.10):

$$B_+ \ni v \to \Gamma(v) = \frac{v}{b(v)} \in V^0; \tag{6.12}$$

here $b(v)$ is the function (6.11). Note that $\|\Gamma(v)\|_0 \to \infty$ as $v \to \partial B_+$.

It is easily seen that the image

$$\Gamma(B_+) = \{y \in V^0 : y = \Gamma(v) \quad \text{for some} \quad v \in B_+\} \tag{6.13}$$

under the mapping Γ splits the space V^0 into two parts,

(6.14)
$$V^0_- = \{v \in V^0 :\ [0,v] \cap \Gamma(B_+) = \varnothing\}, \quad V^0_+ = \{v \in V^0 :\ [0,v) \cap \Gamma(B_+) \neq \varnothing\};$$

here $[0,v]$ is the closed interval in V^0 connecting a point $v \in V^0$ and the origin of the space V^0, and $[0,v)$ is the corresponding half-interval. One clearly has

(6.15)
$$V^0 = V^0_- \cup V^0_+ \cup \Gamma(B_+).$$

We partition the set B_+ into the following two subsets:

(6.16)
$$\begin{aligned} B_{+,f} &= \{v \in B_+ : \quad \text{the maximum in (6.11) is attained for } t_0 < \infty\}, \\ B_{+,\infty} &= \{v \in B_+ : \quad \text{the maximum in (6.11) is attained for } t_0 = \infty\}. \end{aligned}$$

It is readily seen that

(6.17)
$$\Gamma(B_+) = \Gamma(B_{+,f}) \cup \Gamma(B_{+,\infty}).$$

THEOREM 6.1. *The following relations hold:*

(6.18)
$$M_- = V^0_-, \quad M_+ = V^0_+ \cup \Gamma(B_{+,f}), \quad M_g = \Gamma(B_{+,\infty});$$

here V^0_-, V^0_+, $B_{+,f}$, $B_{+,\infty}$ *are the sets defined in* (6.14), (6.16), Γ *is the mapping* (6.12), M_- *is the set of stability,* M_+ *is the set of explosions, and* M_g *is the set of increase* (*see Definitions* 6.1, 6.2, 6.3).

PROOF. Consider, first, the positivity cone of the functional Φ,

(6.19)
$$P\Phi = \{v \in V^0 :\ \exists v_0 \in B_+, \lambda > 0, \text{ such that } v = \lambda v_0\},$$

where B_+ is defined in (6.10), and second, the negativity cone of the functional Φ,

(6.20)
$$N\Phi = \{v \in V^0 :\ \exists v_0 \in A_-, \lambda > 0, \text{ such that } v = \lambda v_0\},$$

the set A_- is defined in (6.9). Clearly,

$$V^0 = P\Phi \cup N\Phi \cup \{0\}.$$

Let $\rho > 0$. Similarly to $\Gamma(v)$, we define the mapping $\Gamma_\rho(v)$:

(6.21)
$$B_+ \ni v \to \Gamma_\rho(v) = \frac{\rho v}{b(v)} \in V^0;$$

the function $b(v)$ is defined in (6.11). It is readily checked that

(6.22)
$$V^0_- = N\Phi \cup \left\{ \bigcup_{\rho \subset (0,1)} \Gamma_\rho(B_+) \right\} \cup \{0\}.$$

Using Lemma 6.1,

(6.23)
$$N\Phi \subset M_-(1).$$

Now assume that $y_0 \in \Gamma_\rho(B_+)$, where $\rho = (\alpha - 1)/\alpha$ with some $\alpha > 1$. Then $y_0 = \Gamma_\rho(v)$ for some $v \in B_+$, and hence, in view of (6.21) and (6.11),

$$\sup_{t>0} \int_0^t \Phi(S(\tau; y_0))\, d\tau = \rho \frac{\sup_{t>0} \int_0^t \Phi(S(\tau; v))\, d\tau}{\sup_{t>0} \int_0^t \Phi(S(\tau; v))\, d\tau} = \frac{\alpha - 1}{\alpha},$$

whence $\Gamma_\rho(B_+) \subset M_-(\alpha)$ for $\rho = (\alpha - 1)/\alpha$ by Lemma 6.1. Hence, by (6.22), (6.23) and (6.3),

(6.24)
$$V^0_- \subset M_-.$$

Let $v \in B_{+,f}$, and let $0 < t_0 < \infty$ be the least time at which the maximum in (6.11) is attained for this v. We set $y_0 = \frac{v}{b(v)} \in \Gamma(B_{+,f})$. Since the operator $S(t;v)$ is linear in v and since the functional $\Phi(v)$ is homogenous in v, it follows in view of (3.2) that

$$y(t,\cdot;y_0) = \frac{S(t;v/b(v))}{1-\int_0^t \Phi(S(\tau;v/b(v)))\,d\tau} = \frac{S(t;v)}{b(v)-\int_0^t \Phi(S(\tau;v))\,d\tau} \tag{6.25}$$

for any $t < t_0$.

Using (6.25), (6.11) it is seen that $\|y(t,\cdot;y_0)\|_0 \to \infty$ as $t \nearrow t_0$, and hence

$$\Gamma(B_{+,f}) \subset M_+ . \tag{6.26}$$

Now let us consider the case $y_0 \in V_+^0$. From the definition of the set V_+^0 it follows that there exists $\rho > 1$ such that $y_0/\rho \in \Gamma(B_+)$ for some $v \in B_+$. But this means that $y_0 = \rho v/b(v)$. Hence, similarly to (6.25),

$$y(t,\cdot;y_0) = \frac{S(t;\rho v/b(v))}{1-\int_0^t \Phi(S(\tau;\rho v/b(v)))\,d\tau} = \frac{\rho S(t;v)}{b(v)-\rho\int_0^t \Phi(S(\tau;v))\,d\tau}. \tag{6.27}$$

We have $\rho > 1$, and hence there exists $0 < t_0 < \infty$ such that for $t < t_0$ the denominator on the right-hand side of the second fraction in (6.27) is positive and tends to zero as $t \to t_0$. As a result, $y_0 \in M_+$, and therefore,

$$V_+^0 \subset M_+. \tag{6.28}$$

Finally, assume that $y_0 = \frac{v}{b(v)} \in \Gamma(B_{+,\infty})$. Then the denominator of the second fraction in (6.25) is positive and tends to zero as $t \to \infty$. In view of (5.8), we have for sufficiently large t,

$$b(v) - \int_0^t \Phi(S(\tau;v))\,d\tau = \int_t^\infty \Phi(S(\tau;v))\,d\tau = e^{-\alpha t}\left(\frac{a_0}{b_0\alpha} + \sum_{m=1}^\infty \frac{c_m}{m+\alpha}e^{-mt}\right),$$

where $\alpha \geqslant 2$, $\frac{a_0}{b_0\alpha} > 0$, the series converging. Hence, using (6.25),

$$\|y(t,\cdot;y_0)\|_0 \asymp O(e^{-t}/e^{-\alpha t}) \to \infty \quad \text{as} \quad t \to \infty;$$

that is, $y_0 \in M_g$, and so

$$\Gamma(B_{+,\infty}) \subset M_g. \tag{6.29}$$

Now (6.18) follows from the inclusions (6.24), (6.26), (6.28), (6.29), equalities (6.15), (6.17), and the clear relations $M_- \cap M_+ = M_- \cap M_g = M_+ \cap M_g = \varnothing$. □

References

[1] Hopf E., Über die Anfangswertaufgabe für die Hydrodynamischen Grundgleichungen.- Math. Nachr., 4, (1950), 213–231. MR0050423 (14:327b)

[2] Ladyzhenskaya, O. A., The mathematical theory of viscous incompressible flow, Moscow (1961), Gordon and Breach, New York (1969). MR0254401 (40:7610)

[3] R. Temam, Navier–Stokes equations. Theory and numerical analysis, Amsterdam–New York, North-Holland (1979); Moscow, Mir (1981). MR644804 (83a:35082)

[4] Fursikov, A.V., On one semilinear parabolic equation of normal type, Proceeding volume "Mathematics and life sciences", De Gruyter v.1, 2012, 147–160. MR3059755

[5] Fursikov, A.V., The simplest semilinear parabolic equation of normal type. Mathematical Control and Related Fields(MCRF), v.2, N2, (2012), 141–170. MR2929443

[6] Fursikov, A.V., On the normal semilinear parabolic equations corresponding to 3D Navier-Stokes system, D. Homberg and F. Troltzsch (Eds.): CSMO 2011, IFIP AICT 391, (2013), 338–347.

[7] Ladyzhenskaya, O. A., Solution in the large of the boundary-value problem for Navier-Stokes equations for two, Doklady SSSR, 123, (1958), 427–429.
[8] V. I. Yudovich, Non-stationary flows of an ideal incompressible fluid, Zh. Vychisl. Mat. Mat. Fiz., 1963, 3 (6), 1032–1066. MR0158189 (28:1415)

M. V. Lomonosov Moscow State University, Faculty of Mechanics and Mathematics, Vorob'evy gory, 119899 Moscow, Russia
E-mail address: fursikov@gmail.com

Amer. Math. Soc. Transl.
Volume **232**, 2014

A discrete Bernoulli free boundary problem

M. del Mar Gonzalez, M. Gualdani, and H. Shahgholian

To the memory of O.A. Ladyzhenskaya

Abstract. In this paper we consider a free boundary problem for the p-Laplace operator which can be seen as a discrete version of the so-called Bernoulli free boundary problem. In this formulation, the classical boundary gradient condition is replaced by a condition on the distance between two different level surfaces of the solution. For suitable scalings the model converges to the classical Bernoulli problem; one of the advantages of this formulation lies in the simplicity of the arguments, since one does not need to consider the boundary gradient.

This problem was studied by Andrew Acker in late 1970s and early 1980s where he used operator methods to prove the existence of a fixed point that was exactly the solution to this discrete Bernoulli problem. However, Acker's operator method uses the global setting of the problem, and even if it seems likely to be applied to a local setting, the complications seem much more than that of the supersolution technique, which our method is based upon, and can be applied to problems in local setting as well as a wider class of problems/questions, such as regularity of the solutions and that of the free boundary.

1. Introduction

For a bounded domain $K \subset \mathbb{R}^N$ $(N \geq 2)$, the well-known (exterior) Bernoulli free boundary problem is to find a domain $\Omega \supset \overline{K}$, with the requirement that the harmonic function u in the region $\Omega \setminus \overline{K}$ with given boundary values satisfies a prescribed Neumann condition on $\partial\Omega$; the boundary $\partial\Omega$ is thus called Bernoulli free boundary. There is a vast literature on the subject, including by-now classical references [**1, 2, 3, 9, 11, 14, 26**]. One may also replace the Laplacian operator by the p-Laplacian operator; this problem arises in various nonlinear flow laws and several physical situations, e.g. electrochemical machining and potential flow in fluid mechanics (see for instance [**9, 23, 29, 31, 32, 47, 13**]).

In [**2, 9, 6, 7**] Andrew Acker approached the Bernoulli free boundary problem using an operator method, where he first proved existence of a discrete version of the free boundary (which is the main topic of this paper) and then showed that the discrete version converges to the Bernoulli problem.

2010 *Mathematics Subject Classification.* Primary 35R35.
Key words and phrases. Free boundary, Bernoulli boundary condition.

In this paper we propose the supersolution technique to solve this problem. It is also noteworthy that the operator method of Acker was done in global setting, even though it is likely that the formulation can be done in local setting. The supersolution technique seems in general more straightforward and has proven to be worked out in local setting as well; see [**18**] and the references therein.

To formulate the problem, let u be the p-capacitary potential $(1 < p < \infty)$ in the annulus $\Omega \setminus \overline{K}$, then we ask for a particular Ω such that

$$\operatorname{dist}(x, \Gamma_l) = \lambda, \quad \forall x \in \partial\Omega, \tag{1.1}$$

where $\Gamma_l := \{x \in \Omega \ : \ u(x) = l\}$ is the l-level set of u and l, λ are two positive given constants.

Condition (1.1) was also considered in [**42, 43, 44**] while studying diffusion problems with stationary isothermic surfaces, and independently by the third author in [**49**], where he considered a discrete version of Serrin's problem. This second approach was inspired by a parabolic free boundary model in finance in which the distance between the free boundary (that is the zero level set for the solution) and the location of a source is prescribed (cf. [**36, 28, 21, 20**] for an overview).

Two different problems can be set up according to where the free boundary lies in the annulus: the exterior problem (P_E), considered in Section 3, and the interior one (P_I), considered in Section 4. It is clear that the geometry of the domains will play a crucial role in both problems.

In this paper we deal with two different geometries: the convex and the star-shaped cases. We address the questions of existence, multiplicity, uniqueness, and regularity for the discrete Bernoulli problem, together with convergence to the usual Bernoulli problem in the convex setting.

The main ideas in our paper follow the approach by sub- and supersolutions introduced by Beurling in [**14**]. For the classical Bernoulli problem in the convex case with p-Laplacian operator, questions of existence and uniqueness were addressed in [**31, 32, 33**]. It is noteworthy that the use of the properties of the distance function (1.1) avoids many of the technicalities: for example, higher regularity for solutions is an immediate consequence from existence theory, which is not the case for the classical Bernoulli problem.

The most standard approach for the Bernoulli problem is either variational or singular perturbation. In the case $p = 2$, the seminal paper by Alt and Caffarelli [**11**] shows existence of a solution using variational techniques without any geometric restriction on the set $\partial\{u > 0\}$. Moreover, it is shown that the free boundary is flat and, consequently, of class $\mathcal{C}^{1,\alpha}$ up to a hypersurface of $\mathcal{H}^{n-1}$-measure zero. For $N = 2$ they prove that the free boundary is globally analytic (see [**35**]). Recently in [**19**] it was shown that there exists no variational solution with singular free boundary for $N = 3$. The question remains open for $4 \leq N \leq 6$, since in [**25**] existence of minimal cones is shown for $N = 7$.

For $p \neq 2$, see [**22**] and [**23**]; in particular regularity is considered in [**23, 24**], where the authors extend the results of [**11**] ($\mathcal{C}^{1,\alpha}$-regularity near flat points). They prove that for $N = 2$ the free boundary is globally analytic for $2-\delta < p < +\infty$ with δ an absolute constant. For $N \neq 2$ not many results are available: Petrosyan in [**47**] shows that for any $N \leq k^* - 1$ and any $2 - \delta < p < 2 + \delta$ the variational solution has globally analytic free boundary, where k^* is the critical dimension above which any variational solution has free boundary with singular points.

If one adds a priori geometrical assumptions, the problem becomes more tractable. If the initial given domain is convex, for any $1 < p < +\infty$ there exists a unique classical solution, and the free boundary is also convex and $\mathcal{C}^{2,\alpha}$ for both interior and exterior case. This is done in [**31, 32, 33**] using Beurling's method on sub- and supersolutions.

Although the literature is not yet exhaustive, one would expect similar results if we relax the geometry of the domain to be star-shaped only (see, for instance [**9**]). The p-capacitary potential in a star-shaped annulus has star-shaped level sets, according to the rearrangement results of [**34**]; see Section 5 for more detailed literature on the star-shaped Bernoulli problem. However, two questions remain open:

(i) Existence of a solution using Beurling's method, which, by uniqueness, would coincide with the variational one.
(ii) The best regularity one expects for the free boundary.

The paper is structured as follows: after recalling in Section 2 some preliminary results that will be used throughout the manuscript, we will consider the exterior and interior problem in the convex setting; see Sections 3 and 4, respectively. Our problem will be then considered for star-shaped domains in Section 5. In Section 6 we further extend the problem to more general cases, such as nonconstant distance condition including a constraint involving mean curvature. In the last section we make some final remarks and show that the proposed model might give insights for new numerical algorithms that could more efficiently approximate solutions to the classical Bernoulli problem.

2. Preliminaries

For $1 < p < \infty$, we define the p-Laplacian operator Δ_p as

$$\Delta_p u := \mathrm{div}(|\nabla u|^{p-2}\nabla u), \quad 1 < p < \infty.$$

Let K, Ω be two domains in $\mathbb{R}^N$, $N \geq 2$, satisfying $\overline{K} \subset \Omega$, Ω bounded. In the sequel we say that u_Ω is the *p-capacitary* potential of the ring shaped region $\Omega \setminus \overline{K}$ if:

$$(\mathrm{P}_\Omega) \qquad \begin{cases} \Delta_p\, u_\Omega = 0 \ \text{ in } \ \Omega \setminus \overline{K}, \\ u_\Omega = 1 \quad \text{on } \overline{K}, \quad u_\Omega = 0 \quad \text{on } \partial\Omega. \end{cases}$$

Let $l \in [0, 1)$. From [**38**] one sees that u_Ω is real analytic and the level sets $\{u_\Omega > l\}$ are convex (with strictly positive principal curvatures) and analytic. Now we deal with boundary regularity:

PROPOSITION 2.1 ([**50, 40, 38**]). *Let Ω, K be two bounded convex domains such that $\overline{K} \subset \Omega$, and denote by u_Ω the p-capacitary potential of $\Omega\setminus\overline{K}$ as given in equation* (P_Ω). *Set $d_0 := \mathrm{dist}(\partial\Omega, K)$.*

i. Suppose that K satisfies the uniform interior cone condition for opening r_0. Then there is a constant $M = M(r_0, d_0)$ such that $\|u_\Omega\|_{\mathcal{C}^\alpha(\overline{\Omega\setminus K})} \leq M$.
ii. Suppose that K satisfies the uniform interior ball condition for radius r_0. Then there is a constant $M = M(r_0, d_0)$ such that $|\nabla u_\Omega| \leq M$ in $\Omega\setminus\overline{K}$.
iii. Suppose that $\Omega\setminus\overline{K}$ satisfies the uniform interior ball condition for radius r_0. There is a constant M_0 such that $|\nabla u_\Omega| \geq M_0$ in $\Omega\setminus\overline{K}$.

To define the gradient of the solution at the boundary of a convex ring, and thus give a notion of classical solutions for the usual Bernoulli problem we need:

PROPOSITION 2.2 (Theorem 1.3 in [**30**]). *Let K and Ω be two open convex domains, let $\overline{K} \subset \Omega$, and denote by u_Ω the p-capacitary potential of $\Omega\backslash\overline{K}$. Suppose also that ∂K and $\partial\Omega$ are $\mathcal{C}^1$. Then ∇u is semi-continuous in $\overline{\Omega\backslash K}$, and nontangentially continuous up to $\partial\Omega \cup \partial K$.*

3. The exterior problem in convex setting

The first aim of this paper is to solve the following problem: given a convex open bounded set $K \subset \mathbb{R}^N$, $N \geq 2$, and $l \in (0,1)$, $\lambda > 0$ positive constants, find a function u and a convex open bounded domain $\Omega \subset \mathbb{R}^N$, $\Omega \supset \overline{K}$, solving

$$(\mathrm{P}_E) \qquad \begin{cases} \Delta_p u = 0 \text{ in } \Omega \setminus \overline{K}, \quad u = 1 \text{ in } \overline{K}, \quad u = 0 \text{ on } \partial\Omega, \\ \operatorname{dist}(x, \Gamma_l) = \lambda \quad \text{for all } x \in \partial\Omega. \end{cases}$$

THEOREM 3.1. *Problem (P_E) has a unique regular solution (u, Ω) with Ω convex, $\partial\Omega \in \mathcal{C}^\infty$, and $u \in \mathcal{C}^{1,\alpha}(\overline{\Omega\backslash K})$ uniformly up to the boundary of Ω.*

The proof is inspired by previous works of one of the authors [**31, 32**] via Beurling's method (see also the classical reference [**14**]). The method is based on constructions of super- and subsolutions for (P_E); hence, existence of a solution for (P_E) follows by the proof of the existence of a *minimal* super-solution which turns out to be a subsolution (or solution) as well.

3.1. Existence. For each convex set Ω, let u_Ω be the p-capacitary potential of $\Omega\backslash\overline{K}$. Let us consider the class $\mathcal{C} := \{\Omega \text{ convex bounded open set in } \mathbb{R}^N, \ \Omega \supset \overline{K}\}$, and define the following subsets (subsolutions, strict subsolutions and supersolutions, respectively):

$$(3.1) \qquad \begin{aligned} \mathcal{A} \quad &= \{\Omega \in \mathcal{C} \mid \sup_{x\in\partial\Omega} \operatorname{dist}(x, \Gamma_l) \leq \lambda\}, \\ \mathcal{A}_0 \quad &= \{\Omega \in \mathcal{C} \mid \inf_{x\in\partial\Omega} \operatorname{dist}(x, \Gamma_l) < \lambda\}, \\ \mathcal{B} \quad &= \{\Omega \in \mathcal{C} \mid \inf_{x\in\partial\Omega} \operatorname{dist}(x, \Gamma_l) \geq \lambda\}. \end{aligned}$$

LEMMA 3.2. *There exist a supersolution and a strict subsolution for problem* (P_E).

PROOF. By translation suppose that the origin of the cartesian system lies inside K. In order to construct a supersolution, take $0 < r < R$ big enough such that $K \subset B_r(0)$. The solution of the capacitor problem for the annulus $B_R(0)\backslash\overline{B_r(0)}$ is explicitly given by the function

$$\overline{u}(|x|) = \frac{\log\left(\frac{R}{|x|}\right)}{\log\left(\frac{R}{r}\right)} \quad \text{for } p = N, \quad \overline{u}(|x|) = \frac{|x|^{\frac{p-N}{p-1}} - R^{\frac{p-N}{p-1}}}{r^{\frac{p-N}{p-1}} - R^{\frac{p-N}{p-1}}} \quad \text{for } p \neq N.$$

It is clear that we can have $\operatorname{dist}(\{\bar{u} = l\}, \{\bar{u} = 0\})$ as big as we wish by taking R big enough. Consider also u_R to be the p-capacitary potential of the set $B_R(0)\backslash\overline{K}$. By the comparison principle, $u_R \leq \bar{u}$ for all $x \in B_R(0)\backslash B_r(0)$. This immediately implies that $\operatorname{dist}(x, \{u_R = l\}) \geq \lambda$ for all $x \in \partial B_R(0)$, which shows that u_R is a supersolution for our problem, i.e, $B_R(0) \in \mathcal{B}$.

Now we seek a strict subsolution. Let $B_R(0)$ and u_R be as above. Since $K \subset B_R(0)$ and u_R is the p-capacitary potential in this annulus, by Proposition 2.1 there exists a small neighborhood U of ∂K and some $\alpha > 0$ such that $|\nabla u_R(x)| \geq \alpha > 0$ for all $x \in U \backslash K$. Choose the level set $\{u_R = 1 - \varepsilon\}$, and set $U_\varepsilon := \{x : u_R(x) > 1 - \varepsilon\} \cap U$. Define now

$$u_\varepsilon(x) := \frac{u_R(x) - (1-\varepsilon)}{\varepsilon}.$$

The function $u_\varepsilon(x)$ is a p-harmonic function in the domain U_ε, $u_\varepsilon(x) = 0$ in $\{u_R(x) - 1 - \varepsilon\}$ and $u_\varepsilon(x) = 1$ in K, so it is the p-capacitary potential of the set U_ε. We need to prove that $u_\varepsilon \in \mathcal{A}_0$, i.e., $\sup_{x \in \{u_\varepsilon = 0\}} \operatorname{dist}(x, \{u_\varepsilon = l\}) < \lambda$. This is easy to see since from the above lower bound for the gradient, for ε small, $\operatorname{dist}(x, \{u_\varepsilon = l\}) \leq \varepsilon l / \alpha < \lambda$, for $x \in \{u_\varepsilon = 0\}$. □

LEMMA 3.3. *Let (u_S, Ω_S) be any supersolution and (u_s, Ω_s) any subsolution for problem* (P_E). *Then $\Omega_s \subset \Omega_S$ and $u_s \leq u_S$.*

PROOF. We use Lavrent'ev's rescaling method. Assume by contradiction that there is a subsolution (u_s, Ω_s) which is not smaller than a supersolution (u_S, Ω_S). Without loss of generality, assume that $0 \in K$. Rescale the function u_s by a parameter $\varepsilon < 1$ small enough, i.e., $u_s^\varepsilon(x) = u_s(\frac{x}{\varepsilon})$, so that u_s^ε is a p-capacitary function in $\Omega_s^\varepsilon \backslash \overline{K_\varepsilon}$ for $\Omega_s^\varepsilon \subset \Omega_S$, $K^\varepsilon \subset K$. Moreover, for all $x \in \partial \Omega_s^\varepsilon$,

$$\operatorname{dist}(x, \{u_s^\varepsilon = l\}) \leq \lambda^\varepsilon < \lambda.$$

Consider the biggest $\varepsilon_0 < 1$ such that $\Omega_s^{\varepsilon_0} \subseteq \Omega_S$ and $\partial \Omega_s^{\varepsilon_0} \cap \partial \Omega_S \neq 0$. By the strong comparison principle $u_s^\varepsilon \leq u_S$ for each $x \in \Omega_s^{\varepsilon_0} \backslash K$. Let $x^0 \in \partial \Omega_s^{\varepsilon_0} \cap \partial \Omega_S$: it holds that $\lambda > \lambda^\varepsilon \geq \operatorname{dist}(x^0, \ \{u_s^\varepsilon = l\}) \geq \operatorname{dist}(x^0, \ \{u_S = l\}) \geq \lambda$, which is a contradiction. □

The next lemma shows convergence of the sets Ω_n.

LEMMA 3.4. *Let $\Omega_1 \supset \Omega_2 \supset \cdots$ be a decreasing sequence of domains in $\mathcal{B}$. Suppose that the set Ω, defined as as interior of the closure $\overline{\bigcap \Omega_k}$, is convex, bounded and $\Omega \supset \overline{K}$. Then $\Omega \in \mathcal{B}$.*

PROOF. Let u_k be the p-capacitary potential of $\Omega_k \backslash \overline{K}$. By construction, $0 \leq u_k \leq 1$, and it is a decreasing sequence. Then we can easily show that on every compact subset of $\Omega := \text{Interior}(\overline{\bigcap \Omega_k})$, u_k converges in $\mathcal{C}^{1,\alpha}$-norm to a p-harmonic function u. However, the lack of control of the gradients at the boundary does not allow us to show higher regularity up to the boundary, but only $\mathcal{C}^\alpha$.

We first prove that u is precisely the solution to the p-capacitary problem in $\Omega \backslash \overline{K}$, denoted by u_Ω. Clearly, $u \equiv 1$ in $\overline{K}$. We also need to check that $u \equiv 0$ outside Ω. For that, let $x \notin \Omega$. Since K is convex, $\Omega_k \backslash \overline{K}$ automatically satisfies the uniform exterior cone condition, and one may use Proposition 2.1 to estimate

$$0 \leq u_k(x) \leq \operatorname{dist}(x, \mathbb{R}^N \setminus \Omega_k)^\alpha \sup \|u_k\|_{\mathcal{C}^\alpha(\overline{\Omega} \backslash K)} \leq M \operatorname{dist}(x, \mathbb{R}^N \setminus \Omega_k)^\alpha \to 0,$$

as $k \to \infty$. This implies that $u(x) = 0$, as desired.

Since the sequence of functions is decreasing, the set $\{u_k = l\}$ converges to the set where $\{u = l\}$. However, the boundary of the set Ω_k may not converge to the set $\{u = 0\}$: that may happen if the gradient of u vanishes at some point of $\partial\Omega$ so that the limit function flattens out, and in this case the distance property would be violated. As a consequence, we need to check that $u(x) > 0$ for all $x \in \Omega$. By the minimum principle, u cannot vanish in the interior, but it could happen that it vanishes at the boundary of some set $\tilde{\Omega} \subset \Omega$. Hence assume by contradiction that $\tilde{\Omega} \neq \Omega$; consider a point $x^0 \in \partial\tilde{\Omega}$, $x^0 \notin \Omega$, and let $P \in \{u = l\}$ be the point where the distance between x^0 and $\{u = l\}$ achieves its minimum. It can happen that $x^0 = P$, but this is not a problem in the argument. Note that the ball $B_{\varepsilon/2}(x^0)$ is completely inside Ω_k for all k (otherwise, the distance property for u_k would be violated). Moreover, $u_k(x) \to 0$ as $k \to \infty$ for all $x \in B_{\varepsilon/2}(x^0) \setminus (B_{\varepsilon/2}(x^0) \cap \tilde{\Omega})$. Cover the segment $\overline{x^0P}$ by a finite number of balls of radius $\varepsilon/2$, and define $S = \bigcup_{i=0}^N B_{\varepsilon/2}(x^i)$. In the set S the Harnack inequality applied to the function u_k implies $l = u(P) \leq u_k(P) \leq \sup_S u_k(x) \leq C \inf_S u_k(x)$, where C depends only on S. But this is a contradiction since, as $k \to \infty$, it holds $l \leq C \inf_S u_k(x) \to 0$. We have shown that $\tilde{\Omega} = \Omega$, so that $u > 0$ on Ω. Then $u = u_\Omega$ is the p-capacitary potential for the set $\Omega\backslash\overline{K}$. Moreover the set Ω belongs to $\mathcal{B}$ since

$$\lambda \leq \operatorname{dist}(\partial\Omega_n, \{u_n = l\}) \to \operatorname{dist}(\partial\Omega, \{u = l\}) \quad \text{as } n \to +\infty. \tag{3.2}$$

Note here that the passage to the limit is justified because $\partial\Omega_n \to \partial\Omega$ and $\{u_n = l\} \to \{u = l\}$ in the Hausdorff distance sense. □

COROLLARY 3.5. *Assume that there exist two domains $\Omega^0 \in \mathcal{A}_0$ and $\Omega^1 \in \mathcal{B}$, with $\Omega^0 \subset \Omega^1$. Set $\mathcal{S} := \{\Omega \in \mathcal{B} : \Omega^0 \subset \Omega\}$. Then there exists a domain Ω in $\mathcal{S}$ which is minimal for the inclusion.*

PROOF. Let $\tilde{\Omega}$ be the intersection of all domains in the class $\mathcal{S}$, and set Ω to be the interior of the closure of $\tilde{\Omega}$, which is still convex. To prove $\Omega \in \mathcal{B}$, we select a sequence of domains $\{U_k\}_{k=1}^\infty$ in $\mathcal{S}$ such that $\bigcap U_k = \tilde{\Omega}$. Consider the sequence of domains defined by $\Omega_1 = U_1$ and $\Omega_{k+1} = \Omega_k \cap U_{k+1}$ for $k \geq 1$. Each Ω_k is convex and belongs to $\mathcal{B}$ (since this class is closed under intersections). Lemma 3.4 completes the proof of the corollary. □

In the following, we always denote by Ω the (nontrivial) minimal element in the class $\mathcal{S}$ defined in the previous proposition. To finalize the our proof, we need to check that this minimal Ω and its corresponding p-capacitor potential u_Ω satisfy the distance property in (P_E).

The proof for the standard Bernoulli problem uses crucially the convexity assumption. Our problem may be solved through a different and much simpler argument, which may be extended to more complex geometries, for instance, the star-shaped case. By contradiction, suppose that there is a point $x^0 \in \partial\Omega$ such that $\operatorname{dist}(\{u_\Omega = l\}, x^0) = \lambda + \varepsilon$, $\varepsilon > 0$. Set $\mathcal{L} := \{u_\Omega > l\}$. Note that $\mathcal{L}\backslash\overline{K}$ is a convex ring. Define $\tilde{\Omega}$ as the set

$$\tilde{\Omega} := \{x \in \Omega : \operatorname{dist}(x, \mathcal{L}) < \lambda\},$$

with $\tilde{u}$ the p-capacitary potential of $\tilde{\Omega}\backslash\overline{K}$. By the hypothesis, $\tilde{\Omega} \subsetneq \Omega$, and hence the comparison principle gives that $\tilde{u} \leq u$. One may estimate, for $x \in \partial\tilde{\Omega}$, that $\operatorname{dist}(x, \{\tilde{u} = l\}) \geq \operatorname{dist}(x, \{u = l\}) \geq \lambda$. Since $\partial\tilde{\Omega}$ lies at fixed distance λ from $\mathcal{L}$, and $\mathcal{L}$ is a convex bounded set, we can conclude that also $\tilde{\Omega}$ is convex. Hence $\tilde{\Omega}$ belongs to the set $\mathcal{B}$; but this contradicts the fact that Ω is the minimal set in $\mathcal{B}$.

LEMMA 3.6. *Assume K is star-shaped. If problem* (P_E) *admits a solution, then it is unique.*

PROOF. It is a straightforward consequence from Lemma 3.3. □

REMARK 3.7. *The uniqueness result together with the constructive construction of Ω as intersection of convex sets give that the solution Ω of problem* (P_E) *must be convex.*

We look now at regularity. It is clear that $u \in \mathcal{C}^\infty(\Omega \backslash \overline{K})$ and is even analytic in the interior of $\Omega \backslash K$, but what about the regularity of the free boundary? First note that $\{u = l\}$ is level set of an analytic function, hence it is analytic. Moreover, all the normal curvatures at the level set are positive (since level sets in a convex rings are also convex). The level set $\{u = 0\}$ is at a fixed distance λ from $\{u = l\}$ and its curvatures are of the form $\frac{\mu_i}{1+\lambda\mu_i}$, where μ_i are the curvatures of $\{u = l\}$; hence we can conclude [**27**] that $\{u = 0\}$ is also $\mathcal{C}^\infty$ and analytic.

The proof of Theorem 3.1 is thus completed.

3.2. Convergence to the Bernoulli problem. Finally, we show that the discrete Bernoulli problem converges, at the limit, to the usual Bernoulli problem:

$$(\mathrm{P}_B) \qquad \begin{cases} \Delta_p u = 0 \text{ in } \Omega \setminus \overline{K}, \quad u = 1 \text{ in } \overline{K}, \quad u = 0 \text{ on } \partial\Omega, \\ |\nabla u| = \omega, \quad \text{for all } \ x \in \partial\Omega. \end{cases}$$

The boundary condition $|\nabla u| = \omega$ is to be understood in the following sense:

$$(3.3) \qquad \liminf_{y \to x, y \in \Omega} |\nabla u(y)| = \limsup_{y \to x, y \in \Omega} |\nabla u(u)| = \omega, \quad \text{for every } x \in \partial\Omega.$$

It is shown in [**32**] that the solution for the Bernoulli problem (P_B) exists, it is unique if K is bounded and has $\mathcal{C}^{2,\alpha}$-boundary $\partial\Omega$.

THEOREM 3.8. *Let l_n and λ_n be two decreasing sequences of positive numbers such that $l_n \to 0$, $\lambda_n \to 0$ as $n \to +\infty$ and $l_n = \lambda_n \omega$. Let (u_n, Ω_n) be the solution of the exterior problem* (P_E) *with constants $l = l_n$ and $\lambda = \lambda_n$. As $n \to \infty$, $\{\Omega_n\}$ is a decreasing sequence of sets converging in Hausdorff distance to some convex set Ω and the function u_n converges in $\mathcal{C}^{1,\alpha}_{\text{loc}}(\Omega \backslash \overline{K}) \cap \mathcal{C}^\alpha(\overline{\Omega \backslash K})$ to the (unique) solution u of* (P_B).

PROOF. Define the rescaled distance function

$$d_n(x) = l_n[1 - \operatorname{dist}(x, \{u_n = l_n\})/\lambda_n].$$

Clearly, when $x \in \partial\Omega_n \cup \{u_n = l_n\}$, we have that $d_n(x) = u_n(x)$. Define also $S_n := \{0 \le u_n \le l_n\}$, a narrow ring. Since the level sets of u_n are convex surfaces, the function d_n is p-superharmonic in the set S_n, i.e., $-\Delta_p d_n \ge 0$. Since d_n and u_n have the same boundary values on ∂S_n, the comparison principle yields that $u_n \le d_n$ everywhere in S_n. In particular, for $x \in \partial\Omega_n$ and $l_m \le l_n$,

$$\operatorname{dist}(x, \{u_n = l_m\}) \ge \operatorname{dist}(x, \{d_n = l_m\}).$$

On the other hand for any $x \in \{0 \le d_n \le l_n\}$, it holds $|\nabla d_n(x)| = \omega$. Since $0 \le l_m \le l_n$,

$$\operatorname{dist}(x, \{d_n = l_m\}) \ge l_m / \sup |\nabla d_n| = l_m/\omega = \lambda_m \quad \text{for all} \quad m \ge n.$$

Thus, u_n is a supersolution for every (P_E) with constants (λ_m, l_m), $m \ge n$, i.e., $\Omega_n \in \mathcal{B}_m$ where

$$\mathcal{B}_m := \{\Omega \in \mathcal{C} \mid \inf_{x \in \partial\Omega} \operatorname{dist}(x, \{u_\Omega = l_m\}) \ge \lambda_m\},$$

for all $m \geq n$. Because of our construction of the solution for the exterior problem with constants (λ_m, l_m) as the minimal set in $\mathcal{B}_m$, this immediately gives that Ω_n contains all the Ω_m for $m \geq n$. Consequently, we have shown that the sequence of $\{\Omega_n\}_{n=0}^{\infty}$ is decreasing and from the maximum principle it follows that $\{u_n\}$ is a decreasing sequence.

Since $u_n \leq d_n$ on S_n, using Hopf's boundary lemma, we arrive at $|\nabla u_n| \leq |\nabla d_n| = \omega$ on $\partial\Omega_n$. This tells us that u_n is a supersolution for the Bernoulli problem with constant ω.

As a consequence, the sequence of sets Ω_n converges (for example in the Hausdorff measure) as $n \to \infty$ toward a convex domain $\Omega \supset \overline{K}$. Moreover, u_n converges in $\mathcal{C}^{1,\alpha}_{\text{loc}}$-norm to a p-harmonic function u: by [**32**, Theorem 3.2] we know that u solves the p-capacitary problem in $\Omega \setminus K$ and is a supersolution, i.e., $|\nabla u| \leq \omega$ for all x in $\partial\Omega$.

The last step consists of showing that (u, Ω) solves the Bernoulli problem (P_B) with constant ω in the sense indicated in (3.3). Again, using comparison to the distance function, we obtain that $|\nabla u_n| \geq \omega$ on the level set $\{u_n = l_n\}$. But [**32**, Lemma 2.2] gives that the inequality $|\nabla u_n| \geq \omega$ is also true for every point in $\{l_n \leq u < 1\}$. Since we have uniform convergence of the gradients away from the boundary and the sequence of sets $\Omega_n \to \Omega$ is decreasing, we are able to pass to the limit and thus $|\nabla u(y)| \geq \omega$ for every point $y \in \Omega \setminus \overline{K}$. The proof of the theorem is completed. □

4. The interior problem in convex setting

We consider now the following problem: given a convex open bounded set $\Omega \subset \mathbb{R}^N$ for $N \geq 2$ and $l \in (0,1)$, $\lambda > 0$, find a function u and a convex open bounded domain $K \subset \Omega$ such that

$$(\mathrm{P}_I) \qquad \begin{cases} \Delta_p u = 0 \text{ in } \Omega \setminus \overline{K}, \quad u = 0 \text{ in } \overline{K}, \quad u = 1 \text{ in } \partial\Omega, \\ \operatorname{dist}(x, \Gamma_l) = \lambda, \quad \text{for all } x \in \partial K. \end{cases}$$

Our first main result for the interior case is the following.

THEOREM 4.1. *There exists a constant $\lambda_{\Omega,\max}$ that depends only on l and Ω such that for any $\lambda \leq \lambda_{\Omega,\max}$ problem (P_I) has a smooth solution (u, K) with K a convex set and $\partial K \in \mathcal{C}^{0,1}$.*

For each convex set K, one can define its associated p-capacitary potential u_K as the solution of the following problem:

$$(\mathrm{P}_K) \qquad \begin{cases} \Delta_p u_K = 0 \text{ in } \Omega \setminus \overline{K}, \\ u_K = 0 \text{ in } \overline{K}, \quad u_K = 1 \text{ on } \partial\Omega. \end{cases}$$

Note that, without risk of confusion, we have switched the notation from problem (P_Ω) in the exterior case.

The proof of Theorem 4.1 goes, as in the exterior case, by considering the maximal element in the class:

$$\mathcal{B}_\lambda = \{K \subset \Omega \text{ convex bounded open} \mid \inf_{x \in \partial K} \operatorname{dist}(x, \{u_K = l\}) \geq \lambda\}. \tag{4.1}$$

In the following we show that for fixed $l \in (0,1)$, the set $\mathcal{B}$ is nonempty if the value of λ is smaller than a certain critical value, which will be denoted by $\lambda_{\Omega,\max}$. We also show that problem (P_I) does not necessarily have a unique solution for certain values of λ.

4.1. A minimal supersolution and the Bernoulli constant. Consider the simple case that Ω is a ball: this particular example shows the existence of a constant $\lambda_{\Omega,\max}$ such that no solution exists for $\lambda > \lambda_{\Omega,\max}$, one or two solutions can occur for $\lambda < \lambda_{\Omega,\max}$, and only one solution exists if $\lambda = \lambda_{\Omega,\max}$. Using Beurling's terminology, we call the unique solution that corresponds to the value $\lambda = \lambda_{\Omega,\max}$ of *parabolic* type, the smallest solution for $\lambda < \lambda_{\Omega,\max}$ of *hyperbolic* type and the biggest solution for $\lambda < \lambda_{\Omega,\max}$ of *elliptic* type. The elliptic solutions form a decreasing family, while the hyperbolic solutions form a increasing family of solutions.

LEMMA 4.2. *Let $\Omega = B_R(0) \subset \mathbb{R}^N$ be a ball of radius R. Fix $l \in (0,1)$ and $p > 1$, we have:*

i. If $p \le N$, then problem (P_I) has a unique solution for $\lambda = \lambda_{\max}$, two solutions if $0 < \lambda < \lambda_{\max}$ and no solutions if $\lambda > \lambda_{\max}$, where

$$\lambda_{\max} = R(1-l)^{1/l}\left(1/l - 1\right)^{-1}.$$

ii. If $p > N$, then (P_I) has a unique solution for $\lambda \in (0, \lambda_{\min}) \cup \{\lambda_{\max}\}$, two solutions if $\lambda \in (\lambda_{\min}, \lambda_{\max})$ and no solutions if $\lambda > \lambda_{\max}$, for

$$\lambda_{\min} = Rl^{\frac{p-1}{p-N}}, \quad \lambda_{\max} = R\left(\frac{l}{1-(1-l)^{\frac{p-1}{N-1}}}\right)^{\frac{p-1}{p-N}} - R\left(\frac{l}{(1-l)^{\frac{N-p}{N-1}} - (1-l)}\right)^{\frac{p-1}{p-N}}.$$

PROOF. It has been shown in [**48**] that all the solutions of the p-capacitary problem with $\Omega = B_R(0)$ are balls centered in the origin. Then one uses the precise formulas for the p-capacitary potential in an annulus to conclude. □

Now we go back to the problem (P_I) for a general convex bounded set Ω in $\mathbb{R}^N$ and a fixed $l \in (0,1)$ (although we may not write it, the dependence in l will be always implicit). Consider the set of supersolutions as defined in (4.1). Since $\mathcal{B}_{\lambda_1} \subset \mathcal{B}_{\lambda_2}$ for $\lambda_1 > \lambda_2$, we can define the Bernoulli constant for our problem as

$$\lambda_{\Omega,\max} := \sup\{\lambda > 0 \mid \mathcal{B}_\lambda \text{ is not empty}\}. \tag{4.2}$$

LEMMA 4.3. *Let $l \in (0,1)$, and let Ω be any open bounded convex set. Then*

$$\lambda_{\Omega,\max} \ge \lambda_{\max},$$

where R is the radius of the largest ball inscribed in Ω.

PROOF. According to Theorem 4.1 and the definition in (4.2), problem (P_I) is solvable for given constants $\lambda > 0$, $l \in (0,1)$, provided that $\mathcal{B}_\lambda \neq \emptyset$. Following the discussion above, for any $\lambda \le \Lambda(r_{\max})$, there exists a unique constant $r_{\lambda,2}$, with $r_{\max} \le r_{\lambda,2} < R$ such that the solution of the p-capacitary problem $u_{r_{\lambda,2}}$ in $B_R(0) \setminus \overline{B_{r_{\lambda,2}}(0)}$ satisfies the distance property,

$$\operatorname{dist}(x, \{u_{r_{\lambda,2}} = l\}) = \lambda \quad \text{for all } x \in \partial B_{r_{\lambda,2}}(0).$$

Note that if $\lambda = D(r_{\max})$, then $r_{\max} = r_{\lambda,2}$.

For any $\lambda \le D(r_{\max})$ consider now the solution $\bar{u}$ of the p-capacitary problem in $\Omega \setminus \overline{B_{r_{\lambda,2}}(0)}$. By the comparison principle applied to the functions $u_{r_{\lambda,2}}$ and $\bar{u}$, the (minimum) distance between the sets $\{\bar{u} = l\}$ and $\partial B_{r_{\lambda,2}}(0)$ is $\ge \lambda$. This shows that $\bar{u}$ is a supersolution for (P_I). □

COROLLARY 4.4. *Fix $l \in (0,1)$, and let Ω be a convex bounded domain in $\mathbb{R}^N$. Then, for each $\lambda \leq \lambda_{\Omega,\max}$, there exists a supersolution (u, K) such that K belongs to the class $\mathcal{B}_\lambda$.*

The proof of Theorem 4.1 follows now as in the exterior case by constructing an extremal supersolution in $\mathcal{B}_\lambda$, so we will not write the details here.

REMARK 4.5. *In contrast to the exterior problem, in the interior discrete Bernoulli problem the free boundary may have corners; consequently, ∂K is only $\mathcal{C}^{0,1}$ and the solution is $\mathcal{C}^{0,\alpha}$ up to the boundary.*

4.2. Convergence to the Bernoulli problem. Before we state the main theorem of the section, let us recall what is known for the interior Bernoulli problem:

THEOREM 4.6 ([**31**]). *Given a convex domain Ω, there exists a constant ω_0 depending on Ω and p such that for all $\omega \geq \omega_0$ the Bernoulli-type problem*

$$(\mathrm{P}_{B_I}) \qquad \begin{cases} \Delta_p v = 0 \text{ in } \Omega \setminus \overline{D}, \quad v = 0 \text{ in } \overline{D}, \quad v = 1 \text{ in } \partial\Omega, \\ |\nabla v| = \omega \quad \text{for all } x \in \partial D, \end{cases}$$

has a solution (v, D) (not necessarily unique). Moreover D is convex. The boundary condition $|\nabla v| = \omega$ is again understood in the sense of (3.3).

REMARK 4.7. *We remind the reader that the solution (v, D) given in the previous theorem is the one constructed as the extremal supersolution for problem (P_{B_I}), although there may be others. The set of supersolutions for each fixed ω is defined in [**31**] as the class of Lipschitz functions in Ω such that $v = 1$ on $\partial\Omega$, $\Delta_p v \leq 0$ in $\{v > 0\} \cap \Omega$ and $|\nabla v| \leq \omega$ on $\partial\{v > 0\} \cap \Omega$.*

Now, the main theorem of this section is:

THEOREM 4.8. *Let Ω be a bounded convex domain in $\mathbb{R}^N$. Let $\omega > \omega_0$, where ω_0 is the Bernoulli constant for the domain Ω given in Theorem 4.6. Fix two decreasing sequences $\{l_n\}$ and $\{\lambda_n\}$ of positive numbers such that $l_n \to 0$, $\lambda_n \to 0$ as $n \to +\infty$ and $l_n = \lambda_n \omega$. Let also $\lambda^n_{\Omega,\max}$ be the corresponding quantity (4.2) to each l_n. Then:*

- *There exists a solution (v, D) for the interior Bernoulli problem (P_{B_I}) with constant ω.*
- *There exists a sequence of solutions (u_n, K_n) of problem (P_I) for each $l = l_n$, $\lambda = \lambda_n$.*
- *The sequence of convex sets $\{K_n\}$ is decreasing and as $n \to \infty$ converges in Hausdorff distance to the convex set D.*
- *The functions u_n converge in $\mathcal{C}^{1,\alpha}_{\mathrm{loc}}(\Omega \backslash \overline{D}) \cap \mathcal{C}^\alpha(\overline{\Omega \backslash D})$ to v.*

The existence of a solution for the Bernoulli problem with constant $\omega > \omega_0$ is the statement of Theorem 4.6. Then the proof of the theorem follows from the following claims:

Claim 1. For each n, there exists a solution (u_n, K_n) to the interior problem (P_I) with constants $\lambda = \lambda_n$ and $l = l_n$. The proof of this claim uses the following lemma on the relation between the solvability of the distance problem and the Bernoulli one:

LEMMA 4.9. *Given a convex bounded domain* Ω *and* $l \in (0,1)$, *the Bernoulli constant defined in* (4.2) *for the distance problem satisfies*

$$\lambda_{\Omega,\max} \geq l/\omega_0, \tag{4.3}$$

where ω_0 *is the Bernoulli constant from Theorem* 4.6.

PROOF. As in the proof of Lemma 4.3, it is enough to show that any solution (v, D) for the Bernoulli problem (P_B) with constant ω_0 is also a supersolution for (P_I) for $\lambda \leq \frac{l}{\omega_0}$. Indeed, the Bernoulli condition for v tells us that $|\nabla v| = \omega_0$ on ∂D, which implies that $|\nabla v| \leq \omega_0$ in $\Omega \backslash D$ because of [**32**, Lemma 2.2]. Then, if $x \in \partial D$,

$$\operatorname{dist}(x, \{v = l\}) \geq l/\sup|\nabla v| = l/\omega_0 \geq \lambda,$$

which implies $D \in \mathcal{B}_\lambda$. □

Using the lemma for the domain Ω and the level l_n we may estimate $\lambda^n_{\Omega,\max} \geq l_n/\omega_0 \geq l_n/\omega = \lambda_n$, which tells us that the distance problem (P_I) with constants (λ_n, l_n) may be solved (Theorem 4.1). This proves the following Claim 2.

Claim 2. $\{K_n\}$ is a decreasing sequence of sets. In particular, $u = \sup_n u_n$ and $K = \bigcap_n K_n$, where (u_n, K_n) is the solution to the distance problem (P_I) with constants $\lambda = \lambda_n$, $l = l_n$.

Fix $m \leq n$, so that $l_n \leq l_m$ and $\lambda_n \leq \lambda_m$. Let us check first that u_n belongs to the set of supersolutions $\mathcal{B}_m$, defined as

$$\mathcal{B}_m := \{K \subset \Omega \mid \inf_{x \in \partial K} \operatorname{dist}(x, \{u_K = l_m\}) \geq \lambda_m\}.$$

For that, consider the distance function d_n between two level sets of u_n, rescaled by a factor ω, so that d_n and u_n have the same boundary values in the set $S_n := \{0 \leq u_n \leq l_n\}$. More precisely,

$$d_n(x) = l_n[1 - \operatorname{dist}(x, \{u_n = l_n\})/\lambda_n]. \tag{4.4}$$

Since the level set $\{u_n = l_n\}$ is smooth, the distance function is smooth (a reference for the regularity of distance functions may be found in [**27**]). Note that in order to have d_n well defined outside S_n, we need to take the distance function to the level set $\operatorname{dist}(x, \{l_n = u_n\})$ with a change of sign once we pass the level set.

Next, since the level sets of u_n are convex surfaces, the distance function defined in the set S_n is p-subharmonic, i.e., $\Delta_p d_n \geq 0$. Moreover, since d_n and u_n share the same boundary values on both boundaries of the set S_n, we can use Hopf's lemma on the smooth curve $\{u_n = l_n\}$, thus obtaining that

$$\omega = |\nabla d_n| \geq |\nabla u_n|, \quad \forall\, x \in \{u_n = l_n\}.$$

Define now the set $A_n := \{l_n \leq u_n \leq 1\}$. From Hopf's lemma we can deduce that the function $f_n := d_n - u_n$ is strictly positive in a annulus M_n (a neighborhood of $\{u_n = l_n\}$) inside A_n. Suppose, by contradiction, that M_n is strictly contained in the neighborhood A_n. Consider f_n in the domain $\tilde{M}_n := S_n \cup M_n$: it holds that $f_n = 0$ on ∂K_n, $f_n > 0$ in $\{0 < u_n < l_n\}$, $f_n = 0$ on $\{u_n = l_n\}$, $f_n > 0$ in M_n and $f_n = 0$ on ∂M_n. But this is a contradiction since f_n is superharmonic and should reach its maximum value at the boundary of $\tilde{M}_n$. Hence $M_n = A_n$. Then in the set A_n one has that $u_n \leq d_n$. As a consequence, for $m \leq n$, i.e., $l_m \geq l_n$, and $x \in \partial K_n = \{u_n = 0\}$,

$$\operatorname{dist}(x, \{u_n = l_m\}) \geq \operatorname{dist}(x, \{d_n = l_m\}) = \lambda_m,$$

and $u_n \in \mathcal{B}_m$ as claimed. In particular, since u_m is constructed as the extremal supersolution in the set $\mathcal{B}_m$, we have that $u_m \leq u_n$, which in particular implies that $K_n \subset K_m$. This is, $K_n \to K$ is a decreasing sequence of sets, and the claim is proved.

Claim 3. Let $K := \bigcap_n K_n$. Then $K \supseteq D$ and is convex. In particular, it is not degenerate.

Following the steps of Lemma 4.9, one can show that the solution (v, D) of problem (P_{B_I}) is a supersolution to the (λ_n, l_n) distance problem. Hence $K_n \supseteq D$ for any $n \in \mathbb{N}$.

Claim 4. There exists a function u such that $u_n \to u$ in the space $\mathcal{C}^{1,\alpha}_{\text{loc}}(\Omega \backslash \overline{K}) \cap \mathcal{C}^{\alpha}(\Omega \backslash K)$.

Claim 5. In addition, $u = 0$ on ∂K and $\partial K \in \mathcal{C}^{0,\alpha}$, so that u is the p-capacitary potential in the annulus $\Omega \backslash K$.

Claims 4 and 5 follow similarly as the exterior case as a consequence of uniform $\mathcal{C}^{\alpha}$ estimates up to the boundary for a convex annulus $\Omega \backslash K$, so we will not make further comment.

Now we remark that the the regularity of the boundary ∂K is slightly worse than in the exterior case (see Remark 4.5); indeed, it is only $\mathcal{C}^{0,1}$, and we may not have gradient bounds up to the boundary. However, from the proof of Claim 3, in the set $A_n := \{l_n \leq u_n < 1\}$,

$$|\nabla u_n| \leq \omega. \tag{4.5}$$

Now we may pass to the limit $n \to \infty$. It is easy to see that $|\nabla u_n|$ converges uniformly to $|\nabla u|$ on compact sets inside $\Omega \backslash \overline{K}$. As a consequence of (4.5) we get the following claim.

Claim 6. $|\nabla u| \leq \omega$ on ∂K. From [**32**, Lemma 2.2] we also get that $|\nabla u| \leq \omega$ in $\overline{\Omega \backslash K}$.

To complete the proof of Theorem 4.8 it suffices to show that $D = K$ and $u = v$. First, in Claim 3 we have shown that $D \subset K$. On the other hand, in Claim 6 we have shown that $|\nabla u| \leq \omega$ on ∂K. This in particular implies that (u, K) is a supersolution for the Bernoulli problem (P_{B_I}) in the sense of Remark 4.7. As a consequence, one gets that $K \subset D$ (see [**31**]).

5. The star-shaped case

A domain $U \subset \mathbb{R}^N$ is called star-shaped with respect to the point $x_0 \in U$ if for all $x \in U$ the segment joining x and x_0 is contained in U. Moreover, the set U is called star-shaped with respect to the ball $B \subset U$ if U is star-shaped with respect to every point $x_0 \in B$. These domains are also known as star-shaped Lipschitz ([**17**]), and they clearly satisfy both the interior and exterior cone conditions. One may also consider star-shaped Lipschitz rings: they are of the form $U = \Omega \backslash \overline{K}$, where Ω and K are star-shaped Lipschitz with respect to the same center.

Given a bounded, star-shaped Lipschitz domain $K \subset \mathbb{R}^N$, we consider the exterior distance problem

$$(\mathrm{P}_*) \qquad \begin{cases} \Delta_p u = 0 \text{ in } \Omega \setminus \overline{K}, & u = 1 \text{ in } \overline{K}, \quad u = 0 \text{ on } \partial\Omega, \\ \operatorname{dist}(x, \Gamma_l) = \lambda & \text{for all } x \in \partial\Omega, \end{cases}$$

for some function u and a star-shaped domain $\Omega \subset \mathbb{R}^N$, with the same center as K. Our main result in this section is stated in the following theorem:

THEOREM 5.1. *Problem* (P_*) *has a unique solution* (u, Ω) *with* $\partial\Omega \in \mathcal{C}^{0,\alpha}$ *and* $\mathcal{C}^{1,1}$ *from inside.*

Before we give the proof of the theorem, let us briefly recall what is known for the (exterior) classical Bernoulli problem (P_B) when K star-shaped with respect to a ball. For the case $p = 2$, Kawohl proved in [**34**] existence and uniqueness of a (variational) solution via rearrangement method. In addition he proved that such a solution has star-shaped Lipschitz level sets. Also Acker and Meyer [**9**] showed that, still for $p = 2$, the solution is regular and that $\partial\Omega \in \mathcal{C}^\infty$ is also starlike with respect to the same ball.

The regularity results from Section 2 still hold for star-shaped rings (see [**39**], for instance). Let $l \in [0, 1)$. Given a (Lipschitz) star-shaped ring $\Omega \backslash \overline{K}$ and its p-capacitary function u_Ω described as above, the level sets $\{u_\Omega > l\}$ are (Lipschitz) star-shaped ([**34**]).

Since, by hypothesis, the initial domain K is star-shaped Lipschitz with respect to some point, then there exists some small ball B_δ centered at that point such that K is star-shaped with respect to that ball. Consider the class

$$\mathcal{C}_* := \{\Omega \text{ open, bounded, star-shaped with respect to } B_\delta,\ \Omega \supset \overline{K}\}.$$

We define the sets $\mathcal{A}$, $\mathcal{A}_0$, $\mathcal{B}$ as subclasses of $\mathcal{C}_*$ analogous to (3.1).

The construction of a minimal supersolution follows the same lines as in the convex regime, but some additional ingredients are needed to compensate for the lack of convexity. One only needs to check that this minimal supersolution u is indeed a solution for the distance problem (P_*), i.e.

$$(5.1) \qquad \operatorname{dist}(x, \Gamma_l) = \lambda \quad \text{for all } x \in \partial\Omega.$$

This follows from a simple argument: Let Ω be the extremal set as constructed above, and set $\mathcal{L} := \{u > l\}$. Note that $\mathcal{L}$ is a star-shaped annulus. By contradiction, assume that there exists $x_0 \in \partial\Omega$ such that $\operatorname{dist}(x_0, \partial\mathcal{L}) > \lambda$. Define $\tilde{\Omega}$ as $\tilde{\Omega} := \{x \in \Omega : \operatorname{dist}(x, \mathcal{L}) < \lambda\}$, and let $\tilde{u}$ be the p-capacitary potential of $\tilde{\Omega}\backslash\overline{K}$. By hypothesis, $\tilde{\Omega} \subsetneq \Omega$. The comparison principle implies that $\tilde{u} \leq u$, and consequently

$$\operatorname{dist}(x, \{\tilde{u} = l\}) \geq \operatorname{dist}(x, \{u = l\}) \geq \lambda, \quad \text{for } x \in \partial\tilde{\Omega}.$$

Since $\partial\tilde{\Omega}$ has a fixed distance λ from Γ_l and Γ_l is a star-shaped surface, we conjecture that $\tilde{\Omega}$ is also star-shaped with respect to the ball B_δ. We proceed by contradiction: Suppose that $\tilde{\Omega}$ is not star-shaped with respect to a point in B_δ. Without loss of generality we can assume this point to be the origin $x = 0$. Note that both sets $\mathcal{L} := \{l \leq u \leq 1\}$ and $\tilde{\Omega}$ are smooth. Then there exists a point $z \in \partial\tilde{\Omega}$ such that the tangent plane Π to $\partial\tilde{\Omega}$ at z passes through the origin.

By construction the point z lies at a distance λ from Γ_l; in particular there exists $x_0 \in \Gamma_l$ where this distance is attained. Since all the sets here are smooth, we know that the line joining x_0 to z is normal to the plane Π. Then one may

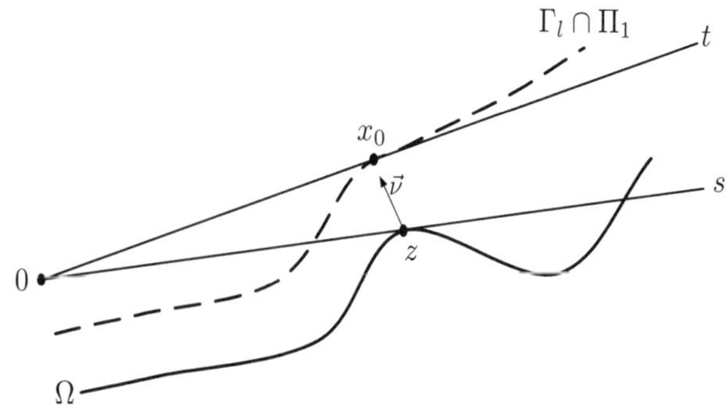

FIGURE 1. Star-shaped domain

just work with the intersection of $\tilde{\Omega}$ with the plane Π_1, where Π_1 is the plane that contains the origin, the point z and the normal vector $\vec{\nu}$ (see Figure 1).

Consider now the line s passing through the points 0 and z and the line t passing through 0 and x_0. Since the set $\mathcal{L}$ is star-shaped with respect to 0, the curve $\Gamma_l \cap \Pi_1$ has to cross the line t exactly at x_0, and after crossing, it lies on the half-plane bounded by t that does not contain the point z. Hence there exist some points in $\tilde{\Omega} \cap \Pi_1$ which are on the half-plane bounded by s (which does not contain x_0) that lie at a distance strictly greater than λ from Γ_l. But this contradicts the definition of $\tilde{\Omega}$.

Hence $\tilde{\Omega}$ is star-shaped and belongs to $\mathcal{B}$; by the minimality of Ω if follows that $\tilde{\Omega} = \Omega$ and (5.1) is proven.

Finally, uniqueness was already shown in Lemma 3.6. For the regularity, just note that the distance property implies that $\partial\Omega$ is $\mathcal{C}^{1,1}$ from inside (i.e., for every point $x_0 \in \partial\Omega$, there exists some ball contained in Ω that touches $\partial\Omega$ exactly at x_0), but further regularity is unknown. The proof of Theorem 5.1 is thus completed.

6. The distance problem with mean curvature condition

The next question one may ask is what happens when we replace the constant λ in the distance property by a nonconstant function $\lambda(x)$. For the Bernoulli problem, this problem was considered in [**30**]. In particular, it is solvable if $\lambda(x)$ is continuous and locally convex. Under the same conditions, the nonconstant distance problem may be solved with appropriate modifications.

We shall now consider the new case where $\lambda = \lambda(x, \kappa(x))$, with $\kappa(x)$ being the mean curvature of $\partial\Omega$ at the point x. The standard Bernoulli problem for harmonic functions (case $p = 2$) with a nonconstant gradient condition depending on the mean curvature has been studied in [**12, 45, 46**] through a variational formulation for the functional

$$\int_D |\nabla u|^2\, dx + \mathrm{Per}(\{v > 0\}).$$

DEFINITION 6.1. *For U a bounded convex domain, assume that $0 \in \partial U$ and that the interior normal ν to ∂U at the origin is in the direction of the e-axis. Let $S_A = \{(x, e) : e = \langle Ax, x\rangle\}$ for any symmetric matrix A.*

i. *The* (interior) *mean curvature of* ∂U *in the viscosity sense is*

$$\kappa(\partial U)(0) := \inf_{A \in \mathfrak{U}} \kappa(S_A)(0),$$

where $\mathfrak{U}$ *is the set of all symmetric matrices* A *such that the set* S_A *locally touches* ∂U *at* 0 *from inside. If no paraboloid touches* ∂U *at* 0 *from inside, we define* $\kappa(\partial U)(0) = \infty$.

ii. *The* (exterior) *mean curvature* $\underline{\kappa}(\partial U)(0)$ *is*

$$\underline{\kappa}(\partial U)(0) := \sup_{A \in \mathfrak{V}} \kappa(S_A)(0),$$

where $\mathfrak{V}$ *is the set of all symmetric matrices* A *such that the set* S_A *(the graph of a quadratic polynomial) locally touches* ∂U *at* 0 *from outside. If the graph of any quadratic polynomial locally touches* ∂U *at* $x = 0$ *from outside, we say that* $\underline{\kappa}(\partial U)(0) = \infty$.

Clearly, for any point $x_0 \in \partial U$ where $\partial U \in \mathcal{C}^2$ in a neighborhood of x_0 it holds that $\underline{\kappa}(\partial U)(x_0) = \kappa(\partial U)(x_0)$.

In this section we study the exterior discrete Bernoulli problem (P'_E) with the distance condition depending on the mean curvature. More precisely, given a convex bounded domain K with $\partial K \in \mathcal{C}^{1,1}$, we prescribe the condition

$$\operatorname{dist}(x, \Gamma_l) = \frac{\tilde{l}}{\kappa(\partial\Omega)(x)} =: \lambda_{\kappa,\tilde{l}}(x), \quad x \in \partial\Omega, \tag{6.1}$$

with $\tilde{l} \in (0,1)$. We do not include $\tilde{l} = 1$ since it is not an admissible value for the case when K is a ball, as explained in Lemma 6.4.

THEOREM 6.2. *Let* $K \subset \mathbb{R}^N$, $N \geq 2$, *be a convex open bounded set and* $l, \tilde{l} \in (0,1)$. *There exist a function* $u \in \mathcal{C}^{1,\alpha}(\Omega)$ *and a convex open bounded domain* $\Omega \subset \mathbb{R}^N$, $\Omega \supset \overline{K}$, *with* $\partial\Omega \in \mathcal{C}^{0,1}$ *which solve*

$$(\mathrm{P}'_K) \qquad \begin{cases} \Delta_p u = 0 \text{ in } \Omega \setminus \overline{K}, \quad u = 1 \text{ in } \overline{K}, \quad u = 0 \text{ on } \partial\Omega, \\ \operatorname{dist}(x, \Gamma_l) \geq \lambda_{\kappa,\tilde{l}}(x) \quad \text{for all } x \in \partial\Omega, \\ \operatorname{dist}(x, \Gamma_l) = \lambda_{\kappa,\tilde{l}}(x) \quad \text{for all } x \in \partial\Omega_\kappa, \end{cases}$$

where $\lambda_{\kappa,\tilde{l}}(x)$ *is defined in* (6.1) *and* $\partial\Omega_\kappa$ *is the set of all points* $x \in \partial\Omega$ *for which at least one of the following assumptions is satisfied:*

(*A1*) x *is an extremal point of* Ω *and* $\partial\Omega$ *is* $\mathcal{C}^2$*-smooth at* x,

(*A2*) x *is not an extremal point, but it can be written as a linear combination of extremal points satisfying* (*A1*).

Moreover the set $\partial\Omega \in \mathcal{C}^{0,1}$ *does not contain points* x *of the form*

(*A3*) x *is an extremal point of* Ω *and* $\kappa(\partial\Omega)(x) = \underline{\kappa}(\partial\Omega)(x) = \infty$,

(*A4*) x *is not an extremal point, but it can be written as a linear combination of extremal points satisfying* (*A3*).

LEMMA 6.3. *Let* Ω_n *be a decreasing sequence of convex bounded domains converging in Hausdorff measure to* Ω. *Let* x_0 *be a point on* $\partial\Omega$. *It holds*

$$\liminf_{\{x_n\}} \kappa(\partial\Omega_n)(x_n) \leq \kappa(\partial\Omega)(x_0), \tag{6.2}$$

where the lim inf *is taken for all sequences* $\{x_n\}$ *satisfying* $x_n \in \partial\Omega_n$ *and* $x_n \to x_0$.

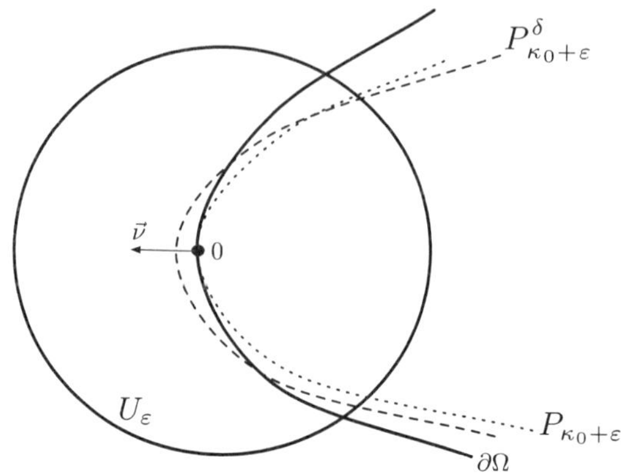

FIGURE 2

PROOF. First note that if $\kappa(\partial\Omega)(x_0) = +\infty$, inequality (6.2) holds. Let x_0 be a regular point with curvature $0 \leq k_0 := \kappa(\partial\Omega)(x_0) < +\infty$. Without loss of generality assume that $x_0 = 0$. For any $\varepsilon > 0$, there exists a paraboloid $P_{k_0+\varepsilon}$ with curvature $\kappa_0 + \varepsilon$ at the point $x = 0$ that touches $\partial\Omega$ from inside at 0 in a neighborhood $U_\varepsilon \ni 0$ (see Figure 2).

Shift now the paraboloid $P_{k_0+\varepsilon}$ by a small positive distance δ in the direction of the outward normal $\vec{\nu}$ to $\partial\Omega$ at 0; denote this shifted ball by $P^{\delta}_{k_0+\varepsilon}$. Clearly, if δ is small enough, $P^{\delta}_{k_0+\varepsilon} \cap \partial\Omega \subset U_\varepsilon$. Since Ω_n are convex sets and $\{\partial\Omega_n\}$ converges to $\partial\Omega$ as $n \nearrow \infty$, there exists at least a set Ω_{n_δ}, with n_δ dependent on δ, such that $\Omega \cup (P^{\delta}_{k_0+\varepsilon} \cap U_\varepsilon) \subseteq \Omega_{n_\delta}$ and $\partial\Omega_{n_\delta} \cap \partial P^{\delta}_{k_0+\varepsilon} = \{x_{n_\delta}\}$; moreover, $P^{\delta}_{k_0+\varepsilon}$ touches tangentially $\partial\Omega_{n_\delta}$ at the point x_{n_δ}. Inequality (6.2) follows since ε and δ can be taken arbitrarily small. □

One may define the set of supersolutions $\mathcal{B}$ by the condition

$$\text{dist}(x, \Gamma_l) \geq \lambda_{\kappa,\tilde{l}}(x). \tag{6.3}$$

Note that if $x_0 \in \partial\Omega$ is a point where $\kappa(\partial\Omega)(x) = +\infty$, then the distance property (6.3) does not impose any restriction. One may define analogously $\mathcal{A}$ (and $\mathcal{A}_0$) to be the set of subsolutions (strict subsolutions) through condition

$$\text{dist}(x, \Gamma_l) \leq (<)\ \lambda_{\kappa,\tilde{l}}(x). \tag{6.4}$$

LEMMA 6.4. *The sets $\mathcal{B}$ and $\mathcal{A}$ are nonempty. Moreover, any supersolution is greater or equal than every admissible subsolution.*

PROOF. A possible supersolution for our problem is the function u_R constructed in Lemma 3.2, for R big enough such that $\left(1 - \left(\frac{r}{R}\right)^l\right) \geq \tilde{l}$. For the subsolution consider the function u_ε constructed in Lemma 3.2. It holds that $\text{dist}(x, \{u_\varepsilon = l\}) \leq \varepsilon\tilde{l}/\alpha$ for all $x \in \{u_\varepsilon = 0\}$, and (6.4) is satisfied by choosing $\varepsilon \leq \alpha / \max_{x \in \{u_R = 1-\varepsilon\}} \kappa(x)$, where $\kappa(x)$ is the mean curvature of the level set $\{u_R = 1 - \varepsilon\}$, which is well defined since level sets are analytic surfaces. □

LEMMA 6.5. *Let $\{\Omega_n\}$ be a decreasing sequence of bounded convex domains in $\mathcal{B}$, and define $\Omega :=$ Interior$(\bigcap \Omega_k)$. Then the p-capacitary potential of $\Omega \backslash \overline{K}$,* *denoted by u_Ω, satisfies* (6.3) *and thus, $\Omega \in \mathcal{B}$. Moreover, there exists a positive constant c_0 such that $\kappa(\partial\Omega)(x) \geq c_0$ for all $x \in \partial\Omega$.*

PROOF. Let u_n be the p-capacitary potential of $\Omega_n \backslash \overline{K}$. The sequence $\{u_n\}$ converges in $\mathcal{C}^{1,\alpha}$ norm to a p-harmonic function u. Convexity immediately implies that $u \equiv 0$ outside Ω. In addition it holds that the set $\{u_n = l\}$ converges to the set $\{u = l\}$ and that $\partial\Omega_n$ converges to $\partial\Omega$ in Hausdorff distance.

The existence of a subsolution (found in Lemma 6.4) implies, using Proposition 2.1, that $|\nabla u_n| \leq M$ for a constant M independent of n. Consequently, for any $n \in \mathbb{N}$, it holds that

$$\text{dist}(x, \{u_n = l\}) \geq l/M, \quad \forall x \in \partial\Omega_n. \tag{6.5}$$

This implies that $K \Subset \Omega$. Unfortunately it does not give any information on uniform bounds from above for $\kappa(\partial\Omega_n)$.

Instead we show that $\kappa(\partial\Omega)$ is uniformly bounded from below by a positive constant. By construction we have that $\Omega_n \subset \Omega_0$ for every n. Then, for every point $x_n \in \partial\Omega_n$,

$$\frac{\tilde{l}}{\kappa(\partial\Omega_n)(x_n)} \leq \text{dist}(x_n, \{u_n = l\}) \leq \text{diam}(\Omega_0), \quad \forall x_n \in \partial\Omega_n. \tag{6.6}$$

In particular $\partial\Omega_n$ cannot have flat parts. Moreover, if we take a sequence $x_n \in \partial\Omega_n$ such that $x_n \to x_0$ as $n \nearrow \infty$ with $x_0 \in \partial\Omega$, from (6.6) and (6.2) (maybe after passing to a subsequence) we get $\frac{\tilde{l}}{\kappa(\partial\Omega)(x_0)} \leq \text{diam}(\Omega_0)$, for all $x_0 \in \partial\Omega$, which shows that $\partial\Omega$ cannot have flat parts either.

Two things are now to be proven: (i) the function u is the p-capacitary potential in $\Omega \backslash \overline{K}$, and (ii) the function u satisfies (6.3). Suppose that $\tilde{\Omega} := \{\text{supp}(u)\} \subset \Omega$; then there exists a point $x_0 \in \partial\tilde{\Omega} \backslash \partial\Omega$ with a neighborhood of radius ε, denoted by $B_\varepsilon(x_0)$, such that $\tilde{\Omega} \cup B_\varepsilon(x_0) \subset \Omega$. Let $P \in \{u = l\}$ be the closest point of $\{u = l\}$ to x_0 and cover the segment $\overline{x_0, P}$ with a finite number of balls if radius ε. Note that this is possible because the segment $\overline{x_0, P}$ has a nonnegative finite length whose bounds are given in (6.5) and (6.6). By mimicking the argument in Lemma 3.4, one can show that Harnack inequality leads to a contradiction. Hence (i) is proven.

Consider now a point $x_0 \in \partial\Omega$ and any sequence $\{x_n \in \partial\Omega_n\}$ such that $x_n \to x_0$ as $n \nearrow \infty$; it holds $\text{dist}(x_n, \{u_n = l\}) \to \text{dist}(x_0, \{u = l\})$, as $n \nearrow \infty$. On the other hand, taking $\limsup_{n\to+\infty}$ on both sides of (6.3) and using (6.2), we obtain

$$\lim_{n\to+\infty} \text{dist}(x_n, \{u_n = l\}) \geq \limsup_{n\to+\infty} \frac{\tilde{l}}{\kappa(\partial\Omega_n)(x_n)} \geq \frac{\tilde{l}}{\kappa(\partial\Omega)(x_0)}. \qquad \square$$

PROOF OF THEOREM 6.2. It remains to prove that the p-capacitary function (u_Ω, Ω) constructed in Lemma 6.5 satisfies the distance property

$$\text{dist}(x, \Gamma_l) = \lambda_{\kappa,\tilde{l}}(x), \tag{6.7}$$

for every x that satisfies either $(A1)$ and $(A2)$ and that $\partial\Omega$ does not have points that satisfy $(A3)$ nor $(A4)$.

Suppose, by contradiction, that there exists an extremal point $x_0 \in \partial\Omega$ such that $\text{dist}(x_0, \{u_\Omega = l\}) \geq \frac{\tilde{l}}{\kappa(\partial\Omega)(x_0)} + \delta, \delta > 0$. Then there exists a neighborhood U

of x_0 (depending on δ) such that for any $x \in U \cap \partial\Omega$ it holds that

$$\operatorname{dist}(x, \{u_\Omega = l\}) \geq \frac{\tilde{l}}{\kappa(\partial\Omega)(x_0)} + \frac{\delta}{2}. \tag{6.8}$$

We first assume that $\partial\Omega$ is $\mathcal{C}^2$ in a neighborhood of x_0, i.e., assumption $(A1)$ is satisfied. We apply translations and rotations so that $x_0 = 0$ and choose coordinates $x = (x', x_n)$ in such a way that $\Omega \subset \{x_n > 0\}$, and we parametrize the surface $\partial\Omega$ near x_0 as a graph $x_n = \varphi(x')$, $x' \in \mathbb{R}^{N-1}$, such that $\varphi(0) = 0$, $\nabla\varphi(0) = 0$. Defined a new function $\varphi_1(x') := (1-\varepsilon)\varphi(x') + \varepsilon^2$. For ε small enough the intersection set $V_\varepsilon := \{(x', x_n), \mid \varphi(x') = \varphi_1(x')\}$ is strictly contained in the small neighborhood U. Consider the new function $\tilde{\varphi} := \max\{\varphi, \varphi_1\}$ and define the set $\tilde{\Omega} = (\Omega \cap U \cap \{(x', x_n) \mid x_n > \tilde{\varphi}(x')\}) \cup (\Omega \cap U^c)$. By construction it holds that $\tilde{\Omega} \subsetneq \Omega$; moreover the mean curvature of $\tilde{\Omega}$ can be estimated in terms of the mean curvature or $\partial\Omega$ at x_0, except maybe at the points in $V_\varepsilon \cap \partial\Omega$, in which case it may be infinity. Hence let $\tilde{u}$ be the p-capacitary potential of $\tilde{\Omega}\backslash\overline{K}$. The comparison principle and (6.8) yield, for ε small enough,

$$\operatorname{dist}(x_1, \{\tilde{u} = l\}) \geq \operatorname{dist}(x_1, \{u_\Omega = l\}) \geq \frac{\tilde{l}}{\kappa(\partial\Omega)(x_0)} + \frac{\delta}{2} \geq \frac{\tilde{l}}{\kappa(\partial\Omega)(x_1)},$$

for $x_1 \in \partial\tilde{\Omega} \cap U$. Note that we may obtain the same conclusion of $x_1 \in V_\varepsilon$ even if the curvature at those points is infinity. The existence of such $\tilde{\Omega} \subsetneq \Omega$ contradicts the minimality of Ω.

Let us assume now that the extremal point x_0 satisfies $(A3)$. Inequality (6.8) reduces to $\operatorname{dist}(x, \{u_\Omega = l\}) \geq \delta/2$. We perturb Ω to $\tilde{\Omega}$ in the following way: Consider, as before, $\partial\Omega$ written as the graph $x_n = \varphi(x')$. By hypothesis, $\overline{\kappa}(\partial\Omega)(x_0) = \infty$, so for each constant $M > 0$, there exists a paraboloid $m(x')$ with curvature equal M at x_0 that touches $\partial\Omega$ from outside at x_0. Define a new function $\varphi_1(x') := (1-\varepsilon)m(x') + \varepsilon^2$, and consider $\tilde{\varphi} := \max\{\varphi, \varphi_1\}$. Define the set $\tilde{\Omega}$ as in the previous case; contradiction follows similarly. Hence in such point x_0 we have that $\operatorname{dist}(x_0, \{u_\Omega = l\}) = 0$. But this is not possible because we have uniform gradient bounds for u_Ω (see Proposition 2.1).

Next we consider the case of x_0 not being an extremal point of $\partial\Omega$: [**46**, Lemma 12] shows that $\frac{1}{\kappa(\partial\Omega)(x)}$ is a locally concave function if Ω is convex. Then we may proceed as in the previous section in order to conclude that property (6.7) is satisfied for each points of the form $(A2)$ and that $\partial\Omega$ does not contain points of the form $(A4)$. □

REMARK 6.6. *Open problems:*

(i) *Full regularity of $\partial\Omega$ for* (P'_K). *For the classical Bernoulli problem full regularity of $\partial\Omega$ is obtained via the well-known regularity theory for almost minimal surfaces (see* [**10**]*). It is an interesting open question to show that such theory can be applied also for* (P'_K).

(ii) *One may also consider the problem with more general distance functions* $\lambda_{\kappa,\tilde{l}} = \lambda(x, \kappa(x))$ *in* (P'_K).

7. Final comments

In this final section we make further remarks and lines of research on the problem. Denote by $\Lambda(\Omega) := \lambda_{\Omega,\max}$ the Bernoulli constant for the (interior) distance problem defined in (4.2). One may show a Brunn–Minkowski inequality for $\Lambda(\Omega)$ in the spirit of [**16, 15, 41**]:

PROPOSITION 7.1. *Fix $l \in (0,1)$. Let Ω_0, Ω_1 be two bounded convex domains in $\mathbb{R}^N$. For each $t \in (0,1)$, define*

$$\Omega_t := (1-t)\Omega_0 + t\Omega_1.$$

Then

$$\Lambda(\Omega_t) \geq (1-t)\Lambda(\Omega_0) + t\Lambda(\Omega_1).$$

We conjecture that equality holds if and only if Ω_0 and Ω_1 are homothetic.

The multiphase Bernoulli problem (cf. [**4, 5, 37**] for $p = 2$, [**8**] for general p), has many applications in studying interfaces, for instance. In a similar flavor, one may formulate and solve the multiphase discrete Bernoulli problem.

Another application is in numerical simulations. One of the major problems in designing numerical algorithms for the (interior and exterior) classical Bernoulli problem, as well as for the formulation presented in this manuscript, is related to the fact that the set Ω (for the exterior case) or K (for the interior case) is unknown and chosen by the solution. Consequently, to impose numerically the gradient or the distance condition to the solution itself at the unknown boundary becomes a nontrivial task.

Consider for a moment the exterior discrete Bernoulli problem (the interior can be treated similarly)

$$(\mathrm{P}_E) \qquad \begin{cases} \Delta_p u = 0 \text{ in } \Omega \setminus \overline{K}, \quad u = 1 \text{ in } \overline{K}, \quad u = 0 \text{ on } \partial\Omega, \\ \operatorname{dist}(x, \{u = \lambda\omega\}) = \lambda \quad \text{for all } x \in \partial\Omega, \end{cases}$$

and the classical one

$$(\mathrm{P}_B) \qquad \begin{cases} \Delta_p u = 0 \text{ in } \Omega \setminus \overline{K}, \quad u = 1 \text{ in } \overline{K}, \quad u = 0 \text{ on } \partial\Omega, \\ |\nabla u| = \omega \quad \text{for all } x \in \partial\Omega. \end{cases}$$

The numerical implementation of a distance condition between two level sets of a p-harmonic function might be easier than the implementation of a gradient condition on a level set. Moreover we have rigorously shown in the previous sections that solutions of (P_E) converge to solutions of (P_B) as $\lambda \to 0$.

Based on these two observations, we propose the following algorithm to numerically approximate solutions to problem (P_E) and consequently solutions to (P_B). Fix the set $K \subset \mathbb{R}^N$ and (u_{n-1}, Ω_{n-1}) solution of (P_Ω) with $\Omega = \Omega_{n-1}$:

1. Let $B_\lambda(x)$ be the ball of radius λ and centrum in x; define

$$\Omega_n := \{u_{n-1} \geq \lambda\omega\} \cup \bigcup_{x \in \Gamma_{n-1}} B_\lambda(x)$$

with $\Gamma_{n-1} = \{x \in \Omega_{n-1}, \mid u_{n-1}(x) = \lambda\omega\}$.

2. If $\Omega_{n-1} = \Omega_n$, the couple (Ω_{n-1}, u_{n-1}) is a solution for (P_E). If $\Omega_{n-1} \neq \Omega_n$ solve (P_Ω) with $\Omega = \Omega_n$.

3. Let (u_n, Ω_n) be a solution of (P_Ω) with $\Omega = \Omega_n$. Restart from step **1.** for $n \to n+1$.

Hence a numerical algorithm for (P_B) consists on solving (P_E) with the scheme $\{\mathbf{1},\mathbf{2},\mathbf{3}\}$ for a decreasing sequence of values for λ.

We remark that estimates showing convergence and stability of algorithm $\{\mathbf{1},\mathbf{2},\mathbf{3}\}$ are not provided in this manuscript but are very interesting open questions.

Acknowledgments

M.d.M Gonzalez is supported by grants MTM2008-06349-C03-01 and MTM-2011-27739-C04-01 (from the Spanish government), and 2009SGR345 (from Generalitat de Catalunya). M.P. Gualdani is supported by NSF-DMS 1109682. H. Shahgholian is supported in part by Swedish Research Council. This work was finalized at KTH with the support from Göran Gustafsson foundation. The authors would like to thank the hospitality of MSRI during the program *Free Boundary Problems, Theory and Applications* in the Spring 2011. The authors thank Andrew Acker for bringing their attention to his work from late 70's.

References

1. A. Acker. *Heat flow inequalities with applications to heat flow optimization problems.* SIAM J. Math. Anal. 8 (1977), no. 4, 604–618. MR0473960 (57:13618)
2. A. Acker. *How to approximate the solutions of certain free boundary problems for the Laplace equation by using the contraction principle.* Z. Angew. Math. Phys. 32 (1981), no. 1, 22–33. MR615552 (82m:35004)
3. A. Acker. *Interior free boundary problems for the Laplace equation.* Arch. Rational Mech. Anal. 75 (1980/81), no. 2, 157–168. MR605527 (82f:35185)
4. A. Acker. *On the multi-layer fluid problem: regularity, uniqueness, convexity, and successive approximation of solutions.* Comm. Partial Differential Equations 16 (1991), no. 4-5, 647–666. MR1113101 (92k:35290)
5. A. Acker. *On the existence of convex classical solutions for multilayer free boundary problems with general nonlinear joining conditions.* Trans. Amer. Math. Soc. 350 (1998), no. 8, 2981–3020. MR1422592 (98k:35209)
6. A. Acker. On the existence of convex classical solutions for multilayer free boundary problems with general nonlinear joining conditions. *Trans. Amer. Math. Soc.*, 350(8):2981–3020, 1998. MR1422592 (98k:35209)
7. A. Acker. On the successive approximation of classical solutions for multi-surface free-boundary problems with general non-linear joining conditions. *Pacific J. Math.*, 188(1):1–51, 1999. MR1680419 (2000a:35258)
8. A. Acker, A. Henrot, M. Poghosyan and H. Shahgholian. *The multi-layer free boundary problem for the p-Laplacian in convex domains.* Interfaces Free Bound. 6 (2004), no. 1, 81–103. MR2047074 (2005e:35256)
9. A. Acker and R. Meyer. *A free boundary problem for the p-Laplacian: uniqueness, convexity, and successive approximation of solutions.* Electron. J. Differential Equations, pages No. 08, approx. 20 pp. (electronic), 1995.
10. F. J., Jr. Almgren. *Existence and regularity almost everywhere of solutions to elliptic variational problems with constraints.* Mem. Amer. Math. Soc. 4 (1976), no. 165, viii+199 pp. MR0420406 (54:8420)
11. H. W. Alt and L. A. Caffarelli. *Existence and regularity for a minimum problem with free boundary.* J. Reine Angew. Math., 325 (1981), 105–144. MR618549 (83a:49011)
12. I. Athanasopoulos, L. A. Caffarelli, C. Kenig, and S. Salsa. *An area-Dirichlet integral minimization problem.* Comm. Pure Appl. Math., 54(4) (2001), 479–499. MR1808651 (2001k:49003)
13. C. Atkinson and C. R. Champion. *Some boundary-value problems for the equation* $\nabla\cdot(|\nabla\varphi|^N\nabla\varphi)=0$. Quart. J. Mech. Appl. Math., 37 (1984), no. 3, 401–419. MR760209
14. A. Beurling. *The collected works of Arne Beurling.* Vol. 1. Complex analysis. Edited by L. Carleson, P. Malliavin, J. Neuberger and J. Wermer. Contemporary Mathematicians. Birkhäuser Boston Inc., Boston, MA, 1989. MR1057613 (92k:01046a)

15. C. Bianchini and P. Salani. *Concavity properties for elliptic free boundary problems.* Nonlinear Anal. 71 (2009), no. 10, 4461–4470. MR2548676 (2011d:35523)
16. C. Borell. *Capacitary inequalities of the Brunn-Minkowski type.* Math. Ann. 263 (1984) 179–184. MR698001 (84e:31005)
17. V. Burenkov. *Sobolev spaces on domains.* Teubner-Texte zur Mathematik [Teubner Texts in Mathematics], 137 (1998). MR1622690 (99g:46040)
18. L. Caffarelli *A Harnack inequality approach to the regularity of free boundaries. III. Existence theory, compactness, and dependence on X.* Ann. Scuola Norm. Sup. Pisa Cl. Sci. (4) 15 (1988), no. 4, 583–602 (1989). MR1029856 (91a:35170)
19. L. Caffarelli, D. Jerison and C. Kenig. *Global energy minimizers for free boundary problems and full regularity in three dimensions.* Noncompact problems at the intersection of geometry, analysis, and topology, 83–97. Contemp. Math., 350, Amer. Math. Soc., Providence, RI, 2004. MR2082392 (2005e:35258)
20. L. Caffarelli, P. Markowich and J-F. Pietschmann. *On a price formation free boundary model by Lasry & Lions.* C. R. Math. Acad. Sci. Paris 349 (2011), no. 11-12, 621–624. MR2817378 (2012g:35396)
21. L. Chayes, M. d. M. González, M. P. Gualdani, and I. Kim. *Global existence, uniqueness and asymptotics of solutions to a model in price formation.* SIAM J. Math. Anal., 41 (2009), no. 5, 2107–2135. MR2578801 (2011m:35382)
22. D. Daners and B. Kawohl. *An isoperimetric inequality related to a Bernoulli problem.* Calc. Var. (2010) 39: 547–555. MR2729312 (2012a:35369)
23. D. Danielli and A. Petrosyan. *A minimum problem with free boundary for a degenerate quasilinear operator.* Calc. Var. 23 (2005), no.1, 97–124. MR2133664 (2006c:35303)
24. D. Danielli and A. Petrosyan. *Fully regularity of the free boundary in two dimensions.* Math. Res. Lett.13 (2006) 4, 667–681. MR2250499 (2007f:35306)
25. D. De Silva, D. Jerison. *A singular energy minimizing free boundary.* J. Reine Angew. Math. 635 (2009), 1–21. MR2572253 (2010k:49003)
26. M. Flucher and M. Rumpf. *Bernoulli's free-boundary problem, qualitative theory and numerical approximation*, J. Reine Angew. Math. **486** (1997), 165–204. MR1450755 (98i:35214)
27. R. Foote, *Regularity of the distance function*, Proc. AMS. 92 (1984), no. 1, 153–155. MR749908 (85m:58024)
28. M.d.M. González and M. Gualdani. *Asymptotics for a free-boundary model in price formation*, Nonlinear Analysis 74 (2011), pp. 3269–3294. MR2793561 (2012d:91087)
29. A. Greco and B. Kawohl. *On the convexity of some free boundaries.* Interfaces Free Bound. 11 (2009), no. 4, 503–514. MR2576214 (2011c:35631)
30. A. Henrot and H. Shahgholian. *The one phase free boundary problem for the p-Laplacian with non-constant Bernoulli boundary condition.* Trans. Amer. Math. Soc. 354 (2002), no. 6, 2399–2416. MR1885658 (2003c:35165)
31. A. Henrot and H. Shahgholian. *Existence of classical solutions to a free boundary problem for the p-Laplace operator. II. The interior convex case.* Indiana Univ. Math. J. 49 (2000), no. 1, 311–323. MR1777029 (2001m:35326)
32. A. Henrot and H. Shahgholian. *Existence of classical solutions to a free boundary problem for the p-Laplace operator. I. The exterior convex case.* J. Reine Angew. Math. 521 (2000), 85–97. MR1752296 (2001f:35442)
33. A. Henrot and H. Shahgholian. *Convexity of free boundaries with Bernoulli type boundary condition.* Nonlinear Anal. 28 (1997), no. 5, 815–823. MR1422187 (97j:35165)
34. B. Kawohl. *On starshaped rearrangement and applications.* Transactions. AMS 296, (1986), no. 1, 377–386. MR837818 (88d:35196)
35. D. Kinderlehrer and L. Nirenberg. *Regularity in free boundary problems.* Ann. Scuola Norm. Sup. Pisa Cl. Sci. (4) 4 (1977), no. 2, 373–391. MR0440187 (55:13066)
36. J.-M. Lasry and P.-L. Lions. *Mean field games.* Japanese Journal of Mathematics, 2 (2007), no. 1, 229–260. MR2295621 (2008k:91034)
37. P. Laurence and E. Stredulinsky. *Existence of regular solutions with convex levels for semilinear elliptic equations with nonmonotone L^1 nonlinearities. I. An approximating free boundary problem.* Indiana Univ. Math. J. 39 (1990), no. 4, 1081–1114. MR1087186 (92m:35279)
38. J.L. Lewis. *Capacitary functions in convex rings.* Arch. Rational Mech. Anal. 66 (1977), no. 3, 201–224. MR0477094 (57:16638)

39. J.L. Lewis and K. Nyström. *Boundary behavior for p-harmonic functions in Lipschitz and starlike Lipschitz ring domains.* Ann. Sci. Ecole Norm. Sup. (4) 40 (2007), no. 5, 765–813. MR2382861 (2009d:35093)
40. G.M. Lieberman. *Boundary regularity for solutions of degenerate elliptic equations.* Nonlinear Anal. 12 (1988), no. 11, 1203–1219. MR969499 (90a:35098)
41. M. Longinetti and P. Salani. *On the Hessian matrix and Minkowski addition of quasiconvex functions.* J. Math. Pures Appl. 9 88(3)(2007), 276–292. MR2355460 (2008k:35163)
42. R. Magnanini and S. Sakaguchi. *Matzoh ball soup: heat conductors with a stationary isothermic surface.* Ann. of Math. (2) 156 (2002), no. 3, 931–946. MR1954240 (2003k:35086)
43. R. Magnanini and S. Sakaguchi. *Stationary isothermic surfaces for unbounded domains.* Indiana Univ. Math. J. 56 (2007), no. 6, 2723–2738. MR2375699 (2009i:35126)
44. R. Magnanini and S. Sakaguchi. *Nonlinear diffusion with a bounded stationary level surface.* Ann. Inst. H. Poincaré Anal. Non Linéaire 27 (2010), no. 3, 937–952. MR2629887 (2011d:35263)
45. F. Mazzone. *A single phase variational problem involving the area of level surfaces.* Comm. Partial Differential Equations, 28(5-6) (2003), 991–1004. MR1986058 (2004m:35062)
46. H. Mikayelyan and H. Shahgholian. *Convexity of the free boundary for an exterior free boundary problem involving the perimeter.* Preprint.
47. A. Petrosyan. *On the full regularity of the free boundary in a class of variational problems.* Proc. Amer. Math. Soc. 136 (2008), no. 8, 2763–2769. MR2399040 (2010b:35501)
48. W. Reichel. *Radial symmetry by moving planes for semilinear elliptic BVPs on annuli and other non-convex domains.* Elliptic and parabolic problems (Pont-a-Mousson, 1994), 164Ð182, Pitman Res. Notes Math. Ser., 325, Longman Sci. Tech., Harlow, 1995. MR1416582 (97k:35083)
49. H. Shahgholian. *Diversifications of Serrin's and related Symmetry problems.* Complex Var. Elliptic Equ. 57 (2012), no. 6, 653–665. MR2916825
50. P. Tolksdorf. *On the Dirichlet problem for quasilinear equations in domains with conical boundary points.* Comm. Partial Differential Equations 8 (1983), no. 7, 773–817. MR700735 (85g:35053)

Universitat Politècnica de Catalunya, ETSEIB - Dept. de MA1, Av. Diagonal 647, 08028 Barcelona, Spain
E-mail address: mar.gonzalez@upc.edu

Department of Mathematics, University of Texas at Austin, 2515 Speedway C1200, Austin, Texas 78712
E-mail address: gualdani@math.utexas.edu

Department of Mathematics, Royal Institute of Technology 100 44 Stockholm, Sweden
E-mail address: henriksh@kth.se

Amer. Math. Soc. Transl.
Volume **232**, 2014

On Plateau's problem in Riemannian manifolds

S. Hildebrandt and F. Sauvigny

Dedicated to the memory of Olga A. Ladyzhenskaya in high admiration

ABSTRACT. Let $\mathcal{C}(\Gamma, M)$ be the class of mappings $X : B \to M$ from the unit disk B into a Riemannian manifold, bounded by a closed Jordan curve Γ in M. At first, we prove a generalization of the fact that for $M \in C^{2,\mu}$ and $\Gamma \in C^{2,\mu}$ the infimum of area in $\mathcal{C}(\Gamma, M)$ is the same as the infimum of energy (i.e., of Dirichlet's integral). Then it is investigated whether a corresponding result holds for relative minimizers of area and energy in $\mathcal{C}(\Gamma, M)$.

1. Introduction

Let $\mathcal{C}(\Gamma, M)$ be the class of mappings $X : B \to M$ from the unit disk B in $\mathbb{C}$ into a Riemannian manifold M which are bounded by a closed Jordan curve Γ in M. If $\mathcal{C}(\Gamma, M)$ is nonempty, $M \in C^{3,\mu}$ with $0 < \mu < 1$, and Γ is rectifiable. Morrey, in [**7**], [**8**], solved Plateau's problem, i.e., he proved the existence of a harmonic mapping $X : B \to M$ which is conformally parametrized, continuous on $\bar{B}$, and bounded, assuming also that M is homogeneously regular. This has been achieved by minimizing the Dirichlet integral $\mathcal{D}$ in $\mathcal{C}(\Gamma, M)$, proving regularity results for the minimizers, and verifying that the infimum of $\mathcal{D}$ in $\mathcal{C}(\Gamma, M)$ is the same as the infimum of the area functional $\mathcal{A}$ in $\mathcal{C}(\Gamma, M)$.

Here we prove $\inf \mathcal{D} = \inf \mathcal{A}$ in $\mathcal{C}(\Gamma, M)$ assuming only that M is of class C^1 and can be embedded into some Euclidean space $\mathbb{R}^N$. No regularity results for minimizers are needed. Instead we simultaneously consider the minimum problems

$$\mathcal{A}^\epsilon := (1-\epsilon)\mathcal{A} + \epsilon\mathcal{D} \to \min \quad \text{in} \quad \mathcal{C}(\Gamma, M) \quad \text{for} \quad 0 < \epsilon \leq 1$$

and use Courant's formula for inner variations of $\mathcal{A}^\epsilon$ which, because of $\partial\mathcal{A}^\epsilon = (1-\epsilon)\partial\mathcal{A} + \epsilon\partial\mathcal{D} = \epsilon\partial\mathcal{D}$, amounts to an expression for $\partial\mathcal{D}$. The idea for this method is taken from [**3**]. In Section 2 we carry out this approach by treating an even more general problem. Moreover, we recall various results on the class $\mathcal{M}(\Gamma, M)$ of minimal surfaces in $\mathcal{C}(\Gamma, M)$.

At the end of Section 2 and in all of Section 3 we investigate the question whether relative minimizers of $\mathcal{A}$ in $\mathcal{M}(\Gamma, M)$ are relative minimizers of $\mathcal{D}$ in $\mathcal{C}(\Gamma, M)$, and vice versa. This converse seems to be quite difficult to prove, and

2010 *Mathematics Subject Classification.* 49Q05, 35J60, 53A10.
Key words and phrases. Plateau's problem, Riemannian manifolds, minimizers of area and energy.

we merely have a weak version which only holds for immersed minimal surfaces that are relative minimizers of $\mathcal{D}$. The main result about this question is stated as Theorem 4. The proof essentially follows the approach used in [**4**] for minimal surfaces in $\mathbb{R}^3$. The basic tool is taken from work of Vekua.

2. Absolute minimizers of area and energy

Let K be a closed set in $\mathbb{R}^N, N > 3$, which contains a closed rectifiable Jordan curve Γ. By $\mathcal{C}(\Gamma, K)$ we denote the class of mappings $X : B \to \mathbb{R}^N$ from the unit disk

$$B := \{w = (u, v) \in \mathbb{R}^2 : |w| < 1\}$$

into $\mathbb{R}^N$, which have the following properties:

(i) $X \in H^{1,2}(B, \mathbb{R}^N) \cap C^0(\partial B, \mathbb{R}^N)$;
(ii) $X(B) \subset K$;
(iii) the continuous representative of the "Sobolev trace $X|_{\partial B}$" maps ∂B monotonically onto Γ.

Condition (ii) means that every representative of X, again denoted by X, satisfies

$$X(w) \in K \quad \text{a.e. on} \quad B.$$

Furthermore, we can assume that Γ is equipped with a fixed orientation and that $X|_{\partial B} : \partial B \to \Gamma$ induces the same orientation of Γ. The assumption "monotonical" is to be understood in the sense of "weakly monotonical", i.e., $(X|_{\partial B})^{-1}(p)$ for $p \in \Gamma$ might be a closed subarc of ∂B.

Throughout this section we suppose that the set $\mathcal{C}(\Gamma, K)$ is nonempty:

$$\mathcal{C}(\Gamma, K) \neq \emptyset . \tag{2.1}$$

For example, this assumption is satisfied for all closed, rectifiable curves Γ in K if the set K is C^1-diffeomorphic to some affine subspace of $\mathbb{R}^N$. If K is an arbitrary differentiable submanifold of $\mathbb{R}^N$, then (2.1) is a topological condition on K, which amounts to the property that Γ is zero-homotopic in K.

We denote the points of $\mathbb{R}^N$ by $x = (x^1, x^2, \ldots, x^N)$, $y = (y^1, y^2, \ldots, y^N)$ etc., and write $\langle x, y\rangle$ for the Euclidean scalar product $\delta_{jk}x^j y^k$ of x, y, and $|x| = \langle x, x\rangle^{1/2}$ for the Euclidean length of x.

If K is a C^r-differentiable submanifold of $\mathbb{R}^N$ with $r \geq 1$, then the Euclidean metric ds_0, given by

$$ds_0^2 = \delta_{jk} dx^j dx^k,$$

induces a Riemannian ds on K, and $M := (K, ds)$ is a Riemannian manifold of class C^r. Whenever a global "Nash embedding theorem" holds, one has this situation.

For mappings $X \in H^{1,2}(B, \mathbb{R}^N)$ we define the *Dirichlet integral* ("energy")

$$D(X) := \frac{1}{2} \int_B (|X_u|^2 + |X_v|^2)\, dudv$$

and the *area functional*

$$A(X) := \int_B \sqrt{|X_u|^2\, |X_v|^2 - \langle X_u, X_v\rangle^2}\, dudv\, .$$

Then we have

$$A(X) \leq D(X) \quad \text{for all} \quad X \in H^{1,2}(B, \mathbb{R}^N),$$

and

$$A(X) = D(X) \quad \text{if and only if } X \text{ satisfies the equations}$$

$$|X_u|^2 = |X_v|^2 \,, \langle X_u, X_v \rangle = 0, \tag{2.2}$$

the so-called *conformality relations.* For any real ϵ with $0 < \epsilon \le 1$, we define the functional A^ϵ on $H^{1,2}(B, \mathbb{R}^N)$ by

$$A^\epsilon(X) := (1-\epsilon)A(X) + \epsilon D(X) \,, \quad X \in H^{1,2}(B, \mathbb{R}^N). \tag{2.3}$$

Then

$$A \le A^\epsilon \le D \tag{2.4}$$

and

$$A(X) = A^\epsilon(X) = D(X) \quad \text{if and only if (2.2) holds.} \tag{2.5}$$

Both D and A are sequentially weakly semicontinuous on $H^{1,2}(B, \mathbb{R}^N)$, cf. [**1**, p. 300, Lemma 1]. Therefore, also A^ϵ has this property, i.e., $X_j \rightharpoonup X$ in $H^{1,2}(B, \mathbb{R}^N)$ for $j \to \infty$ implies

$$A^\epsilon(X) \le \liminf_{j\to\infty} A^\epsilon(X_j).$$

Next we choose three points $w_1, w_2, w_3 \in \partial B$ and three points $P_1, P_2, P_3 \in \Gamma$ such that the representation $\gamma : \partial B \to \Gamma$ of the oriented curve Γ satisfies $\gamma(w_j) = P_j$, $j = 1, 2, 3$. Then the set $\mathcal{C}^*(\Gamma, K)$, defined by

$$\mathcal{C}^*(\Gamma, K) := \{X \in \mathcal{C}(\Gamma, K) : \, X(w_j) = P_j \,, j = 1, 2, 3\},$$

is sequentially weakly closed. This follows from the Courant–Lebesgue lemma due to [**1**] and from the fact that $X_j \rightharpoonup X$ in $H^{1,2}(B, \mathbb{R}^N)$ implies $X_j \to X$ in $L^2(B, \mathbb{R}^N)$, and therefore $X_{j_k}(w) \to X(w)$ a.e. on B for some subsequence $\{X_{j_k}\}$ of $\{X_j\}$, taking also into account that the countable union of sets of measure zero is again a set of measure zero.

PROPOSITION 1. *For each $\epsilon \in (0, 1]$ there is an $X^\epsilon \in \mathcal{C}^*(\Gamma, K)$ such that*

$$A^\epsilon(X^\epsilon) = \inf\{A^\epsilon(X) : X \in \mathcal{C}(\Gamma, K)\} =: a(\epsilon) \tag{2.6}$$

and

$$|X_u^\epsilon|^2 = |X_v^\epsilon|^2 \,, \, \langle X_u^\epsilon, X_v^\epsilon \rangle = 0. \tag{2.7}$$

PROOF: Since A^ϵ is conformally invariant, it follows that

$$a(\epsilon) = \inf\{A^\epsilon(X) : X \in \mathcal{C}^*(\Gamma, K)\}.$$

Hence, there is a sequence $X_j \in \mathcal{C}(\Gamma, K)$, $j = 1, 2, \ldots$, such that

$$A^\epsilon(X_j) \to a(\epsilon) \quad \text{as} \quad j \to \infty \,,$$

whence

$$A^\epsilon(X_j) \le c_0(\epsilon) \quad \text{for all} \quad j \in \mathbb{N}$$

and some constant $c_0(\epsilon)$. Then

$$D(X_j) \le \epsilon^{-1} c_0(\epsilon) \quad \text{for all} \quad j \in \mathbb{N}.$$

Since $X_j(\partial B) = \Gamma$, a suitable Poincaré inequality yields

$$\|X_j\|^2_{L^2(B,\mathbb{R}^N)} \le c_1 D(X_j) \quad \text{for all} \quad j \in \mathbb{N}.$$

Thus we have

$$\|X_j\|^2_{H^{1,2}(B,\mathbb{R}^N)} = \|X_j\|^2_{L^2(B,\mathbb{R}^N)} + 2D(X_j) \le [c_1+2]\epsilon^{-1}c_0(\epsilon)\,.$$

Then a subsequence $\{X_{j_k}\}$ of $\{X_j\}$ and some $X^\epsilon \in H^{1,2}(B,\mathbb{R}^N)$ exist such that $X_{j_k} \rightharpoonup X^\epsilon$ in $H^{1,2}(B,\mathbb{R}^N)$ as $k \to \infty$ holds true. Since $\mathcal{C}^*(\Gamma, K)$ is sequentially weakly closed in $H^{1,2}(B,\mathbb{R}^N)$, one obtains $X^\epsilon \in \mathcal{C}^*(\Gamma,K)$ and therefore

$$a(\epsilon) \le A^\epsilon(X^\epsilon).$$

On the other hand, the sequential weak lower semicontinuity of A^ϵ on $H^{1,2}(B,\mathbb{R}^N)$ implies

$$A^\epsilon(X^\epsilon) \le \lim_{k\to\infty} A^\epsilon(X_{j_k}) = a(\epsilon),$$

and $a(\epsilon) = A^\epsilon(X^\epsilon)$ follows.

Using this minimum property of X^ϵ and applying "inner variations of B", one obtains

$$0 = \partial A^\epsilon(X^\epsilon,\lambda) = (1-\epsilon)\partial A(X^\epsilon,\lambda) + \epsilon\partial D(X^\epsilon,\lambda)$$

for any $\lambda \in C^1(\overline{B},\mathbb{R}^2)$ such that $\lambda|_{\partial B}$ is tangential to ∂B. Since $A(X\circ\tau) = A(X)$ for any $X \in H^{1,2}(B,\mathbb{R}^N)$ and any diffeomorphism $\tau : \overline{B} \to \overline{B}$ of the unit disk onto itself, it follows $\partial A(X^\epsilon,\lambda) = 0$. Because of $\epsilon > 0$, we then obtain

$$\partial D(X^\epsilon,\lambda) = 0 \quad \text{for all} \quad \lambda \in C^1(\overline{B},\mathbb{R}^2) \quad \text{with} \quad \lambda|_{\partial B} \text{ tangential to } \partial B.$$

A computation due to Courant (cf. [**1**, Section 4.5]) yields

$$\int_B [\alpha\cdot(\mu_u - \nu_v) + \beta(\mu_v + \nu_u)]\,dudv = 0 \tag{2.8}$$

for $\alpha^\epsilon := |X^\epsilon_u|^2 - |X^\epsilon_v|^2, \beta^\epsilon := \langle X^\epsilon_u, X^\epsilon_v\rangle$ and for all $\lambda = (\mu,\nu) \in C^1(\overline{B},\mathbb{R}^2)$ such that $\lambda|_{\partial B}$ is tangential to B. Using (2.8) for all $\lambda \in C^1_c(B,\mathbb{R}^2)$ one infers that $f^\epsilon := \alpha^\epsilon - i\beta^\epsilon$ is holomorphic in $w \in B$ (cf. [**1**, pp. 302–303]). Finally, another reasoning by Courant implies $f^\epsilon = 0$, using the full variational formula (2.8) and Poisson's integral formula (see [**1**, pp. 271–272]). □

PROPOSITION 2. *Let $X^\epsilon \in \mathcal{C}^*(\Gamma,K)$ be solutions of (2.6) as determined in Proposition 1, i.e.*

$$A^\epsilon(X^\epsilon) = a(\epsilon) \quad \textit{for} \quad 0<\epsilon\le 1,$$
$$a(\epsilon) := \inf_{\mathcal{C}(\Gamma,K)} A^\epsilon,$$

and set

$$a(\Gamma,K) := \inf_{\mathcal{C}(\Gamma,K)} A, \quad d(\Gamma,K) := \inf_{\mathcal{C}(\Gamma,K)} D. \tag{2.9}$$

Then one obtains

$$a(\Gamma,K) = d(\Gamma,K) = A(X^\epsilon) = D(X^\epsilon) = A^\epsilon(X^\epsilon) \quad \textit{for all} \quad \epsilon \in (0,1]. \tag{2.10}$$

PROOF: The conformality relations (2.7) together with (2.5) yield

$$A(X^\epsilon) = A^\epsilon(X^\epsilon) = D(X^\epsilon) \quad \text{for} \quad 0<\epsilon\le 1. \tag{2.11}$$

Then

$$d(\Gamma,K) \le D(X^\epsilon) = A^\epsilon(X^\epsilon) \quad \text{for} \quad 0<\epsilon\le 1,$$

and, since X^ϵ minimizes A^ϵ in $\mathcal{C}(\Gamma,K)$, it follows also that

$$A^\epsilon(X^\epsilon) \le A^\epsilon(Z) = (1-\epsilon)A(Z) + \epsilon D(Z) \quad \text{for any} \quad Z \in \mathcal{C}(\Gamma,K).$$

Therefore, we see

$$d(\Gamma, K) \leq (1-\epsilon)A(Z) + \epsilon D(Z) \quad \text{for} \quad 0 < \epsilon \leq 1.$$

Letting $\epsilon \to +0$, one obtains

$$d(\Gamma, K) \leq A(Z) \quad \text{for all} \quad Z \in \mathcal{C}(\Gamma, K),$$

whence

$$d(\Gamma, K) \leq a(\Gamma, K).$$

On the other hand, $A \leq A^\epsilon \leq D$ and (2.11) imply

$$a(\Gamma, K) \leq a(\epsilon) = A^\epsilon(X^\epsilon) = A(X^\epsilon) = D(X^\epsilon) \leq d(\Gamma, K);$$

therefore

$$a(\Gamma, K) = A(X^\epsilon) = A^\epsilon(X^\epsilon) = D(X^\epsilon) = d(\Gamma, K) \quad \text{for} \quad 0 < \epsilon \leq 1. \qquad \square$$

REMARK 1. *We thank Professor Guofang Wang at Freiburg for pointing out this modification of our original proof for Proposition* 2, *which used a reasoning presented in* [**1**, pp. 303–304].

Combining our Propositions 1 and 2 and choosing X as some of the minimizers X^ϵ with $0 < \epsilon \leq 1$, one arrives at

THEOREM 1. *There is a mapping $X \in \mathcal{C}(\Gamma, K)$ which simultaneously minimizes the area functional A, the Dirichlet integral D, and their convex combinations $A^\epsilon = (1-\epsilon)A + \epsilon D$ within $\mathcal{C}(\Gamma, K)$.*

REMARK 2. *The proof of Theorem* 1 *via Propositions* 1 *and* 2 *uses only variations in the parameter domain B and no variations in the target space K. That means no Euler equations for any simultaneous minimizer of A, A^ϵ, D are employed. In fact, any Euler equations for A and A^ϵ cannot be expected to hold since the integrand*

$$[|X_u|^2 \, |X_v|^2 - \langle X_u, X_v \rangle^2]^{1/2} = |X_u \wedge X_v| \tag{2.12}$$

of $A(X)$ can vanish for mappings $X \in H^{1,2}(B, \mathbb{R}^N)$. The artifice of Courant and Morrey for obtaining minimizers of A consisted in proving the existence of minimizers of D which are minimal surfaces, i.e., harmonic mappings satisfying the conformality relations, and then verifying $a(\Gamma, K) = d(\Gamma, K)$ for $K = \mathbb{R}^N$ or K representing a positively homogeneous Riemannian manifold. This artifice is in our case replaced by the trick of using the convex combinations $A^\epsilon = (1-\epsilon)A + \epsilon D, 0 < \epsilon \leq 1$. Note that one cannot directly operate with A since the inner variation ∂A of A vanishes identically and, therefore, does not provide any useful information. Compared with Morrey's approach in [**7**] *or* [**8**], *the advantage of our method is that it does not need any regularity theorems for minimizers of D in $\mathcal{C}(\Gamma, K)$ and for weak solutions of the corresponding Euler equations. Therefore, we can apply our approach to fairly general K, say, differentiable manifolds in $\mathbb{R}^N$ which become Riemannian manifolds $M = (K, ds)$, when we equip K with the metric ds induced by the Euclidean metric ds_0 of $\mathbb{R}^N$.*

REMARK 3. *Let M be an n-dimensional Riemannian manifold of class C^1. Using local coordinates $x = (x^1, \dots, x^n)$ in a coordinate patch of M, the local representation of the metric ds of M is*

$$ds^2 = g_{jk}(x)dx^j dx^k$$

with g_{jk} locally in C^0. If $M \in C^r$ or $C^{r,\mu}$ with $r \geq 1, 0 < \mu < 1$, then $g_{jk} \in C^{r-1}$ or $g_{jk} \in C^{r-1,\mu}$ locally.

If $(X^1, ..., X^n)$ denotes the local representation of a mapping $X : B \to M$ in such a coordinate patch, we obtain the Gaussian symbols

(2.13)

$$\mathcal{E}(X) := g_{jk}(X)X_u^j X_u^k, \quad \mathcal{F}(X) := g_{jk}(X)X_u^j X_v^k, \quad \mathcal{G}(X) := g_{jk}(X)X_v^j X_v^k.$$

Then the area of $X \in C^1(B, M)$ is given by

$$\mathcal{A}(X) := \int_B \sqrt{\mathcal{E}(X)\mathcal{G}(X) - \mathcal{F}^2(X)}\, dudv, \tag{2.14}$$

and the Dirichlet integral of X appears as

$$\mathcal{D}(X) := \frac{1}{2}\int_B [\mathcal{E}(X) + \mathcal{G}(X)]\, dudv. \tag{2.15}$$

We suppose that there exists a global isometric C^1-embedding

$$e : (M, ds) \to (\mathbb{R}^N, ds_0)$$

of M into $\mathbb{R}^N$. Then $X \in C^1(B, M)$ is transformed into $\mathcal{X} := e \circ X \in C^1(B, \mathbb{R}^N)$, and

$$\mathcal{E}(X) = |\mathcal{X}_u|^2, \quad \mathcal{F}(X) = \langle \mathcal{X}_u, \mathcal{X}_v \rangle, \quad \mathcal{G}(X) = |\mathcal{X}_v|^2,$$

as well as

$$\mathcal{A}(X) = A(\mathcal{X}), \quad \mathcal{D}(X) = D(\mathcal{X}).$$

We say that $X \in \mathcal{C}(\Gamma, M)$ holds true if and only if $\mathcal{X} = e \circ X \in \mathcal{C}(\Gamma', K)$, where $\Gamma' := e(\Gamma)$ is a closed, rectifiable Jordan curve in $K := e(M)$; cf. also [**5**, Appendix].

By Proposition 2 we have $a(\Gamma', K) = d(\Gamma', K)$ if (2.1) holds. Thus we obtain for every closed, rectifiable Jordan curve Γ in M with $\mathcal{C}(\Gamma, M) \neq \emptyset$ that

$$\inf_{\mathcal{C}(\Gamma,M)} \mathcal{A} = \inf_{\mathcal{C}(\Gamma,M)} \mathcal{D}. \tag{2.16}$$

An immediate consequence of (2.16) is formulated in

THEOREM 2. *Any minimizer of $\mathcal{D}$ in $\mathcal{C}(\Gamma, M)$ minimizes also $\mathcal{A}$ in $\mathcal{C}(\Gamma, M)$. Conversely, any minimizer X of $\mathcal{A}$ in $\mathcal{C}(\Gamma, M)$ is a minimizer of $\mathcal{D}$ in $\mathcal{C}(\Gamma, M)$ provided that X satisfies the conformality relations*

$$\mathcal{E}(X) = \mathcal{G}(X), \quad \mathcal{F}(X) = 0. \tag{2.17}$$

PROOF:

(i) If $X \in \mathcal{C}(\Gamma, M)$ satisfies $\mathcal{D}(X) = \inf_{\mathcal{C}(\Gamma,M)} \mathcal{D}$, then $\mathcal{A} \leq \mathcal{D}$ implies

$$\mathcal{A}(X) \leq \mathcal{D}(X) = \inf_{\mathcal{C}(\Gamma,M)} \mathcal{D} = \inf_{\mathcal{C}(\Gamma,M)} \mathcal{A} \leq \mathcal{A}(X).$$

Then $\mathcal{A}(X) = \mathcal{D}(X) = \inf_{\mathcal{C}(\Gamma,M)} \mathcal{A}$, and therefore also (2.17).

(ii) If $X \in \mathcal{C}(\Gamma, M)$ fulfils (2.17), then $\mathcal{A}(X) = \mathcal{D}(X)$. Therefore $\mathcal{A}(X) = \inf_{\mathcal{C}(\Gamma,M)} \mathcal{A}$ implies

$$\inf_{\mathcal{C}(\Gamma,M)} \mathcal{D} \leq \mathcal{D}(X) = \mathcal{A}(X) = \inf_{\mathcal{C}(\Gamma,M)} \mathcal{A} = \inf_{\mathcal{C}(\Gamma,M)} \mathcal{D}$$

whence

$$\mathcal{D}(X) = \inf_{\mathcal{C}(\Gamma,M)} \mathcal{D}. \qquad \square$$

The above result raises the question whether a similar result holds for relative minimizers of $\mathcal{A}$ and $\mathcal{D}$. We shall discuss this problem in the next section.

REMARK 4. *The Euler equations of* $\mathcal{D}$ *are*

$$\Delta X^l + \Gamma^l_{jk}(X)[X^j_u X^k_u + X^j_v X^k_v] = 0, \quad l = 1, \ldots, n, \tag{2.18}$$

in B *where* Γ^l_{jk} *are the Christoffel symbols of the second kind with respect to* $ds^2 = g_{jk}(x)dx^j dx^k$. *Solutions* $X : B \to M$ *of* (2.18) *are called harmonic mappings from* B *into* M, *and they are said to be disk-type minimal surfaces in* M *if they also satisfy* (2.17).

For "homogeneously regular" M *of class* $C^{r,\mu}$ *with* $r \geq 3$, $0 < \mu < 1$, *and* $\mathcal{C}(\Gamma, M) \neq \emptyset$, *C.B. Morrey* [**7**] *and* [**8**] *has proved the existence of minimal surfaces* $X : B \to M$ *which lie in* $\mathcal{C}(\Gamma, M)$, *minimize both* $\mathcal{A}$ *and* $\mathcal{D}$ *in* $\mathcal{C}(\Gamma, M)$, *and are of class* $C^0(\overline{B}, M) \cap C^{r,\mu}(B, M)$. *Actually, using a result of Tomi* [**10**], *it can be seen that Morrey's theorem also holds for* $r = 2$, $0 < \mu < 1$. *In fact, Tomi's result implies the following: Let* $B' \Subset B$ *be a subdisk of* B *such that* X *is of class* $H^{1,2}(B', \mathbb{R}^n) \cap C^0(B', \mathbb{R}^n)$ *and* $\Gamma^l_{jk} \in C^0(\Omega)$ *for some open set* Ω *with* $\overline{X(B')} \subset \Omega$. *Then it follows* $X \in C^{1,\gamma}(B', \mathbb{R}^n)$, $0 < \gamma < 1$. *Therefore* $M \in C^{2,\mu}$ *yields* $g_{jk} \in C^{1,\mu}, \Gamma^l_{jk} \in C^{0,\mu}$, *and consequently* $X \in C^{2,\mu}$ *on* B. *If also* $\Gamma \in C^{2,\mu}$ *holds true, then one obtains even* $X \in C^{2,\mu}(\overline{B}, M)$; *cf. Heinz and Hildebrandt* [**2**]. *Concerning a priori estimates, we refer to Jost* [**5**].

3. Relative minimizers of area and energy

Now we want to investigate whether there holds an analogous result to Theorem 2 for relative minimizers of area $\mathcal{A}$ and energy $\mathcal{D}$. For the sake of technical simplicity, we assume that the n-dimensional manifold M is homeomorphic to $\mathbb{R}^n$ and homogeneously regular. Then M can be identified with $\mathbb{R}^n$, and the metric ds of M is bounded by the Euclidean metric ds_0 as follows: There are constants m_1 and m_2 with $0 < m_1 \leq m_2$ such that

$$m_1 |\xi|^2 \leq g_{jk}(x)\xi^j \xi^k \leq m_2 |\xi|^2 \quad \text{for all} \quad x, \xi \in \mathbb{R}^n,$$

which means $m_1\, ds_0^2 \leq ds^2 \leq m_2\, ds_0^2$. We also assume that M is of class $C^{2,\mu}$, i.e., $g_{jk} \in C^{1,\mu}(\mathbb{R}^n)$ with $0 < \mu < 1$, and that Γ is a closed Jordan curve in $\mathbb{R}^n$ of class $C^{2,\mu}$. Let $\mathcal{M}(\Gamma, M)$ be the class of (disk-type) minimal surfaces $X : \overline{B} \to M$ bounded by Γ in the sense that $X \in \mathcal{C}(\Gamma, \mathbb{R}^n) \cong \mathcal{C}(\Gamma, M)$. Then we can interpret any $X \in \mathcal{M}(\Gamma, M)$ as a mapping

$$X : \overline{B} \to \mathbb{R}^n \quad \text{with} \quad X(u,v) = (X^1(u,v), \ldots, X^n(u,v)),\ w = (u,v) \in B,$$

which is of class $\mathcal{C}(\Gamma, \mathbb{R}^n) \cap C^{2,\mu}(\overline{B}, \mathbb{R}^n)$ and satisfies (2.17) and (2.18) on $\overline{B}$. Furthermore, the special structure of $M \cong \mathbb{R}^n$ allows us to interpret the

C^s- and $C^{s,\mu}$-distances of mappings $X, Y : \overline{B} \to M$ as the corresponding distances $\|X - Y\|_{C^s(\overline{B},\mathbb{R}^n)}$ and $\|X - Y\|_{C^{s,\mu}(\overline{B},\mathbb{R}^n)}$ of mappings $\overline{B} \to \mathbb{R}^n$.

Then we obtain the following result that is easy to deduce:

THEOREM 3. *If $X \in \mathcal{M}(\Gamma, M)$ is a relative minimizer of $\mathcal{A}$ in $\mathcal{C}(\Gamma, M)$ with respect to the $C^{2,\mu}(\overline{B}, \mathbb{R}^n)$-norm $\|.\|_{2,\mu}$, i.e.,*

$$\mathcal{A}(X) \leq \mathcal{A}(Y) \quad \text{for all} \quad Y \in \mathcal{C}(\Gamma, M) \quad \text{with} \quad \|X - Y\|_{2,\mu} < \epsilon$$

for some $\epsilon > 0$, then

$$\mathcal{D}(X) \leq \mathcal{D}(Y) \quad \text{for all} \quad Y \in \mathcal{C}(\Gamma, M) \quad \text{with} \quad \|X - Y\|_{2,\mu} < \epsilon,$$

i.e., X is also a relative minimizer of $\mathcal{D}$ in $\mathcal{C}(\Gamma, M)$ with respect to the norm $\|.\|_{2,\mu}$.

PROOF:

$$\mathcal{D}(X) = \mathcal{A}(X) \leq \mathcal{A}(Y) \leq \mathcal{D}(Y) \quad \text{for} \quad Y \in \mathcal{C}(\Gamma, M) \quad \text{with} \quad \|Y - X\|_{2,\mu} < \epsilon. \quad \square$$

REMARK 5. *The same reasoning holds for $\|.\|$-relative minimizers of $\mathcal{A}$ with respect to any suitable norm $\|.\|$, say*

$$\|.\|_{C^0} + \sqrt{\mathcal{D}(.)}, \quad \|.\|_{C^1}, \quad \|.\|_{C^{1,\nu}}, \quad \|.\|_{C^2}, \quad etc.$$

It is natural to conjecture that also the converse statement of Theorem 3 holds true. We can prove such a result only for immersed minimal surfaces, however, in a slightly weakened form:

THEOREM 4. *Suppose that $X \in \mathcal{M}(\Gamma, M)$ is immersed, i.e. $\mathcal{E}(X)(w) > 0$ for $w \in \overline{B}$, and that X satisfies*

$$\text{(3.1)} \quad \mathcal{D}(X) \leq \mathcal{D}(Z) \quad \text{for all} \quad Z \in \mathcal{C}(\Gamma, M) \cap C^2(\overline{B}, \mathbb{R}^n) \quad \text{with} \quad \|Z - X\|_{C^2} < \epsilon$$

for some $\epsilon > 0$. Then there is a number $\delta(\epsilon) > 0$ such that
(3.2)

$$\mathcal{A}(X) \leq \mathcal{A}(Y) \quad \text{for all} \quad Y \in \mathcal{C}(\Gamma, M) \cap C^{2,\mu}(B, \mathbb{R}^n) \quad \text{with} \quad \|Y - X\|_{C^{2,\mu}} < \delta(\epsilon).$$

PROOF: We follow the reasoning of [**4**] and proceed in three steps.

Step 1. Fix some $X \in \mathcal{M}(\Gamma, M)$, $\Gamma \in C^{2,\mu}$, $M \in C^{2,\mu}$, which is immersed. Then $X \in C^{2,\mu}(\overline{B}, \mathbb{R}^n)$ and $\mathcal{E}(X)(w) > 0$ on $\overline{B}$; the line element $d\sigma$ on X, given by

$$d\sigma^2 := g_{jk}(X) dX^j dX^k = \mathcal{E}(X)[du^2 + dv^2],$$

satisfies $\mathcal{E}(X) \geq \delta_0(X) > 0$ for some positive constant $\delta_0(X)$. Consider a "perturbation" Y of X with $Y \in C^{2,\mu}(\overline{B}, \mathbb{R}^n)$ and

$$\|Y - X\|_{C^{2,\mu}} < \eta, \quad \eta \in (0, 1).$$

The line element ds of Y, given by $ds^2 = g_{jk}(Y) dY^j dY^k$, can be written as

$$ds^2 = a\, du^2 + 2b\, du dv + c\, dv^2.$$

Then $a, b, c, \Delta := ac - b^2 \in C^{1,\mu}(\overline{B})$ holds, and there are constants $\kappa(X) > 0, \eta(X) \in (0, 1)$, and $\lambda_0(X) > 0$ such that $\gamma_{11} := a, \gamma_{22} := c, \gamma_{12} = \gamma_{21} := b$ satisfy

$$\|\gamma_{jk} - \mathcal{E}(X)\delta_{jk}\|_{C^{1,\mu}(\overline{B})} \leq \kappa(X)\eta(X)$$

and

$$\sqrt{\Delta} \geq \lambda_0(X) > 0.$$

This implies $\sqrt{\Delta} \in C^{1,\mu}(\overline{B})$ and

$$\left\| \sqrt{\Delta} - \mathcal{E}(X) \right\|_{C^{1,\mu}(\overline{B})} \leq \kappa'(X)\eta(X)$$

for some constant $\kappa'(X) > 0$. We also note that

$$\sqrt{\Delta} = \mathcal{E}(X) \quad \text{if } Y = X \text{ holds true.}$$

Using a method by I. N. Vekua [**11**], as presented in Sauvigny's textbook [**9**, Chap. XII], we shall construct a diffeomorphism $f := \overline{B} \to \overline{\Omega}$ of class $C^{2,\mu}(\overline{B}, \mathbb{R}^2)$ such that its inverse $g : \Omega \to B$ pulls ds^2 back into the form

$$g^* ds^2 = \lambda(x,y)\{dx^2 + dy^2\}, \quad (x,y) \in \overline{\Omega}, \tag{3.3}$$

where f is close to the identical map if $\eta(X) \ll 1$. Clearly, for fixed X, the mappings f, g and the domain $\Omega = f(B)$ depend on Y:

$$f = f(Y), \quad g = g(Y), \quad \Omega = \Omega(Y).$$

To carry out Vekua's method, we identify $\mathbb{R}^2$ with the complex plane and B as well as Ω with domains in $\mathbb{C}$, and we write $w = u + iv$, $z = x + iy$, $f : w \mapsto z = f(w)$. For solving (3.3) by $g = f^{-1}$ via a diffeomorphism f close to $id_{\overline{B}}$, one has to treat the *complex Beltrami equation*

$$f_{\overline{w}} - q(w) f_w = 0 \quad \text{in} \quad \overline{B} \tag{3.4}$$

with

$$q := \frac{a - \sqrt{\Delta} + ib}{a + \sqrt{\Delta} - ib} = \frac{a - c + 2ib}{a + c + 2\sqrt{\Delta}} \quad \in \quad C^{1,\mu}(\overline{B}, \mathbb{C})$$

by the ansatz

$$f = id_{\overline{B}} + T_B[\psi] \quad \text{with} \quad \psi \text{ solving } \psi = q + q\Pi_B[\psi]. \tag{3.5}$$

Here T_B means the Cauchy operator $\psi \mapsto T_B[\psi]$ for $\psi \in C^{1,\mu}(\overline{B}, \mathbb{C})$ given by

$$T_B[\psi](w) := -\frac{1}{\pi} \int_B \frac{\psi(z)}{z - w}\, dxdy, \quad w = u + iv \in \overline{B}, \quad z = x + iy \in \mathbb{C},$$

and $\Pi_B[\psi]$ denotes Vekua's integral operator $\psi \mapsto \Pi_B[\psi]$ which is defined by the Cauchy principal value:

$$\Pi_B[\psi](w) := \lim_{\rho \to +0} \left\{ -\frac{1}{\pi} \int_{B \setminus B_\rho(w)} \frac{\psi(z)}{(z - w)^2}\, dxdy \right\}, \quad w \in \overline{B}.$$

On account of Vekua's work [**11**], the mapping represents a bounded linear operator on $E := C^{1,\mu}(\overline{B}, \mathbb{C})$ with the norm $\|.\|_{1,\mu} := \|.\|_{C^{1,\mu}(\overline{B},\mathbb{C})}$. Hence, we have a constant $m(\mu) > 1$ such that

$$\|\Pi_B[\psi]\|_{1,\mu} \leq m(\mu)\, \|\psi\|_{1,\mu} \quad \text{for all} \quad \psi \in C^{1,\mu}(\overline{B}, \mathbb{C}).$$

Suppose now that q satisfies $m(\mu)\, \|q\|_{1,\mu} < 1/4$. Then the mapping $L_q : E \to E$ defined by

$$L_q[\psi] := q + q\Pi_B[\psi]$$

yields a contraction on the Banach space $(E, \|.\|_{1,\mu})$, and consequently L_q possesses a uniquely determined fixed point $\psi \in E$ with

$$\|\psi\|_{1,\mu} \leq \|q\|_{1,\mu} \cdot [1 - m(\mu)\, \|q\|_{1,\mu}]^{-1}.$$

Thus $L_q[\psi] = \psi$ holds true or

$$(0+\psi) - q\{1 + \Pi_B[\psi]\} = 0.$$

Since we have

$$(T_B[\psi])_w = \Pi_B[\psi], \quad (T_B[\psi])_{\overline{w}} = \psi,$$

we see that $f(w) = w + T_B[\psi], w \in \overline{B}$ from (3.5) satisfies equation (3.4), using $T_B[\psi] \in C^{2,\mu}(\overline{B}, \mathbb{C})$ if $\psi \in C^{1,\mu}(\overline{B}, \mathbb{C})$. The mapping f represents a local diffeomorphism, since its Jacobian J_f can be estimated from below by

$$\begin{aligned} J_f &= |f_w|^2 - |f_{\overline{w}}|^2 = |1 + \Pi_B[\psi]|^2 - |\psi|^2 \\ &\geq \left|1 - m(\mu)\,\|\psi\|_{1,\mu}\right|^2 - \|\psi\|_{1,\mu}^2 > \left|1 - \frac{1}{3}\right|^2 - \frac{1}{3^2} = \frac{1}{3} \text{ on } \overline{B}. \end{aligned}$$

Furthermore, $f : \overline{B} \to \overline{\Omega}$ represents a one-to-one mapping. In fact, there exists a constant $c_0(\mu)$ with $0 < c_0(\mu) < 1/[2m(\mu)]$ such that for $\|q\|_{1,\mu} \leq c_0(\mu)$ the fixed point ψ of L_q satisfies

$$|T_B[\psi](w) - T_B[\psi](w')| \leq 2^{-1}\,|w - w'| \quad \text{for} \quad w, w' \in B,$$

and consequently

$$\begin{aligned} |f(w) - f(w')| &\geq |w - w'| - |T_B[\psi](w) - T_B[\psi](w')| \\ &\geq 2^{-1}\,|w - w'| \quad \text{for all} \quad w, w' \in \overline{B}. \end{aligned}$$

Thus f from (3.5) yields a $C^{2,\mu}$-diffeomorphism from $\overline{B}$ onto $\overline{\Omega}$ with $\Omega = f(B)$. Since

$$\|T_B[\psi]\|_{2,\mu} \quad \leq \quad \frac{1}{2} c_1(\mu)\,\|\psi\|_{1,\mu}$$

holds true for some constant $c_1(\mu)$ and

$$\|\psi\|_{1,\mu} \quad \leq \quad 2\,\|q\|_{1,\mu} \quad \text{if} \quad \|q\|_{1,\mu} \leq c_0(\mu),$$

we obtain

$$\|f - id_{\overline{B}}\|_{2,\mu} \quad \leq \quad c_1(\mu)\,\|q\|_{1,\mu} \quad \text{if} \quad \|q\|_{1,\mu} \leq c_0(\mu). \tag{3.6}$$

Summarizing the above results, we have found

PROPOSITION 3. *There are constants $c_0(\mu), c_1(\mu) > 0$ such that for any $q \in C^{2,\mu}(\overline{B}, \mathbb{C})$ with $\|q\|_{1,\mu} \leq c_0(\mu)$ the Tricomi equation $\psi - q\Pi_B[\psi] = q$ has exactly one solution $\psi \in C^{1,\mu}(\overline{B}, \mathbb{C})$ satisfying $\|\psi\|_{1,\mu} \leq 2\,\|q\|_{1,\mu}$, and then $f := id_{\overline{B}} + T_B[\psi] \in C^{2,\mu}(\overline{B}, \mathbb{C})$ fulfils* (3.4). *Furthermore, f is a $C^{2,\mu}$-diffeomorphism from $\overline{B}$ onto $\overline{\Omega}$ where Ω is a simply connected domain in $\mathbb{C}$ bounded by a regular closed Jordan curve $\gamma \in C^{2,\mu}$, and f satisfies* (3.6). *The inverse $g = f^{-1} \in C^{2,\mu}(\overline{\Omega}, \mathbb{C})$ transfers ds^2 into the isothermal form*

$$g^* ds^2 = \lambda(x,y)\{dx^2 + dy^2\}, \quad z = x + iy \widehat{=} (x,y) \in \overline{\Omega}, \tag{3.7}$$

where $\lambda \in C^{1,\mu}(\overline{\Omega})$.

Moreover, if f is close to $id_{\overline{B}}$ with respect to the $C^{2,\mu}(\overline{B}, \mathbb{C})$-norm, then g is close to $id_{\overline{\Omega}}$ with respect to the $C^{2,\mu}(\overline{\Omega}, \mathbb{C})$-norm. Therefore, we obtain

PROPOSITION 4. *There exists an* $\epsilon_2^* \in (0,1)$ *with the following property: For any* ϵ_2 *with* $0 < \epsilon_2 \le \epsilon_2^*$ *there is a number* $\eta_2(X, \epsilon_2)$ *such that*

$$\|f(Y) - id_{\overline{B}}\|_{C^{2,\mu}(\overline{B},\mathbb{C})} < \epsilon_2, \quad \left\|g(Y) - id_{\overline{\Omega}(Y)}\right\|_{C^{2,\mu}(\overline{B},\mathbb{C})} < \epsilon_2,$$

$$\|\lambda(x,y) - 1\|_{C^{1,\mu}(\overline{B},\mathbb{C})} < \epsilon_2,$$

provided that $\|Y - X\|_{C^{2,\mu}(\overline{B},\mathbb{R}^2)} < \eta_2(X, \epsilon_2)$.

Step 2. Applying work of Courant (cf. [**1**, 4.3, Theorem 4]), Kellogg [**6**], and Warschawski [**12**]–[**14**], one proves, similarly as in [**4**] (cf. Proposition 4), the following perturbation result:

PROPOSITION 5. *There exists a number* $\epsilon_3^* \in (0,1)$ *with the following property: For* ϵ_3 *with* $0 < \epsilon_3 < \epsilon_3^*$ *we have a number* $\eta_3(X, \epsilon_3) > 0$ *such that for any* Y *with* $\|Y - X\|_{C^{2,\mu}} < \eta_3(X, \epsilon_3)$ *there is a diffeomorphism* $\varphi(Y) = \varphi(Y, .) \in C^{2,\mu}(\overline{B}, \mathbb{C})$ *from* $\overline{B}$ *onto* $\overline{\Omega}(Y)$ *which maps* $\overline{B}$ *conformally onto* $\overline{\Omega(Y)}$ *and satisfies*

$$\|\varphi(Y) - id_{\overline{B}}\|_{C^2(\overline{B},\mathbb{C})} < \epsilon_3.$$

Combining the results of Propositions 3–5 and introducing the mapping $\tau(Y) : \overline{B} \to \overline{B}$ by

$$\tau(Y) := g(Y) \circ \varphi(Y)$$

one obtains

PROPOSITION 6. *There exists a number* $\epsilon_4^* \in (0,1)$ *such that for any* $\epsilon_4 \in (0, \epsilon_4^*)$, *we have another number* $\eta_4(X, \epsilon_4) > 0$ *with the following property: for any* Y *with* $\|Y - X\|_{C^{2,\mu}(\overline{B},\mathbb{R}^n)} < \eta_4(X, \epsilon_4)$, *one can find a diffeomorphism* $\tau(Y) \in C^2(\overline{B}, \mathbb{C})$ *from* $\overline{B}$ *onto itself such that*

$$\|\tau(Y) - id_{\overline{B}}\|_{C^2(\overline{B},\mathbb{C})} < \epsilon_4, \tag{3.8}$$

and $Z := Y \circ \tau(Y)$ *satisfies the conformality relations*

$$\mathcal{E}(Z) = \mathcal{G}(Z), \quad \mathcal{F}(Z) = 0. \tag{3.9}$$

Step 3. Now we verify the assertion (3.2) of Theorem 4 assuming (3.1): For any $Y \in \mathcal{C}(\Gamma, M)$ with $\|Y - X\|_{C^{2,\mu}} < \eta(X, \epsilon_4)$, at first, we choose $\tau(Y)$ as in Proposition 6 and form $Z = Y \circ \tau(Y) \in \mathcal{C}(\Gamma, M)$. Second, if $\epsilon > 0$ denotes the number in (3.1), a straightforward estimation yields that choosing ϵ_4 sufficiently small, one can find a number $\delta(\epsilon) \in (0, \epsilon/2)$ such that

$$\|Z - Y\|_{C^2} < \epsilon/2 \quad \text{if} \quad \|Y - X\|_{C^{2,\mu}} < \delta(\epsilon).$$

(Here and below we abbreviate $C^2 = C^2(\overline{B}, \mathbb{R}^n)$, $C^{2,\mu} = C^{2,\mu}(\overline{B}, \mathbb{R}^n)$.) Then we see

$$\begin{aligned}\|Z - X\|_{C^2} &\le \|Z - Y\|_{C^2} + \|Y - X\|_{C^2} \\ &\le \|Z - Y\|_{C^2} + \|Y - X\|_{C^{2,\mu}} < \epsilon/2 + \epsilon/2 = \epsilon\end{aligned}$$

for $\|Y - X\|_{C^{2,\mu}} < \delta(\epsilon)$, and by (3.1) it follows that

$$\mathcal{D}(X) \le \mathcal{D}(Z).$$

Moreover, (2.17) and (3.9) yield

$$\mathcal{A}(X) = \mathcal{D}(X) \quad \text{and} \quad \mathcal{A}(Z) = \mathcal{D}(Z),$$

and the parameter invariance of $\mathcal{A}$ implies

$$\mathcal{A}(Z) = \mathcal{A}(Y).$$

Since $\mathcal{A}(Y) \le \mathcal{D}(Y)$ holds true, we finally obtain

$$\mathcal{A}(X) = \mathcal{D}(X) \le \mathcal{D}(Z) = \mathcal{A}(Z) = \mathcal{A}(Y) \quad \text{for} \quad \|Y - X\|_{C^{2,\mu}} < \delta(\epsilon),$$

and (3.2) is proved. □

References

[1] Dierkes, U.; Hildebrandt, S.; Sauvigny, F.: *Minimal Surfaces.* Grundlehren der mathematischen Wissenschaften **339**, Springer-Verlag, Berlin 2010. MR2566897 (2012b:49002)

[2] Heinz, E.; Hildebrandt, S.: *Some remarks on minimal surfaces in Riemannian manifolds.* Comm. Pure Appl. Math. **23**, 371-377 (1970). MR0259765 (41:4398)

[3] Hildebrandt, S.; von der Mosel, H.: *On Lichtenstein's theorem about globally conformal mappings.* Calc. Var. **23**, 415-424 (2005). MR2153031 (2006g:49071)

[4] Hildebrandt, S.; Sauvigny, F.: *Relative minimizers of energy are relative minimizers of area.* Calc. Var. **37**, 475-483 (2010) . MR2592983 (2011a:58022)

[5] Jost, J.: *Two-dimensional geometric variational problems.* Wiley, New York 1990. MR1100926 (92h:58045)

[6] Kellogg, O.D.: *Harmonic functions and Green's integral.* Trans. Amer. Math. Soc. **13**, 109-132 (1912). MR1500909

[7] Morrey, C.B.: *The problem of Plateau on a Riemannian manifold.* Ann. Math **49**, 807-851 (1948). MR0027137 (10:259f)

[8] Morrey, C.B.: *Multiple integrals in the calculus of variations.* Grundlehren der mathematischen Wissenschaften **130**, Springer-Verlag, Berlin 1966. MR0202511 (34:2380)

[9] Sauvigny, F.: *Partial Differential Equations 2.* Universitext, Springer, Berlin 2006; Second revised and enlarged edition: Springer, London 2012. MR2254750 (2007c:35002)

[10] Tomi, F.: *Ein einfacher Beweis eines Regularitätssatzes für schwache Lösungen gewisser elliptischer Systeme.* Math. Z. **112**, 214-218 (1969). MR0257549 (41:2199)

[11] Vekua, I.N.: *Verallgemeinerte analytische Funktionen.* Akademie-Verlag, Berlin 1963. MR0158085 (28:1312)

[12] Warschawski, S.: *Über einen Satz von O. D. Kellogg.* Nachr. Akad. Wiss. Göttingen (1932), 73-86.

[13] Warschawski, S.: *Über das Randverhalten der Abbildungsfunktion bei konformer Abbildung.* Math. Z. **35**, 321-456 (1932) MR1545302

[14] Warschawski, S.: *On the higher derivatives at the boundary in conformal mapping.* Trans. Amer. Math. Soc. **38**, 310-340 (1935). MR1501813

Mathematisches Institut der Universität Bonn, Endenicher Allee 60, D-53115 Bonn, Germany

Mathematisches Institut der Brandenburgischen Technischen Universität, Platz der Deutschen Einheit 1, D-03046 Cottbus, Germany

E-mail address: sauvigny@math.tu-cottbus.de

Amer. Math. Soc. Transl.
Volume **232**, 2014

The boundary Harnack principle for second order elliptic equations in John and uniform domains

H. Kim and M. Safonov

In memory of Olga A. Ladyzhenskaya

ABSTRACT. Consider second order uniformly elliptic equations $Lu := a_{ij}D_{ij}u + b_iD_iu = 0$ in a domain $\Omega \subset \mathbb{R}^n$ with bounded measurable a_{ij} and $b_i \in \mathbb{L}^n(\Omega)$. We give direct proofs of the boundary Harnack principle for the ratios of positive solutions in John domains, and the Hölder regularity near the boundary in uniform domains.

1. Introduction

We consider strong solutions to second order elliptic equations in *non-divergence* form

$$Lu := a_{ij}D_{ij}u + b_iD_iu = 0 \quad \text{in an open set} \quad \Omega \subset \mathbb{R}^n, \tag{1.1}$$

and the corresponding inequalities $Lu \le 0$ or $Lu \ge 0$ in Ω. This means that $u \in W(\Omega) := W^{2,n}_{loc}(\Omega) \cap C(\overline{\Omega})$, i.e., u, $D_iu := \partial u/\partial x_i$, and $D_{ij}u := D_iD_ju$ belong to $\mathbb{L}^n(\Omega_0)$ for every bounded open subset $\Omega_0 \subset \overline{\Omega_0} \subset \Omega$, and the relations $Lu = 0$, $Lu \le 0$, and $Lu \ge 0$ are understood in the sense of almost everywhere (a.e.) in Ω. Throughout this article, we adopt the summation convention over repeated indices. We assume that $a_{ij} = a_{ij}(x)$ are measurable functions satisfying the *uniform ellipticity condition*

$$a_{ij} = a_{ji}, \qquad \nu|\xi|^2 \le a_{ij}\xi_i\xi_j \le \nu^{-1}|\xi|^2 \quad \text{for all} \quad \xi \in \mathbb{R}^n, \tag{1.2}$$

with the *ellipticity constant* $\nu \in (0,1]$, and $b_i \in \mathbb{L}^n(\Omega)$. Note that the quantity

$$S := \int_\Omega |\mathbf{b}(x)|^n dx < \infty, \quad \textit{where} \quad \mathbf{b} := (b_1, \dots, b_n), \tag{1.3}$$

is invariant with respect to rescaling $x \to \text{const} \cdot x$ in (1.1).

In a simplified formulation, the *boundary Harnack principle* states that if $x_0 \in \partial\Omega$, and $u, v \in W(\Omega)$ are positive solutions of (1.1) in $\Omega_R := \Omega \cap B_R(x_0)$ which

---

2010 *Mathematics Subject Classification.* Primary 35B05; Secondary 35J15.

Key words and phrases. Boundary Harnack principle, Carleson type estimates, elliptic equations with measurable coefficients.

vanish on $(\partial\Omega)\cap B_R(x_0)$, then

$$\sup_{\Omega_{R/2}} \frac{u}{v} \le N \cdot \inf_{\Omega_{R/2}} \frac{u}{v}, \tag{1.4}$$

with a constant N depending only on the prescribed quantities. If this constant does not depend on R, then by a simple iteration, one can get the Hölder continuity of the ratio u/v and further applications to the Martin boundary, H^p and BMO spaces. A systematic exposition of these and other results for harmonic functions in non-tangentially accessible (NTA) domains was presented by Jerison and Kenig in [**JK82**]. The assumptions on NTA domains include the exterior measure condition, which guarantees the boundary regularity of solutions. Bass and Burdzy [**BB91, BB92**] proved the estimate (1.4) for solutions to second order elliptic equations in the *divergence* form

$$D_i(a_{ij}D_j u) = 0 \quad \text{in} \quad \Omega, \tag{1.5}$$

assuming that Ω is a twisted Hölder domain of order $\alpha \in (1/2, 1]$, and they showed that this estimate fails in general if $\alpha \in (0, 1/2)$. In our recent works [**KmS11, KmS11a**], we have extended their result to the non-divergence equation (1.1) and also gave a counterexample for the boundary case $\alpha = 1/2$ (Example 5.1 in [**KmS11**]). See the above references for definitions, history, and more applications.

The bounded twisted Hölder domains of order $\alpha = 1$ are also called the twisted cone domains, or *John domains*. They can be defined in many equivalent ways. We will use the definition in terms of lengths of curves $\gamma(a,x) \subset \Omega$ connecting the points a and x in Ω. All the curves under consideration are rectifiable.

DEFINITION 1.1. A bounded domain $\Omega \subset \mathbb{R}^n$ is a *John domain with center* $z_0 \in \Omega$ and constant $K \ge 1$ if for each $a \in \Omega\backslash\{z_0\}$, there is a curve $\gamma = \gamma(a, z_0) \subset \Omega$, connecting a and z_0, and such that

$$|\gamma(a,x)| \le Kd(x) \qquad \forall x \in \gamma(a, z_0), \tag{1.6}$$

where $d(x) := \operatorname{dist}(x, \partial\Omega)$, and $|\gamma(a,x)|$ denotes the length of the subcurve $\gamma(a,x) \subset \gamma(a,z_0)$ which connects a and x.

Note that from (1.6) with $x = z_0$ it follows that

$$|\gamma| = |\gamma(a, z_0)| \le R_0 := Kd(z_0) \qquad \forall a \in \Omega. \tag{1.7}$$

Since $|a - z_0| \le |\gamma(a, z_0)|$, we get $\Omega \subset B_{R_0}(z_0)$, and

$$\operatorname{diam}\Omega \le 2R_0 = 2Kd(z_0) \le K \cdot \operatorname{diam}\Omega. \tag{1.8}$$

In [**KmS11, KmS11a**], the John domains were treated as special cases of twisted Hölder domains, so that the proofs of the boundary Harnack principle and other related results were rather involved. In the present article, we give more direct proofs of the boundary Harnack principle for the ratios u/v of positive solutions to equation (1.1) in John domains and its natural consequences, such as the Hölder continuity of u/v in uniform domains. These direct proofs are more transparent and flexible for application to other situations, e.g., to equations on manifolds.

One of our main results is the following theorem.

THEOREM 1.2 (Boundary Harnack principle). *Let Ω be a John domain with center $z_0 \in \Omega$ and constant $K \ge 1$ in (1.6). Let functions $u, v \in W(\Omega)$ be such that*

$$u > 0, \quad v > 0; \quad Lu = Lv = 0 \qquad \text{in} \quad \Omega_R := \Omega \cap B_R(x_0), \tag{1.9}$$

and $u \equiv 0$ *on* $\Gamma_R := (\partial\Omega) \cap B_R(x_0)$, *where* $x_0 \in \partial\Omega$ *and* $R > 0$ *satisfies* $\operatorname{diam}\Omega \le K_0 R$ *with* $K_0 = \text{const} > 1$. *Then*

$$\sup_{\Omega_{R/2}} \frac{u}{v} \le N_1 \cdot \frac{u(z_0)}{v(z_0)}, \tag{1.10}$$

where the constant $N_1 = N_1(n, \nu, S, K, K_0) > 1$.

In this theorem, we do not assume that $v \equiv 0$ on Γ_R. Under this additional assumption, the estimate (1.4) easily follows.

COROLLARY 1.3. *In addition to the assumptions in Theorem* 1.2, *let* $v \equiv 0$ *on* Γ_R. *Then the estimate* (1.4) *holds true with* $N := N_1^2$, *where* N_1 *is the constant in* (1.10).

PROOF. Interchanging u and v in (1.8), we get

$$\inf_{\Omega_{R/2}} \frac{u}{v} = \left(\sup_{\Omega_{R/2}} \frac{v}{u}\right)^{-1} \ge \left(N_1 \cdot \frac{v(z_0)}{u(z_0)}\right)^{-1} = N_1^{-1} \cdot \frac{u(z_0)}{v(z_0)},$$

and (1.4) follows with $N := N_1^2$. □

In Section 2, we discuss some known properties and equivalent definitions of John domains and closely related *uniform* domains. Our definitions of c-uniform and c-John domains below are taken from [**MS78, V88, NV91, GNV94**], in a simplified form, so that the same constant c appears in both properties (1.11) and (1.12).

DEFINITION 1.4. A domain $\Omega \subset \mathbb{R}^n$ is *c-uniform* with $c = \text{const} \ge 1$, if each pair of distinct points $a, b \in \Omega$ can be connected by a curve $\gamma = \gamma(a, b) \subset \Omega$ satisfying

$$|\gamma(a, x)| \wedge |\gamma(b, x)| \le c\, d(x) \qquad \forall x \in \gamma(a, b), \tag{1.11}$$

where $A \wedge B := \min\{A, B\}$, and

$$|\gamma(a, b)| \le c \cdot |a - b|. \tag{1.12}$$

According to Definitions 1.1 and 1.4, every John domain is bounded, whereas c-uniform domains can be unbounded. In this context, it is natural to treat c-uniform domains as subclass of c-John domains, which can also be unbounded.

DEFINITION 1.5. A domain $\Omega \subset \mathbb{R}^n$ is *c-John domain* with $c = \text{const} \ge 1$, if each pair of distinct points $a, b \in \Omega$ can be connected by a curve $\gamma = \gamma(a, b) \subset \Omega$ satisfying (1.11).

It is known that the class of John domains coincide with the class of *bounded* c-John domains (Theorem 2.3). Having in mind this relation, we have the following strict inclusions between different classes of domains:

$$\text{Lipschitz} \quad \subset \quad \text{NTA} \quad \subset \quad \text{Uniform} \quad \subset \quad \text{John}. \tag{1.13}$$

For more classes, cf. [**JK82**, p.92] and [**Ak05**, p.42].

There is extensive literature on this subject; see Section 4.3 in the book [**EE04**], the articles [**MS78, V88, NV91, GNV94, Ak05, DRS10**], and references therein. It includes numerous equivalent definitions of such kind of domains, which may be more convenient in other applications. In particular, one can get six equivalent definitions of John domains (with different constants K or c) by taking $\operatorname{diam}\gamma(a,x)$ or $|a-x|$ in (1.6) or (1.11) in place of $|\gamma(a,x)|$ (cf.[**NV91**, 2.16]). The corresponding relations are easier to verify, because $|a-b| \le \operatorname{diam}\gamma \le |\gamma|$ for $\gamma = \gamma(a,b)$. We provide the proofs of equivalence of different definitions, which are related to our technique, i.e., some arguments in the proofs are used in our constructions. These include the equivalence of definitions of John domains in terms of different characteristics of curves (Theorem 2.6).

In Section 3, we review the main tools in our techniques: a *growth lemma* (Lemma 3.2) and the interior *Harnack inequality.* So-called growth lemmas were originally introduced by Landis [**L71**] in his proof of the Harnack inequality and Hölder regularity of solutions to second order uniformly elliptic and parabolic equations in *divergence* form. The growth lemmas were also successfully used in the proof of the Harnack inequality for equations (1.1) with $b_i \in \mathbb{L}^\infty$ in [**KS80, S80, K85**] and in its recent extension to $b_i \in \mathbb{L}^n$ in [**S10**].

The proof of Theorem 1.2 is given in Section 4. We precede it with a particular case $v \equiv 1$ (Theorem 4.2). It is called a Carleson type estimate, because it first appeared in the work by Carleson [**C61**] for harmonic functions in Lipschitz domains.

Finally, in Section 5 we derive the Hölder continuity of ratios of solutions in uniform domains based on the boundary Harnack principle in John domains of small size. The boundary Harnack principle and Hölder continuity of ratios u/v of harmonic functions in uniform domains was proved by Aikawa [**Ak01**]. His technique uses the estimates for Green's functions which are not available in the case of non-divergence equations (1.1). There are easy examples (e.g., Example 3.2 in [**KmS11a**]) which show that in general there is no continuity of u/v near the boundary of a John domain.

REMARK 1.6. For uniformly elliptic equations in *divergence form*

$$D_i(a_{ij}D_ju) = 0 \quad \text{in} \quad \Omega,$$

a growth lemma in a formulation close to our Lemma 3.2 is contained in [**L71**, Ch. 2, Lemma 3.4]. Since our technique uses only this growth lemma and the maximum principle, the results in this article remain valid for such equations. However, most of them were already obtained by different means in [**BB91, BB92**]. Some new growth lemmas were recently proved by Nazarov and Uraltseva [**NU12**] for elliptic and parabolic equations in divergence form with additional lower order terms b_iD_iu with coefficient b_i in appropriate Lebesgue spaces satisfying $D_ib_i = 0$, so that $b_iD_iu = D_i(b_iu)$, and the whole equation has divergence structure. These are Lemma 2.2 and most closely, Corollary 3.4 in [**NU12**] in the elliptic and parabolic cases, correspondingly. Combining these facts with classical methods in [**LU, LSU**], they proved the Harnack inequality and other related facts for such equations.

We use the following notations:

$x = (x_1, x_2, \dots, x_n)$ are vectors or points in $\mathbb{R}^n$, $|x|$ is the length of $x \in \mathbb{R}^n$.

We also denote $|\gamma|$ the length of a rectifiable curve $\gamma \subset \mathbb{R}^n$, and $|V|$ the Lebesgue measure of an open set $V \subset \mathbb{R}^n$.

$B_r(y) := \{x \in \mathbb{R}^n : |x-y| < r\}$ is a ball of radius $r > 0$ centered at $y \in \mathbb{R}^n$.

$\Omega_r(y) := \Omega \cap B_r(y)$ for $r > 0$ and $y \in \mathbb{R}^n$, and a given open set $\Omega \subset \mathbb{R}^n$.

$d(x) := \operatorname{dist}(x, \partial\Omega)$ for $x \in \Omega$.

$\operatorname{diam}\Omega := \sup\{|x-y| : x, y \in \Omega\}$.

$[a, b]$ is a line segment connecting the points a and b in $\mathbb{R}^n$.

$[a_0, a_1, \ldots, a_m] := [a_0, a_1] \cup [a_1, a_2] \cup \cdots \cup [a_{m-1}, a_m]$ is a polygonal line in $\mathbb{R}^n$.

$c_1 \vee c_2 := \max\{c_1, c_2\}$, $c_1 \wedge c_2 := \min\{c_1, c_2\}$ for $c_1, c_2 \in \mathbb{R}^1$.

$[c] := \max\{\text{integers } k \le c\}$ is the integer part of $c \in \mathbb{R}^1$, $c_+ := \max\{c, 0\}$, $c_- := \max\{-c, 0\}$.

The notation $A := B$, or $B =: A$ means "$A = B$ by definition". Throughout the paper, N (with indices or without) denotes different constants depending only on the prescribed quantities such that n, ν, etc. This dependence is indicated in parentheses: $N = N(n, \nu, \ldots)$.

2. John domains and uniform domains

In this section, we prove the equivalence between different definitions of John domains and establish some relations between John and uniform domains. These relations are of a qualitative nature, i.e., in most cases, the constants are far from optimal. In our notations and some details of proofs, we partially follow [**MS78, V88, NV91**].

Originally, *John domains* were introduced by F. John [**J61**, Definition on p. 402] in terms of the inner and outer radii, which can be described as follows.

DEFINITION 2.1. We say that a bounded domain Ω has an *inner radius* $r_0 > 0$ and an *outer radius* $R_0 > 0$ if there exists a point $z_0 \in \Omega$, and for each $a \in \Omega \setminus \{z_0\}$, there is a curve $\gamma = \gamma(a, z_0)$ of length $|\gamma| \le R_0$ satisfying the property

$$|\gamma(a, x)| \le \frac{|\gamma|}{r_0} \cdot d(x) \qquad \forall x \in \gamma. \tag{2.1}$$

THEOREM 2.2. *Definitions* 1.1 *and* 2.1 *are equivalent in the following sense.*

(I) *If Ω is a John domain with center z_0 and constant $K \ge 1$, then it has inner radius $r_0 := (2K)^{-1} d(z_0)$ and outer radius $R_0 := K d(z_0)$.*

(II) *If Ω has inner radius $r_0 > 0$ and outer radius $R_0 > 0$, then it is a John domain with center $z_0 \in \Omega$ and constant $K := R_0 / r_0 \ge 1$.*

In both definitions, the point (center) $z_0 \in \Omega$ is the same.

PROOF. (I) Let Ω be a John domain in Definition 1.1. Fix a point $a \in \Omega \setminus \{z_0\}$, and let $\gamma = \gamma(a, z_0) \subset \Omega$ be a curve satisfying (1.6). In order to verify the property (2.1), consider separately the cases (i) and (ii).

(i) $|\gamma| = |\gamma(a, z_0)| < \rho := d(z_0)/2 = K r_0$. In this case, $\gamma \subset B_\rho(z_0)$, and for every $x \in \gamma$,

$$d(x) := \operatorname{dist}(x, \partial\Omega) \ge d(z_0) - \rho = \rho, \tag{2.2}$$

and

$$|\gamma(a, x)| \le |\gamma| \le \frac{|\gamma|}{\rho} \cdot d(x) \le \frac{|\gamma|}{r_0} \cdot d(x).$$

(ii) $|\gamma| \ge \rho$. Then by (1.6),

$$|\gamma(a, x)| \le K d(x) = \frac{\rho}{r_0} \cdot d(x) \le \frac{|\gamma|}{r_0} \cdot d(x) \qquad \forall x \in \gamma.$$

Therefore, (2.1) holds true in any case.

(II) Let Ω, r_0, R_0, z_0, a, and let $\gamma = \gamma(a, z_0)$ be as in Definition 2.1. Then from (2.1) and $|\gamma| \le R_0$ it follows (1.6) that

$$|\gamma(a,x)| \le \frac{R_0}{r_0} \cdot d(x) =: Kd(x) \qquad \forall x \in \gamma.$$

The theorem is proved. □

THEOREM 2.3. *For bounded domains $\Omega \subset \mathbb{R}^n$, Definitions 1.1 and 1.5 are equivalent in the following sense.*

(I) *If Ω is a John domain with center $z_0 \in \Omega$ and constant $K \ge 1$, then it is a c-John domain with $c := K$.*

(II) *If Ω is a c-John domain with $c = \text{const} \ge 1$, then it is a John domain with center $z_0 \in \Omega$ and constant $K := 4c^3$.*

PROOF. (I) Let $\Omega \subset \mathbb{R}^n$ be a John domain with center $z_0 \in \Omega$ and constant $K \ge 1$. Then for arbitrary $a, b \in \Omega$, $a \ne b$, we can take the curves $\gamma(a, z_0)$ and $\gamma(b, z_0)$ according to (1.6), and define

$$\gamma(a,b) := \gamma(a,z_0) \cup \gamma(b,z_0) \subset \Omega.$$

If $x \in \gamma(a, z_0)$, then by (1.6),

$$|\gamma(a,x)| \wedge |\gamma(b,x)| \le |\gamma(a,x)| \le Kd(x).$$

By symmetry, we also have similar inequalities for $x \in \gamma(b, z_0)$, and (1.6) holds true with $c := K$.

(II) Let Ω be a bounded c-John domain with $c = \text{const} \ge 1$. For fixed $a_0, b_0 \in \Omega$, $a_0 \ne b_0$, take a curve $\gamma = \gamma(a_0, b_0)$ by Definition 1.4, and choose a point $z_0 \in \gamma$ such that $|\gamma(a_0, z_0)| = |\gamma(b_0, z_0)| = |\gamma|/2$. Using (1.11) with $x = z_0$, we get

$$|a_0 - b_0| \le |\gamma(a_0,b_0)| = 2\big(|\gamma(a_0,z_0)| \wedge |\gamma(b_0,z_0)|\big) \le 2cd(z_0). \tag{2.3}$$

By an easy compactness argument, there is a point $z_0 \in \Omega$ such that

$$\operatorname{diam}\Omega := \sup\big\{|a_0 - b_0| : \ a_0, b_0 \in \Omega\big\} \le 2cd(z_0),$$

which in turn implies

$$d(z) \le \frac{1}{2} \cdot \operatorname{diam}\Omega \le cd(z_0) \qquad \forall z \in \Omega. \tag{2.4}$$

Now fix $a \in \Omega \setminus \{z_0\}$ and take $\gamma = \gamma(a, z_0) \subset \Omega$ satisfying (1.11) with $b = z_0$, i.e.,

$$|\gamma(a,x)| \wedge |\gamma(z_0,x)| \le cd(x) \qquad \forall x \in \gamma(a,z_0). \tag{2.5}$$

Choose a point $z \in \gamma(a, z_0)$ such that $|\gamma(a, z)| = |\gamma(z_0, z)| = |\gamma(a, z_0)|/2$. Similarly to (2.3), we have $|\gamma(a, z_0)| \le 2cd(z)$. Together with (2.4), this yields

$$|\gamma(a,z_0)| \le 2c^2\, d(z_0). \tag{2.6}$$

For each $x \in \gamma(a, z_0)$, we have one of three possible options:

(i) $|\gamma(a, x)| \le |\gamma(z_0, x)|$. Then by (2.5), $|\gamma(a, x)| \le cd(x)$.

(ii) $|\gamma(a, x)| > |\gamma(z_0, x)| \ge \rho := d(z_0)/2$. In this case, from (2.5) it follows $d(z_0) = 2\rho \le 2cd(x)$.

(iii) $|\gamma(a, x)| \wedge \rho > |\gamma(z_0, x)|$. As in (2.2), $\gamma(z_0, x) \subset B_\rho(z_0)$ and $d(z_0) = 2\rho \le 2d(x)$.

In both cases (ii) and (iii), we have $d(z_0) \le 2cd(x)$. Combining this inequality with (2.6), we obtain

$$|\gamma(a,x)| \le |\gamma(a,z_0)| \le 2c^2 d(z_0) \le 4c^3 d(x).$$

We see that in each of these three cases, the desired inequality (1.6) holds true with $K := 4c^3$. The theorem is proved. □

For curves $\gamma = \gamma(a,b) \subset \Omega$ connecting a and b, $a \ne b$, one can compare the following characteristics:

(i) $l_1(\gamma(a,b)) := |\gamma(a,b)|$, the *length* of $\gamma(a,b)$;
(ii) $l_2(\gamma(a,b)) := \operatorname{diam} \gamma(a,b)$, the *diameter* of $\gamma(a,b)$;
(iii) $l_3(\gamma(a,b)) := |a-b|$, the *distance* between the ends of $\gamma(a,b)$.

Obviously,

$$(2.7) \qquad |\gamma(a,b)| \ge \operatorname{diam} \gamma(a,b) \ge |a-b|, \quad \text{i.e.,} \quad l_1(\gamma) \ge l_2(\gamma) \ge l_3(\gamma).$$

As in [**MS78, V88, DRS10**], consider the following objects.

DEFINITION 2.4. Let $\gamma = \gamma(a,b), a \ne b$, be a curve in $\mathbb{R}^n$. For $j = 1,2,3$ and $K > 0$, $c > 0$, define the following:

(i) *K-carrots with core* γ:

$$(2.8) \qquad \operatorname{car}_j(\gamma, K) := \bigcup_{x\in\gamma} \left\{ B_{r(x)}(x) : \, r(x) = K^{-1} \cdot l_j\big(\gamma(a,x)\big) \right\};$$

(ii) *c-cigars with core* γ:

$$(2.9) \qquad \operatorname{cig}_j(\gamma, c) := \bigcup_{x\in\gamma} \left\{ B_{r(x)}(x) : \, r(x) = c^{-1} \cdot \Big(l_j\big(\gamma(a,x)\big) \wedge l_j\big(\gamma(b,x)\big)\Big) \right\}.$$

Note that for fixed $\gamma = \gamma(a,z_0) \subset \Omega$ and $x \in \gamma$, the inequality in (1.6) holds true if and only if $r(x) := K^{-1} \cdot |\gamma(a,x)| \le d(x)$, i.e., $B_{r(x)}(x) \subset \Omega$. Therefore, for any individual curve $\gamma = \gamma(a,z_0) \in \Omega$, the property (1.6) is equivalent to $\operatorname{car}_1(\gamma,K) \subset \Omega$. Similarly, (1.11) is equivalent to $\operatorname{cig}_1(\gamma,c) \subset \Omega$. For these relations, we do not need the assumptions $K \ge 1$ and $c \ge 1$. However, these restrictions are needed in Definitions 1.1, 1.4, and 1.5, in which (1.6) and (1.11) hold true uniformly for all $a \in \Omega$ and $b \in \Omega$. For example, from (1.6) with $x = z_0$, it follows that

$$d(z_0) \le \sup_{a\in\Omega} |a - z_0| \le \sup_{a\in\Omega} |\gamma(a,z_0)| \le K d(z_0),$$

so that $K \ge 1$. Under these assumptions, we also have the following elementary facts.

LEMMA 2.5. *Let $\gamma = \gamma(a,b)$ be a curve in $\mathbb{R}^n$ with $a \ne b$.*

(i) *If $K = \text{const} \ge 1$, then $\operatorname{car}_1(\gamma,K)$ is a John domain with center b and constant K.*

(ii) *If $c = \text{const} \ge 1$, then $\operatorname{cig}_1(\gamma,c)$ is a c-John domain.*

PROOF. For fixed $\gamma = \gamma(a,b), a \ne b$, and $K \ge 1$, denote $\Omega_1 := \operatorname{car}_1(\gamma,K)$ according to (2.8) with $l_1\big(\gamma(a,x)\big) := |\gamma(a,x)|$. For each $p \in \Omega_1$, there is a point $q \in \gamma$ depending on p and such that $p \in B_r(q) \subset \Omega_1$, where $r = r(q) := K^{-1} \cdot |\gamma(a,q)|$. Set

$$\gamma_p := \gamma_p(p,b) := [p,q] \cup \gamma(q,b), \quad \text{where} \quad \gamma(q,b) \subset \gamma(a,b).$$

If $x \in [p, q]$, then

$$|\gamma_p(p, x)| = |p - x| \le \operatorname{dist}\big(x, \partial B_r(q)\big) \le d_1(x) := \operatorname{dist}(x, \partial\Omega_1).$$

Note that $|p - q| \le r \le |\gamma(a, q)|$ because $K \ge 1$. Therefore, if $x \in \gamma(q, b)$, then

$$|\gamma_p(p, x)| = |p - q| + |\gamma(q, x)| \le |\gamma(a, q)| + |\gamma(q, x)| = |\gamma(a, x)| = Kr(x) \le Kd_1(x).$$

Hence

$$|\gamma_p(p, x)| \le Kd_1(x) = K \cdot \operatorname{dist}(x, \partial\Omega_1) \qquad \forall x \in \gamma_p(p, b).$$

Since p is an arbitrary point in Ω_1, the part (i) is proved. The proof of (ii) is quite similar. □

The equivalence of definitions of John and c-John domains in the diameter sense, i.e., the analogue of Theorem 2.3(II) with $\operatorname{diam}\gamma$ instead of $|\gamma|$ was proved in [**V88**] with $K := 8c^2$. In the following theorem, we bring together three equivalent definitions (D_1), (D_2), and (D_3) of John domains Ω, which are based correspondingly on the characterizations $l_1(\gamma)$, $l_2(\gamma)$, and $l_3(\gamma)$ for curves $\gamma = \gamma(a, z_0) \subset \Omega$. The equivalence between (D_1) and (D_2) was proved in [**MS78**, 2.7], and between (D_2) and (D_3)—in [**V88**, Theorem 2.18]. The implication $(D_1) \Rightarrow (D_2) \Rightarrow (D_3)$ follows automatically from (2.7). By modifying the argument in [**V88**], we show that $(D_3) \Rightarrow (D_1)$, hence all these definitions are equivalent. Note that our "unifying" estimates (2.11) do not cover the results in [**V88**], because in fact the bound for K_2 in terms of K_3 does not depend on the dimension n.

THEOREM 2.6. *Let Ω be a bounded domain in $\mathbb{R}^n$. We claim that for $j = 1, 2, 3$, and fixed $z_0 \in \Omega$, the following properties (D_j) are equivalent.*

(D_j) There is a constant $K_j \ge 1$ such that for each $a \in \Omega \setminus \{z_0\}$, there exists a curve $\gamma_j = \gamma_j(a, z_0) \subset \Omega$ connecting a and z_0, such that

$$l_j\big(\gamma_j(a, x)\big) \le K_j d(x) \qquad \forall x \in \gamma_j(a, z_0), \tag{2.10}$$

where $\gamma_j(a, x)$ is a subcurve of $\gamma_j(a, z_0)$ connecting a and x. Moreover, if we choose the minimal possible constants $K_j \ge 1$ satisfying (D_j), then

$$K_1 \ge K_2 \ge K_3, \quad \text{and} \quad K_1 \le 4 \cdot (8K_3^2 + 2)^n. \tag{2.11}$$

We precede the proof of this theorem with a simple lemma.

LEMMA 2.7. *For each $j = 1, 2, 3$, from the property (2.10) it follows that*

$$d(a) \le 2K_j d(x) \qquad \forall x \in \gamma_j(a, z_0). \tag{2.12}$$

PROOF. For each $x \in \gamma_j(a, z_0)$, we have one of the following two cases.
(i) $|x - a| \le d(a)/2$. In this case,

$$d(x) \ge d(a) - |x - a| \ge \frac{d(a)}{2}, \quad \text{so that} \quad d(a) \le 2d(x) \le 2K_j d(x).$$

(ii) $|x - a| > d(a)/2$. Then by virtue of (2.10),

$$d(a) < 2|x - a| \le 2|\gamma_j(a, x)| \le 2K_j d(x).$$

Therefore, (2.12) holds true in any case. □

PROOF OF THEOREM 2.6. From inequalities $l_1(\gamma) \ge l_2(\gamma) \ge l_3(\gamma)$ in (2.7) it follows that $(D_1) \Rightarrow (D_2) \Rightarrow (D_3)$ with the minimal constants $K_1 \ge K_2 \ge K_3$. We only need to show that if Ω and $z_0 \in \Omega$ satisfy (D_3) with a constant $K := K_3 \ge 1$, then they also satisfy (D_1) with $K_1 := 4 \cdot (8K^2+2)^n$. Under these assumptions, fix a point $a \in \Omega \setminus \{z_0\}$ and $\gamma = \gamma(a, z_0) := \gamma_3(a, z_0) \subset \Omega$ satisfying (2.10), which in turn implies (2.12), so that we have

$$(2.13) \qquad |a - x| \le Kd(x), \qquad d(a) \le 2Kd(x) \qquad \forall x \in \gamma(a, z_0).$$

We will construct a polygonal line $\gamma_1 = \gamma_1(a, z_0)$ satisfying the desired property

$$(2.14) \qquad |\gamma_1(a, x)| \le K_1 d(x) \qquad \forall x \in \gamma_1(a, z_0).$$

Set $r_0 := (4K)^{-1} d(a) > 0$. By (2.13), we have $d(x) \ge 2r_0$ on $\gamma(a, z_0)$. Therefore,

$$(2.15) \qquad d(x) \ge r_0 \quad \text{on} \quad [p, q] \quad \text{if} \quad p, q \in \gamma(a, z_0) \quad \text{and} \quad |p - q| \le 2r_0.$$

The curve γ can be parametrized by the arc length

$$\gamma = \gamma(a, z_0) =: \{g(s) : \ 0 \le s \le S := |\gamma|\}.$$

Starting from the point $a_0 = a = g(0)$, introduce a finite sequence

$$(2.16) \qquad \{a_j = g(s_j) : \ 0 = s_0 < s_1 < \cdots < s_{k-1} < s_k := S\} \subset \gamma(a, z_0)$$

by induction as follows. For $j \ge 1$, if $a_{j-1} \ne z_0$ is already selected and $|a_{j-1} - z_0| \le 2r_0$, we set $a_j := a_k := z_0$, and the construction is complete. Otherwise, we have

$$|g(s_{j-1}) - g(S)| = |a_{j-1} - z_0| > 2r_0.$$

Using continuity of $g(s)$, choose s_j as the maximal possible value of s in (s_{j-1}, S) for which $|g(s_{j-1}) - g(s)| = 2r_0$. Then $a_j := g(s_j)$ satisfies $|a_{j-1} - a_j| = 2r_0$. Moreover, we also have

$$|g(s_i) - g(s_j)| = |a_i - a_j| \ge 2r_0 \quad \text{if} \quad 0 \le i \le j-2.$$

Indeed, if this is not the case, then

$$|g(s_i) - g(s_j)| < 2r_0 < |g(s_i) - g(S)|,$$

and by continuity, there exists $s^* \in (s_j, S)$ such that $|g(s_i) - g(S)| = 2r_0$. Here $s_i < s_{i+1} < s_j < s^*$, which contradicts to the choice of s_{i+1} as the maximal value of s satisfying $|g(s_i) - g(s)| = 2r_0$. This argument shows that $a_j := g(s_j)$ satisfy

$$(2.17) \qquad |a_i - a_j| \ge 2r_0 \quad \text{if} \quad 0 \le i, j \le k-1, \quad i \ne j,$$

and the balls $B_{r_0}(a_j)$ are disjoint for $0 \le j \le k-1$.

Set $m := m_0^n$, where $m_0 := [8K^2] + 2$. Here $[\cdots]$ denotes the integer part. If $k \le m$, then we take $\gamma_1 = \gamma_1(a, z_0) := P_0 := [a_0, a_1, \ldots, a_k]$—a polygonal line of length

$$|P_0| = |a_1 - a_0| + |a_2 - a_1| + \cdots + |a_k - a_{k-1}| \le 2kr_0 \le 2mr_0.$$

In addition, by the property (2.15), we also have $d(x) \ge r_0$ on P_0, so that the inequality (2.14) holds true with $K_1 := 2m$.

If $m < k$, then the ball $B_{m_0 r_0}(a)$ cannot contain $m + 1 = m_0^n + 1$ disjoint balls of radius r_0, hence one can choose $j \in \{0, 1, \ldots, m\}$ such that $B_{r_0}(a_j) \not\subseteq B_{m_0 r_0}(a)$. This yields

$$|a - a_j| \ge (m_0 - 1)r_0 \ge 8K^2 r_0 = 2Kd(a).$$

On the other hand, by virtue of (2.13), $|a - a_j| \le Kd(a_j)$. Thus,

$$d(a_j) \ge 2d(a) \quad \text{for some} \quad j \le m, \quad \text{provided} \quad m < k. \tag{2.18}$$

Fix such j and set $P_0 := [a_0, a_1, \dots, a_j]$. Then starting from the point $a^{(1)} := a_j$, take a curve $\gamma = \gamma(a^{(1)}, z_0) := \gamma_3(a^{(1)}, z_0) \subset \Omega$ satisfying (2.10), and repeat the previous constructions with $a^{(1)}$ and $r_1 := (4K)^{-1}d(a^{(1)}) \ge 2r_0$ instead of a and r_0. In this way, we get a polygonal line P_1 of length $|P_1| \le 2mr_1$ which connects $a^{(1)}$ either to z_0 or to another point $a^{(2)}$ with $d(a^{(2)}) \ge 2r_1$, and also $d(x) \ge r_1$ on P_1.

Proceeding in a similar way, we construct a sequence of polygonal lines $P_0, P_1, \dots$, with the properties

$$|P_i| \le 2mr_i, \quad d(x) \ge r_i \quad \text{on} \quad P_i, \quad r_i \ge 2r_{i-1} \quad \text{for} \quad i \ge 1. \tag{2.19}$$

Note that P_i is a portion of a polygonal line which connects $a^{(i)}$ to z_0, and along the whole line (including z_0) we have $d(x) \ge r_i \ge 2^i r_0$. Thus we get a finite sequence P_i, $i \le s$, with $2^s r_0 \le d(z_0)$.

The remaining part of the proof is straightforward. We set $\gamma = \gamma(a, z_0) := P_0 \cup P_1 \cup \dots \cup P_s$. Then for each $x \in \gamma$, we have $x \in P_i$ for some $i \le s$. By the properties (2.19), $d(x) \ge r_i$, and

$$\begin{aligned} |\gamma(a,x)| &\le |P_0| + |P_1| + \dots + |P_i| \le 2m \cdot (r_0 + r_1 + \dots + r_i) \\ &\le 2m \cdot \left(2^{-i} + 2^{1-i} + \dots + 1\right) r_i < 4mr_i \le 4md(x). \end{aligned}$$

Since $m \le (8K^2 + 2)^n$, where $K := K_3$, the desired estimate (2.10) for $j = 1$ holds true with $K_1 := 4 \cdot (8K_3^2 + 2)^n$. The theorem is proved. □

The following lemma is a shorter version of Lemma 3.3 in [**KmS11a**], in which there was an additional restriction $d(z_0) = r$. In our applications, this restriction is not needed, because $d(z_0)$ and $\operatorname{diam}\Omega$ are comparable by virtue of (1.8).

LEMMA 2.8. *Let Ω be a c-uniform domain with a constant $c \ge 1$ in (1.11)–(1.12). Let $x_0 \in \partial\Omega$ and $R > 0$ be such that $\overline{\Omega} \cap (\partial B_R(x_0))$ is not empty. Then for each $r \in (0, R/3c]$, there is a John domain J_r with center $z_0 \in \Omega \cap (\partial B_r(x_0))$ and constant $K := 4c^3 \ge 4$, such that*

$$\Omega_r(x_0) := \Omega \cap B_r(x_0) \subset J_r \subset \Omega_{3cr}(x_0) \subseteq \Omega_R. \tag{2.20}$$

PROOF. Fix an arbitrary $r \in (0, R/3c]$. As in (2.3) with a_0 close to x_0 and $b_0 \in \Omega \cap (\partial B_{2r}(x_0))$, we get the existence of a point $z_0 \in \Omega \cap (\partial B_r(x_0))$ with $r \le cd(z_0)$. Then fix a point $a \in \Omega_r(x_0)$ and take $\gamma = \gamma(a, z_0) \subset \Omega$ by Definition 1.4 with $b = z_0$. Then by the property (1.12),

$$|\gamma(a, z_0)| \le c \cdot |a - z_0| \le 2cr \le 2c^2 d(z_0), \tag{2.21}$$

i.e., the same estimate as in (2.6). The remaining part of the proof of Theorem 2.3 (II) also remains the same. It shows that

$$|\gamma(a,x)| \le Kd(x) := 4c^3 d(x) \qquad \forall x \in \gamma(a, z_0). \tag{2.22}$$

For each $a \in \Omega_r(x_0)$, we have a curve $\gamma(a, z_0)$ satisfying (2.21), which is equivalent to $\Omega^a := \operatorname{car}_1(\gamma(a, z_0), K) \subset \Omega$. Finally, we claim that

$$J_r := \Omega_r(x_0) \cup \Big(\bigcup_{a \in \Omega_r} \Omega^a \Big) \tag{2.23}$$

satisfies all the desired properties. Indeed, by Lemma 2.5(i), each Ω^a in this expression is a John domain with center z_0 and constant K. This means that if $p = a \in \Omega_r$, then

$$|\gamma(p,x)| \le K \cdot \operatorname{dist}(x,\Omega^a) \le K \cdot \operatorname{dist}(x,\partial J_r) \qquad \forall x \in \gamma(p,z_0), \tag{2.24}$$

and also if $p \in \Omega^a \setminus \Omega_r$, then there exists a curve $\gamma(p,z_0) \subset \Omega^a \subset J_r$ satisfying (2.24). This proves that J_r is a John domain with center z_0 and constant K.

Finally, since by (2.21) we have $|\gamma(a,z_0)| \le 2cr$, where $|a - x_0| < r$, $|z_0 - x_0| = r$, then by Definition 2.4(i), the set $\Omega^a \subset \Omega_{(c+2)r}(x_0) \subset \Omega_{3cr}(x_0)$ for each $a \in \Omega$. Hence J_r in (2.23) satisfies (2.20). The lemma is proved. □

3. General properties of solutions

We deal with non-divergence elliptic operators $L := a_{ij}D_{ij} + b_iD_i$ which are applied to the functions $u \in W(\Omega) := W^{2,n}_{loc}(\Omega) \cap C(\overline{\Omega})$, where Ω is an open set in $\mathbb{R}^n$. The coefficients a_{ij} satisfy the uniform ellipticity conditions (1.2), and $b_i \in \mathbb{L}^n(\Omega)$ with a constant S in (1.3). In this section we describe the properties of solutions in terms of constants n, ν, and S, without any additional assumptions on Ω. Note that the constants ν and S are invariant with respect to rescaling $x \to \text{const} \cdot x$, i.e., after substitution $x = \text{const} \cdot y$, the equation $Lu = 0$ in a domain Ω can be rewritten as a similar equation with the same ν, S in the corresponding domain (cf. [**S10**, Remark 1.4]).

From the maximum principle of A.D. Aleksandrov [**A63**, Section 5] (cf. also [**GT83**, Section 9.1]) we have

THEOREM 3.1 (Comparison principle). *Let Ω be a bounded open set in $\mathbb{R}^n$, and let functions $u_1, u_2 \in W(\Omega)$ satisfy $Lu_1 \ge Lu_2$ in Ω and $u_1 \le u_2$ on $\partial\Omega$. Then $u_1 \le u_2$ in Ω.*

We will also use a growth lemma in the following form.

LEMMA 3.2 (Growth lemma). *Let $u \in W(\Omega)$, $x_0 \in \Omega$, and $r > 0$ be such that*

$$u > 0, \quad Lu \ge 0 \quad \text{in} \quad \Omega; \qquad u = 0 \quad \text{on} \quad (\partial\Omega) \cap B_r(x_0), \tag{3.1}$$

and the Lebesgue measure

$$|\Omega_r(x_0)| := |\Omega \cap B_r(x_0)| \le \mu \cdot |B_r|, \quad \text{where} \quad \mu = \text{const} \in (0,1). \tag{3.2}$$

Then

$$u(x_0) \le \beta \cdot \sup_{\Omega_r(x_0)} u, \quad \text{where} \quad \beta = \beta(n,\nu,S,\mu) < 1. \tag{3.3}$$

PROOF. In [**S10**, Lemmas 2.4 and 2.5], this lemma was proved in the following equivalent form. Let $u \in W(\Omega)$ satisfy

$$u > 0, \quad Lu \ge 0 \quad \text{in} \quad \Omega; \qquad u = 0 \quad \text{on} \quad (\partial\Omega) \cap B_{2r}(x_0), \tag{3.1$'$}$$

and let Ω satisfy (3.2). Extend $u \equiv 0$ on $B_{2r}(x_0) \setminus \Omega$. Then

$$\sup_{B_r(x_0)} u \le \beta' \cdot \sup_{B_{2r}(x_0)} u, \quad \text{where} \quad \beta' = \beta'(n,\nu,S,\mu) < 1. \tag{3.3$'$}$$

We need a few elementary additional steps in order to prove it in our formulation.

Step 1. Set

$$\mu_0 := \frac{1-\mu}{2}, \quad \varepsilon_0 := \frac{\mu_0}{n}, \quad \text{and} \quad F := B_{(1-\varepsilon_0)r}(x_0) \setminus \Omega.$$

Then

$$\left|B_r \setminus B_{(1-\varepsilon_0)r}\right| = \left(1-(1-\varepsilon_0)^n\right)\cdot |B_r| \le n\varepsilon_0 \cdot |B_r| = \mu_0 \cdot |B_r|,$$

and from (3.2) it follows that

$$|F| \ge \left|B_r(x_0)\setminus\Omega\right| - \left|B_r \setminus B_{(1-\varepsilon_0)r}\right| \ge (1-\mu)\cdot|B_r| - \mu_0\cdot|B_r| = \mu_0\cdot|B_r|.$$

Step 2. Next, take $\rho := \varepsilon_0 r/2$, $h := \rho/\sqrt{n}$. One can cover F by a finite family of disjoint cubes $\{Q_j\}$ with side length h, so that each Q_j is obtained from a standard cube $Q_0 := [0,h)^n$ by parallel translation in $\mathbb{R}^n$. We consider only Q_j for which the intersections $F \cap Q_j$ are not empty. Since $\operatorname{diam} Q_j = \rho$ does not exceed the distance between $F \subset B_{(1-\varepsilon_0)r}(x_0)$ and $\partial B_r(x_0)$, we have

$$F \subset \bigcup_j Q_j \subset B_r(x_0),$$

and

$$\sum_j |F\cap Q_j| = |F| \ge \mu_0 \cdot |B_r| \ge \mu_0 \cdot \sum_j |Q_j|.$$

Hence we can pick a cube $Q := Q_j$ satisfying $|F\cap Q| \ge \mu_0 \cdot |Q|$. Then for an arbitrary point $y_0 \in F\cap Q$,

(3.4) $$Q \subset B_\rho(y_0) \subset B_{2\rho}(y_0) \subset B_r(x_0).$$

Since Q contains a ball of radius $h/2$, we have rough estimates $|Q| \ge |B_{h/2}| = (2\sqrt{n})^{-n}\cdot|B_\rho|$, and

(3.5) $$|\Omega_\rho(y_0)| \le |B_\rho| - |F\cap Q| \le |B_\rho| - \mu_0\cdot|Q| \le \mu'\cdot|B_\rho|,$$

where

(3.6) $$\mu' = \mu'(n,\mu) := 1-(2\sqrt{n})^{-n}\mu_0 = 1-2^{-n-1}n^{-n/2}\cdot(1-\mu) \in (0,1).$$

By virtue of (3.4) and (3.5), we have the properties (3.1′) and (3.2) with y_0, ρ, and μ' instead of x_0, r, and μ, correspondingly. Assuming that u is extended as $u \equiv 0$ on $B_r(x_0)\setminus\Omega$, from (3.3′) and (3.4) it follows that

(3.7) $$\sup_{B_\rho(y_0)} u \le \beta_0 \cdot \sup_{B_{2\rho}(y_0)} u \le \beta_0\cdot M, \quad \text{where} \quad M := \sup_{B_r(x_0)} u,$$

and $\beta_0 = \beta_0(n,\nu,S,\mu) \in (0,1)$.

Step 3. Set $m := [2/\varepsilon_0] \ge 4$—the integer part of $2/\varepsilon_0$. Since $y_0 \in F \subset B_{(1-\varepsilon_0)r}(x_0)$ and $r = 2\cdot\varepsilon_0^{-1}\rho$, we have

$$|y_0 - x_0| < (1-\varepsilon_0)\,r = 2\,(\varepsilon_0^{-1}-1)\,\rho < m\rho.$$

Divide the segment $[y_0, x_0]$ into m equal parts by the point $y_0, y_1, \dots, y_{m-1}, y_m := x_0$, i.e.,

(3.8) $$|y_j - y_{j+1}| = \frac{|y_0-x_0|}{m} < \rho \quad \text{for} \quad j = 0,1,\dots,m-1.$$

We claim that

(3.9) $$\sup_{B_\rho(y_j)} u \le M_j := \left(1-(1-\beta_0)^{j+1}\right)\cdot M.$$

In the case $j = 0$, this estimate is contained in (3.7). Suppose that it is true for some $j \in \{0, 1, \dots, m-1\}$, and extend to $j+1$ in place of j as follows. Consider the function

$$u_j := (u - M_j) \vee 0 \quad \text{in} \quad \Omega^j := \{x \in \Omega : \ u(x) > M_j\}. \tag{3.10}$$

By (3.8), $B_\rho(y_{j+1}) \subset B_{2\rho}(y_{j+1}) \subset B_r(x_0)$, and the intersection $B_\rho(y_{j+1}) \cap B_\rho(y_j)$ contains a ball of radius $\rho/2$. Having in mind (3.9) and (3.6), we conclude that

$$\left|\Omega^j \cap B_\rho(y_{j+1})\right| \le \left|B_\rho(y_{j+1}) \setminus B_\rho(y_j)\right| \le (1 - 2^{-n}) \cdot |B_\rho| < \mu' \cdot |B_\rho|.$$

Similarly to (3.7),

$$\sup_{B_\rho(y_{j+1})} u_j \le \beta_0 \cdot \sup_{B_r(x_0)} u_j = \beta_0 \cdot (M - M_j),$$

so that from (3.10) and (3.9) it follows that

$$\sup_{B_\rho(y_{j+1})} u \le M_j + \beta_0 \cdot (M - M_j) = M_{j+1}.$$

Therefore, by induction, we have (3.9) for all $j \in \{0, 1, \dots, m-1\}$.

Finally, since $x_0 = y_m \in B_\rho(y_{m-1})$, the estimate (3.3) holds true with $\beta := 1 - (1 - \beta_0)^m \in (0, 1)$. The lemma is proved. □

The following *interior Harnack inequality* for linear elliptic equations with $b_i \in \mathbb{L}^n$ was proved in [**S10**, Theorem 3.1]. Note that the constant N in (3.11) does not depend on r.

THEOREM 3.3 (Interior Harnack inequality). *Let $x_0 \in \mathbb{R}^n$, $r > 0$, and a function $u \in W(B_{8r}) := W(B_{8r}(x_0))$ be such that*

$$u > 0 \quad \text{and} \quad Lu := a_{ij} D_{ij} u + b_i D_i u = 0 \quad \text{in} \quad B_{8r}.$$

Then

$$\sup_{B_r} u \le N \cdot \inf_{B_r} u, \quad \text{where} \quad N = N(n, \nu, S) > 1, \quad S := \int_{B_{8r}} |\mathbf{b}|^n dx. \tag{3.11}$$

COROLLARY 3.4. *Let Ω be an open set in $\mathbb{R}^n$, and let $u \in W(\Omega)$ be such that $u > 0$ and $Lu = 0$ in Ω. Suppose that two points $a, b \in \Omega$, $a \ne b$, are connected by a curve $\gamma = \gamma(a, b) \subset \Omega$ such that its length*

$$|\gamma| \le K_0 d(x) := K_0 \cdot \operatorname{dist}(x, \partial\Omega) \qquad \forall x \in \gamma, \tag{3.12}$$

where $K_0 = \text{const} > 0$. Then

$$u(a) \le N_0 u(b), \quad \text{where} \quad N_0 = N_0(n, \nu, S, K_0) \ge 1. \tag{3.13}$$

PROOF. Set $m := [8K_0] + 1$, the minimal integer satisfying $m > 8K_0$, and divide $\gamma = \gamma(a, b)$ into m parts by the points

$$a = x_0, x_1, \dots, x_m = b, \quad \text{such that } |\gamma(x_j, x_{j+1})| = \frac{|\gamma|}{m} \quad \text{for all } j = 0, 1, \dots, m-1.$$

Then

$$r := \frac{|\gamma|}{8K_0} > \frac{|\gamma|}{m} = |\gamma(x_j, x_{j+1})| \ge |x_j - x_{j+1}|,$$

and $8r = |\gamma|/K_0 \le d(x_j)$. This means that

$$x_{j+1} \in B_r(x_j) \subset B_{8r}(x_j) \subset \Omega \quad \text{for} \quad j = 0, 1, \dots, m-1.$$

By Theorem 3.3,

$$u(a) = u(x_0) \le Nu(x_1) \le N^2 u(x_2) \le \cdots \le N^m u(x_m) = N^m u(b),$$

and the inequality (3.13) follows with $N_0 := N^m$. □

4. Boundary Harnack principle

In this section, we discuss the properties of positive solutions of the equation $Lu = 0$ in a John domain Ω with center $z_0 \in \Omega$ and constant $K \ge 1$. By Lemma 2.7, for each $a \in \Omega$, the curve $\gamma(a, z_0) \subset \Omega$ in Definition 1.1 satisfies

$$(4.1) \qquad |\gamma(a,x)| \le Kd(x), \qquad d(a) \le 2Kd(x) \qquad \forall x \in \gamma(a, z_0),$$

where $d(x) := \operatorname{dist}(x, \partial\Omega)$. In the following theorem, we do not impose any boundary conditions on solutions of $Lu = 0$.

THEOREM 4.1. *Let Ω be a John domain with center $z_0 \in \Omega$ and constant $K \ge 1$ in* (1.6), *and let $u \in W(\Omega)$ be a function satisfying $u > 0$ and $Lu = 0$ in Ω. Then*

$$(4.2) \qquad N_1^{-1} d^{c_1}(a) u(z_0) \le u(a) \le N_1 d^{-c_1}(a) u(z_0) \qquad \forall x \in \Omega,$$

with constants $c_1 = c_1(n, \nu, S, K) > 0$ and $N_1 = N_1(n, \nu, S, K, \operatorname{diam}\Omega) \ge 1$.

PROOF. For fixed $a \in \Omega \setminus \{z_0\}$, take a curve $\gamma(a, z_0) \subset \Omega$ according to (4.1). Take an arbitrary point

$$(4.3) \qquad b \in \gamma(a, z_0) \quad \text{such that} \quad d(b) \le 2d(a).$$

Then from (4.1) it follows that

$$|\gamma(a,b)| \le Kd(b) \le 2Kd(a) \le 4K^2 d(x) \qquad \forall x \in \gamma(a,b),$$

i.e., we have (3.12) with $K_0 := 4K^2$. Therefore, by Corollary 3.4, in which one can interchange a and b,

$$(4.4) \qquad N_0^{-1} u(b) \le u(a) \le N_0 u(b),$$

where the constant $N_0 \ge 1$ depends only on n, ν, S, and K.

We claim that the inequalities (4.2) hold true with $c_1 := \log_2 N_0$ and $N_1 := (\operatorname{diam}\Omega)^{c_1}$. If $d(z_0) \le 2d(a)$, then we can use (4.2)–(4.3) with $b = z_0$. From the equality $2d(a) \le \operatorname{diam}\Omega$, it follows that $N_0 = 2^{c_1} \le N_1 d^{-c_1}(a)$, which in turn implies (4.2).

In the remaining case $2d(a) < d(z_0)$, since $d(x)$ is continuous, one can choose $b \in \gamma(a, z_0)$ such that $d(b) = 2d(a)$. By (4.3),

$$d^{c_1}(a) u(a) \le N_0 d^{c_1}(a) u(b) = 2^{c_1} d^{c_1}(a) u(b) = d^{c_1}(b) u(b),$$

and similarly, $d^{-c_1}(a)u(a) \ge d^{-c_1}(b)u(b)$.

Now observe that the subcurve $\gamma(b, z_0) \subset \gamma(a, z_0)$ satisfies (1.6) with the same constant $K \ge 1$, which means that inequalities (4.1) hold true with $a_1 = b$ in place of a. By continuing this procedure, we get a finite sequence of points $a_1, a_2, \ldots, a_k$ in $\gamma(a, z_0)$, such that $d(a_j) = 2^j d(a)$ for each j, and $d(a_k) < d(z_0) \le 2d(a_k)$. Since we already know that the desired estimates hold true for a_k, we finally get

$$d^{c_1}(a)u(a) \le d^{c_1}(a_1)u(a_1) \le \cdots \le d^{c_1}(a_k)u(a_k) \le N_1 u(z_0),$$

and similarly, $d^{-c_1}(a)u(a) \ge N_1^{-1} u(z_0)$. In other words, we have the desired inequalities (4.2). □

The following theorem can be considered as a particular case of Theorem 1.2 with $v \equiv 1$.

THEOREM 4.2 (Carleson type estimate). *Let Ω be a John domain with center $z_0 \in \Omega$ and constant $K \geq 1$. Let $u \in W(\Omega)$ be a function satisfying $u > 0$ and $Lu = 0$ in $\Omega_R(x_0) := \Omega \cap B_R(x_0)$, and moreover,*

$$u = 0 \quad on \quad \Gamma_R := (\partial\Omega) \cap B_R(x_0), \tag{4.5}$$

where $x_0 \in \partial\Omega$ and $R > 0$ satisfies $\operatorname{diam}\Omega \leq K_0 R$ with $K_0 = \mathrm{const} > 1$. Then

$$\sup_{\Omega_{R/2}(x_0)} u \leq N_2 u(z_0), \tag{4.6}$$

where the constant $N_2 = N_2(n, \nu, S, K, K_0) > 1$.

For the proof of this theorem, we will use a technical lemma.

LEMMA 4.3. *Let Ω be a John domain with center $z_0 \in \Omega$ and constant $K \geq 1$. Let $a \in \Omega$ and $h = \mathrm{const} > 0$ satisfy*

$$d(a) \leq h < Ah \leq d(z_0), \quad where \quad A := 2K + 1. \tag{4.7}$$

Then there is a point b such that

$$B_h(b) \subset \big(\Omega^h \cap B_{Ah}(a)\big), \quad where \quad \Omega^h := \{x \in \Omega : \ d(x) > h\}. \tag{4.8}$$

PROOF. By Definition 1.1 of a John domain, there is a curve $\gamma = \gamma(a, z_0)$ such that $|\gamma(a, x)| \leq K d(x)$ for every $x \in \gamma$. From (4.7) it follows that

$$|a - z_0| \geq d(z_0) - d(a) \geq 2Kh,$$

so that γ must intersect the sphere $\partial B_{2Kh}(a)$ at some point b, which satisfies

$$2Kh = |a - b| \leq |\gamma(a, b)| \leq K d(b).$$

Hence $d(b) \geq 2h$, and (4.8) follows. □

PROOF OF THEOREM 4.2. Without loss of generality, we can assume that $x_0 = 0$ and $u(z_0) = 1$. Moreover, since all the conditions in this theorem are invariant with respect to rescaling $x \to \mathrm{const} \cdot x$, we also assume $R = 2$.

By Theorem 4.1, we have

$$u \leq N_1 d^{-c_1} \quad \text{in} \quad \Omega \tag{4.9}$$

with some prescribed constants $c_1 = c_1(n, \nu, S, K) > 0$ and $N_1 \geq 1$. Note that in the proof of this theorem, $N_1 := (\operatorname{diam}\Omega)^{c_1}$, and in our case $\operatorname{diam}\Omega \leq K_0 R = 2K_0$. Therefore, (4.9) holds true with the constant $N_1 := (2K_0)^{c_1} > 1$, which depends only on n, ν, S, K, and K_0.

Now consider

$$M_0 := \sup_{\Omega} d_0^{c_1} u \quad \text{where} \quad d_0 = d_0(x) := \operatorname{dist}\big(x, (\partial\Omega) \setminus \Gamma_2\big). \tag{4.10}$$

Obviously, $d_0 \geq 1$ and $u \leq M_0$ on $\Omega_1(0) := \Omega \cap B_1(0)$. Therefore, for the proof of (4.6), it suffices to obtain the estimate $M_0 \leq N_2$.

Note that $d_0^{c_1} u = 0$ on $\partial\Omega$, so that

$$M_0 = d_0^{c_1}(a_0)\, u(a_0) \quad \text{at some point} \quad a_0 \in \Omega. \tag{4.11}$$

We concentrate on the case

$$u(a_0) > C_0 := N_1 h^{-c_1}, \quad \text{where} \quad h := \varepsilon_0 d_0(a_0), \tag{4.12}$$

c_1 and N_1 are constants in (4.9), and $\varepsilon_0 = \varepsilon_0(n, \nu, S, K) > 0$ is a small constant which satisfies

$$2K(2K+1)\varepsilon_0 \le 1, \tag{4.13}$$

and another restriction (4.19) below.

Set $A := 2K + 1$, and consider the function

$$u_0 := u - C_0 \quad \text{in} \quad \Omega_0 := \Omega \cap B_{Ah}(a_0) \cap \{u > C_0\}. \tag{4.14}$$

We have

$$a_0 \in \Omega_0, \qquad Ah = (2K+1)\varepsilon_0 d_0(a_0) \le d_0(a_0),$$

hence

$$u_0 > 0, \quad Lu_0 = 0 \quad \text{in} \quad \Omega_0; \qquad \text{and} \quad u_0 = 0 \quad \text{on} \quad (\partial\Omega_0) \cap B_{Ah}(a_0). \tag{4.15}$$

From (4.9), (4.12), and (4.14) it follows that

$$\Omega_0 \subset \{x \in \Omega : \ d(x) < h\} = \Omega \setminus \overline{\Omega^h}. \tag{4.16}$$

By (1.8), we also have

$$h := \varepsilon_0 d_0(a_0) < \varepsilon_0 \operatorname{diam} \Omega \le 2K\varepsilon_0 d(z_0).$$

By our assumption (4.13), the point $a = a_0$ satisfies (4.7) in Lemma 4.3. Therefore, the set $\Omega^h \cap B_{Ah}(a_0)$ contains a ball $B_h(b)$, which does not intersect Ω_0 because of (4.16). Then

$$|\Omega_0| \le |B_{Ah}| - |B_h| = (1 - A^{-n}) \cdot |B_{Ah}|.$$

This means that u_0 satisfies assumptions (3.1) and (3.2) in Lemma 3.2 with

$$\Omega := \Omega_0, \quad B_r(x_0) := B_{Ah}(a_0), \quad \text{and} \quad \mu := 1 - A^{-n}.$$

By this lemma,

$$u_0(a_0) \le \beta \cdot \sup_{\Omega_0} u_0, \quad \text{where} \quad \beta = \beta(n, \nu, S, K) < 1. \tag{4.17}$$

Using the maximum principle and (4.14), (4.15), we get

$$u_0(a_0) \le \beta u_0(b_0), \quad \text{where} \quad |a_0 - b_0| = Ah.$$

Then $u = u_0 + C_0$ satisfies

$$u(a_0) \le \beta u(b_0) + C_0. \tag{4.18}$$

From the triangle inequality it follows that

$$d_0(b_0) \ge d_0(a_0) - |a_0 - b_0| = d_0(a_0) - Ah.$$

By definition of M_0 in (4.10),

$$\begin{aligned} u(b_0) &\le d_0^{-c_1}(b_0) \cdot M_0 \le \big(d_0(a_0) - Ah\big)^{-c_1} M_0 \\ &= (1 - A\varepsilon_0)^{-c_1} d_0^{-c_1}(a_0) M_0. \end{aligned}$$

Combining this estimate together with (4.11), (4.18), and (4.12), we derive

$$\begin{aligned} M_0 &= d_0^{c_1}(a_0)\, u(a_0) \le d_0^{c_1}(a_0) \cdot \big(\beta u(b_0) + C_0\big) \\ &\le (1 - A\varepsilon_0)^{-c_1} \beta M_0 + \varepsilon_0^{-c_1} N_1. \end{aligned}$$

Since $\beta < 1$, we can choose a small constant $\varepsilon_0 > 0$ satisfying (4.13) and such that

$$\beta_0 := (1 - A\varepsilon_0)^{-c_1} \beta < 1. \tag{4.19}$$

The choice of ε_0 and β_0 depends only on n, ν, S, and K. Then the preceding inequality can be rewritten as $M_0 \le \beta_0 M_0 + \varepsilon_0^{-c_1} N_1$, and the desired estimate is

$$M_0 \le N_2 := (1-\beta_0)^{-1} \varepsilon_0^{-c_1} N_1. \tag{4.20}$$

This estimate was proved under the assumption $u(a_0) > C_0$ in (4.12). If $u(a_0) \le C_0$, it follows immediately from (4.11) that

$$M_0 = d_0^{c_1}(a_0)\, u(a_0) \le C_0 d_0^{c_1}(a_0) = \varepsilon_0^{-c_1} N_1 < N_2.$$

The theorem is proved. □

Now we are ready to prove the boundary Harnack principle.

PROOF OF THEOREM 1.2. We can assume that $x_0 = 0, R = 2, u(z_0) = 1$, and also $v(z_0) = 1$. We claim that

$$u \le N_3 \quad \text{in} \quad \Omega_{3R/4}(0) = \Omega_{3/2}(0) := \Omega \cap B_{3/2}(0) \tag{4.21}$$

with a constant $N_3 = N_3(n, \nu, S, K, K_0) \ge 1$. Indeed, for an arbitrary $x \in \Omega_{3/2}$, we have one of two cases: (i) $d(x) := \operatorname{dist}(x, \partial\Omega) < 1/6$, or (ii) $d(x) \ge 1/6$.

In the case (i), there is a point $\xi_0 \in \partial\Omega$ such that $d(x) = |x - \xi_0| < 1/6$. Then

$$|\xi_0| < \frac{3}{2} + \frac{1}{6} = \frac{5}{3}, \quad \text{and} \quad x \in \Omega_{1/6}(\xi_0) \subset \Omega_{1/3}(\xi_0) \subset \Omega_2(0).$$

Here we can use Theorem 4.2 with $R := 1/3$, which yields $u(x) \le N_2$.

In the case (ii), as in (4.9),

$$u(x) \le N_1 d^{-c_1}(x) \le 6^{c_1} N_1, \quad \text{where} \quad N_1 = N_1(n, \nu, S, K, K_0) > 1.$$

In any case, (4.21) holds true with $N_3 := (6^{c_1} N_1) \vee N_2$. In fact, the explicit expression for N_2 in (4.20) automatically satisfies $N_2 > 6^{c_1} N_1$, because from (4.13) it follows $\varepsilon_0 \le 1/6$. Therefore, we can take $N_3 := N_2$ in this expression.

For integers $j \ge 1$, introduce the sets

$$T_j := \Omega_{r_j}(0) \cap \{d < h_j\}, \quad U_j := \Omega_{r_j}(0) \cap \{d \ge h_j\}, \quad \text{where } r_j := 1 + \frac{1}{j}, h_j := 2^{-j}. \tag{4.22}$$

From the estimates (4.2) for functions u and v it follows that

$$\frac{u}{v} \le N_1^2 d^{-2c_1} \le N_1^2 4^{c_1 j} \quad \text{on} \quad U_j. \tag{4.23}$$

Since $r_j \searrow 1$ and $h_j \searrow 0$ as $j \to \infty$, each point in Ω_1 belongs to U_j for some j. Therefore, for the proof of the desired estimate $u/v \le N$ in $\Omega_{R/2}(0) = \Omega_1(0)$, it suffices to show that

$$M_j := \sup_{U_j} \frac{u}{v} \le N = N(n, \nu, S, K, K_0) \tag{4.24}$$

uniformly for all $j \ge 1$.

Fix an integer $j_0 \ge 4$ depending only on K, such that

$$K(2K+1)h_j \le 1, \quad (2K+1)j(j+1)h_j \le 1 \qquad \forall j \ge j_0, \tag{4.25}$$

and suppose that $M_j \le N_j = \text{const}$ for some $j \ge j_0$. Consider the function

$$w_j := u - N_j v \quad \text{on the set} \quad W_j := \Omega_{r_j}(0) \cap \{w_j > 0\}. \tag{4.26}$$

If the intersection $W_j \cap \Omega_{r_{j+1}}(0)$ is empty, then $w_j \le 0$ on the set $\Omega_{r_{j+1}}(0)$ which contains U_{j+1}, so that $M_{j+1} \le N_j$.

Now consider the case when $W_j \cap \Omega_{r_{j+1}}(0)$ is not empty. Since $w_j \le 0$ on $U_j \cup (\partial\Omega)$, we have $W_j \subset T_j \subset \{d < h_j\}$, and $w_j = 0$ on a portion of ∂W_j which lies inside $B_{r_j}(0)$. Moreover, $Lw_j = 0$ in W_j, hence by the maximum principle, the function

$$\lambda_j(\rho) := \sup_{\Omega_\rho(0)} w_j \quad \text{is positive and non-decreasing on} \quad [r_{j+1}, r_j].$$

Note that by (1.8) we have $1 = R/2 \le (\operatorname{diam}\Omega)/2 \le Kd(z_0)$. From (4.25) it follows that

$$s_j := (2K+1)h_j \le K(2K+1)h_j d(z_0) \le d(z_0) \quad \text{and} \quad s_j \le \frac{1}{j(j+1)} = r_j - r_{j+1}. \tag{4.27}$$

Therefore, there are values ρ satisfying $r_{j+1} \le \rho < \rho + s_j \le r_j$ (e.g., $\rho = r_{j+1}$). Fix such ρ and observe that by the maximum principle, $\lambda_j(\rho) = w_j(a)$ at some point $a \in \partial\Omega_\rho(0)$, for which we must have $|a| = \rho$ and $d(a) := \operatorname{dist}(a, \partial\Omega) < h_j$. This inequality together with $s_j \le d(z_0)$ guarantees that we can apply Lemma 4.3 with $h = h_j$, so that there is a ball

$$B_{h_j}(b) \subset \big(\Omega^{h_j} \cap B_{s_j}(a)\big) \subset \big(B_{s_j}(a) \setminus W_j\big),$$

and the measure

$$|W_j \cap B_{s_j}(a)| \le |B_{s_j}| - |B_{h_j}| = \mu \cdot |B_{s_j}|, \quad \text{where} \quad \mu := 1 - (2K+1)^{-n} < 1.$$

Moreover, since $\rho + s_j \le r_j$, we also have

$$\big(W_j \cap B_{s_j}(a)\big) \subset B_{\rho+s_j}(0) \subset B_{r_j}(0), \quad \text{and} \quad w_j = 0 \quad \text{on} \quad (\partial W_j) \cap B_{s_j}(a).$$

Similarly to (4.17), we get the estimate

$$\lambda_j(\rho) = w_j(a) \le \beta \cdot \sup_{W_j \cap B_{s_j}(a)} w_j \le \beta \cdot \sup_{\Omega_{\rho+s_j}(a)} w_j \le \beta\lambda_j(\rho + s_j), \tag{4.28}$$

where $\beta = \beta(n, \nu, S, K) < 1$. Starting from $\rho = r_{j+1}$, one can iterate this inequality:

$$\lambda_j(r_{j+1}) \le \beta\lambda_j(r_{j+1} + s_j) \le \cdots \le \beta^{m_j}\lambda_j(r_{j+1} + m_j s_j) \le \beta^{m_j}\lambda_j(r_j),$$

where $m_j := [d_j]$—the integer part of $d_j := (r_j - r_{j+1})/s_j$. By (4.27), we have $d_j \ge 1$, which implies

$$m_j \ge \frac{d_j}{2} = \frac{2^{j-1}}{(2K+1)j(j+1)} \ge \frac{c_2 2^j}{j^2} \qquad \forall j \ge 4$$

with a prescribed constant $c_2 = c_2(K) > 0$. Together with the rough estimate $w_j \le u \le N_3$ in (4.21), we obtain $\lambda_j \le N_3$, and

$$\sup_{U_{j+1}} w_j \le \sup_{\Omega_{r_{j+1}}(0)} w_j = \lambda_j(r_{j+1}) \le N_3\beta^{m_j} \le N_3 \exp\big(-c_3 \cdot j^{-2}2^j\big), \tag{4.29}$$

where $c_3 = c_3(n, \nu, S, K) := -c_2 \ln\beta > 0$.

By Theorem 4.1 applied to the function v, we also have

$$\inf_{U_{j+1}} v \ge N_1^{-1}h_{j+1}^{c_1} = N_1^{-1}2^{-(j+1)c_1} \ge N_1^{-1} \cdot \exp\big(-c_4 \cdot (j+1)\big),$$

where $c_4 = c_4(n, \nu, S, K) := c_1 \cdot \ln 2 > 0$. Therefore,

$$\sup_{U_{j+1}} \frac{w_j}{v} \le \varepsilon_j := N_1 N_3 \cdot \exp\big(c_4 \cdot (j+1) - c_3 \cdot j^{-2}2^j\big).$$

This is the estimate for $w_j := u - N_j v$ under the assumption that in (4.24) we have $M_j \le N_j$, and $j \ge j_0 \ge 4$ satisfies (4.25). From this estimate we derive

$$M_{j+1} := \sup_{U_{j+1}} \frac{u}{v} = N_j + \sup_{U_{j+1}} \frac{w_j}{v} \le N_{j+1} := N_j + \varepsilon_j.$$

The series $\sum \varepsilon_j$ obviously converges, because $c_4 \cdot (j+1) - c_3 \cdot j^{-2} 2^j < -j$ for large j. According to (4.23), we have $M_{j_0} \le N_{j_0} := N_1^2 4^{c_1 j_0}$, and then by induction,

$$M_j \le N_j := N_{j_0} + \varepsilon_{j_0} + \varepsilon_{j_0+1} + \cdots + \varepsilon_{j-1} \quad \text{for} \quad j > j_0.$$

Finally, since $\Omega_1(0)$ is covered by $\{U_j\}$,

$$\sup_{\Omega_1(0)} \frac{u}{v} \le \sup_{j \ge 1} M_j \le N := N_{j_0} + \sum_{j=j_0}^{\infty} \varepsilon_j.$$

Here the constant N depends only on n, ν, S, K, and K_0. Theorem 1.2 is proved. □

In the following statement, we do not assume the positivity of u in $\Omega_R(x_0)$.

COROLLARY 4.4. *Let Ω be a John domain with center $z_0 \in \Omega$ and constant $K \ge 1$. Let functions $u, v \in W(\Omega)$ be such that*

$$v > 0; \quad Lu = Lv = 0 \quad \text{in} \quad \Omega_R := \Omega \cap B_R(x_0), \tag{4.30}$$

and $u \equiv 0$ on $\Gamma_R := (\partial\Omega) \cap B_R(x_0)$, where $x_0 \in \partial\Omega$ and $R > 0$ satisfies $\operatorname{diam}\Omega \le K_0 R$ *with* $K_0 = \text{const} > 1$. *Then*

$$\sup_{\Omega_{R/2}(x_0)} \frac{u}{v} \le \frac{N_1}{v(z_0)} \cdot \sup_{\Omega_{3R/4}(x_0)} u_+, \tag{4.31}$$

where $N_1 = N_1(n, \nu, S, K, K_0) \ge 1$ is the constant in Theorem 1.2.

PROOF. In the proof of Theorem 1.2, we set

$$N_3 := \sup_{\Omega_{3R/4}(x_0)} u_+$$

instead of (4.21). Then we have $u/v \le N_1 N_3 \cdot 2^{c_1 j}$ on U_j instead of (4.23), and the rest of proof remains the same. □

5. Ratios of solutions in uniform domains

In this section, we assume that Ω is a c-uniform domain according to Definition 1.4 with a given constant $c \ge 1$. We consider the functions $u, v \in W(\Omega)$ satisfying

$$v > 0, \quad Lu = Lv = 0 \quad \text{in} \quad \Omega. \tag{5.1}$$

Our goal is to show that the interior and boundary Harnack inequalities imply the Hölder continuity of the ratio u/v near a portion of boundary $\Gamma \subset \partial\Omega$, on which both u and v vanish. This result is very standard and the most essential part of the technique goes back to J. Moser [**M61**]. However, in many sources (cf. [**JK82**], [**Ak01**], and others) it is assumed that both u and v are positive in Ω, while the positivity of u is not needed for the estimates such as (4.31). We also want to show the explicit dependence of the Hölder exponent $\alpha > 0$ for u/v in terms of the given quantities. Therefore, here we put together a few standard statements.

The following lemma follows immediately from the maximum principle. In this lemma and also in Theorem 5.2 below, the John property of Ω is not needed.

LEMMA 5.1. *Let Ω be a bounded domain in $\mathbb{R}^n$, and let functions $u, v \in W(\Omega)$ satisfy* (5.1), *and $u = v = 0$ on a closed subset $\Gamma \subset \partial\Omega$. Then*

$$\sup_{\Omega} \frac{u}{v} = \sup_{(\partial\Omega)\setminus\Gamma} \frac{u}{v}, \qquad \inf_{\Omega} \frac{u}{v} = \inf_{(\partial\Omega)\setminus\Gamma} \frac{u}{v}. \tag{5.2}$$

PROOF. Let M denote the right-hand side of the first equality in (5.2). Obviously, we can assume $M < \infty$. Then the function $w := u - Mv$ satisfies

$$Lw = 0 \quad \text{in} \quad \Omega; \qquad \text{and} \quad w \le 0 \quad \text{on} \quad \partial\Omega.$$

By the maximum principle, $w \le 0$ on $\partial\Omega$, and the first equality in (5.2) follows. The proof of the second one is quite similar. □

THEOREM 5.2. *Let functions $u, v \in W(\Omega)$ satisfy* (5.1). *Then for every ball $B_R := B_R(y) \subseteq \Omega$, the oscillation over concentric balls,*

$$\omega(r) := \operatorname*{osc}_{B_r} \frac{u}{v} \le \left(\frac{8r}{R}\right)^{\alpha_0} \omega(R) \quad \text{for} \quad 0 < r \le R, \tag{5.3}$$

where

$$\alpha_0 = \alpha_0(n, \nu, S) := \log_8 N_0 > 0, \qquad N_0 = N_0(n, \nu, S) := \frac{N^2 + 1}{N^2 - 1} > 1, \tag{5.4}$$

and $N = N(n, \nu, S) > 1$ is the constant in the Harnack inequality (3.11).

PROOF. We can write $\omega(r) = M(r) - m(r)$, where

$$M(r) := \sup_{B_r} \frac{u}{v}, \qquad m(r) := \inf_{B_r} \frac{u}{v}.$$

By Lemma 5.1 applied to the balls B_r, the functions $M(r)$ and $m(r)$ are monotone on $(0, R]$. If $0 < 8r \le R$ and $u \ge 0$ in B_{8r}, then from the interior Harnack inequality (Theorem 3.3) it follows that

$$\sup_{B_r} u \le N \cdot \inf_{B_r} u, \quad \inf_{B_r} v \ge N^{-1} \sup_{B_r} v, \quad \text{therefore} \quad M(r) \le N^2 m(r).$$

Applying this estimate to functions $u - m(8r) \cdot v$ and $M(8r) \cdot v - u$, which are non-negative in B_{8r}, we obtain

$$\begin{aligned} M(r) - m(8r) &\le N^2 \cdot \big(m(r) - m(8r)\big), \\ M(8r) - m(r) &\le N^2 \cdot \big(M(8r) - M(r)\big). \end{aligned}$$

After summation of both sides, we get

$$\omega(8r) + \omega(r) \le N^2 \cdot \big(\omega(8r) - \omega(r)\big).$$

Using notations (5.4), one can rewrite the last inequality in the form $N_0\omega(r) \le \omega(8r)$, and

$$\omega(r) \le \frac{N^2 - 1}{N^2 + 1} \cdot \omega(8r) = N_0 \omega(8r) = 8^{-\alpha_0} \omega(8r) \quad \text{for} \quad 0 < 8r \le R. \tag{5.5}$$

For each $r \in (0, R]$, there is a unique integer $k \geq 0$ such that $8^{-k-1}R < r \leq 8^{-k}R$. Iterating the inequality (5.5) and using monotonicity of $\omega(r)$, we finally get

$$\omega(r) \leq 8^{-\alpha_0}\omega(8r) \leq \cdots \leq 8^{-k\alpha_0}\omega(8^k r) \leq \left(\frac{8r}{R}\right)^{\alpha_0} \omega(R),$$

which is the desired property (5.3). □

THEOREM 5.3. *Let Ω be a c-uniform domain in $\mathbb{R}^n$ with $c = \text{const} \geq 1$, Let functions $u, v \in W(\Omega)$ be such that*

$$v > 0; \quad Lu = Lv = 0 \quad \text{in} \quad \Omega_R(x_0) := \Omega \cap B_R(x_0), \tag{5.6}$$

and $u \equiv 0$, $v \equiv 0$ on $\Gamma_R := (\partial\Omega) \cap B_R(x_0)$, where $x_0 \in \partial\Omega$ and $R > 0$. Then

$$\omega(r) := \operatorname*{osc}_{\Omega_r(x_0)} \frac{u}{v} \leq \left(\frac{6cr}{R}\right)^{\alpha_1} \omega(R) \quad \text{for} \quad 0 < r \leq R, \tag{5.7}$$

where

$$\alpha_1 = \alpha_1(n, \nu, S, c) := \log_{6c}\left(\frac{N_1^2 + 1}{N_1^2 - 1}\right) > 0 \tag{5.8}$$

and $N_1 > 1$ is the constant in the boundary Harnack inequality (1.10) *evaluated for $K := 4c^3$ and $K_0 := 6c$.*

PROOF. Fix $r \in (0, R/3c]$. By Lemma 2.8, there is a John domain J_r with center $z_0 \in J_r$ and constant $K := 4c^3$, such that $\Omega_r(x_0) \subset J_r \subset \Omega_{3cr}(x_0)$. In particular, $\operatorname{diam} J_r \leq 6cr$, and the assumption $\operatorname{diam}\Omega \leq K_0 R$ in Theorem 1.2 is satisfied with J_r, r, and $6c$ in place of Ω, R, and K_0, correspondingly. Therefore, if $u \geq 0$ in $\Omega_{3cr}(x_0)$, then by Corollary 1.3,

$$\sup_{\Omega_{r/2}(x_0)} \frac{u}{v} \leq N_1^2 \cdot \inf_{\Omega_{r/2}(x_0)} \frac{u}{v}.$$

Now we can proceed as in the previous proof with B_r and B_{8r} being replaced by $\Omega_{r/2}(x_0)$ and $\Omega_{3cr}(x_0)$ correspondingly, so that the ratio 8 is replaced by $6c$. This gives us the estimate

$$\omega(r/2) \leq \frac{N_1^2 - 1}{N_1^2 + 1} \cdot \omega(3cr) = (6c)^{-\alpha_1}\omega(3cr) \quad \text{for} \quad 0 < 3cr \leq R$$

instead of (5.5). By iteration, the desired estimate (5.7) follows. □

Combining Theorems 5.2 and 5.3, we get the Hölder estimates of ratios of solutions u/v near the boundary.

THEOREM 5.4. *Under the assumptions of Theorem 5.3,*

$$\left|\frac{u(x)}{v(x)} - \frac{u(y)}{v(y)}\right| \leq N \cdot \left(\frac{|x - y|}{R}\right)^{\alpha} \cdot \operatorname*{osc}_{\Omega_R(x_0)} \frac{u}{v} \qquad \forall x \neq y \in \Omega_{R/2}(x_0), \tag{5.9}$$

where the constant $N = N(n, \nu, S, c) > 1$, and $\alpha = \alpha(n, \nu, S, c) := \alpha_0 \wedge \alpha_1 > 0$, the minimum of constants α_0 in (5.4) *and α_1 in* (5.8).

PROOF. Note that $(r/R)^{\alpha_{0,1}} \leq (r/R)^{\alpha}$ for $0 < r \leq R$. Therefore, we can use a simplified interior estimate (5.3) in the form

$$\omega(r) \leq A_0 \left(\frac{r}{R}\right)^{\alpha} \omega(R) \quad \text{for} \quad 0 < r \leq R, \quad \text{where} \quad A_0 := 8^{\alpha_0}, \tag{5.10}$$

and a simplified boundary estimate (5.7) in the form

$$\omega(r) \le A_1\left(\frac{r}{R}\right)^{\alpha}\omega(R) \quad \text{for} \quad 0<r\le R, \quad \text{where} \quad A_1 := (6c)^{\alpha_1}. \tag{5.11}$$

Fix $x, y \in \Omega_{R/2}(x_0)$, $x \ne y$, and denote $\rho := |x-y|/2 \in (0, R/2)$, $\xi := (x+y)/2$, the center of the segment $[x, y]$, and $d := d(\xi) := \operatorname{dist}(\xi, \partial\Omega)$. This notation makes sense and $x, y \in \overline{\Omega_\rho(\xi)}$ even if $\xi \notin \Omega$. Therefore, the desired estimate (5.9) will follow from

$$\omega(\xi,\rho) := \operatorname*{osc}_{\Omega_\rho(\xi)} \frac{u}{v} \le N\left(\frac{\rho}{R}\right)^{\alpha}\omega(x_0, R). \tag{5.12}$$

In order to prove this estimate, we consider four different cases:

(i) $R/6 \le \rho < R/2$. Then $\Omega_\rho(\xi) \subset \Omega_R(x_0)$, which implies $\omega(\xi,\rho) \le \omega(x_0, R)$, and we have (5.12) if $N \ge 6^\alpha$.

(ii) $\rho < R/6 \le d := d(\xi)$. In this case we must have $\xi \in \Omega$ and $B_\rho(\xi) \subset B_{R/6}(\xi) \subset \Omega_R(x_0)$. Then by the interior estimate (5.10),

$$\omega(\xi,\rho) \le A_0\left(\frac{\rho}{R/6}\right)^{\alpha}\omega(\xi, R/6) \le 6^\alpha A_0\left(\frac{\rho}{R}\right)^{\alpha}\omega(x_0, R).$$

In this case, (5.12) is true if $N \ge 6^\alpha A_0$.

(iii) $d \le \rho < R/6$. Take $\xi_0 \in \partial\Omega$ such that $d = |\xi - \xi_0|$. Then

$$\Omega_\rho(\xi) \subset \Omega_{2\rho}(\xi_0) \subset \Omega_{R/3}(\xi_0) \subset \Omega_R(x_0),$$

by the boundary estimate (5.11),

$$\omega(\xi,\rho) \le \omega(\xi_0, 2\rho) \le A_1\left(\frac{2\rho}{R/3}\right)^{\alpha}\omega(\xi_0, R/3) \le 6^\alpha A_1\left(\frac{\rho}{R}\right)^{\alpha}\omega(x_0, R),$$

and we can take $N \ge 6^\alpha A_1$ in (5.12).

(iv) $\rho < d < R/6$. In this case, we first use the interior estimate (5.10) with $R := d$:

$$\omega(\xi,\rho) \le A_0\left(\frac{\rho}{d}\right)^{\alpha}\omega(\xi, d). \tag{5.13}$$

Next, take $\xi_0 \in \partial\Omega$ with $|\xi - \xi_0| = d$. Then

$$B_d(\xi) \subset \Omega_{2d}(\xi_0) \subset \Omega_{R/3}(\xi_0) \subset \Omega_R(x_0),$$

and as in (iii), with d instead of ρ,

$$\omega(\xi,\rho) \le \omega(\xi_0, 2d) \le 6^\alpha A_1\left(\frac{d}{R}\right)^{\alpha}\omega(x_0, R).$$

Together with (5.13), this estimate implies (5.12) with $N := 6^\alpha A_0 A_1$. This choice of N is also valid for the previous cases (i)–(iii). The theorem is proved. □

References

[A63] A. D. Aleksandrov, Uniqueness conditions and estimates for the solution of the Dirichlet problem, [in Russian], *Vestn. Leningr. Univ.* **18**, No. 3, 5–29 (1963); English transl.: *AMS Transl. (2)* **68**, 89–119 (1968). MR0164135 (29:1434)

[Ak01] H. Aikawa, Boundary Harnack principle and Martin boundary for a uniform domain, *J. Math. Soc. Japan* **53**, No. 1, 119–145 (2001). MR1800526 (2001m:31007)

[Ak05] H. Aikawa, Martin boundary and boundary Harnack principle for non-smooth domains. In: *Selected papers on differential equations and analysis, AMS Transl. (2)*, **215**, 33–55 (2005). MR2186550

[BB91] R. F. Bass and K. Burdzy, A boundary Harnack principle in twisted Hölder domain, *Ann. Math.* **134**, 253–276 (1991). MR1127476 (92m:31006)

[BB92] R. F. Bass and K. Burdzy, Lifetimes of conditioned diffusions, *Probab. Th. Rel. Fields* **91**, 405–443 (1992). MR1151804 (93e:60155)

[C61] L. Carleson, On the existence of boundary values for harmonic functions in several variables, *Ark. Mat.* **4**, 339–343 (1961). MR0159013 (28:2232)

[DRS10] L. Diening, M. Ružicka, and K. Schumacher, A decomposition technique for John domains, *Ann. Acad. Sci. Fenn. Math.* **35**, No. 1, 87-114 (2010). MR2643399 (2011i:26020)

[EE04] D. E. Edmunds and W. D. Evans, *Hardy Operators, Function Spaces and Embeddings*, Springer-Verlag, Berlin–Heidelberg–New York, 2004. MR2091115 (2005g:46068)

[GNV94] M. Ghamsari, R. Näkki, and J. Väisalä, John disks and dxtension of maps, *Mh. Math.* **117**, 63–94 (1994). MR1266774 (95e:30023)

[GT83] D. Gilbarg and N.S. Trudinger, *Elliptic Partial Differential Equations of Second Order*, 2nd ed., Springer-Verlag, Berlin–Heidelberg–New York–Tokyo, 1983. MR737190 (86c:35035)

[JK82] D. S. Jerison and C. E. Kenig, Boundary behavior of harmonic functions in non-tangentially accessible domains, *Adv. Math.* **46**, 80–147 (1982). MR676988 (84d:31005b)

[J61] F. John, Rotation and strain, *Comm. Pure Appl. Math.* **14**, 391–413 (1961). MR0138225 (25:1672)

[KmS11] H. Kim and M. V. Safonov, Carleson type estimates for second order elliptic equations with unbounded drift, *J. Math. Sci., New York* **176**, No. 6, 928–944 (2011). MR2838983 (2012j:35061)

[KmS11a] H. Kim and M. V. Safonov, Boundary Harnack principle for second order elliptic equations with unbounded drift, *J. Math. Sci., New York* **179**, No. 1, 128–144 (2011). MR3014101

[K85] N. V. Krylov, *Nonlinear Elliptic and Parabolic Equations of Second Order* [in Russian], "Nauka", Moscow (1985); English transl.: D. Reidel, Dordrecht, Holland (1987). MR815513 (87h:35002)

[KS80] N. V. Krylov and M. V. Safonov, A certain property of solutions of parabolic equations with measurable coefficients, [in Russian] *Izv. Akad. Nauk SSSR, Ser. Mat.* **44**, 161–175 (1980); English transl.: *Math. USSR Izvestija* **16**, 151–164 (1981). MR563790 (83c:35059)

[L71] E. M. Landis, *Second Order Equations of Elliptic and Parabolic Type* [in Russian], "Nauka", Moscow (1971); English transl.: Transl. Math. Monographs **171**, AMS, Providence, RI (1997). MR0320507 (47:9044)

[LSU] O. A. Ladyzhenskaya, V. A. Solonnikov and N. N. Ural'tseva, *Linear and Quasilinear Equations of Parabolic Type*, Nauka, Moscow, 1967 in Russian; English transl.: Amer. Math. Soc., Providence, RI, 1968. MR0241822 (39:3159b)

[LU] O. A. Ladyzhenskaya and N. N. Ural'tseva, *Linear and Quasilinear elliptic Equations*, Nauka, Moscow, 1964 in Russian; English transl.: New York, Academic Press, 1968. 2nd Russian ed. 1973. MR0244627 (39:5941)

[M61] J. Moser, On Harnack's theorem for elliptic differential equation, *Comm. Pure Appl. Math.*, **14**, 577–591 (1961). MR0159138 (28:2356)

[MS78] O. Martio and J. Sarvas, Injectivity theorems in plane and space, *Ann. Acad. Sci. Fenn. Ser. A I Math.* **4** 383–401 (1978-79). MR565886 (81i:30039)

[NU12] A. I. Nazarov and N. N. Uraltseva, The Harnack inequality and related properties for solutions of elliptic and parabolic equations with divergence-free lower-order coefficients, [in Russian] *Algebra i Analiz* **28**, No. 1 (2011); English transl.: *St. Petersburg Math. J.* **23**, No. 1, 93–115 (2012). MR2760150 (2012d:35026)

[NV91] R. Näkki and J. Väisälä, John disks, *Expo. Math.* **9:1**, 3–43 (1991). MR1101948 (92i:30021)

[S80] M. V. Safonov, Harnack inequality for elliptic equations and the Hölder property of their solutions, [in Russian] *Zap. Nauchn. Semin. LOMI* **96**, 272–287 (1980); MR579490 (82b:35045)

[S10] M. V. Safonov, Non-divergence elliptic equations of second order with unbounded drift, In: *Nonlinear Partial Differential Equations and Related Topics, AMS Transl. (2)* **229**, 211–232 (2010). MR2667641 (2011i:35040)

[V88] J. Väisälä, Uniform domains, *Tôhoku Math. J.* **40**, 101–118 (1988). MR927080 (89d:30027)

Department of Mathematics, University of Michigan, Ann Arbor, Michigan 48109
E-mail address: khyejin@umich.edu

School of Mathematics, University of Minnesota, Minneapolis, Minnesota 55455
E-mail address: safonov@math.umn.edu

Amer. Math. Soc. Transl.
Volume **232**, 2014

Oblique derivative problem for non-divergence parabolic equations with time-discontinuous coefficients

V. Kozlov and A. Nazarov

In memory of O.A. with reverence

ABSTRACT. We consider an oblique derivative problem for non-divergence parabolic equations with discontinuous in t coefficients in a half-space. We obtain weighted coercive estimates of solutions in anisotropic Sobolev spaces. We also give an application of this result to linear parabolic equations in a bounded domain. In particular, if the boundary is of class $\mathcal{C}^{1,\delta}$, $\delta \in (0,1]$, then we present a coercive estimate of solutions in weighted anisotropic Sobolev spaces, where the weight is a power of the distance to the boundary.

1. Introduction

Consider the parabolic equation

$$(\mathcal{L}_0 u)(x,t) \equiv \partial_t u(x,t) - a^{ij}(t) D_i D_j u(x,t) = f(x,t) \tag{1}$$

for $x \in \mathbb{R}^n$ and $t \in \mathbb{R}$. Here and elsewhere D_i denotes the operator of differentiation with respect to x_i and $\partial_t u$ is the derivative of u with respect to t.

The only assumptions about the coefficients in (1) are that a^{ij} are measurable real-valued functions of t satisfying $a^{ij} = a^{ji}$ and

$$\nu|\xi|^2 \le a^{ij}\xi_i\xi_j \le \nu^{-1}|\xi|^2, \qquad \xi \in \mathbb{R}^n, \quad \nu = \text{const} > 0. \tag{2}$$

It was proved by Krylov [**2, 3**] that for $f \in L_{p,q}(\mathbb{R}^n \times \mathbb{R})$ with $1 < p, q < \infty$, equation (1) in $\mathbb{R}^n \times \mathbb{R}$ has a unique solution such that $\partial_t u$ and $D_i D_j u$ belong to $L_{p,q}(\mathbb{R}^n \times \mathbb{R})$ and

$$\|\partial_t u\|_{p,q} + \sum_{ij} \|D_i D_j u\|_{p,q} \le C\|f\|_{p,q}\,. \tag{3}$$

2010 *Mathematics Subject Classification.* Primary 35K20.

Key words and phrases. Non-divergence parabolic equations, discontinuous in time coefficients, oblique derivative problem, coercive estimates, anisotropic Sobolev spaces.

The first author was supported by the Swedish Research Council (VR).

The second author was supported by RFBR grant 12-01-00439 and by St. Petersburg University grant 6.38.670.2013. He is also acknowledges the Linköping University for the financial support of his visit in February 2012.

Here $L_{p,q}(\Omega \times I) = L_q(I \to L_p(\Omega))$ is the space of functions on $\Omega \times I$ with finite norm

$$\|f\|_{p,q} = \Big(\int_I \Big(\int_\Omega |f(x,t)|^p dx \Big)^{\frac{q}{p}} dt \Big)^{\frac{1}{q}}$$

(with natural change in the case $p = \infty$ or $q = \infty$).

In the authors' paper [**4**] estimate (3) was supplemented by a similar one in the space $\widetilde{L}_{p,q}(\mathbb{R}^n \times \mathbb{R})$,

$$|||\partial_t u|||_{p,q} + \sum_{ij} |||D_i D_j u|||_{p,q} \le C |||f|||_{p,q}.$$

Here $\widetilde{L}_{p,q}(\Omega \times I) = L_p(\Omega \to L_q(I))$ is the space of functions on $\Omega \times I$ with finite norm

$$|||f|||_{p,q} = \Big(\int_\Omega \Big(\int_I |f(x,t)|^q dt \Big)^{\frac{p}{q}} dx \Big)^{\frac{1}{p}}$$

(with natural change in the case $p = \infty$ or $q = \infty$). This space arises naturally in the theory of quasilinear non-divergence parabolic equations (see [**9**]). Note that for $p = q$, we have

$$\widetilde{L}_{p,p}(\Omega \times I) = L_{p,p}(\Omega \times I) = L_p(\Omega \times I); \qquad |||f|||_{p,p} = \|f\|_{p,p} = \|f\|_p.$$

The homogeneous Dirichlet problem for (1) in $\mathbb{R}^n_+ \times \mathbb{R}$, where $\mathbb{R}^n_+$ the half-space $\{x = (x', x_n) \in \mathbb{R}^n : x_n > 0\}$, was considered in [**2, 4**]. It was proved that its solution satisfies the weighted coercive estimate

$$\|x_n^\mu \partial_t u\|_{p,q} + \sum_{ij} \|x_n^\mu D_i D_j u\|_{p,q} \le C \|x_n^\mu f\|_{p,q}\,, \tag{4}$$

where $1 < p, q < \infty$ and $\mu \in (-\frac{1}{p}, 2 - \frac{1}{p})$ (in [**2**] this estimate was proved only for $\mu \in (1 - \frac{1}{p}, 2 - \frac{1}{p})$). An analog of estimate (4), where the norm $\|\cdot\|_{p,q}$ is replaced by $|||\cdot|||_{p,q}$, is also proved in [**4**].

In the paper [**5**] the homogeneous Dirichlet problem for (1) in cones and wedges was considered, and coercive estimates for solutions were obtained in the scales of weighted $L_{p,q}$ and $\widetilde{L}_{p,q}$ spaces, where the weight is a power of the distance to the vertex (edge).

Let us turn to the oblique derivative problem in the half-space $\mathbb{R}^n_+$. Now equation (1) is satisfied for $x_n > 0$ and $\frac{\partial u}{\partial \gamma} = 0$ for $x_n = 0$. Here γ is a constant vector field with $\gamma_n > 0$.

By changing the spatial variables, one can reduce the boundary condition to the case

$$D_n u = 0 \qquad \text{for} \quad x_n = 0. \tag{5}$$

One of the main results of this paper is the proof of estimate (4) and its analog for the norm $|||\cdot|||_{p,q}$, for solutions of the oblique derivative problem (1), (5) with arbitrary $p, q \in (1, \infty)$ and for μ satisfying

$$-\frac{1}{p} < \mu < 1 - \frac{1}{p}.$$

In the case of time independent coefficients, such estimates for the Neumann problem were proved in [**9**].

We use an approach based on the study of Green functions. In Section 2 we collect (partially known) results on the estimate of the Green function of the Dirichlet problem for equation (1). Section 3 is devoted to the estimates of the Green function of problem (1), (5).

In Section 4 we apply the obtained estimates to the oblique derivative problem for linear non-divergence parabolic equations with discontinuous in time coefficients in cylinders $\Omega \times (0,T)$, where Ω is a bounded domain in $\mathbb{R}^n$. We prove solvability results in weighted $L_{p,q}$ and $\widetilde{L}_{p,q}$ spaces, where the weight is a power of the distance to the boundary of Ω. The smoothness of the boundary is characterized by smoothness of local isomorphisms in neighborhoods of boundary points, which flatten the boundary. In particular, if the boundary is of the class $\mathcal{C}^{1,\delta}$ with $\delta \in (0,1]$, then for solutions to the equation (1)[1] in $\Omega \times (0,T)$ with zero initial and boundary conditions the following coercive estimate is proved in Theorem 4 (see Remark 1):

$$\|(\widehat{d}(x))^{\mu}\partial_t u\|_{p,q} + \sum_{ij} \|(\widehat{d}(x))^{\mu} D_i D_j u\|_{p,q} \le C\|(\widehat{d}(x))^{\mu} f\|_{p,q},$$

$$|\!|\!|(\widehat{d}(x))^{\mu}\partial_t u|\!|\!|_{p,q} + \sum_{ij} |\!|\!|(\widehat{d}(x))^{\mu} D_i D_j u|\!|\!|_{p,q} \le C|\!|\!|(\widehat{d}(x))^{\mu} f|\!|\!|_{p,q},$$

where μ, p, q and δ satisfy $1 < p,q < \infty$, $1-\delta-\frac{1}{p} < \mu < 1-\frac{1}{p}$.

Let us recall some notation: $x = (x_1,\dots,x_n) = (x',x_n)$ is a point in $\mathbb{R}^n$; $Du = (D_1u,\dots,D_nu)$ is the gradient of u.

We denote

$$Q_R(x^0,t^0) = \{(x,t) : |x-x^0| < R,\ 0 < t^0 - t < R^2\};$$

$$Q_R^+(x^0,t^0) = \{(x,t) : |x-x^0| < R,\ x_n > 0,\ 0 < t^0 - t < R^2\}.$$

The last notation will be used only for $x^0 \in \overline{\mathbb{R}^n_+}$.

Set

$$\mathcal{R}_x = \frac{x_n}{x_n + \sqrt{t-s}}, \qquad \mathcal{R}_y = \frac{y_n}{y_n + \sqrt{t-s}}.$$

In what follows we denote by the same letter the kernel and the corresponding integral operator, i.e.,

$$(\mathcal{K}h)(x,t) = \int_{-\infty}^{t}\int_{\mathbb{R}^n} \mathcal{K}(x,y;t,s)h(y,s)\,dyds.$$

Here we expand functions $\mathcal{K}$ and h by zero to whole space-time if necessary.

We adopt the convention regarding summation from 1 to n with respect to repeated indices. We use the letter C to denote various positive constants. To indicate that C depends on some parameter a, we sometimes write C_a.

[1]Here the coefficients a^{ij} may depend on x (namely, we assume $a^{ij} \in \mathcal{C}(\Omega \to L_\infty(0,T))$).

2. Preliminary results

2.1. The estimates in the whole space and in the half-space under the Dirichlet boundary condition. Let us consider equation (1) in the whole space $\mathbb{R}^n$. Using the Fourier transform with respect to x, one can obtain the following representation of solution through the right-hand side:

$$u(x,t) = \int_{-\infty}^{t} \int_{\mathbb{R}^n} \Gamma(x,y;t,s) f(y,s)\, dy\, ds, \tag{6}$$

where Γ is the Green function of the operator $\mathcal{L}_0$ given by

$$\Gamma(x,y;t,s) = \frac{\det\left(\int_s^t A(\tau)d\tau\right)^{-\frac{1}{2}}}{(4\pi)^{\frac{n}{2}}} \exp\left(- \frac{\left(\left(\int_s^t A(\tau)d\tau\right)^{-1}(x-y), (x-y)\right)}{4}\right)$$

for $t > s$ and 0 otherwise. Here $A(t)$ is the matrix $\{a^{ij}(t)\}_{i,j=1}^n$. The above representation implies, in particular, the following evident estimates.

PROPOSITION 1. *Let α and β be two arbitrary multi-indices. Then*

$$|D_x^\alpha D_y^\beta \Gamma(x,y;t,s)| \le C\,(t-s)^{-\frac{n+|\alpha|+|\beta|}{2}} \exp\left(-\frac{\sigma|x-y|^2}{t-s}\right)$$

and

$$|\partial_s D_x^\alpha D_y^\beta \Gamma(x,y;t,s)| \le C\,(t-s)^{-\frac{n+|\alpha|+|\beta|}{2}-1} \exp\left(-\frac{\sigma|x-y|^2}{t-s}\right)$$

for $x,y \in \mathbb{R}^n$ and $s < t$. Here σ depends only on the ellipticity constant ν and C may depend on ν, α and β.

In the next proposition we present solvability results for equation (1) in the whole space.

PROPOSITION 2. *Let $p,q \in (1,\infty)$.*

(i) *If $f \in L_{p,q}(\mathbb{R}^n \times \mathbb{R})$, then the solution of equation* (1) *given by* (6) *satisfies*

$$\|\partial_t u\|_{p,q} + \sum_{ij} \|D_i D_j u\|_{p,q} \le C\,\|f\|_{p,q},$$

where C depends only on ν, p, q.

(ii) *If $f \in \widetilde{L}_{p,q}(\mathbb{R}^n \times \mathbb{R})$, then the solution of equation* (1) *given by* (6) *satisfies*

$$|\!|\!|\partial_t u|\!|\!|_{p,q} + \sum_{ij} |\!|\!|D_i D_j u|\!|\!|_{p,q} \le C\,|\!|\!|f|\!|\!|_{p,q}, \tag{7}$$

where C depends only on ν, p, q.

The first assertion is proved in [**2**] and the second one in [**4**].

We denote by $\Gamma^{\mathcal{D}}(x,y;t,s)$ the Green function of the operator $\mathcal{L}_0$ in the half-space $\mathbb{R}^n_+$ subject to the homogeneous Dirichlet boundary condition on the boundary $x_n = 0$.

The next statement is proved in [**4**, Theorem 3.6].

PROPOSITION 3. *For $x, y \in \mathbb{R}^n_+$ and $t > s$ the following estimate is valid:*

$$|D_x^\alpha D_y^\beta \Gamma^{\mathcal{D}}(x,y;t,s)| \le C\,\frac{\mathcal{R}_x^{2-\alpha_n-\varepsilon}\mathcal{R}_y^{2-\beta_n-\varepsilon}}{(t-s)^{\frac{n+|\alpha|+|\beta|}{2}}}\exp\left(-\frac{\sigma|x-y|^2}{t-s}\right), \tag{8}$$

where σ is a positive number depending on ν and n, ε is an arbitrary small positive number and C may depend on ν, α, β and ε. If $\alpha_n \le 1$ (or $\beta_n \le 1$), then $2-\alpha_n-\varepsilon$ ($2-\beta_n-\varepsilon$) must be replaced by $1-\alpha_n$ ($1-\beta_n$), respectively, in the corresponding exponents.

Since $(\partial_s + a_{ij}(-s)D_{y_i}D_{y_j})\Gamma^{\mathcal{D}}(x,y;t,s) = 0$ for $s < t$, we obtain

COROLLARY 1. *For $x, y \in \mathbb{R}^n_+$ and $t > s$,*

$$|D_x^\alpha D_y^\beta \partial_s \Gamma^{\mathcal{D}}(x,y;t,s)| \le C\,\frac{\mathcal{R}_x^{2-\alpha_n-\varepsilon}\mathcal{R}_y^{-\beta_n-\varepsilon}}{(t-s)^{\frac{n+2+|\alpha|+|\beta|}{2}}}\exp\left(-\frac{\sigma|x-y|^2}{t-s}\right). \tag{9}$$

If $\alpha_n \le 1$, then $2-\alpha_n-\varepsilon$ must be replaced by $1-\alpha_n$.

2.2. Coercive estimates for weak solutions to the Dirichlet problem in the half-space. We formulate two auxiliary results on estimates of integral operators. The first statement is a particular case $m = 1$ of [**4**, Lemmas A.1 and A.3 and Remark A.2]; see also [**9**, Lemmas 2.1 and 2.2].

PROPOSITION 4. *Let $1 \le p \le \infty$, $\sigma > 0$, $0 < r \le 2$, $\lambda_1 + \lambda_2 > -1$, and let*

$$-\frac{1}{p} - \lambda_1 < \mu < 1 - \frac{1}{p} + \lambda_2. \tag{10}$$

Suppose also that the kernel $\mathcal{T}(x,y;t,s)$ satisfies the inequality

$$|\mathcal{T}(x,y;t,s)| \le C\,\frac{\mathcal{R}_x^{\lambda_1+r}\mathcal{R}_y^{\lambda_2}}{(t-s)^{\frac{n+2-r}{2}}}\,\frac{x_n^{\mu-r}}{y_n^{\mu}}\exp\left(-\frac{\sigma|x-y|^2}{t-s}\right),$$

for $t > s$. Then the integral operator $\mathcal{T}$ is bounded in $L_p(\mathbb{R}^n \times \mathbb{R})$ and in $\widetilde{L}_{p,\infty}(\mathbb{R}^n \times \mathbb{R})$.

The next proposition is a particular case $m = 1$ of [**4**, Lemma A.4]; see also [**9**, Lemma 3.2].

PROPOSITION 5. *Let $1 < p < \infty$, $\sigma > 0$, $\varkappa > 0$, $0 \le r \le 2$, $\lambda_1 + \lambda_2 > -1$, and let μ be subject to (10). Also let the kernel $\mathcal{T}(x,y;t,s)$ satisfy the inequality*

$$|\mathcal{T}(x,y;t,s)| \le C\,\frac{\mathcal{R}_x^{\lambda_1+r}\mathcal{R}_y^{\lambda_2}}{(t-s)^{\frac{n+2-r}{2}}}\,\frac{x_n^{\mu-r}}{y_n^{\mu}}\left(\frac{\delta}{t-s}\right)^{\varkappa}\exp\left(-\frac{\sigma|x-y|^2}{t-s}\right)$$

for $t > s + \delta$. Then for any $s^0 > 0$ the norm of the operator

$$\mathcal{T} \;:\; L_{p,1}(\mathbb{R}^n \times \;(s^0-\delta, s^0+\delta)) \;\to\; L_{p,1}(\mathbb{R}^n \times \;(s^0+2\delta, \infty))$$

does not exceed a constant C independent of δ and s^0.

Now we consider the problem

$$Lu = f_0 + \operatorname{div}(\mathbf{f}) \qquad \text{in} \quad \mathbb{R}^n_+ \times \mathbb{R} \tag{11}$$

(here $\mathbf{f} = (f_1, \dots, f_n)$) with the boundary condition

$$u = 0 \qquad \text{for} \quad x_n = 0. \tag{12}$$

THEOREM 1. *Let* $1 < p, q < \infty$ *and* $\mu \in (-\frac{1}{p}, 1 - \frac{1}{p})$.

(i) *Suppose that* $x_n^{\mu+1} f_0$, $x_n^\mu \mathbf{f} \in \widetilde{L}_{p,q}(\mathbb{R}^n_+ \times \mathbb{R})$. *Then the function*

$$(13) \qquad u(x,t) = \int_{-\infty}^{t} \int_{\mathbb{R}^n_+} \Big(\Gamma^{\mathcal{D}}(x,y;t,s) f_0(y,s) - D_y \Gamma^{\mathcal{D}}(x,y;t,s) \cdot \mathbf{f}(y,s)\Big)\, dyds$$

gives a weak solution of problems (11) *and* (12), *and satisfies the estimate*

$$(14) \qquad |\!|\!| x_n^\mu Du |\!|\!|_{p,q} + |\!|\!| x_n^{\mu-1} u |\!|\!|_{p,q} \le C(|\!|\!| x_n^{\mu+1} f_0 |\!|\!|_{p,q} + |\!|\!| x_n^\mu \mathbf{f} |\!|\!|_{p,q}).$$

(ii) *Suppose that* $x_n^{\mu+1} f_0$, $x_n^\mu \mathbf{f} \in L_{p,q}(\mathbb{R}^n_+ \times \mathbb{R})$. *Then the function* (13) *gives a weak solution of problems* (11) *and* (12), *and satisfies the estimate*

$$(15) \qquad \| x_n^\mu Du \|_{p,q} + \| x_n^{\mu-1} u \|_{p,q} \le C(\| x_n^{\mu+1} f_0 \|_{p,q} + \| x_n^\mu \mathbf{f} \|_{p,q}).$$

PROOF. First, function (13) obviously solves problems (11) and (12) in the sense of distributions. Thus, it is sufficient to prove estimates (14) and (15).

Put

$$\mathcal{K}_0(x,y;t,s) = \frac{x_n^{\mu-1}}{y_n^{\mu+1}} \Gamma^{\mathcal{D}}(x,y;t,s); \qquad \mathcal{K}_1(x,y;t,s) = \frac{x_n^{\mu-1}}{y_n^{\mu}} D_y \Gamma^{\mathcal{D}}(x,y;t,s);$$

$$\mathcal{K}_2(x,y;t,s) = \frac{x_n^{\mu}}{y_n^{\mu+1}} D_x \Gamma_{\mathcal{D}}(x,y;t,s); \quad \mathcal{K}_3(x,y;t,s) = \frac{x_n^{\mu}}{y_n^{\mu}} D_x D_y \Gamma_{\mathcal{D}}(x,y;t,s).$$

(i) By Proposition 3 the kernels $\mathcal{K}_0$ and $\mathcal{K}_1$ satisfy the conditions of Proposition 4 with $r = 1$ and

with $\lambda_1 = -1$, $\lambda_2 = 1$ and μ replaced by $\mu + 1$ for the kernel $\mathcal{K}_0$;

with $\lambda_1 = \lambda_2 = 0$ for the kernel $\mathcal{K}_1$, respectively.

This implies that for $\mu \in (-\frac{1}{p}, 1 - \frac{1}{p})$,

$$(16) \qquad \| x_n^{\mu-1} u \|_p \le C(\| x_n^{\mu+1} f_0 \|_p + \| x_n^\mu \mathbf{f} \|_p)$$

and

$$(17) \qquad |\!|\!| x_n^{\mu-1} u |\!|\!|_{p,\infty} \le C(|\!|\!| x_n^{\mu+1} f_0 |\!|\!|_{p,\infty} + |\!|\!| x_n^\mu \mathbf{f} |\!|\!|_{p,\infty}).$$

Interpolating (16) and (17) we arrive at

$$(18) \qquad |\!|\!| x_n^{\mu-1} u |\!|\!|_{p,q} \le C(|\!|\!| x_n^{\mu+1} f_0 |\!|\!|_{p,q} + |\!|\!| x_n^\mu \mathbf{f} |\!|\!|_{p,q}),$$

for $1 < p \le q < \infty$ and $\mu \in (-\frac{1}{p}, 1 - \frac{1}{p})$. Now duality argument gives (18) for all $1 < p, q < \infty$ and for the same interval for μ.

To estimate the first term in the left-hand side of (14), we use local estimates. We put

$$B_{\rho,\vartheta}(\xi) = \{x \in \mathbb{R}^n \,:\, |x' - \xi'| < \rho, \frac{\rho}{\vartheta} < x_n < \rho\}.$$

Localization of estimate (7) using an appropriate cut-off function, which is equal to 1 on $B_{\rho,2}$ and 0 outside $B_{2\rho,8}$, gives

$$\int_{B_{\rho,2}(\xi)} \Big(\int_{\mathbb{R}} |Du|^q dt\Big)^{\frac{p}{q}} dx \le C \int_{B_{2\rho,8}(\xi)} \Big(\int_{\mathbb{R}} (|u|^q \rho^{-q} + \rho^q |f_0|^q + |\mathbf{f}|^q dt\Big)^{\frac{p}{q}} dx.$$

Using a proper partition of unity in $\mathbb{R}^n_+$, we arrive at

$$\int\limits_{\mathbb{R}^n_+}\Big(\int\limits_{\mathbb{R}}|Du|^q dt\Big)^{p/q} x_n^{\mu p}dx \le C\Big(\int\limits_{\mathbb{R}^n_+}\Big(\int\limits_{\mathbb{R}}|u|^q\,dt\Big)^{p/q} x_n^{\mu p-p}dx$$
$$+\int\limits_{\mathbb{R}^n_+}\Big(\int\limits_{\mathbb{R}}|\mathbf{f}|^q\,dt\Big)^{p/q} x_n^{\mu p}dx+\int\limits_{\mathbb{R}^n_+}\Big(\int\limits_{\mathbb{R}}|f_0|^q\,dt\Big)^{p/q} x_n^{\mu p+p}dx\Big).$$

This immediately implies (14) with regard of (18).

(ii) To deal with the scale $L_{p,q}$, we need the following lemma.

LEMMA 1. *Let a function h be supported in the layer $|s-s^0|\le\delta$ and satisfy $\int h(y;s)\,ds\equiv 0$. Also let $p\in(1,\infty)$ and $\mu\in(-\frac{1}{p},1-\frac{1}{p})$. Then the operators $\mathcal{K}_j$, $j=0,1,2,3$, satisfy*

$$\int\limits_{|t-s^0|>2\delta}\|(\mathcal{K}_j h)(\cdot;t)\|_p\,dt\le C\,\|h\|_{p,1},$$

where C does not depend on δ and s^0.

PROOF. By $\int h(y;s)\,ds\equiv 0$, we have

$$(\mathcal{K}_j h)(x;t)=\int\limits_{-\infty}^{t}\int\limits_{\mathbb{R}^n}\Big(\mathcal{K}_j(x,y;t,s)-\mathcal{K}_j(x,y;t,s^0)\Big)\,h(y;s)\,dy\,ds \tag{19}$$

(we recall that all functions are assumed to be extended by zero).

We choose $\varepsilon>0$ such that

$$-\frac{1}{p}<\mu<1-\frac{1}{p}-\varepsilon. \tag{20}$$

For $|s-s^0|<\delta$ and $t-s^0>2\delta$, estimate (9) implies

$$\big|\mathcal{K}_j(x,y;t,s)-\mathcal{K}_j(x,y;t,s^0)\big|\le\int\limits_{s^0}^{s}|\partial_\tau\mathcal{K}_j(x,y;t,\tau)|\,d\tau$$
$$\le C\,\frac{\mathcal{R}_x^{\ell_1}\mathcal{R}_y^{\ell_2-\varepsilon}}{(t-s)^{\frac{n+2-r}{2}}}\,\frac{x_n^{\ell_3}}{y_n^{\ell_4}}\,\frac{\delta}{t-s}\,\exp\Big(-\frac{\sigma|x-y|^2}{t-s}\Big),$$

with $r=2$, $\ell_1=1$, $\ell_2=0$, $\ell_3=\mu-1$, $\ell_4=\mu+1$ for the kernel $\mathcal{K}_0$;
with $r=1$, $\ell_1=1$, $\ell_2=-1$, $\ell_3=\mu-1$, $\ell_4=\mu$ for the kernel $\mathcal{K}_1$;
with $r=1$, $\ell_1=0$, $\ell_2=0$, $\ell_3=\mu$, $\ell_4=\mu+1$ for the kernel $\mathcal{K}_2$;
with $r=0$, $\ell_1=0$, $\ell_2=-1$, $\ell_3=\mu$, $\ell_4=\mu$ for the kernel $\mathcal{K}_3$.

On the other hand, estimate (8) implies

$$\big|\mathcal{K}_j(x,y;t,s)-\mathcal{K}_j(x,y;t,s^0)\big|\le C\,\frac{\mathcal{R}_x^{\ell_1}\mathcal{R}_y^{\ell_2+1}}{(t-s)^{\frac{n+2-r}{2}}}\,\frac{x_n^{\ell_3}}{y_n^{\ell_4}}\,\exp\Big(-\frac{\sigma|x-y|^2}{t-s}\Big).$$

Combination of these estimates gives

$$\big|\mathcal{K}_j(x,y;t,s)-\mathcal{K}_j(x,y;t,s^0)\big|\le\frac{C\delta^{\varkappa}\,\mathcal{R}_x^{\ell_1}\mathcal{R}_y^{\ell_2+1-\varepsilon}}{(t-s)^{\frac{n+2-r}{2}+\varkappa}}\,\frac{x_n^{\ell_3}}{y_n^{\ell_4}}\,\exp\Big(-\frac{\sigma|x-y|^2}{t-s}\Big),$$

where $\varkappa = \frac{\varepsilon}{1+\varepsilon}$. Thus, the kernels in (19) satisfy the assumptions of Proposition 5
with $\lambda_1 = -1$, $\lambda_2 = 1-\varepsilon$ and μ replaced by $\mu+1$ for kernels $\mathcal{K}_0$ and $\mathcal{K}_2$;
with $\lambda_1 = 0$, $\lambda_2 = -\varepsilon$ for kernels $\mathcal{K}_1$ and $\mathcal{K}_3$, respectively.

Inequality (20) becomes (10), and the Lemma follows. □

We continue the proof of the second statement of Theorem 1. Estimate (14) for $q = p$ provides boundedness of the operators $\mathcal{K}_j$, $j = 0,1,2,3$, in $L_p(\mathbb{R}^n \times \mathbb{R})$, which gives the first condition in [**1**, Theorem 3.8]. Lemma 1 is equivalent to the second condition in this theorem. Therefore, [**1**, Theorem 3.8] ensures that these operators are bounded in $L_{p,q}(\mathbb{R}^n \times \mathbb{R})$ for any $q \in (1,p)$. For $q \in (p,\infty)$ this statement follows by duality arguments. This implies estimate (15). □

3. Oblique derivative problem

3.1. The Green function.

THEOREM 2. *There exists a Green function $\Gamma^{\mathcal{N}} = \Gamma^{\mathcal{N}}(x,y;t,s)$ of problems (1) and (5), and for arbitrary $x, y \in \mathbb{R}^n_+$ and $t > s$ it satisfies the estimate*

$$|D_x^\alpha D_y^\beta \Gamma^{\mathcal{N}}(x,y;t,s)| \le C \frac{\mathcal{R}_x^{\widehat{\alpha}_n} \mathcal{R}_y^{\widehat{\beta}_n}}{(t-s)^{\frac{n+|\alpha|+|\beta|}{2}}} \exp\left(-\frac{\sigma|x-y|^2}{t-s}\right), \tag{21}$$

$$|D_x^\alpha D_y^\beta \partial_s \Gamma^{\mathcal{N}}(x,y;t,s)| \le C \frac{\mathcal{R}_x^{\widehat{\alpha}_n} \mathcal{R}_y^{-1-\beta_n-\varepsilon}}{(t-s)^{\frac{n+2+|\alpha|+|\beta|}{2}}} \exp\left(-\frac{\sigma|x-y|^2}{t-s}\right), \tag{22}$$

where

$$\widehat{\alpha}_n = \begin{cases} 0, & \alpha_n = 0; \\ 2-\alpha_n, & \alpha_n = 1,2; \\ 3-\alpha_n-\varepsilon, & \alpha_n \ge 3; \end{cases} \qquad \widehat{\beta}_n = \begin{cases} 0, & \beta_n = 0; \\ 1-\beta_n-\varepsilon, & \beta_n \ge 1. \end{cases}$$

Here σ is a positive number depending on ν and n, ε is an arbitrary small positive number and C may depend on ν, α, β, and ε.

PROOF. Let u be a solution of problems (1) and (5). Then the derivative $D_n u$ obviously satisfies the Dirichlet problems (11) and (12) with $f_0 = 0$ and $\mathbf{f} = (0,\dots,0,f)$. Therefore,

$$D_n u = -\int\limits_{-\infty}^{t} \int\limits_{\mathbb{R}^n_+} D_{y_n} \Gamma^{\mathcal{D}}(x,y;t,s) f(y;s)\, dyds,$$

and we can write solution to problems (1) and (5) as

$$u(x;t) = \int\limits_{-\infty}^{t} \int\limits_{\mathbb{R}^n_+} \Gamma^{\mathcal{N}}(x,y,t,s) f(y,s)\, dyds, \tag{23}$$

where

$$\Gamma^{\mathcal{N}}(x,y;t,s) = \int\limits_{x_n}^{\infty} D_{y_n} \Gamma^{\mathcal{D}}(x',z_n,y;t,s)\, dz_n.$$

Since $D_{x_n}\Gamma^{\mathcal{N}}(x,y;t,s) = -D_{y_n}\Gamma^{\mathcal{D}}(x,y;t,s)$, we derive from (8) that

$$|D_x^\alpha D_y^\beta D_{x_n}\Gamma^{\mathcal{N}}(x,y;t,s)| \le C\,\frac{\mathcal{R}_x^{2-\alpha_n-\varepsilon}\mathcal{R}_y^{1-\beta_n-\varepsilon}}{(t-s)^{\frac{n+1+|\alpha|+|\beta|}{2}}}\exp\left(-\frac{\sigma|x-y|^2}{t-s}\right), \tag{24}$$

where $2-\alpha_n-\varepsilon$ must be replaced by $1-\alpha_n$ if $\alpha_n \le 1$ and $1-\beta_n-\varepsilon$ by 0 if $\beta_n = 0$. Estimate (21) with $\alpha_n \ge 1$ follows from (24).

In a similar way we derive from (9) that

$$\begin{aligned} &|D_x^\alpha D_{x_n} D_y^\beta \partial_s \Gamma^{\mathcal{N}}(x,y;t,s)| \\ &\quad \le C\,\frac{\mathcal{R}_x^{2-\alpha_n-\varepsilon}\mathcal{R}_y^{-1-\beta_n-\varepsilon}}{(t-s)^{\frac{n+3+|\alpha|+|\beta|}{2}}}\exp\left(-\frac{\sigma|x-y|^2}{t-s}\right), \end{aligned} \tag{25}$$

where $2-\alpha_n-\varepsilon$ must be replaced by $1-\alpha_n$ if $\alpha_n \le 1$. Estimate (22) with $\alpha_n \ge 1$ follows from (25).

To estimate derivatives with respect to x', we consider two cases.

Case 1: $|x_n - y_n| \le \sqrt{t-s}$. Then (24) implies

$$\begin{aligned} |D_{x'}^{\alpha'} D_y^\beta \Gamma^{\mathcal{N}}(x,y;t,s)| &\le \int_{x_n}^{\infty} |D_{x'}^{\alpha'} D_y^\beta D_{z_n}\Gamma^{\mathcal{N}}(x',z_n,y;t,s)|\,dz_n \\ &\le \frac{C\,\mathcal{R}_y^{\widehat{\beta}_n}}{(t-s)^{\frac{n+|\alpha'|+|\beta|}{2}}}\exp\left(-\frac{\sigma|x'-y'|^2}{t-s}\right)\int_{\mathbb{R}}\exp\left(-\frac{\sigma|z_n-y_n|^2}{t-s}\right)\frac{dz_n}{\sqrt{t-s}} \\ &\le \frac{C\,\mathcal{R}_y^{\widehat{\beta}_n}}{(t-s)^{\frac{n+|\alpha'|+|\beta|}{2}}}\exp\left(-\frac{\sigma|x-y|^2}{t-s}+\sigma\right) \end{aligned}$$

(the last inequality is due to $\frac{|x_n-y_n|^2}{t-s} \le 1$), which gives (21) with $\alpha_n = 0$ in Case 1. In a similar way we derive estimate (22) with $\alpha_n = 0$ in Case 1 from (25).

Case 2: $|x_n - y_n| > \sqrt{t-s}$. Then we rewrite equation $\mathcal{L}_0\Gamma^{\mathcal{N}} = 0$ as

$$\begin{aligned} \mathcal{L}_0'\Gamma^{\mathcal{N}} &\equiv \partial_t\Gamma^{\mathcal{N}} - \sum_{i,j=1}^{n-1} a_{ij}(t) D_{x_i} D_{x_j}\Gamma^{\mathcal{N}} \\ &= \mathcal{F} \equiv \Bigl(2\sum_{j=1}^{n-1} a_{jn} D_{x_j} D_{x_n}\Gamma^{\mathcal{N}} + a_{nn} D_{x_n} D_{x_n}\Gamma^{\mathcal{N}}\Bigr). \end{aligned} \tag{26}$$

From (24) and (26) it follows that

$$|D_{x'}^{\alpha'} D_y^\beta \mathcal{F}(x,y;t,s)| \le C\,\frac{\mathcal{R}_y^{\widehat{\beta}_n}}{(t-s)^{\frac{n+2+|\alpha'|+|\beta|}{2}}}\exp\left(-\frac{\sigma|x-y|^2}{t-s}\right). \tag{27}$$

Let $\Gamma'(x',y';t,s)$ be the Green function of the operator $\mathcal{L}_0'$ in $\mathbb{R}^{n-1}\times\mathbb{R}$. Then solving (26), we get

$$\Gamma^{\mathcal{N}}(x,y;t,s) = \int_s^t \int_{\mathbb{R}^{n-1}} \Gamma'(x',z';t,\tau)\mathcal{F}(z',x_n,y;\tau,s)\,dz'd\tau.$$

Since $\Gamma'(x',z';t,\tau)$ depends only on the difference $x'-z'$, we obtain

$$D_{x'}^{\alpha'}\Gamma^{\mathcal{N}}(x,y;t,s)=\int_s^t\int_{\mathbb{R}^{n-1}}\Gamma'(x',z';t,\tau)D_{z'}^{\alpha'}\mathcal{F}(z',x_n,y;\tau,s)\,dz'd\tau. \tag{28}$$

Using Proposition 1 for Γ' we get from (28) and (27)

$$|D_{x'}^{\alpha'}D_y^{\beta}\Gamma^{\mathcal{N}}(x,y;t,s)|\le\int_s^t\int_{\mathbb{R}^{n-1}}\frac{C}{(t-\tau)^{\frac{n-1}{2}}}\exp\left(-\frac{\sigma|x'-z'|^2}{t-\tau}\right)$$
$$\times\frac{\mathcal{R}_y^{\widehat{\beta}_n}}{(\tau-s)^{\frac{n+2+|\alpha'|+|\beta|}{2}}}\exp\left(-\frac{\sigma|(z',x_n)-y|^2}{\tau-s}\right)dz'd\tau.$$

We observe that $\mathcal{R}_y$ here has non-standard time argument: $\tau-s$ instead of $t-s$. However, since $\widehat{\beta}_n\le 0$, we can estimate a "non-standard" $\mathcal{R}_y^{\widehat{\beta}_n}$ by a standard one.

Integrating with respect to z', herewith using Fourier transform, we get

$$|D_{x'}^{\alpha'}D_y^{\beta}\Gamma^{\mathcal{N}}(x,y;t,s)|\le\frac{C\,\mathcal{R}_y^{\widehat{\beta}_n}}{(t-s)^{\frac{n-1}{2}}}\exp\left(-\frac{\sigma|x'-y'|^2}{t-s}\right)$$
$$\times\int_s^t\frac{1}{(\tau-s)^{\frac{3+|\alpha'|+|\beta|}{2}}}\exp\left(-\frac{\sigma(x_n-y_n)^2}{\tau-s}\right)d\tau.$$

Substituting $\theta=\frac{t-\tau}{\tau-s}$, we arrive at

$$|D_{x'}^{\alpha'}D_y^{\beta}\Gamma^{\mathcal{N}}(x,y;t,s)|\le\frac{C\,\mathcal{R}_y^{\widehat{\beta}_n}}{(t-s)^{\frac{n+|\alpha'|+|\beta|}{2}}}\exp\left(-\frac{\sigma|x'-y'|^2}{t-s}\right)$$
$$\times\int_0^\infty(\theta+1)^{\frac{|\alpha'|+|\beta|-1}{2}}\exp\left(-\frac{\sigma(x_n-y_n)^2}{t-s}(\theta+1)\right)d\theta.$$

Since $\frac{|x_n-y_n|^2}{t-s}>1$, this implies

$$|D_{x'}^{\alpha'}D_y^{\beta}\Gamma^{\mathcal{N}}(x,y;t,\tau)|\le\frac{C\,\mathcal{R}_y^{\widehat{\beta}_n}}{(t-s)^{\frac{n+|\alpha'|+|\beta|}{2}}}\exp\left(-\frac{\sigma|x-y|^2}{t-s}\right)$$
$$\times\int_0^\infty(\theta+1)^{\frac{|\alpha'|+|\beta|-1}{2}}\exp(-\sigma\theta)\,d\theta,$$

which gives (21) with $\alpha_n=0$ in Case 2.

In a similar way we derive the estimate (22) with $\alpha_n=0$ in Case 2, and the proof is complete. □

3.2. Coercive estimates in $\widetilde{L}_{p,q}$ and in $L_{p,q}$.

THEOREM 3. *Let $1<p,q<\infty$ and $\mu\in(-\frac{1}{p},1-\frac{1}{p})$.*

(i) *If $f\in\widetilde{L}_{p,q}(\mathbb{R}_+^n\times\mathbb{R})$, then solution (23) to problems (1) and (5) satisfies*

$$\|x_n^{\mu}\partial_t u\|_{p,q}+\|x_n^{\mu}D(Du)\|_{p,q}\le C\,\|x_n^{\mu}f\|_{p,q}. \tag{29}$$

(ii) *If* $f \in L_{p,q}(\mathbb{R}^n_+ \times \mathbb{R})$, *then solution* (23) *to problems* (1) *and* (5) *satisfies*

$$\|x_n^\mu \partial_t u\|_{p,q} + \|x_n^\mu D(Du)\|_{p,q} \le C\ \|x_n^\mu f\|_{p,q}. \tag{30}$$

The constant C *depends only on* ν, μ, p, *and* q.

PROOF. First, we recall that the function $D_n u$ satisfies the Dirichlet problems (11) and (12) with $f_0 = 0$ and $\mathbf{f} = (0, \dots, 0, f)$. Thus, Theorem 1 gives

$$|\!|\!| x_n^\mu D(D_n u) |\!|\!|_{p,q} \le C |\!|\!| x_n^\mu f |\!|\!|_{p,q}, \tag{31}$$

$$\|x_n^\mu D(D_n u)\|_{p,q} \le C \|x_n^\mu f\|_{p,q}. \tag{32}$$

To estimate the derivatives $D'D'u$ in $\widetilde{L}_{p,q}$-norm, we proceed similarly to Theorem 2. We rewrite equation (1) as in (26):

$$\mathcal{L}_0' u = \widetilde{f} \equiv f + 2 \sum_{j=1}^{n-1} a_{jn} D_j D_n u + a_{nn} D_n D_n u.$$

Using Proposition 2(ii) in $\mathbb{R}^{n-1}$ we obtain

$$|\!|\!| D'D'u(\cdot, x_n) |\!|\!|_{p,q} \le C |\!|\!| \widetilde{f}(\cdot, x_n) |\!|\!|_{p,q} \tag{33}$$

almost for all $x_n > 0$. Multiplying both sides of (33) by x_n^μ and taking the L_p-norm with respect to x_n, we arrive at

$$|\!|\!| x_n^\mu D'D'u |\!|\!|_{p,q} \le C |\!|\!| x_n^\mu \widetilde{f} |\!|\!|_{p,q} \le C |\!|\!| x_n^\mu f |\!|\!|_{p,q}, \tag{34}$$

where we have used estimate (31). The first term in (29) is estimated by using (31), (34) and equation (1), and the statement (i) follows.

For $L_{p,q}$-norm of $D'D'u$ this approach fails, so we proceed as in part (ii) of Theorem 1. Let us introduce the kernels

$$\mathcal{K}_4(x, y; t, s) = \frac{x_n^\mu}{y_n^\mu} D'_{x'} D'_{x'} \Gamma_{\mathcal{N}}(x, y; t, s); \qquad \mathcal{K}_4^*(x, y; t, s) = \mathcal{K}_4(y, x; s, t).$$

Estimate (34) with $q = p$ means that the operator $\mathcal{K}_4$ is bounded in $L_p(\mathbb{R}^n \times \mathbb{R})$. Choose $\varepsilon > 0$ such that relation (20) holds. Using estimates (21) and (22), it is easy to check that $\mathcal{K}_4$ satisfies the same estimates as the kernel $\mathcal{K}_3$ in Theorem 1. Verbatim repetition of arguments shows that this operator is bounded in $L_{p,q}(\mathbb{R}^n \times \mathbb{R})$ for any $q \in (1, p)$.

Further, by duality the operator $\mathcal{K}_4^*$ is bounded in $L_{p'}(\mathbb{R}^n \times \mathbb{R})$. Using (21) and relation $(\partial_s - a_{ij}(s) D_{y_i} D_{y_j}) \Gamma^{\mathcal{N}}(y, x; s, t) = 0$ for $s > t$, we obtain

$$|\partial_s \mathcal{K}_4^*(x, y; t, s)| \le \frac{C}{(s-t)^{\frac{n+4}{2}}} \frac{x_n^{-\mu}}{y_n^{-\mu}} \exp\left(-\frac{\sigma |x - y|^2}{t - s}\right).$$

For $|s - s^0| < \delta$ and $s^0 - t > 2\delta$ this implies

$$\left|\mathcal{K}_4^*(x, y; t, s) - \mathcal{K}_4^*(x, y; t, s^0)\right| \le \frac{C\delta}{(t-s)^{\frac{n+2}{2}+1}} \frac{x_n^{-\mu}}{y_n^{-\mu}} \exp\left(-\frac{\sigma |x - y|^2}{t - s}\right).$$

The last estimate allows us to apply Proposition 5 with $\varkappa = 1$, $r = 0$, $\lambda_1 = \lambda_2 = 0$, and p replaced by p'. Therefore, [**1**, Theorem 3.8] ensures that for any $q \in (p, \infty)$, the operator $\mathcal{K}_4^*$ is bounded in $L_{p',q'}(\mathbb{R}^n \times \mathbb{R})$. By duality the operator $\mathcal{K}_4$ is bounded in $L_{p,q}(\mathbb{R}^n \times \mathbb{R})$.

Thus, we have

$$\|x_n^\mu D'D'u\|_{p,q} \le C \|x_n^\mu f\|_{p,q} \tag{35}$$

for all $1 < q < \infty$. The first term in (30) is estimated by (32), (35) and equation (1), and the statement (ii) also follows. □

4. Solvability of the oblique derivative problem in a bounded cylinder

Let Ω be a bounded domain in $\mathbb{R}^n$ with boundary $\partial\Omega$. For a cylinder $Q = \Omega \times (0,T)$, we denote by $\partial'' Q = \partial\Omega \times (0,T)$ its lateral boundary.

We introduce two scales of functional spaces: $\mathbb{L}_{p,q,(\mu)}(Q)$ and $\widetilde{\mathbb{L}}_{p,q,(\mu)}(Q)$, with norms

$$\|f\|_{p,q,(\mu),Q} = \|(\widehat{d}(x))^{\mu} f\|_{p,q,Q} = \Big(\int\limits_0^T \Big(\int\limits_\Omega (\widehat{d}(x))^{\mu p} |f(x,t)|^p dx\Big)^{\frac{q}{p}} dt\Big)^{\frac{1}{q}}$$

and

$$|||f|||_{p,q,(\mu),Q} = |||(\widehat{d}(x))^{\mu} f|||_{p,q,Q} = \Big(\int\limits_\Omega \Big(\int\limits_0^T (\widehat{d}(x))^{\mu q} |f(x,t)|^q dt\Big)^{\frac{p}{q}} dx\Big)^{\frac{1}{p}},$$

respectively, where $\widehat{d}(x)$ stands for the distance from $x \in \Omega$ to $\partial\Omega$. For $p = q$ these spaces coincide, and we use the notation $\mathbb{L}_{p,(\mu)}(Q)$ and $\|\cdot\|_{p,(\mu),Q}$.

We denote by $\mathbb{W}^{2,1}_{p,q,(\mu)}(Q)$ and $\widetilde{\mathbb{W}}^{2,1}_{p,q,(\mu)}(Q)$ the set of functions with the finite seminorms

$$\|\partial_t u\|_{p,q,(\mu),Q} + \sum_{ij} \|D_i D_j u\|_{p,q,(\mu),Q}$$

and

$$|||\partial_t u|||_{p,q,(\mu),Q} + \sum_{ij} |||D_i D_j u|||_{p,q,(\mu),Q},$$

respectively. These seminorms become norms on the subspaces defined by $u|_{t=0} = 0$. For $p = q$ we write $\mathbb{W}^{2,1}_{p,(\mu)}(Q)$.

We say $\partial\Omega \in \mathcal{W}^2_{p,(\mu)}$ if for any point $x^0 \in \partial\Omega$ there exists a neighborhood $\mathcal{U}$ and a diffeomorphism Ψ mapping $\mathcal{U} \cap \Omega$ onto the half-ball B_1^+ and satisfying

$$(\widehat{d}(x))^{\mu} D^2\Psi \in L_p(\mathcal{U} \cap \Omega), \qquad x_n^{\mu} D^2 \Psi^{-1} \in L_p(B_1^+),$$

where corresponding norms are uniformly bounded with respect to x^0.

We set $\widehat{\mu}(p,q) = 1 - \frac{n}{p} - \frac{2}{q}$.

We consider the initial boundary value problem

$$\text{(36)} \qquad \begin{aligned} &\mathcal{L}u \equiv \partial_t u - a^{ij}(x,t) D_i D_j u + b^i(x,t) D_i u = f(x,t) \quad \text{in} \quad Q; \\ &\gamma^i(x,t) D_i u|_{\partial'' Q} = 0, \qquad u|_{t=0} = 0. \end{aligned}$$

The matrix of leading coefficients $a^{ij} \in \mathcal{C}(\overline{\Omega} \to L_\infty(0,T))$ is symmetric and satisfies the ellipticity condition (2). The vector field γ is assumed non-tangent to $\partial\Omega$:

$$\text{(37)} \qquad \gamma^i(x,t)\mathbf{n}_i(x) \geq \gamma_0, \qquad (x,t) \in \partial'' Q, \quad \gamma_0 = \text{const} > 0$$

(here $\mathbf{n}(x)$ stands for the unit exterior normal vector to $\partial\Omega$ at the point x).

THEOREM 4. *Let $1 < p, q < \infty$ and $\mu \in (-\frac{1}{p}, 1 - \frac{1}{p})$. Assume that the components γ^i belong to the anisotropic Hölder space $\mathcal{C}^{0,1;\frac{1}{2}}(\partial''Q)$.*

1. *Let $b^i \in \mathbb{L}_{\overline{p},\overline{q},(\overline{\mu})}(Q) + \mathbb{L}_{\infty,(\overline{\overline{\mu}})}(Q)$, where $\overline{p}$ and $\overline{q}$ are subject to*

$$\overline{p} \geq p; \qquad \left[\begin{array}{ll} \overline{q} = q, & \widehat{\mu}(\overline{p},\overline{q}) > 0, \\ q < \overline{q} < \infty, & \widehat{\mu}(\overline{p},\overline{q}) = 0, \end{array} \right.$$

while $\overline{\mu}$ and $\overline{\overline{\mu}}$ satisfy

$$\overline{\mu} = \min\{\mu, \max\{\widehat{\mu}(p,q), 0\}\}; \qquad \overline{\overline{\mu}} < \mu + \tfrac{1}{p}. \tag{38}$$

Suppose also that either $\partial\Omega \in \mathcal{W}^2_{\infty,(\overline{\overline{\mu}})}$ or $\partial\Omega \in \mathcal{W}^2_{\overline{p},(\overline{\mu})}$. Then, for any $f \in \mathbb{L}_{p,q,(\mu)}(Q)$, the initial boundary value problem (36) *has a unique solution $u \in \mathbb{W}^{2,1}_{p,q,(\mu)}(Q)$. Moreover, this solution satisfies*

$$\|\partial_t u\|_{p,q,(\mu)} + \sum_{ij} \|D_i D_j u\|_{p,q,(\mu)} \leq C \|f\|_{p,q,(\mu)},$$

where the positive constant C does not depend on f.

2. *Let $b^i \in \widetilde{\mathbb{L}}_{\overline{p},\overline{q},(\overline{\mu})}(Q) + \mathbb{L}_{\infty,(\overline{\overline{\mu}})}(Q)$, where $\overline{p}$ and $\overline{q}$ are subject to*

$$\overline{q} \geq q; \qquad \left[\begin{array}{ll} \overline{p} = p, & \widehat{\mu}(\overline{p},\overline{q}) > 0, \\ p < \overline{p} < \infty, & \widehat{\mu}(\overline{p},\overline{q}) = 0, \end{array} \right.$$

while $\overline{\mu}$ and $\overline{\overline{\mu}}$ satisfy (38). *Suppose also that $\partial\Omega$ satisfies the same conditions as in part* **1**. *Then, for any $f \in \widetilde{\mathbb{L}}_{p,q,(\mu)}(Q)$, problem* (36) *has a unique solution $u \in \widetilde{\mathbb{W}}^{2,1}_{p,q,(\mu)}(Q)$. Moreover, this solution satisfies*

$$|||\partial_t u|||_{p,q,(\mu)} + \sum_{ij} |||D_i D_j u|||_{p,q,(\mu)} \leq C |||f|||_{p,q,(\mu)},$$

where the positive constant C does not depend on f.

REMARK 1. *It is well known (see, e.g.,* [**7**]*) that if $\partial\Omega \in \mathcal{C}^{1,\delta}$ for some $\delta \in (0,1]$, then $\partial\Omega \in \mathcal{W}^2_{\infty,(1-\delta)}$. In this case the second inequality in* (38) *implies solvability of the problem* (36) *for $1 - \delta - \frac{1}{p} < \mu < 1 - \frac{1}{p}$.*

PROOF. The standard scheme (see [**6**, Ch.IV, §9]) including the partition of unity, local flattening of $\partial\Omega$, and coefficients freezing reduces the proof to the coercive estimates for the model problems to equation (1) in the whole space and in the half-space. These estimates are obtained in [**3**, Theorem 1.1] and our Theorem 3. By the Hölder inequality and the embedding theorems (see, e.g., [**1**, Theorems 10.1 and 10.4]), the assumptions on b^i guarantee that the lower-order terms in (36) belong to desired weighted spaces, $\mathbb{L}_{p,q,(\mu)}(Q)$ and $\widetilde{\mathbb{L}}_{p,q,(\mu)}(Q)$, respectively. By the same reasons, the requirements on $\partial\Omega$ imply $\partial\Omega \in \mathcal{C}^1$ and ensure the invariance of assumptions on b^i under flattening of the boundary.

Next, after the flattening of $\partial\Omega$, we can assume without loss of generality that $\gamma^i(0) = \delta^n_i$ and rewrite the boundary condition as

$$D_n u|_{x_n=0} = \varphi \equiv (\delta^n_i - \gamma^i(x,t)) D_i u. \tag{39}$$

The inhomogeneity in boundary condition (39) will be removed if we subtract from u some function satisfying the same boundary condition. By assumption $\gamma^i \in \mathcal{C}^{0,1;\frac{1}{2}}(\partial''Q)$, the function φ has the same differential properties as Du. Therefore, such a subtraction leaves the space $\mathbb{L}_{p,q,(\mu)}(Q)$ (respectively, $\widetilde{\mathbb{L}}_{p,q,(\mu)}(Q)$) of the right-hand side in (36). This completes the proof. □

The assumption $\gamma^i \in \mathcal{C}^{0,1;\frac{1}{2}}(\partial''Q)$ is not optimal. The sharp assumption here is that multiplication by the vector field γ should keep the space of traces of gradients of functions from $\mathbb{W}^{2,1}_{p,q,(\mu)}(Q)$ (respectively, from $\widetilde{\mathbb{W}}^{2,1}_{p,q,(\mu)}(Q)$). In other words, γ should belong to space $\mathbf{MT}D\mathbb{W}^{2,1}_{p,q,(\mu)}(Q)$ (respectively, $\mathbf{MT}D\widetilde{\mathbb{W}}^{2,1}_{p,q,(\mu)}(Q)$) of multipliers of traces of gradients of weighted Sobolev functions.

Unfortunately, to the best of our knowledge, these spaces are not described yet. In the isotropic case $p = q$, we can give rather sharp sufficient conditions in terms of the Besov spaces (the notation of the Besov spaces corresponds to [**1**, Ch.IV]). The following result can be extracted from the proofs of [**1**, Theorems 18.13 and 18.14], [**10**], and [**8**, 4.4.3].

THEOREM 5. *Let* $1 < p < \infty$ *and* $\mu \in (-\frac{1}{p}, 1 - \frac{1}{p})$.
Suppose that $b^i \in \mathbb{L}_{\overline{p},(\overline{\mu})}(Q) + \mathbb{L}_{\infty,(\overline{\overline{\mu}})}(Q)$, *where* $\overline{p}$, $\overline{\mu}$ *and* $\overline{\overline{\mu}}$ *are subject to*

$$\overline{p} = \max\{p, n+2\}, \quad \text{if} \quad p \neq n+2; \qquad \overline{p} > n+2, \quad \text{if} \quad p = n+2;$$

$$\overline{\mu} = \min\{\mu, \max\{1 - \tfrac{n+2}{p}, 0\}\}; \qquad \overline{\overline{\mu}} < \mu + \tfrac{1}{p}.$$

Suppose also that either $\partial\Omega \in \mathcal{W}^2_{\infty,(\overline{\overline{\mu}})}$ *or* $\partial\Omega \in \mathcal{W}^2_{\overline{p},(\overline{\mu})}$. *Finally, we assume that the components* γ^i *belong to the Besov space* $B^{\boldsymbol{\lambda}}_{\overline{\overline{p}},\theta}(\partial''Q)$ *with parameters*

$$\boldsymbol{\lambda} \equiv (\lambda_x^1, \ldots, \lambda_x^{n-1}, \lambda_t) = \Big(1 - \frac{1}{p}, \ldots, 1 - \frac{1}{p}, \frac{1}{2} - \frac{1}{2p}\Big); \qquad \theta = p;$$

$$\overline{\overline{p}} = \max\Big\{p, \frac{n+1}{1-\mu-\frac{1}{p}}\Big\}, \quad \text{if} \quad p \neq \frac{n+2}{1-\mu}; \quad \overline{\overline{p}} > \frac{n+2}{1-\mu}, \quad \text{if} \quad p = \frac{n+2}{1-\mu}.$$

Then, for any $f \in \mathbb{L}_{p,(\mu)}(Q)$, *the initial boundary value problem* (36) *has a unique solution* $u \in \mathbb{W}^{2,1}_{p,(\mu)}(Q)$. *Moreover, this solution satisfies*

$$\|\partial_t u\|_{p,(\mu)} + \sum_{ij} \|D_i D_j u\|_{p,(\mu)} \leq C \|f\|_{p,(\mu)},$$

where the positive constant C *does not depend on* f.

References

[1] Besov, O.V.; Il'in, V.P.; Nikol'skii, S.M., *Integral representations of functions and imbedding theorems*, ed.2, Moscow, Nauka, 1996 (Russian); English transl. of the 1st ed.: Vol. 1. Scripta Series in Math. Edited by M.H. Taibleson. V. H. Winston & Sons, Washington, D.C.; Halsted Press, New York-Toronto, Ont.-London, 1978. MR519341 (80f:46030a)

[2] Krylov, N.V., *The heat equation in* $L_q((0,T), L_p)$*-spaces with weights*, SIAM J. Math. Anal. **32** (2001), N5, 1117–1141. MR1828321 (2002a:35092)

[3] Krylov, N.V. *The Calderón-Zygmund theorem and its applications to parabolic equations*, Alg. Anal. **13** (2001), N4, 1–25 (Russian); English transl.: St. Petersburg Math. J. **13** (2002), N4, 509–526. MR1865493 (2002g:35033)

[4] Kozlov, V.A.; Nazarov, A.I., *The Dirichlet problem for non-divergence parabolic equations with discontinuous in time coefficients*, Math. Nachr. **282** (2009), N9, 1220–1241. MR2561181 (2011b:35189)

[5] Kozlov, V.A.; Nazarov, A.I., *The Dirichlet problem for non-divergence parabolic equations with discontinuous in time coefficients in a wedge*, preprint available at http://arxiv.org/abs/1112.3031. 37p.

[6] Ladyzhenskaja, O.A.; Solonnikov, V.A.; Ural'tseva, N.N., *Linear and quasi-linear equations of parabolic type*, Moscow, Nauka, 1968 (Russian); English transl.: Transl. of Math. Monographs, Vol. 23. AMS, Providence, R.I., 1967. MR0241821 (39:3159a)

[7] Lieberman, G., *Regularized distance and its applications*, Pacific J. Math., **117** (1985), 329–352. MR779924 (87j:35101)

[8] Maz'ya, V.G.; Shaposhnikova, T.O., *Theory of Sobolev Multipliers With Applications to Differential and Integral Operators*, Comprehensive Studies in Math., Vol. 337. Springer-Verlag, Berlin–Heidelberg, 2009. MR2457601 (2010d:46039)

[9] Nazarov, A.I., *L_p-estimates for a solution to the Dirichlet problem and to the Neumann problem for the heat equation in a wedge with edge of arbitrary codimension*, Probl. Mat. Anal., **22** (2001), 126–159 (Russian); English transl.: J. Math. Sci., **106** (2001), N3, 2989–3014. MR1906030 (2003d:35118)

[10] Uspenskii, S.V., *Embedding theorems for classes with weights*, Tr. Mat. Inst. Steklova, **60** (1961), 282–303 (Russian); English transl.: AMS Transl., **87** (1970), 121–145. MR0136980 (25:440)

Department of Mathematics, University of Linköping, SE-581 83 Linköping, Sweden
E-mail address: vladimir.kozlov@liu.se

St.-Petersburg Department of Steklov Mathematical Institute, Fontanka, 27, St.-Petersburg, 191023, Russia, and St.-Petersburg State University, Universitetskii pr. 28, St.-Petersburg, 198504, Russia
E-mail address: al.il.nazarov@gmail.com

Amer. Math. Soc. Transl.
Volume **232**, 2014

Singular solutions of Navier–Stokes equations

V. V. Pukhnachev

To the memory of Olga Aleksandrovna Ladyzhenskaya

ABSTRACT. Stationary solutions of Navier–Stokes equations with singularities in the flow domain or on its boundary are considered. This class includes well-known Jeffery–Hamel, Landau, Goldshtik, and other self-similar solutions. The history of self-similar solutions, including recent results, is reviewed. The second part of the paper contains the author's original results on flows in plane or axisymmetric domains with singularities of the source or sink type. The velocity field in these problems possesses an infinite Dirichlet integral. Some sufficient conditions for solvability of these problems, based on a priori estimates of the regular component of the solution, are formulated. A few unsolved problems are given in the last part of the paper. One of them is a two-dimensional problem of a point vortex in a domain bounded by a solid wall. The list of references contains 61 items.

1. Introduction

We consider stationary solutions $\mathbf{v}$ and p of the Navier–Stokes equations

$$\mathbf{v} \cdot \nabla \mathbf{v} = -\nabla p + \nu \Delta \mathbf{v}, \quad \nabla \cdot \mathbf{v} = 0. \tag{1}$$

It is assumed that these solutions have singularities either in a domain $\Omega \subset \mathbb{R}^n$ $(n = 2, 3)$ or on its boundary. Here $\mathbf{v}$ is the velocity vector, p is the ratio of the pressure to the constant density of the fluid, and $\nu = const > 0$ is the kinematic viscosity coefficient.

It is well known that (1) admits solutions with a potential velocity field

$$\mathbf{v} = \nabla \phi, \quad p = C - |\nabla \phi|^2/2, \tag{2}$$

where ϕ is an arbitrary harmonic function in $\mathbb{R}^n$ and C is a constant. Choosing a fundamental solution of the Laplace equation or its derivatives as ϕ, we obtain for (1) a family of solutions with pointwise singularities. These solutions are defined in the entire $\mathbb{R}^n$.

Consider the restriction of (2) to a simply connected bounded domain $\Omega \subset \mathbb{R}^2$ and its perturbation by a variation of the velocity vector on the boundary $\partial\Omega$. If $\phi = (2\pi)^{-1} q \log |x|$, where $q = const$, then the singularity is characterized by a source of intensity q at the origin. Another singular solution corresponds to the

2010 *Mathematics Subject Classification.* Primary 35J62, 35Q30, 76D06.
Key words and phrases. Navier–Stokes equations, stationary problems, self-similar solutions, sources and sinks, a priori estimates.

case $\phi = (2\pi)^{-1}\Gamma \arctan(x_2/x_1)$, where $\Gamma = const$. Here a vortex with circulation Γ is placed at the origin.

An axisymmetric analog of the problem with a source is discussed in Section 4. Solutions of (1) invariant under the similarity transformation are discussed in Section 2; these solutions are called self-similar solutions in mechanics. Some open questions of the theory of singular solutions of Navier–Stokes equations are stated in Section 5.

The theory of stationary boundary-value problems for the Navier–Stokes equations was developed in a large degree of completeness in the classical monograph by Ladyzhenskaya [1]. The proof of global solvability of these equations is based on the a priori estimate of the Dirichlet integral

$$I = \int_{\Omega} \nabla \mathbf{v} : \nabla \mathbf{v}\, dx. \tag{3}$$

The perturbation problem for the velocity field generated by a source falls beyond the standard analysis because the Dirichlet integral I in this problem is infinite. Some other problems in Sections 3 and 4 have similar specific features too.

An analogous situation arises if the velocity field on the boundary of Ω has discontinuities. The solvability of such a problem in a special function space was established by Solonnikov [2]. Another example is the problem of a moving line of a three-phase contact, where the Dirichlet integral divergence is caused by incompatibility of the no-slip conditions on the solid wall and the conditions on the free boundary [3].

The theory of solutions of elliptic equations with various singularities has been developed to a good level of completeness. Recent achievements in this theory were described in [4].

2. Self-similar solutions of (1)

System (1) is invariant under the similarity transformations $\mathbf{v}(x) \to \lambda \mathbf{v}(\lambda x)$ and $p(x) \to \lambda^2 p(\lambda x)$. This property allows us to construct self-similar solutions of (1).

In what follows, r, φ, and z are the cylindrical coordinates, while u, v, and w are the corresponding velocity components. The spherical coordinates are denoted by R, θ, and φ, while U, V, and W are the projections of the velocity vector to the axes R, θ, and φ. The solution of (1) is called rotationally symmetric if u, v, w, and p are independent of φ in some cylindrical coordinate system r, φ, z. If, in addition, $v = 0$, then the general form of solution is axisymmetric. Sometimes it is convenient to consider rotationally symmetric solutions in a spherical coordinate system. The self-similar steady-state rotationally symmetric solutions of (1) in spherical coordinates is given by the formulas

$$U = \frac{F'(s)}{R}, \quad V = \frac{F(s)}{R\sin\theta}, \quad W = \frac{\Omega(s)}{R\sin\theta},$$

where $s = \cos\theta$ (see Slezkin [5]). The functions F and Ω satisfy the system of ordinary differential equations

$$\nu(1-s^2)F'''' - 4\nu s F''' + FF''' + 3F'F'' = -2\Omega\Omega'/(1-s^2), \tag{4}$$

$$\nu(1-s^2)\Omega'' + F\Omega' = 0. \tag{5}$$

System (4), (5) has a family of solutions with $\Omega = 0$. They correspond to axisymmetric flows. Equation (4) with $\Omega = 0$ admits the triple integration

$$\nu(1-s^2)F' + 2\nu sF + F^2/2 = C_0 + C_1 s + C_2 s^2 \tag{6}$$

(C_0, C_1, and C_2 are constants) and subsequent linearization by the Riccati transformation, which allows one to express the general solution of equation (6) in terms of hypergeometric functions [6]. The simplest nontrivial solution of equation (6) was found by Landau [7]. It has the form

$$F = 2\nu\frac{1-s^2}{a-s}, \tag{7}$$

where $a > 1$ is a constant. Some detailed analysis of the Landau solution can be found in various textbooks [8, 9]. The velocity field in this solution is a regular in the entire space except the origin. The singularity at the origin is interpreted as an impulse point source; the flow rate of the fluid through each sphere with the center at the origin is zero. The extraordinary character of the Landau solution is confirmed by the fact that it is the only solution regular in $\mathbb{R}^3\backslash\{0\}$ among all self-similar (not necessarily axisymmetric) solutions of (1) (Šverák [10]). This fact was obtained by Šverák on the basis of the analogy he found between self-similar solutions of (1) and conformal geometry of a two-dimensional sphere.

If we consider solutions of equation (6) that are unbounded at one of the points $s = \pm 1$, we can match the self-similar solutions with the no-slip conditions on a plane or conical solid surface. In this connection some unexpected results for the interaction problem of a distributed source with a solid surface were obtained [11, 12].

Thus, let the fluid fill a half-space $z > 0$ whose boundary is a solid surface. Some sources or sinks with a linear density $q = const$ initiating the flow are distributed over the positive half-axis z. To describe such a flow, we have to solve equation (6) under the conditions

$$F(0) = 0, \quad F'(0) = 0, \tag{8}$$

$$\nu F \to q/2\pi, \quad F' \text{ is bounded as } s \to 1. \tag{9}$$

Equalities (8) guarantee that no-slip conditions are satisfied on the plane $z = 0$. The first condition in (9) defines the source (sink) intensity, and the second condition makes the axial component of velocity bounded as $r \to 0$ ($s \to 1$). It can be easily seen that the solution of (6), (8), (9) depends only on the one dimensionless parameter $Re = q/2\pi\nu$ (an analog of the Reynolds number). The cases of $Re > 0$ and $Re < 0$ correspond to a distributed source and a distributed sink, respectively.

Golubinskii and Sychev [11] constructed solutions of (6), (8), (9) that are analytic at the singular point $s = 1$. It turned out that these solutions exist only if the Reynolds number satisfies $-\infty < Re < Re_* \approx 0.847$. The analytic solution is unique if $Re \leq 0$. In the interval $0 < Re < Re_*$, there are two analytic solutions with essentially different qualitative properties. Goldshtik et al. [12] did not require the solution to be analytic, which allowed them to discover that the problem under discussion is solvable for all $Re > 0$. There are infinitely many nonanalytical solutions. To find the unique solution of this family at $Re > 0$, it is necessary to define the magnitude of the streamwise velocity on the symmetry axis. Thus, the self-similar solutions of the interaction problem of a distributed source or sink with a solid surface exist for all values of Re, both positive and negative. In the interval $(0, Re_*)$, there are three solutions to this problem.

Let us assume that the plane $z = 0$ is a free boundary [13]. In this case, instead of equation (8), we have

$$F(0) = 0, \quad F''(0) = 0, \tag{10}$$

and the second condition of (9) is not needed here: if $Re \geq 0$, it is satisfied automatically; if $Re < 0$, then the requirement of boundedness of F' as $s \to 1$ makes the problem unsolvable. It was proved in [13] that (6), (10), (9′) (i.e., the first condition of (9)) has a solution and this solution is unique if $Re \geq -2$ and has no solutions if $Re < -2$. In the latter case, each solution of equation (6) satisfying condition (9′) has a sequence of singularities converging to $s = 1$. The analytic solutions of the problem discussed here exist only for three values of the Reynolds number approximately equal to 1.07, 3.34, and 5.9.

Nonanalytic solutions have a strong singularity of the streamwise velocity on the symmetry axis if $Re \in [-2, 0)$ and a weak singularity if $Re > 0$. When Re is increasing, this singularity is shifted toward the higher-order derivatives of F. The smoothness properties of nonanalytic solutions of (6), (8), (9) and (6), (9′), (10) depending on the parameter are fairly similar, but there is a significant difference in the behavior of solutions of these two problems as $Re \to \infty$. The boundary layer in the first problem has an angular thickness of order $Re^{-2/3}$, whereas in the second problem, the boundary layer thickness is proportional to $Re^{-1}\ln Re$. It is typical for both solutions that both components of dimensional velocity in the flow core have the order Re as $Re \to \infty$. But the radial velocity vanishes when the solid surface is approached, owing to the no-slip condition. Conversely, in the case of the free boundary, the radial velocity drastically increases in the boundary layer. On the surface, the radial velocity has the form $u = q(q + 8\pi\nu)/16\pi\nu r$, so that $u \to \infty$ as $\nu \to 0$. It is of interest that the vorticity ω in each cone with the axis $z > 0$ and the aperture smaller than $\pi/2$ tends to the nonzero limit

$$\lim \omega = \frac{q\sqrt{2}}{4\pi R^2} \cdot \frac{\sin\theta}{(1 + \cos^2\theta)^{3/2}} \quad \text{as} \quad \nu \to 0$$

uniformly with respect to θ, $0 \leq \theta \leq \delta < \pi/2$.

It is of interest to compare the asymptotic solutions as $Re \to -\infty$ and $Re \to \infty$ in the problems (6), (8), (9), where the solution exists for all Re. In the first case, the classical Prandtl boundary layer is formed near the wall; the angular size of this boundary layer has the order $|Re^{-1/2}|$ as $Re \to -\infty$ [11, 12]. In the second case, the flow structure is more complicated because of the nonuniform character of the pressure distribution. The pressure is expressed by the formula $p = p_\infty + q^2/(8\pi^2 r^2)$ on the wall, and the pressure far from the plane where the flow is close to potential is expressed as $p \to p_\infty - q^2/(8\pi^2 r^2)$. The favorable pressure gradient on the wall prevents the boundary layer from separation and leads to the formation of a near-wall jet whose structure was studied in [12].

We leave aside other examples of axisymmetric self-similar flows (see more information on them in [12, 14]) and consider the motion with rotational symmetry. In this case, we have a system of sixth-order quasilinear equations (4), (5) with singularities in the coefficients at $s = \pm 1$. Let us impose the following boundary conditions for the solution of the system (4), (5):

$$F(0) = 0, \quad F'(0) = 0, \quad \Omega(0) = 0, \tag{11}$$

$$\nu\Omega \to \Gamma/2\pi \quad \text{as} \quad s \to 1. \tag{12}$$

It can be interpreted as describing the interaction of a semi-infinite vortex thread $z > 0$ with the solid surface $z = 0$. The parameter Γ defines the vortex intensity, and $Re = \Gamma/2\pi\nu$ (the Reynolds number) is the only governing parameter of this problem. Without loss of generality, we can assume that $\Gamma > 0$. This problem was considered for the first time by Goldshtik [15]. The main result of that work is as follows: if the axial velocity w on the symmetry axis is bounded, then the solution of the problems (4), (5), (11), (12) exists only for $Re \in [0, Re^*]$, $Re^* \approx 5.53$.

Goldshtik's paradox stimulated further studies of the problem of vortex interaction with a plane [16, 17]. It was shown [16] that the problem becomes solvable for all $Re > 0$ if the class of solutions is extended by admitting a logarithmic singularity of the function w as $r \to 0$. However, the coefficient at this singularity remains undetermined.

Up to now, we were discussing the self-similar solutions of system (1) that can be reduced to a system of ordinary differential equations. If the assumption of rotational symmetry of the solution is not valid, then the problem becomes much more difficult. The first results for three-dimensional self-similar stationary solutions of system (1) were obtained by Aristov [18]. They were based the following representation formulas:

$$U = \frac{\nu}{R}\Delta_s\Phi, \quad V = \frac{\nu}{R\sin\theta}\frac{\partial\Psi}{\partial\varphi} - \frac{\nu}{R}\frac{\partial\Phi}{\partial\theta}, \quad W = -\frac{\nu}{R}\frac{\partial\Psi}{\partial\theta} - \frac{\nu}{R\sin\theta}\frac{\partial\Phi}{\partial\varphi}, \quad p = \frac{\rho\nu^2}{R^2}P,$$

where Δ_s is the Laplace–Beltrami operator on a unit sphere. In this case, the continuity equation is satisfied automatically. Aristov [18] found two families of exact solutions of the system of equations for the functions Φ, Ψ, and P. One of them is described by the formulas $\Psi = \pm\Phi$, $P = -\nabla_s\Phi \cdot \nabla_s\Phi$ (∇_s is the surface gradient) with the function Φ satisfying the equation $\Delta_s\Phi = A\exp(-\Phi)$, where A is an arbitrary constant. In the solutions of the second family, $\Psi = 0$, and Φ satisfies the equation

$$\nabla_s(\Delta_s\Phi + 2) \cdot \nabla_s[\ln(\Delta_s\Phi + 2) + \Phi] = Q, \tag{13}$$

where Q is an arbitrary constant. It was shown in [19] that the general solution of equation (13) with the condition of 2π-periodicity with respect to the variable φ admits in the case $Q = 0$ a representation in terms of harmonic functions on a plane. The set of solutions of equation (13) includes the Landau solution, the rotationally symmetric solutions studied in [20], and a number of new solutions. One of them describes the flow generated by the system of an arbitrary number of linear sources with a constant linear density. For nonzero values of the parameter Q, a family of solutions of the form $\Phi = m\log\sin\theta + \Lambda(\varphi)$ was found, where $m = const$, and Λ obeys a fourth-order equation that admits double integration. The analysis and physical interpretation of these solutions can be found in [19]. Chauviere [21] extended the family of Aristov's solutions. He also constructed an asymptotic solution of the interaction problem of a vortex thread with a plane slightly deviatory from the horizontal one for high Reynolds numbers.

Let us consider self-similar solutions of (1) in the two-dimensional case. Let r and φ be polar coordinates in $\mathbb{R}^2$, and let u and v be the corresponding velocity components. The class of such solutions was found by Jeffery [22] and Hamel [23]. The solutions are given by

$$u = \frac{\nu}{r} f(\varphi), \quad v = const, \quad p = \frac{\nu^2}{r^2}\left(2f - \frac{b+v^2}{2}\right),$$

where b is a constant and f is the solution of the equation

$$f'' - vf' + f^2 + 4f = b.$$

Of major interest are the solutions with $v = 0$. The Jeffery–Hamel problem reduces to finding the constant b and the solution of the equation

$$f'' + f^2 + 4f = b \tag{14}$$

in the interval $|\varphi| < \beta \leq \pi$ satisfying the conditions

$$f(\beta) = f(-\beta) = 0, \tag{15}$$

$$\int_{-\beta}^{\beta} f(\varphi) d\varphi = Re, \tag{16}$$

where $Re = q/\nu$ is the Reynolds number that characterizes the ratio of the fluid flow rate q through the cross-section of the flow domain $\Omega = \{r, \varphi : r > 0, |\varphi| < \beta\}$ to the viscosity coefficient ν. Equality (15) means that the no-slip conditions are satisfied on the boundaries of the sector Ω, and the latter may be identified with solid walls. The parameter Re can take both positive and negative values. In the first case, the solution of (14)–(16) describes the flow in a diffuser; in the second case, this solution describes the flow in a confuser. The qualitative behavior of the solutions is different for flows of the diffuser and confuser types; it also depends to a large extent on the angle β.

The Jeffery–Hamel problem has a long story (see the survey [24] and the references therein). The theorem of existence and uniqueness of the solution of (14)–(16) for small values of $|Re|$ and $\beta \neq \beta^*$, where $\beta^* \approx 2.25$ is the root of the equation $\tan\beta = \beta$, was proved in [25]. The case of $\beta = \beta^*$ corresponds to the existence of nontrivial solutions of the linearized homogeneous problem ($b = 0$, $Re = 0$). For large values of $|Re|$, the uniqueness of the problem can be violated.

Let us first consider the confuser flow, and let $\beta < \pi/2$. If $Re < 0$ and $|Re|$ is sufficiently large, the Jeffery–Hamel problem has a solution with the flow close to a potential flow with $f = const$ outside the boundary layers adjacent to the confuser walls $\varphi = \pm\beta$. The solution inside these domains asymptotically approaches the self-similar solution of the Prandtl boundary layer equations [26]. The problem is a rare (and, possibly, the first) example of a situation where not only the asymptotic closeness of solutions of boundary-value problems for Navier–Stokes and Prandtl equations is proved, but also effective estimates of the difference between two solutions as $|Re| \to \infty$ are given [27]. It should be noted that the function f here is even and takes only negative values in the interval $(-\beta, \beta)$.

The situation becomes qualitatively different if $\beta > \pi/2$. By means of careful calculations, Akulenko and Nesterov [28] found that there are sign-variable solutions of (14)–(16) for high values of $|Re|$. An even more unexpected property of the solutions of the Jeffery–Hamel problem is the presence of asymmetric confuser-type solutions [29]. It was earlier believed that the existence of such solutions is the privilege of diffuser-type flows with high values of βRe.

Let us consider the diffuser flow ($Re > 0$). In this case, the nonuniqueness of the solution of (14)–(16) was found long ago [23, 30], but a detailed analysis of the situation was performed only recently [31, 32]. The existence of a countable set of adjacent intervals on the semi-axis ($Re > 0$) was established, with a bifurcation of solutions in each interval. Allowing $Re \to \infty$ and the corresponding increase in the interval number results in an unlimited increase of the interval length and of the number of modes of admissible solutions. A purely diverging solution is possible only if $Re < Re_0(\beta)$. It turned out that $\beta Re_0 \to 3\pi$ as $\beta \to 0$. The function $\beta Re_0(\beta)$ decreases monotonically on the interval $(0, \pi/2)$ from 3π to zero when the flow domain turns into a half-plane.

Although the solution of the Jeffery–Hamel problem is of interest by itself, it can also be used to describe the asymptotic behavior of the fluid flow in domains with sectoral outputs to infinity [25, 33]. Rivkind and Solonnikov [25] considered the case in which there are several outputs of the kind, and the total flow rate of the fluid is small as compared to viscosity. The question is still open whether it is possible to eliminate the condition of small flow rates in formulating Theorem 2 in this paper. It is difficult to answer this question for the more reasons that the solution of the Jeffery–Hamel problem is not unique at the high values of the flow rate.

For equation (14), it is also of interest to consider another problem where the condition of solution periodicity is imposed instead of (15) [12, 23]. The solution of such a problem yields an example of solutions to the stationary Navier–Stokes equations that are defined and regular in $\mathbb{R}^2\backslash\{0\}$. Šverák [10] performed a detailed study of a periodic problem for equation (14) with an additional condition of zero flow rate $Re = 0$. At least for small forces, we may hope to describe the principal term of the asymptotic of the solution to the inhomogeneous system of Navier–Stokes equations with a solenoidal field of external forces as $x \to \infty$ in terms of the solution of this problem. (This possibility comes true in a more favorable case of $n = 3$, where the Landau solution (7) with a given value of the parameter $a > 1$ is unique [10].)

Šverák proved in [10] that there exists a countable set of nontrivial solutions of the periodic problem for equation (14) with the condition of the zero flow rate $f(\varphi) = \tilde{f}_k(\varphi)$, where the function $\tilde{f}_k$ for each $k = 3, 4, \dots$ is a periodic function of φ with the minimal period $2\pi/k$. The corresponding values of $b = b_k$ are expressed by $b_k = \frac{1}{2\pi}\int_0^{2\pi} \tilde{f}_k^2 d\varphi$. The functions $\tilde{f}_k$ can be expressed in terms of the classical elliptic functions. The amplitude of oscillations of $\tilde{f}_k$ has the order k^2.

The solutions of (14)–(16) and the periodic problem are examples of self-similar solutions of system (1), which describe flows with straight-line trajectories. It is possible to propose another physical interpretation of such solutions if we consider the motion in the sector $0 < \varphi < \gamma$, with one or both straight half-lines bounding this sector being free [34, 35]. In contrast to the Jeffery–Hamel problem, the constant b in this case should be equal to zero. (Without writing the statement of problems with a free boundary for system (1) in the general case, we propose that the reader to read the monographs [35, 36].)

The first of our problems of interest is stated as follows: find $\gamma \in (0, 2\pi)$ and the solution $f(\varphi)$ of the equation

$$f'' + f^2 + 4f = 0 \tag{17}$$

in the interval $0 < \varphi < \gamma$ which satisfies (16) and

(18) $$f(0) = f'(\gamma) = 0.$$

In this problem, the line $\varphi = 0$ is the solid wall, and the line $\varphi = \gamma$ is the free boundary. The "free" character of the boundary means that the sector angle γ is not given, but is determined as a function of the Reynolds number Re by solving (16)–(18). Depending on Re, this problem has from one to four solutions if $Re > Re_1 \approx -22.65$. Among these solutions, only one does not change its sign. If $Re > Re_2 \approx 1.66$, the problem solution is unique, and $f > 0$ for $0 < \varphi \leq \gamma$ [34, 35].

The behavior of the solution of (16)–(18) as $Re \to \infty$ is surprising:

$$\gamma = \frac{1}{2Re}\left(1 + \sum_{k=1}^{\infty} c_k Re^{-2k}\right), \quad f(\varphi) = \frac{1}{\gamma^2}\sum_{k=0}^{\infty} \gamma^{2k} y_k\left(\frac{\varphi}{\gamma}\right).$$

The functions y_k and the constants c_k are determined from recurrent relations. It should be noted that the angular size Re^{-1} of the flow domain at the high values of Re is much smaller than the thickness of the near-wall boundary layers in the Jeffery–Hamel confuser flow [26, 27], which have the order $|Re|^{-1/2}$. The structure of the resultant asymptotics shows that there are no narrow zones adjacent to the solid wall and the free boundary in which the solution is described by boundary-layer-type functions and a "core" between them where the effect of viscosity is negligibly small as $Re \to \infty$.

Let us impose the following boundary conditions for the solution of equation (17):

(19) $$f'(0) = f'(\gamma) = 0.$$

For a given value of Re, we consider the following problem: find the function $f(\varphi)$ and the constant $\gamma \in (0, 2\pi)$ such that relations (16), (17), and (19) are satisfied. It was found in [34, 35] that this problem has at least one solution if $-8\pi < Re \leq 0$ and has no solutions if $Re > 0$ or $Re \leq -8\pi$. For small values of $|Re|$, this problem has seven solutions disappearing by pairs as $|Re|$ increases. Finally, for Re close to -8π, there remains only one solution $f = -4$. It is the only solution in which $f(\varphi)$ does not change its sign in the interval $[0, \gamma]$. For low Reynolds numbers the nonuniqueness of solutions to steady problems with a free boundary for Navier–Stokes equations is also observed in other problems, including multidimensional problems [37]. In this aspect, they differ from the first boundary-value problem for system (1) in a fixed domain where the smallness of the Reynolds number guarantees unique solvability [1].

To conclude this section, we should mention Solonnikov's paper [38], where he clearly demonstrated the role of the exact solutions of Navier–Stokes equations as the limiting regimes of a viscous fluid flow in extended domains. He considered a planar steady problem of a heavy fluid flow down an infinite inclined wall into a basin of infinite depth. Let x_1 and x_2 be the horizontal and vertical coordinates, the gravity force is directed vertically downward, and the free boundary in the basin approaches the zero level as $x_1 \to \infty$. Solonnikov [38] proved the existence of the problem solutions at sufficiently low Reynolds numbers $Re = q/\nu$, where $q > 0$ is the flow rate of the fluid through the cross-section of the layer adjacent to the wall. He also found that the problem solution tends to the Nusselt solution with a constant layer thickness h and straight trajectories parallel to the wall as the distance along the wall increases [8]; at $x_2 < 0$ and $(x_1^2 + x_2^2)^{1/2} = r \to \infty$, the solution is close to the positive solution of problem (16)–(18). This solution is unique for low Reynolds numbers $Re > 0$. The asymptotic thickness of the layer h is uniquely determined by the slope of the wall, gravity acceleration, and the values of q and ν.

3. A planar problem with a singularity in the flow domain

Let Σ be a smooth Jordan curve on a plane which bounds the domain Ω containing the origin O. A source or sink of intensity q is located at O. We have to find a solution of (1) in $\Omega\backslash O$ which satisfies the conditions

$$\mathbf{v} = \mathbf{a}(x), \quad x \in \Sigma, \tag{20}$$

$$v_r = \frac{q}{2\pi r}[1 + o(1)], \quad v_\varphi = o(1) \quad \text{as } r \to 0, \tag{21}$$

where $\mathbf{a}(x)$ is a given function and v_r and v_φ are the radial and circular velocity components. Let us assume that the origin is the only singular point of the velocity field. Then the continuity equation $div\,\mathbf{v} = 0$ and conditions (20), (21) yield the equality

$$\int_\Sigma \mathbf{n} \cdot \mathbf{a}\, d\Sigma = q \tag{22}$$

($\mathbf{n}$ is the unit vector of the outer normal to the curve Σ).

Problems (1), (20), (21) are the limiting cases of the problem of a stationary flow of a viscous incompressible fluid in a curvilinear ring with permeability conditions on its boundaries. In the general form, the percolation problem means finding a solution of (1) in $\Omega \in \mathbb{R}^n$ $(n = 2, 3)$ with the boundary $\partial\Omega = \Sigma_1 \cup \Sigma_2 \cup \ldots \cup \Sigma_m$, which satisfies the boundary conditions

$$\mathbf{v} = \mathbf{a}_i(x), \quad x \in \Sigma_i, \quad i = 1, 2, \ldots, m. \tag{23}$$

By the continuity equation, the vector-functions $\mathbf{a}_i$ should satisfy the condition

$$\sum_{i=1}^{m} \int_{\Sigma_i} \mathbf{a}_i \cdot \mathbf{n}_i\, d\Sigma_i = 0, \tag{24}$$

which expresses the equality to zero of the total flow rate of the incompressible fluid through the boundary of the flow domain (here $\mathbf{n}_i$ is the unit outer normal vector to the surface or to the curve Σ_i). If the boundary of the domain Ω is connected (i.e., $m = 1$) and certain smoothness conditions are satisfied, then the solvability of (1), (23) follows from the classical results by Leray [39] and Ladyzhenskaya [1]. This result is also valid if the partial flow rates q_i have zero values:

$$\int_{\Sigma_i} \mathbf{a}_i \cdot \mathbf{n}_i\, d\Sigma_i \equiv q_i = 0, \quad i = 1, 2, \ldots, m. \tag{25}$$

The question whether the necessary condition (3) is sufficient for solvability of (1), (2) remained open until recently.

Note that Leray's proof proceeds by contradiction; it does not contain an a priori estimate of the Dirichlet integral (3). Such an estimate for solutions of (1), (23) with the additional condition (25) was obtained for the first time by Hopf [40]. His considerations were based on the following lemma proved by himself.

LEMMA 1. *Assume that $\Sigma_i \in C^\infty$; $(\mathbf{a}_i)_k \in H^{1/2}(\Sigma_i)$; $k = 1, 2, 3$ $i = 1, \ldots, m$. If condition* (25) *is satisfied, then, for any $\varepsilon > 0$, there exists a solenoidal continuation $\mathbf{b}_i(x) \in \mathbf{H}(\Omega)$ of $\mathbf{a}_i$ to Ω, such that for all $\mathbf{u} \in \mathbf{H}(\Omega)$, we have*

$$\left| \int_\Omega \mathbf{b}_i \cdot \mathbf{u} \cdot \nabla \mathbf{u} \, dx \right| \leq \varepsilon ||\nabla \mathbf{u}||^2_{L^2(\Omega)}, \quad i = 1, 2, \ldots, m. \tag{26}$$

Here $\mathbf{H}(\Omega)$ means the closure of the set of solenoidal infinitely differentiable vector-functions that are compactly supported in the domain Ω in the norm of the Dirichlet integral. The requirement $\Sigma_i \in C^\infty$ can be substantially weakened up to the condition of piecewise smoothness of the surfaces Σ_i (see more details in [1]).

Takeshita [41] proved that (25) is not only sufficient but also necessary for the possibility of the vector field $\mathbf{a}_i$ to be continued to Ω so that (26) be satisfied for arbitrary value $\varepsilon > 0$. Note that violation of (25) does not lead to principal difficulties in the nonstationary flow problem for the Navier–Stokes equations [1]. Concerning the stationary problem, the issue of its global solvability in the three-dimensional case is far from being solved.

Finn [42] was apparently the first researcher who noticed that (25) can be replaced by the condition of the smallness of partial flow rates. Later, numerous results on solvability of steady flow problems of local character were obtained (see, e.g., [43] and the references therein). We should also note the paper by Borchers and Pileckas [44] who studied (1), (23) in the case where the boundary of the domain Ω is diffeomorphic to a spherical layer (or is a curvilinear ring in a planar case). Borchers and Pileckas [44] obtained estimates from above for the ratio $|q_i|/\nu$ ($i = 1, 2$) in terms of the radii $R_1 < R_2$ of spheres (or circumferences) including the domain Ω, which guarantees the solvability of this problem.

The first nonlocal result belongs to Amick [45] who proved the solvability of the problem with no constraints on the magnitude of the values of $|q_i|$ for a flow in a curvilinear ring with an additional condition on the symmetry of the domain and with the velocity vector field given on the boundary of this domain with respect to a certain straight line. It was also proved by contradiction. Sazonov [46], Fujita [47], and Morimoto [48] proposed other proofs, ensuring the effective estimates of the Dirichlet integral of the velocity field in Amick's problem. The solvability of a number of axisymmetric flow problems without conditions on the smallness of the partial flow rates q_i was established in [49]–[51]. A survey of publications on this topic can be found in [49, 52].

A specific feature of (1), (20), (21) is the fact that the Dirichlet integral of the vector function $\mathbf{v}(x)$ is infinite by (21); moreover, its solution possesses an infinite energy. We can hope however that the regularized problem after identification of the singular components of $\mathbf{v}$ and p allows us to obtain an a priori estimate which guarantees the solvability of the problem. (Problems (1), (20), (21) were considered earlier in [53].) Let us introduce the new functions

$$w_r = v_r - \frac{q}{2\pi r}, \quad w_\varphi = v_\varphi, \quad \bar{p} = p + \frac{q^2}{8\pi^2 r^2}. \tag{27}$$

By equations (1) and (27), the vector $\mathbf{w}$ and the function $\bar{p}$ satisfy the system of equations

$$w_r\frac{\partial w_r}{\partial r}+\frac{1}{r}w_\varphi\frac{\partial w_r}{\partial \varphi}-\frac{1}{r}w_\varphi^2+\frac{q}{2\pi r}\frac{\partial w_r}{\partial r}-\frac{q}{2\pi r^2}w_r$$
$$=-\frac{\partial \bar{p}}{\partial r}+\nu\left(\frac{\partial^2 w_r}{\partial r^2}+\frac{1}{r}\frac{\partial w_r}{\partial r}+\frac{1}{r^2}\frac{\partial^2 w_r}{\partial \varphi^2}-\frac{1}{r^2}w_r-\frac{2}{r^2}\frac{\partial w_\varphi}{\partial \varphi}\right)$$
$$w_r\frac{\partial w_\varphi}{\partial r}+\frac{1}{r}w_\varphi\frac{\partial w_\varphi}{\partial \varphi}+\frac{1}{r}w_r w_\varphi+\frac{q}{2\pi r}\frac{\partial w_\varphi}{\partial r}+\frac{q}{2\pi r^2}w_\varphi$$
$$=-\frac{1}{r}\frac{\partial \bar{p}}{\partial \varphi}+\nu\left(\frac{\partial^2 w_\varphi}{\partial r^2}+\frac{1}{r}\frac{\partial w_\varphi}{\partial r}+\frac{1}{r^2}\frac{\partial^2 w_\varphi}{\partial \varphi^2}-\frac{1}{r^2}w_\varphi+\frac{2}{r^2}\frac{\partial w_r}{\partial \varphi}\right),$$

(28) $$\frac{\partial w_r}{\partial r}+\frac{1}{r}w_r+\frac{1}{r}\frac{\partial w_\varphi}{\partial \varphi}=0.$$

Condition (20) may be rewritten as

(29) $$\mathbf{w}=\tilde{\mathbf{a}}(x),\quad x\in\Sigma,$$

where the components of $\tilde{\mathbf{a}}$ have the form $\tilde{a}_r=a_r-q/2\pi r$, $\tilde{a}_\varphi=a_\varphi$. By equations (22) and (27), this vector satisfies the condition

(30) $$\int_\Sigma \mathbf{n}\cdot\tilde{\mathbf{a}}\,d\Sigma=0.$$

Condition (30) allows one to reformulate (28)–(29) in terms of the stream function $\psi(r,\varphi)$ related to the components of the vector $\mathbf{w}$ by the expressions

(31) $$w_r=\frac{1}{r}\frac{\partial\psi}{\partial\varphi},\qquad w_\varphi=-\frac{\partial\psi}{\partial r}.$$

(Previously, before identifying the singular component of the velocity field, the transition to the stream function was not productive because the stream function of a point source in a planar problem is a multivalued function.) The stream function satisfies the equation

(32) $$\nu\Delta^2\psi-\frac{1}{r}\frac{\partial(\Delta\psi,\psi)}{\partial(r,\varphi)}-\frac{q}{2\pi r}\frac{\partial\Delta\psi}{\partial r}=0.$$

The boundary conditions in terms of the stream function have the form

(33) $$\psi=a(x),\quad \frac{\partial\psi}{\partial n}=b(x),\quad x\in\Sigma.$$

The functions a and b are expressed in terms of the components of the vector $\tilde{\mathbf{a}}$. If they belong to $H^{1/2}(\Sigma)$, then $a\in H^{3/2}(\Sigma)$ and $b\in H^{1/2}(\Sigma)$. Below we will assume that these conditions are satisfied:

(34) $$a\in H^{3/2}(\Sigma),\ b\in H^{1/2}(\Sigma).$$

Let us pass in (32), (33) to the new function $\chi=\psi-f$ such that

(35) $$f=a(x),\quad \frac{\partial f}{\partial n}=b(x),\quad x\in\Sigma.$$

We will choose f later. The function χ is the solution of the boundary-value problem

(36) $$\nu\Delta^2\chi-\frac{1}{r}\frac{\partial(\Delta\chi,\chi)}{\partial(r,\varphi)}-\frac{1}{r}\frac{\partial(\Delta\chi,f)}{\partial(r,\varphi)}-\frac{1}{r}\frac{\partial(\Delta f,\chi)}{\partial(r,\varphi)}-\frac{q}{2\pi r}\frac{\partial\Delta\chi}{\partial r}=g,\quad x\in\Omega\backslash\{0\};$$

(37) $$\chi = 0, \quad \frac{\partial \chi}{\partial n} = 0, \quad x \in \Sigma.$$

Here we denote

(38) $$g = -\nu \Delta^2 f + \frac{1}{r}\frac{\partial(\Delta f, f)}{\partial(r,\varphi)} + \frac{q}{2\pi r}\frac{\partial \Delta f}{\partial r}.$$

Let us choose the number $\gamma, 0 < \gamma < dist(\Sigma, \{0\})$ and denote by Ω_γ a doubly connected domain bounded by the curve Σ and the circumference $r = \gamma$. Let us use $\overset{\circ}{H}{}^2(\Omega_\gamma)$ to denote the Hilbert space obtained by closure of the set of functions from $C_0^\infty(\Omega_\gamma)$ in the norm

$$||\eta||^2_{H^2} = \int\limits_{\Omega_\gamma} \left(\eta^2_{rr} + \frac{2}{r^2}\eta^2_{r\varphi} + \frac{1}{r^4}\eta^2_{\varphi\varphi} + \frac{1}{r^2}\eta^2_r\right) r dr d\varphi.$$

For $\gamma \to 0$, we obtain the Hilbert space which is denoted here by $\overset{\circ}{H}{}^2_0(\Omega)$. The norm in this space is determined by the previous equality in which the domain of integration is replaced by Ω. The functions $\eta \in \overset{\circ}{H}{}^2_0(\Omega)$ are continuous in the domain $\bar{\Omega}$ and vanish at $r = 0$. Thus, we obtain the estimate $|\eta| \leq C_0 r^\alpha ||\eta||_{H^2}$, $(r,\varphi) \in \bar{\Omega}$, $0 < \alpha < 1$ with a constant $C_0 > 0$ depending on Ω and α. The subscript in the notation of $\overset{\circ}{H}{}^2_0(\Omega)$ reminds us that the definition of this space is a proper subspace of $\overset{\circ}{H}{}^2(\Omega)$. The issue of the structure of the orthogonal complement of $\overset{\circ}{H}{}^2_0(\Omega)$ to the space $\overset{\circ}{H}{}^2(\Omega)$ is not considered here. A function $\chi \in \overset{\circ}{H}{}^2_0(\Omega)$ is called a generalized solution of (36), (37) if the following integral identity is satisfied for any $\eta \in \overset{\circ}{H}{}^2_0(\Omega)$:

(39) $$\nu \int\limits_\Omega \left(\chi_{rr}\eta_{rr} + \frac{2}{r^2}\chi_{r\varphi}\eta_{r\varphi} + \frac{1}{r^4}\chi_{\varphi\varphi}\eta_{\varphi\varphi} + \frac{1}{r^2}\chi_r\eta_r\right) r dr d\varphi$$
$$+ \int\limits_\Omega \left[\frac{\Delta\chi}{r}\frac{\partial(\chi,\eta)}{\partial(r,\varphi)} + \frac{\Delta\chi}{r}\frac{\partial(f,\eta)}{\partial(r,\varphi)} + \frac{\Delta f}{r}\frac{\partial(\chi,\eta)}{\partial(r,\varphi)} + \frac{q\Delta\chi}{2\pi r}\eta_r\right] r dr d\varphi = \int\limits_\Omega g\eta r dr d\varphi.$$

Our next goal is to obtain an a priori estimate of a generalized solution of problem (36), (37). It is obtained by using a special choice of f.

LEMMA 2. *Let $\Sigma \in C^\infty$ be the Jordan curve including the origin inside. Assume that the functions a and b specified on this curve satisfy* (34). *Then there exists $f \in H^2(\Omega)$ such that* (35) *are satisfied and*

(40) $$\left|\int\limits_\Omega \Delta\chi \frac{\partial(f,\chi)}{\partial(r,\varphi)} dr\, d\varphi\right| \leq \varepsilon ||\chi||^2_{H^2}$$

is valid for all $\varepsilon > 0$, and an arbitrary $\chi \in \overset{\circ}{H}{}^2_0(\Omega)$.

PROOF. Let us choose a point A on the curve Σ and parametrize the functions a and b by the values of the arc length s counted from this point. The functions a and b are periodic functions of s with the period l equal to the length of the curve Σ. By the conditions of Lemma 2, we have $a \in H^{3/2}[0,l]$ and $b \in H^{1/2}[0,l]$. Let us introduce the local curvilinear coordinates s and n fitted to the curve Σ (the

Mises variables [26]). The coordinate s is introduced above, and the coordinate n is chosen as the distance from $(r, \varphi) \subset \Delta_\delta$ to Σ. Here $\Delta_\delta \subset \Omega$ is a curvilinear strip of width δ adjacent to Σ. The value of δ is chosen to be so small that the point on the curve Σ nearest to (r, φ) is unique and the domain Δ_δ does not contain the origin of the coordinate system. Let us divide the interval $[0, l]$ into the N segments $\sigma_k = \{s : s_{k-1} \le s \le s_k;\ k = 1, \dots, N;\ s_0 = 0,\ s_N = l\}$ whose length does not exceed τ. As a result, the strip Δ_δ is divided into the N parts $\Delta_{\delta,k}$. Let us denote the rectangle $\Pi_k = \{s, n : s \in \sigma_k,\ 0 \le n \le \delta\}$ by Π_k. We choose $\tau > 0$ to be so small that the mapping $(s, n) \to (r, \varphi)$ is a diffeomorphism in $\bar{\Pi}_k$ $(k = 1, \dots, N)$.

Consider the cut-off function $\zeta_\kappa(t)$ depending on the parameter κ $(0 < \kappa \le 1/4)$ and possessing the following properties:

$$\zeta_\kappa \in C^\infty(\mathbb{R}_+);\ 0 \le \zeta_\kappa \le 1\ (t \ge 0);\ \zeta_\kappa(0) = 1\ (0 \le t \le \kappa);\ \zeta_\kappa(0) = 0\ (t \ge 1 - \kappa);$$

$$\sup_{t \in [0,1]} |t\zeta_\kappa'(t)| \to 0\ (\kappa \to 0) \tag{41}$$

(see more detail in [1, 47, 54]). Let us assume that

$$f = h(r, \varphi)\ \zeta_\kappa(\delta^{-1} n), \tag{42}$$

where h is the solution of the problem

$$\Delta^2 h = 0, \quad x \in \Omega; \quad h = a(x), \quad \frac{\partial h}{\partial n} = b(x), \quad x \in \Sigma. \tag{43}$$

Because of equation (34), the following relations are valid: $f \in H^2(\Omega)$ and *supp* $f \subset \bar{\Delta}_\delta$. Changing the variables on the right-hand side of (40), we get

$$\int_\Omega \Delta\chi \frac{\partial(f, \chi)}{\partial(r, \varphi)}\, dr\, d\varphi = \int_{\Delta_\delta} (f_s\chi_n - f_n\chi_s)\tilde{\Delta}\chi\, ds\, dn \equiv I_1 \tag{44}$$

(we retain the previous notation for f and χ written in the new variables). In equation (44) the expression

$$\tilde{\Delta}\chi = \frac{\partial}{\partial s}\left(\frac{H_n}{H_s}\chi_s\right) + \frac{\partial}{\partial n}\left(\frac{H_s}{H_n}\chi_n\right)$$

stands for the Laplacian of χ in the variables s and n, while

$$H_s = 1 + \frac{n}{\rho(s)}, \quad H_n = 1,$$

are the corresponding Lame coefficients, and $\rho(s)$ is the curvature radius of Σ.

If

$$I_1 = \sum_{k=1}^{n} \int_{\Delta_{\delta,k}} (f_s\chi_n - f_n\chi_s)\tilde{\Delta}\chi\, ds\, dn,$$

then it suffices to estimate each of the terms

$$\int_{\Delta_{\delta,k}} (f_s\chi_n - f_n\chi_s)\tilde{\Delta}\chi\, ds\, dn = I_{1,k}$$

in order to obtain (39).

By equation (41), the expression for $I_{1,k}$ takes the form

$$I_{1,k} = \int_{\Delta_{\delta,k}} [h_s\zeta_\kappa\chi_n - (h_n\zeta_\kappa + h\zeta_\kappa')\chi_s]\tilde{\Delta}\chi\, ds\, dn$$

(the prime means differentiation of the function $\zeta_\kappa(\delta^{-1}n)$ with respect to its argument).

The quantity $I_{1,k}$ is the sum of three integrals; the absolute value of each integral is estimated separately. Thus,

$$\left| \int\limits_{\Delta_{\delta,k}} h_s \zeta_\kappa \chi_n \tilde{\Delta}\chi \, ds\, dn \right| \le C_1 \delta ||h||_{H^2} ||\chi||^2_{H^2}, \tag{45}$$

where C_1 depends only on Ω. (Here and in what follows, C_k, $k = 1, 2, \dots$, are positive constants.) To obtain (45), we have used the inequality

$$||\chi_n||^4_{L^4} \le C_2 ||\chi_n||^2_{L^2} ||\chi_n||^2_{H^2}$$

that follows from the well-known inequality by Ladyzhenskaya [1], the Poincaré–Steklov inequality for the function χ_n vanishing on $n = 0$, the coercive estimates of the solution of (43), conditions (34), and the embedding theorem of $L^4(\Omega)$ in $H^1(\Omega)$. In a similar manner we obtain

$$\left| \int\limits_{\Delta_{\delta,k}} h_n \zeta_\kappa \chi_s \tilde{\Delta}\chi \, ds\, dn \right| \le C_3 \delta ||h||_{H^2} ||\chi||^2_{H^2}. \tag{46}$$

The estimate of the integral containing $\zeta'_\kappa(\delta^{-1}n)$ is based on an appropriate choice of the parameter κ and the Hardy-type inequality [1, 47]. Choose the value of κ to be so small that

$$\sup_{t\in[0,1]} |t\zeta'_\kappa(t)| = \delta$$

(in this case, κ should have the order $\exp(-1/\delta)$ [1]). Then

(47)

$$\begin{aligned}
\left| \int\limits_{\Delta_{\delta,k}} h\zeta'_\kappa \chi_s \tilde{\Delta}\chi \, ds\, dn \right| &\le \delta \int\limits_{\Delta_{\delta,k}} \left| h \frac{\chi_s}{n} \tilde{\Delta}\chi \right| ds\, dn \\
&\le C_4 \delta \max_{\bar{\Omega}} |h| \int\limits_{\Delta_{\delta,k}} \left| \frac{\chi_s}{n} \tilde{\Delta}\chi \right| ds\, dn \\
&\le C_4 \delta \max_{\bar{\Omega}} |h| \Bigg(\int\limits_{\Delta_{\delta,k}} \frac{\chi_s^2}{n^2} ds\, dn \Bigg)^{1/2} \Bigg(\int\limits_{\Delta_{\delta,k}} (\tilde{\Delta}\chi)^{1/2} \, ds\, dn \Bigg)^{1/2} \\
&\le 2C_4 \delta \max_{\bar{\Omega}} |h| \Bigg(\int\limits_{\Delta_{\delta,k}} \chi_{ss}^2 \, ds\, dn \Bigg)^{1/2} \Bigg(\int\limits_{\Delta_{\delta,k}} (\tilde{\Delta}\chi)^{1/2} \, ds\, dn \Bigg)^{1/2} \\
&\le C_5 \delta ||h||_{H^2} ||\chi||^2_{H^2}.
\end{aligned}$$

Application of the Hardy-type inequality to $\chi \in \overset{\circ}{H}{}^2_0(\Omega)$ is justified by the fact that its derivative χ_s can be approximated in the metric of $H^1(\Omega)$ by smooth functions vanishing near Σ. Summing up (45)–(47), we conclude that (39) is satisfied if δ is chosen in such a way that

$$N(C_1 + C_3 + C_5)\delta ||h||_{H^2} = \varepsilon.$$

The proof of Lemma 2 is complete. □

LEMMA 3. *Let the assumptions of Lemma 2 be satisfied. If, in addition, the value of q is such that*

$$|q| < 2\pi\nu, \tag{48}$$

then, for each generalized solution χ of (36), (37), *the estimate*

$$||\chi||_{H^2_r} \le C_6 \tag{49}$$

holds with the constant $C_6 = C_6(\nu, q, \Omega, a, b)$.

PROOF. Assume that $\eta = \chi$ in (39). Then it takes the form

$$\int\limits_\Omega \left[\nu\left(\chi_{rr}^2 + \frac{2}{r^2}\chi_{r\varphi}^2 + \frac{1}{r^4}\chi_{\varphi\varphi}^2 + \frac{1}{r^2}\chi_r^2\right) + \frac{q}{2\pi}\left(\frac{\chi_r^2}{r^2} - \frac{\chi_\varphi^2}{r^4}\right) + \frac{\Delta\chi}{r}\frac{\partial(f,\chi)}{\partial(r,\varphi)}\right] r\,dr\,d\varphi$$
$$= \int\limits_\Omega g\,\chi\; r\,dr\,d\varphi.$$

From here, by choosing f in accordance with the structure used in the proof of Lemma 2 and taking into account representation formula (38), we obtain the inequality

$$\int\limits_\Omega \left[\nu\left(\chi_{rr}^2 + \frac{2}{r^2}\chi_{r\varphi}^2 + \frac{1}{r^4}\chi_{\varphi\varphi}^2 + \frac{1}{r^2}\chi_r^2\right) - \frac{|q|}{2\pi}\left(\frac{\chi_r^2}{r^2} + \frac{\chi_\varphi^2}{r^4}\right)\right] r\,dr\,d\varphi$$
$$\le \varepsilon||\chi||^2_{H^2} + (\nu||f||_{H^2} + C_7||f||^2_{H^2})||\chi||_{H^2},$$

where C_7 depends only on Ω. To conclude the proof of Lemma 3, it is sufficient to assume that

$$\varepsilon = \frac{\nu}{2} - \frac{|q|}{4\pi},$$

which yields the required a priori estimate of the norm of χ,

$$||\chi||_{H^2} \le 4\pi(2\pi\nu - |q|)^{-1}(\nu||f||_{H^2} + C_7||f||^2_{H^2}) = C_6. \tag{50}$$ □

THEOREM 1. *Let the assumptions of Lemmas 2 and 3 be satisfied. Then problem* (36), (37) *have at least one generalized solution $\chi \in \overset{\circ}{H}{}^2_0(\Omega)$ satisfying estimate* (50).

PROOF. The proof of Theorem 1 is based on the natural regularization of (36), (37). Let us briefly describe it. We consider an auxiliary problem: find a solution of equation (32) in Ω_γ satisfying (33) and

$$\psi = 0, \quad \frac{\partial\psi}{\partial r} = 0, \quad x \in C_\gamma, \tag{33$'$}$$

where C_γ is the circumference of radius γ with the center at the origin of the coordinate system. Let us pass in (32), (33), (33$'$) to the new sought function $\chi^{(\gamma)} = \psi - f^{(\gamma)}$. The function $f^{(\gamma)}$ satisfies the conditions

$$f^{(\gamma)} = a(x), \ \frac{\partial f^{(\gamma)}}{\partial n} = b(x), \ x \in \Sigma; \ f^{(\gamma)} = 0, \ \frac{\partial f^{(\gamma)}}{\partial r} = 0, \ x \in C_\gamma. \tag{35$'$}$$

The function $\chi^{(\gamma)}$ is the solution of the equation

$$(36') \quad \begin{aligned} &\nu\Delta^2\chi^{(\gamma)} - \frac{1}{r}\frac{\partial(\Delta\chi^{(\gamma)},\chi^{(\gamma)})}{\partial(r,\varphi)} - \frac{1}{r}\frac{\partial(\Delta\chi^{(\gamma)},f^{(\gamma)})}{\partial(r,\varphi)} - \frac{1}{r}\frac{\partial(\Delta f^{(\gamma)},\chi^{(\gamma)})}{\partial(r,\varphi)} \\ &\quad - \frac{q}{2\pi r}\frac{\partial\Delta\chi^{(\gamma)}}{\partial r} = g^{(\gamma)}, \end{aligned}$$

satisfying the boundary conditions

$$(37') \quad \chi^{(\gamma)} = 0, \quad \frac{\partial\chi^{(\gamma)}}{\partial n} = 0, \quad x \in \Sigma; \quad \chi^{(\gamma)} = 0, \quad \frac{\partial\chi^{(\gamma)}}{\partial r} = 0, \quad x \in C_r,$$

where

$$(38') \quad g^{(\gamma)} = -\nu\Delta^2 f^{(\gamma)} + \frac{1}{r}\frac{\partial(\Delta f^{(\gamma)},f^{(\gamma)})}{\partial(r,\varphi)} + \frac{q}{2\pi r}\frac{\partial\Delta f^{(\gamma)}}{\partial r}.$$

The function $\chi^{(\gamma)} \in \overset{\circ}{H}{}^2(\Omega_\gamma)$ is called a generalized solution of (36′), (37′), if the following integral identity is satisfied for all $\eta^{(\gamma)} \in \overset{\circ}{H}{}^2(\Omega_\gamma)$:

$$(39') \quad \begin{aligned} &\nu\int\limits_{\Omega_\gamma}\left(\chi_{rr}^{(\gamma)}\eta_{rr}^{(\gamma)} + \frac{2}{r^2}\chi_{r\varphi}^{(\gamma)}\eta_{r\varphi}^{(\gamma)} + \frac{1}{r^4}\chi_{\varphi\varphi}^{(\gamma)}\eta_{\varphi\varphi}^{(\gamma)} + \frac{1}{r^2}\chi_r^{(\gamma)}\eta_r^{(\gamma)}\right) r\,dr\,d\varphi \\ &+ \int\limits_{\Omega_\gamma}\left[\frac{\Delta\chi^{(\gamma)}}{r}\frac{\partial(\chi^{(\gamma)},\eta^{(\gamma)})}{\partial(r,\varphi)} + \frac{\Delta\chi^{(\gamma)}}{r}\frac{\partial(f^{(\gamma)},\eta^{(\gamma)})}{\partial(r,\varphi)}\right. \\ &+ \left.\frac{\Delta f^{(\gamma)}}{r}\frac{\partial(\chi^{(\gamma)},\eta^{(\gamma)})}{\partial(r,\varphi)} + \frac{q\Delta\chi^{(\gamma)}}{2\pi r}\eta_r^{(\gamma)}\right] r\,dr\,d\varphi \\ &= \int\limits_{\Omega_\gamma} g^{(\gamma)}\eta^{(\gamma)} r\,dr\,d\varphi. \end{aligned}$$

Consider a family of (36′), (37′), where $\gamma \in (0,\gamma_0]$. Solvability of each problem is proved by the available methods (see, e.g., [1, 55]). The proof is based on analogs of Lemmas 2 and 3. For $\gamma \in (0,\gamma_0]$, there exists $f^{(\gamma)} \in H^2(\Omega_\gamma)$ satisfying (35′) such that, for all $\varepsilon > 0$, the inequality

$$(40') \quad \left|\int\limits_{\Omega_\gamma}\Delta\chi^{(\gamma)}\frac{\partial(f^{(\gamma)},\chi^{(\gamma)})}{\partial(r,\varphi)}dr\,d\varphi\right| \le \varepsilon\|\chi^{(\gamma)}\|_{H^2}^2$$

holds for arbitrary $\chi^{(\gamma)} \in H^2(\Omega_\gamma)$. The proof of (40′) is almost identical to the proof of Lemma 2. We only have to choose the value of γ_0 to be so small that the support $\bar{\Delta}_\delta$ of $f^{(\gamma)}$ does not intersect the circumference $C_{2\gamma_0}$. It is important to note that the value of ε in (40′) can be chosen identically for all $\gamma \in (0,\gamma_0]$.

Further, we prove a statement similar to Lemma 3. Let the assumptions of Lemma 2 be satisfied, and let q satisfy inequality (48). Then any generalized solution $\chi^{(\gamma)}$ of problem (36′), (37′) admits the estimate

$$(49') \quad \|\chi^{(\gamma)}\|_{H^2} \le C_6$$

with the constant C_6 independent of γ if $0 < \gamma < \gamma_0$. In (36′), (37′) we have to pass to a limit as $\gamma \to 0$. For this purpose, we extend $\chi^{(\gamma)}$ and $f^{(\gamma)}$ by the zero to the circle of radius γ with center at the origin, retaining the previous notation of these functions. The extended functions are defined in the entire domain Ω and have finite norms in $H^2(\Omega)$. By (49$'$), the norms of $\chi^{(\gamma)}$ $(0 < \gamma \leq \gamma_0)$ are bounded in this space. There exists a weak limit χ of the family of $\{\chi^{(\gamma)}\}$, $\gamma \in (0, \gamma_0]$, as $\gamma \to 0$. The limiting function belongs to $\overset{\circ}{H}{}_0^2(\Omega)$. Being weakly compact in $\overset{\circ}{H}{}^2(\Omega)$, the set of the functions $\{\chi^{(\gamma)}\}$, $\gamma \in [0, \gamma_0]$, is compact in $\overset{\circ}{W}{}^{1,4}(\Omega)$. This allows us to perform the passage to the limit in (39$'$) as $\gamma \to 0$. The limiting function χ satisfies (39); thus, the proof of Theorem 1 is complete. □

The issue of smoothness of a generalized solution near the boundary $\Sigma \in C^\infty$ of Ω is resolved by methods [1]. If the components of $\tilde{\mathbf{a}}$ in (29) belong to the Hölder class $C^{2+\alpha}(\Sigma)$, $0 < \alpha < 1$, then w_r, $w_\varphi \in C^{2+\alpha}(\bar{\Omega}')$, where Ω' is an arbitrary subdomain of Ω that does not contain the origin, and $\mathbf{w} = (w_r, w_\varphi)$ is the regular component of the solution of (1), (20), (21). The behavior of the vector $\mathbf{w}$ near the singular point $r = 0$ of (28) requires some special study, in particular, an adequate choice of the weighted class of functions. (Possibly, the classes of Kondratiev [56] will be suitable here.) The issue of uniqueness of the generalized solution of the steady problem for Navier–Stokes equations is extremely complicated, and the regularity of the solution does not guarantee its uniqueness if the Reynolds number is high. For this reason, we should note that even the uniqueness of the radial stationary flow in a circular ring has not been proved, although it is known that this solution is isolated [57].

4. An axisymmetric problem with sources on the symmetry axis

The velocity vector $\mathbf{v} = (v_r, v_z)$ and the pressure p of a steady flow of a viscous incompressible fluid form the solution of the following system of equations:

$$(51)\qquad v_r\frac{\partial v_r}{\partial r} + v_z\frac{\partial v_r}{\partial z} = -\frac{\partial p}{\partial r} + \nu\left(\frac{\partial^2 v_r}{\partial r^2} + \frac{1}{r}\frac{\partial v_r}{\partial r} + \frac{\partial^2 v_r}{\partial z^2} - \frac{v_r}{r^2}\right),$$

$$v_r\frac{\partial v_z}{\partial r} + v_z\frac{\partial v_z}{\partial z} = -\frac{\partial p}{\partial z} + \nu\left(\frac{\partial^2 v_z}{\partial r^2} + \frac{1}{r}\frac{\partial v_z}{\partial r} + \frac{\partial^2 v_z}{\partial z^2}\right), \quad \frac{1}{r}\frac{\partial (r v_r)}{\partial r} + \frac{\partial v_z}{\partial z} = 0.$$

This system is considered in a bounded domain $\Omega \in \mathbb{R}^3$ whose boundary is the connected surface of revolution $\Sigma \in C^\infty$. Let us denote the points of intersection of the surface Σ with the axis z by $A_1 = (0, z_1)$ and $A_2 = (0, z_2)$. We assume that sources or sinks with a constant linear density q are distributed on the segment $L = \{r,\ z :\ r = 0,\ z_1 < z < z_2\}$, and an axisymmetric distribution of velocity is prescribed on Σ:

$$(52)\qquad \mathbf{v} = \mathbf{a}(x), \quad x \in \Sigma.$$

Moreover, we have

$$(53)\qquad \int_\Sigma \mathbf{n}\cdot\mathbf{a}\, d\Sigma = q(z_2 - z_1),$$

where $\mathbf{n}$ is the unit vector of the outer normal to Σ. We have to find a solution of (51) in $\Omega\backslash L$ satisfying (52), (53) and

$$(54)\qquad v_r = \frac{q}{2\pi r}[1 + o(1)], \quad v_z = o(1) \quad \text{for } r \to 0, \quad z_1 < z < z_2.$$

Let us introduce the new functions

$$w_r = v_r - \frac{q}{2\pi r}, \quad w_z = v_z, \quad \bar{p} = p + \frac{q^2}{8\pi^2 r^2}. \tag{55}$$

(The symbols $\mathbf{v}, \mathbf{a}, \mathbf{w}, \tilde{\mathbf{a}}, \Omega$, and Σ denote different objects in Sections 3 and 4, but this should not lead to any misunderstanding.) By equations (51) and (55), the vector $\mathbf{w}$ and the function $\bar{p}$ satisfy the system of equations

$$w_r \frac{\partial w_r}{\partial r} + w_z \frac{\partial w_r}{\partial z} + \frac{q}{2\pi r}\frac{\partial w_r}{\partial r} - \frac{q}{2\pi r^2} w_r = -\frac{\partial \bar{p}}{\partial r} + \nu \left(\frac{\partial^2 w_r}{\partial r^2} + \frac{1}{r}\frac{\partial w_r}{\partial r} + \frac{\partial^2 w_r}{\partial z^2} - \frac{w_r}{r^2} \right),$$

$$w_r \frac{\partial w_z}{\partial r} + w_z \frac{\partial w_z}{\partial z} - \frac{q}{2\pi r^2} w_z = -\frac{\partial \bar{p}}{\partial r} + \nu \left(\frac{\partial^2 w_z}{\partial r^2} + \frac{1}{r}\frac{\partial w_z}{\partial r} + \frac{\partial^2 w_z}{\partial z^2} \right), \tag{56}$$

$$\frac{1}{r}\frac{\partial (r v_r)}{\partial r} + \frac{\partial v_z}{\partial z} = 0$$

in $\Omega \backslash L$ and the boundary condition

$$\mathbf{w} = \tilde{\mathbf{a}}(x), \quad x \in \Sigma, \tag{57}$$

where the components of $\tilde{\mathbf{a}}$ have the form $\tilde{a}_r = a_r - q/2\pi r$, $\tilde{a}_z = a_z$. By equations (53) and (55), this vector satisfies the condition of the zero flow rate

$$\int_\Sigma \mathbf{n} \cdot \tilde{\mathbf{a}} \, d\Sigma = 0. \tag{58}$$

Let us define the function space $\mathbf{H}_r(\Omega)$ as the closure of the set of solenoidal axisymmetric vector-functions $\mathbf{u} = (u_r, u_z)$ such that $u_r, u_z \in C_0^\infty$ in the norm of the Dirichlet integral

$$||\mathbf{u}||^2_{\mathbf{H}_r} = \int_\Omega \left[\left(\frac{\partial u_r}{\partial r}\right)^2 + \left(\frac{\partial u_r}{\partial z}\right)^2 + \left(\frac{\partial u_z}{\partial r}\right)^2 + \left(\frac{\partial u_z}{\partial z}\right)^2 + \frac{u_r^2}{r^2} \right] r \, dr \, dz.$$

Let us assume that there exists a solenoidal axisymmetric extension $\mathbf{b}$ of the vector field $\tilde{\mathbf{a}}$ to Ω such that the Dirichlet integral of the vector function $\mathbf{b}$ is finite. This assumption imposes some requirements of smoothness on the components of the vector $\tilde{\mathbf{a}}$ and their behavior near the endpoints of the segment L (see [51]). Such an extension is carried out by using the stream function $\Psi(r, z)$ of the axisymmetric flow related to the components c_r and c_z of the velocity vector $\mathbf{c}$ by the expressions

$$c_r = \frac{1}{r}\frac{\partial \Psi}{\partial z}, \quad c_z = -\frac{1}{r}\frac{\partial \Psi}{\partial r}. \tag{59}$$

The function Ψ satisfies the equation

$$\left(\frac{\partial^2}{\partial r^2} - \frac{1}{r}\frac{\partial}{\partial r} + \frac{\partial^2}{\partial z^2} \right)^2 \Psi = 0, \quad (r, z) \in \Omega \backslash L, \tag{60}$$

and the boundary conditions

$$\Psi = a(s), \quad \frac{\partial \Psi}{\partial n} = b(s), \quad (r, z) \in \gamma. \tag{61}$$

In conditions (61), γ means a curve formed by the meridional cross section of the surface Σ, and s is the arc length counted from $A_1 = (0, z_1)$. The functions a and b are expressed in terms of the components of $\tilde{\mathbf{a}}$. Let us require that the vector $\mathbf{c}$ with the components c_r and c_z calculated by equations (59) via solving (60), (61) should have a finite Dirichlet integral. Actually, this is the regularity condition, which is imposed onto the boundary values of the solution of problems (56), (57). Let us assume that this condition is satisfied. Then we can construct the required vector field $\mathbf{b}$ with a finite Dirichlet integral such that the following is satisfied for all $\varepsilon > 0$:

$$\left| \int_\Omega \mathbf{b} \cdot \mathbf{u} \cdot \nabla \mathbf{u}\, r\, dr\, dz \right| \le \varepsilon ||\mathbf{u}||^2_{\mathbf{H}_r} \tag{62}$$

and an arbitrary value of $\mathbf{u} \in \mathbf{H}_r(\Omega)$. For this purpose, it is sufficient to put

$$b_r = \frac{1}{r}\frac{\partial(\zeta_\kappa \Psi)}{\partial z}, \quad b_z = -\frac{1}{r}\frac{\partial(\zeta_\kappa \Psi)}{\partial r},$$

where $\zeta_\kappa(\delta^{-1}n)$ is Hopf's cut-off function of the previous section.

Let us pass in (56) to the new sought functions $u_r = w_r - b_r$ and $u_z = w_z - b_z$ and denote $\mathbf{u} = (u_r, u_z)$. The functions u_r, u_z, and $\bar{p}$ satisfy the equations

$$\begin{aligned} &u_r \frac{\partial u_r}{\partial r} + u_z \frac{\partial u_r}{\partial z} + \frac{q}{2\pi r}\frac{\partial u_r}{\partial r} - \frac{q}{2\pi r^2} u_r + b_r \frac{\partial u_r}{\partial r} + b_z \frac{\partial u_r}{\partial z} + u_r \frac{\partial b_r}{\partial r} + b_z \frac{\partial u_r}{\partial z} \\ &\quad = -\frac{\partial \bar{p}}{\partial r} + \nu\left(\frac{\partial^2 u_r}{\partial r^2} + \frac{1}{r}\frac{\partial u_r}{\partial r} + \frac{\partial^2 u_r}{\partial z^2} - \frac{u_r}{r^2}\right) + g_r, \\ &u_r \frac{\partial u_z}{\partial r} + u_z \frac{\partial u_z}{\partial z} + \frac{q}{2\pi r}\frac{\partial u_z}{\partial r} + b_r \frac{\partial u_z}{\partial r} + b_z \frac{\partial u_z}{\partial z} + u_r \frac{\partial b_z}{\partial r} + b_z \frac{\partial u_z}{\partial z} = \\ &\quad = -\frac{\partial \bar{p}}{\partial r} + \nu\left(\frac{\partial^2 w_r}{\partial r^2} + \frac{1}{r}\frac{\partial w_r}{\partial r} + \frac{\partial^2 w_r}{\partial z^2}\right) + g_z, \\ &\frac{1}{r}\frac{\partial (r u_r)}{\partial r} + \frac{\partial u_z}{\partial z} = 0 \end{aligned} \tag{63}$$

in $\Omega \backslash L$ and the boundary conditions

$$u_r = u_z = 0, \quad (r, z) \in \Sigma. \tag{64}$$

The vector $\mathbf{g} = (g_r, g_z)$ has the form

$$\mathbf{g} = \nu \Delta \mathbf{b} - \mathbf{b} \cdot \nabla \mathbf{b} - \frac{q}{2\pi} \nabla \left(\frac{\mathbf{b}}{r}\right).$$

The vector function $\mathbf{u} \in \mathbf{H}_r(\Omega)$ is referred to as a generalized solution of problem (63), (64) if the integral identity

$$\begin{aligned} &\int_\Omega \left[\nu\left(\nabla u_r \cdot \nabla h_r + \nabla u_z \cdot \nabla h_z + \frac{u_r h_r}{r^2}\right) - \mathbf{u} \cdot \mathbf{u} \cdot \nabla \mathbf{h} - \mathbf{u} \cdot \mathbf{b} \cdot \nabla \mathbf{h} - \mathbf{b} \cdot \mathbf{u} \cdot \nabla \mathbf{h} \right. \\ &\quad \left. + \frac{q}{2\pi r}\left(\frac{\partial \mathbf{b}}{\partial r} \cdot \mathbf{h} - \frac{u_r h_r}{r}\right)\right] r\, dr\, dz = \int_\Omega \mathbf{g} \cdot \mathbf{h}\, r\, dr\, dz \end{aligned} \tag{65}$$

is satisfied for any $\mathbf{h} \in \mathbf{H}_r(\Omega)$. Assuming that $\mathbf{h} = \mathbf{u}$ in (65) and taking into account (64) and the solenoidality of $\mathbf{u}$ and $\mathbf{b}$, we obtain

$$\begin{aligned} &\int_\Omega \left[\nu\left(|\nabla u_r|^2 + |\nabla u_z|^2 + \frac{u_r h_r}{r^2}\right) - \frac{q u_r^2}{2\pi r^2} - \mathbf{b} \cdot \mathbf{u} \cdot \nabla \mathbf{u}\right] r\, dr\, dz \\ &\quad = \int_\Omega \mathbf{g} \cdot \mathbf{h}\, r\, dr\, dz. \end{aligned} \tag{66}$$

We choose $\mathbf{b}$ in such a way that (62) is satisfied with $\varepsilon = (4\pi)^{-1}(2\pi\nu - q)$ and assume that

$$q < 2\pi\nu. \tag{67}$$

Then equation (66) yields the a priori estimate

$$||\mathbf{u}||_{\mathbf{H}_r} \leq C_8(\nu||\mathbf{b}||_{\mathbf{H}_r} + C_9||\mathbf{b}||^2_{\mathbf{H}_r}), \tag{68}$$

where $C_9 > 0$ depends only on Ω; as for C_8, we can take $C_8 = 4\pi(2\pi\nu - q)^{-1}$ if $0 \leq q < 2\pi\nu$ and $2/\nu$ if $q \leq 0$. From this estimate, Theorem 2 follows.

THEOREM 2. *Let the vector* $\tilde{\mathbf{a}}$ *in* (57) *be such that the solenoidal vector field* $\mathbf{c}$ *constructed for this vector has a finite Dirichlet integral. If the value of* q *satisfies* (67), *then there exists a generalized solution* $\mathbf{u} \in \mathbf{H}_r(\Omega)$ *of* (63), (64), *and its norm admits* (68).

The proof of Theorem 2 is similar to the proof of Theorem 1. Therefore, we do not describe it here. It is of interest that a one-sided constraint is imposed here on the parameter q: the intensity of distributed sinks can have an arbitrary value. At the same time, the solvability of a planar problem with a source or sink was established under constraint (48) on the absolute value of q. On the other hand, there are no constraints on the norms of the functions $\tilde{\mathbf{a}}$ in both problems.

5. Open problems

One of the yet unsolved problems is the problem of a point vortex located inside a bounded domain $\Omega \in \mathbb{R}^2$. In the simplest version, it consists of finding a solution of (1) in $\Omega \backslash O$, satisfying the conditions

$$\mathbf{v} = 0, \quad x \in \partial\Omega = \Sigma, \tag{69}$$

$$v_\varphi = \frac{\Gamma}{2\pi r}[1 + o(1)], \quad v_r = o(1) \quad \text{as } r \to 0, \tag{70}$$

where Γ is a given circulation of the vortex. We can expect that problems (1), (69), (70) have a solution if the ratio $|\Gamma|/\nu$ is sufficiently small, without making constrains on the properties of Ω, except on the smoothness of its boundary. It seems natural to apply the approach of Section 3 to this problem: the identification of the singular part of the solution and the estimation of the Dirichlet integral of the regular part. The spatial analog of this problem is more complicated. Here the flow domain $\Omega \in \mathbb{R}^3$ is bounded by a smooth surface of revolution Σ. The no-slip condition is satisfied on the boundary, and the flow is induced by a vortex thread with circulation Γ, which is located on the symmetry axis and crosses Σ at two points. The solution of system (1) is sought in the class of rotationally symmetric vector fields. Contrary to the previous problem, there are additional singularities of the velocity field at the points of intersection of the surface Σ with the symmetry axis. Can we expect that the behavior of velocity near these points is approximated by Goldshtik's solution [15], that is unique in a suitable class of self-similar solutions for small values of $|\Gamma|/\nu$?

Let us again consider planar flows with singularities, but now these singularities are located on the domain boundary Σ rather than inside the flow domain. More particularly, let the curve Σ have two points A^+ and A^- where a source and a sink with intensities q and $-q$ are located. The remaining part of the boundary is subject to the no-slip conditions. Does this problem for (1) have a solution if the value of $|q/\nu|$ is small? As first step, we can consider the case where the segments of the curve Σ near A^+ and A^- are segments of straight lines crossing each other at an angle β. Then the singular part of the solution is described by the solution of the Jeffery–Hamel problem localized near corner points. This solution is unique at the small values of $|q/\nu| = Re$ and $\beta \neq \beta^* \approx 2.25$ [25]. After separating out the singular part of the velocity field, we obtain a problem similar to that in Section 3, but this problem is more complicated because of the presence of corner points of the boundary. Basing on results of [58], we can hope that the Dirichlet integral of the velocity field in this problem is finite for sufficiently small angles β and Reynolds numbers Re.

The Landau solution was already mentioned in Section 2. It describes an idealized situation where the impulse point source acts in an unbounded space. Rumer [59] considered the problem of a submerged jet escaping from a thin jet into an ambient space filled with the fluid and found its asymptotic solution far from the orifice. It turned out that the Landau solution yields the principal term of the formal expansion of the solution of Navier–Stokes equations in terms of the small ratio d/s, where d is the orifice diameter and s is the distance from the place of jet exhaustion. The next term of the expansion is proportional to the flow rate through the tube cross-section. This result was obtained "at the physical level of rigorousness." It would be desirable to give a rigorous proof as well.

To conclude, let us consider the problem of perturbation of a point source or sink in a bounded domain of a three-dimensional space by the velocity field on the domain boundary. In the planar problem, the source intensity q and the kinematic viscosity coefficient ν have an identical dimension, and their ratio can be identified with the Reynolds number; in the three-dimensional problem, however, the source intensity Q has the dimension νl, where l is a certain linear size.

Let us assume that a source or sink of the intensity Q is located at the origin. The cases of $Q > 0$ and $Q < 0$ correspond to the source and sink, respectively. For simplicity, we assume that the flow domain is a sphere of radius l with the center at the origin. Moreover, we consider for simplicity only axisymmetric flows. It is convenient to analyze them in spherical coordinates R, θ. The velocity components of the axisymmetric flow are expressed via the stream function Ψ by the formulas (see [60])

$$v_R = -\frac{1}{R^2 \sin\theta}\frac{\partial\Psi}{\partial\theta}, \quad v_\theta = \frac{1}{R\sin\theta}\frac{\partial\Psi}{\partial R}.$$

The stream function Ψ_Q of the point source or sink has the form

$$\Psi_Q = -\frac{Q}{4\pi}(1 - \cos\theta).$$

Let us denote the stream function of the perturbed flow by $\Phi = \Psi - \Psi_Q$. The equation for the function Φ can be easily derived from the formulas given in the book [60]. This equation has the form

$$\text{(71)} \qquad \nu E^2\Phi - \sin\theta\frac{\partial[(R\sin\theta)^{-2}E\Phi, \Phi]}{\partial(R,\theta)} + \frac{Q\sin^2\theta}{4\pi}\frac{\partial[(R\sin\theta)^{-2}E\Phi]}{\partial R} = 0,$$

where

$$E = \frac{\partial^2}{\partial R^2} + \frac{\sin\theta}{R^2}\frac{\partial}{\partial\theta}\left(\frac{1}{\sin\theta}\frac{\partial}{\partial\theta}\right).$$

Equation (71) is completed by the boundary conditions

$$\Phi = Qa(\theta), \quad \frac{\partial \Phi}{\partial R} = \frac{Q}{l} b(\theta), \quad R = l, \tag{72}$$

where the functions a and b are independent of the dimensional parameters of the problem ν, Q, and l, satisfy the natural conditions of smoothness and decay as $\theta \to 0$ and $\theta \to \pi$. The last condition is necessary to ensure the regularity of the solution of problems (71), (72) near the symmetry axis. In addition, it is assumed that the function a has a zero mean value on the segment $[0, \pi]$. This assumption guarantees conservation of the flow rate in the problem of a point source.

Problems (71), (72) are significantly more complicated than the problems of sources considered in Sections 3 and 4. We hope that it will be possible to prove its solvability for small values of the dimensionless parameter $(\nu l)^{-1}|Q|$, which plays the role of the Reynolds number, even though the choice of a suitable function space is not obvious here. Another open question is the asymptotic behavior of the solution as $R \to 0$, that may be expected to be different in the cases of a source and a sink. Finally, it is of interest to consider the limiting case of problems (71), (72) as $l \to \infty$. In the limit, we have a problem of a point source (sink) in an unbounded fluid. By scaling, equation (71) can be reduced to a form that does not contain parameters. The following hypothesis seems to be valid: a radially symmetric solution of the problem of a point source in an unbounded fluid is unique in the class of stream functions Φ such that the velocity field corresponding to this stream function possesses a finite Dirichlet integral. The question of uniqueness of the radially symmetric solution of the problem of a point sink remains open. It is of interest to note that the plane analog of the problem of a sink in an unbounded fluid has solutions different from radial ones [12].

This work was performed within the framework of the Joint Integration Project of the Siberian Branch, Ural Branch, and Far-East Branch of the Russian Academy of Sciences (2012–2014). It was also supported by the Russian Foundation for Basic Research (Grant No. 13–01–00526). The author is grateful to the reviewer of the paper for valuable recommendations that improved the text and to Professor V.S. Belonosov for his attention to this work.

As far as the author knows, the result of Theorem 1 was independently obtained by A. Russo and A. Tartaglioni [**61**].

References

[1] O.A. Ladyzhenskaya, The Mathematical Theory of Viscous Incompressible Flow (Gordon and Breach, New York, 1969). MR0254401 (40:7610)

[2] V.A. Solonnikov, "On boundary value problems for the Navier–Stokes equations with discontinuous boundary data", in: Partial differential equations and related subjects. Pitman Research Notes, Math. Series, 1992, Vol. 269, pp. 227-237. MR1190943 (94b:35216)

[3] V.V. Pukhnachev and V.A. Solonnikov, "On the problem of dynamic contact angle", J. Appl. Math. Mech., 1982, Vol 46, No. 6, pp. 771–779. MR726121 (85h:76025)

[4] A.A. Kovalevsky, I.I. Skrypnik, and A.E. Shishkov, Singular Solutions of Nonlinear Elliptic and Parabolic Equations (Naukova Dumka, Kyiv, 2010). MR2858689 (2012i:35005)

[5] N.A. Slezkin, "On one case of integrability of full differential equations of motion of a viscous fluid", Uch. Zap. Mosk. Gos. Univ., 1934, No. 2, pp. 89-90 (in Russian).

[6] V.I. Yatseev, "On one class of exact solutions of equations of motion of a viscous fluid", Zh. Eksp. Teor. Fiz., 1950, Vol. 20, No. 11, pp. 1031-1034. MR0039449 (12:552a)

[7] L.D. Landau, "On one new exact solution of Navier–Stokes equations", Dokl. Akad. Nauk SSSR, 1944, Vol. 43, No. 7, pp. 299-301.

[8] G.K. Batchelor, An Introduction to Fluid Dynamics (Cambridge University Press, Cambridge, 2000). MR1744638 (2000j:76001)
[9] L.D. Landau and E.M. Lifshits, Fluid Mechanics (Pergamon Press, Oxford, 1987). MR961259 (89i:00006)
[10] V.O. Šhverák, "On Landau solutions of Navier–Stokes equations", Problems of Mathematical Analysis (Tamara Rozhkovskaya, Novosibirsk, 2011), No. 61, pp. 173-191 (n Russian).
[11] A.A. Golubinskii and V.V. Sychev, "On oneself similar solution of Navier–Stokes equations", Uch. Zap. TsAGI, 1976, Vol. 7, No. 6, pp. 11-17 (in Russian). MR0489293 (58:8734)
[12] M.A. Goldshtik, V.N. Shtern, and N.I. Yavorskii, Viscous Flows with Paradoxical Properties (Nauka, Novosibirsk, 1989) (in Russian). MR1046234 (92e:76009)
[13] V.V. Pukhnachev, "Interaction between a distributed source and the free surface of a viscous fluid", Fluid Dynamics, 1996, No. 2, pp. 53-65. MR1407617 (97g:76024)
[14] C.Y. Wang, "Exact solution of the steady Navier–Stokes equations", Annual Rev. Fluid Mech., 1991, Vol. 23, pp. 159-177. MR1090331 (92a:76030)
[15] M.A. Goldshtik, "On one paradoxical solution of Navier–Stokes equations", Prikl. Mat. Mekh., 1960, Vol. 24, No. 4, pp. 610-621.
[16] J. Serrin, "The swirling vortex", Phyl. Trans. Royal Soc., 1972, Vol. 271, No. 1214, pp. 325-360.
[17] R. Ferrnandez-Faria, J.F. De La Mora, M. Perez-Saborid, and A. Barrero, "Conically similar swirling flows at high Reynolds number", Quart. J. Mech. Appl. Math., 1999, Vol. 62, Part 1, pp. 1-53. MR1681456 (2000b:76038)
[18] S.N. Aristov, "An exact solution to the point-source problem", Doklady Physics, 1995, Vol. 40, No. 7, pp. 346–348. MR1353985 (96i:76031)
[19] S.N. Aristov, "Three-dimensional conical viscous incompressible fluid flows", Fluid Dynamics, 1998, Vol. 33, No. 6, pp. 929–932. MR1689324 (2000b:76025)
[20] R. Paull and A.F. Pillow, "Conically similar viscous flows. Part 2. One-parameter swirl-free flows", J. Fluid Mech., 1985, Vol. 155, pp. 343-358. MR801090 (87h:76044b)
[21] A. Chauviere, "Non-axisymmetric boundary layer induced by a swirling downdraft", C. R. Acad. Sci. Paris, 2000, Vol. 321, Ser. II b. pp. 679-684.
[22] G.B. Jeffery, "The two-dimensional steady motion of a viscous fluid", Phil. Mag., 1915, Ser. Vol. 29, No. 172, pp. 455-465.
[23] Hamel G., "Spiralformige Bewegung zahen Flussigkeiten", Jahresber. Deutsch. Mathem. Vereindung. 1917. Bd. 25. S. 34-60.
[24] V.V. Pukhnachev, "Symmetries in Navier–Stokes equations", Usp. Mekhaniki, 2006, Vol. 4, No. 1, pp. 6-76 (in Russian).
[25] L. Rivkind and V.A. Solonnikov, "Jeffery–Hamel steady-state Navier–Stokes flow with sector-like outlets to infinity", J. Math. Fluid Mech., 2000, Vol. 2, No. 4, pp. 324-352. MR1814221 (2002i:76033)
[26] N.E. Kochin, I.A. Kibel, and N.V. Rose, Theoretical Hydromechanics, Vol. 2 (Interscience Publisher, New York, 1964).
[27] J. Serrin, "On the mathematical basis for Prandtl boundary layer theory: an example", Arch. Rational Mech. Anal., 1968, Vol. 28, pp. 217-225. MR0227634 (37:3218)
[28] L.D. Akulenko and S.V. Nesterov, "An efficient numerical-analytical solution to variational problem in mechanics", Doklady Physics, 2000, Vol. 45, pp. 543–546. MR1798953 (2001h:49049)
[29] L.D. Akulenko, D.V. Georgievskii, and S.A. Kumakshev, "The flow of a viscous fluid inside a confuser with a large opening angle", Doklady Physics, 2002, Vol. 47, pp. 698–701. MR1931124
[30] L. Rosenhead, "The steady two-dimensional radial flow of viscous fluid between two inclined plane walls", Proc. Royal Soc. London, Ser. A, 1940, Vol. 175, No. 963, pp. 436-467.
[31] L.D. Akulenko, S.A. Kumakshev, and S.V. Nesterov, "Effective numerical-analytical solution of isoperimetric variational problem of mechanics by an accelerated convergence method", J. Appl. Math. Mech., 2002, Vol. 66, No. 5, pp. 693–708. MR1964609 (2003k:49068)
[32] L.D. Akulenko and S.A. Kumakshev, "Multimode bifurcation of the flow of a viscous fluid in a plane diffuser", Doklady Physics, 2004, Vol. 49, pp. 751–755. MR2115523
[33] G.P. Galdi, M. Padula, and V.A. Solonnikov, "Existence, uniqueness and asymptotic behaviour of solution of steady Navier–Stokes equations in a plane aperture domain", Indiana Univ. Math. J., 1996, Vol. 45, No. 4, pp. 961-995. MR1444475 (98a:35106)

[34] V.V. Pukhnachev, "Plane stationary flow of a viscous incompressible fluid with straightline free boundaries", in: Numerical Methods of Mechanics of Continuous Media, 1971, Vol. 2, No. 4, pp. 67-75 (in Russian). MR0342016 (49:6762)

[35] V.K. Andreev, O.V. Kaptsov, V.V. Pukhnachev, and A.A. Rodionov, Applications of Group-Theoretical Methods in Hydrodynamics (Kluwer Academic Publisher, Dordrecht, 1998). MR1663416 (2000f:76103)

[36] V.K. Andreev, Yu.A. Gaponenko, O.N. Goncharova, and V.V. Pukhnachev, Mathematical Models of Convection (De Gruyter Publisher, Berlin, 2012). MR2961939

[37] V.V. Pukhnachev, "Branching of rotationally symmetric solutions describing flows of a viscous liquid with a free surface", J. Appl. Mech. Tech. Phys., 1973, Vol. 14, No. 2, pp. 253–258.

[38] V.A. Solonnikov, "Solvability of a problem of the flow of a viscous incompressible fluid into an infinite open basin", Proceedings of the Steklov Institute of Mathematics, 1989, Vol. 179, No. 2, pp. 193–226. MR964918 (90k:35208)

[39] J. Leray, "Etude de diverses equations integrals nonlineaires et de quelques problemes que pose l'hydrodynamique", J. Math. Pure. Appl., 1933, Vol. 12, pp. 1-82.

[40] E. Hopf, "Ein allgemeiner Endlichkeitssatz der Hydrodynamik", Math. Ann., 1941, Vol. 117, pp. 764-775. MR0005003 (3:92a)

[41] A. Takeshita, "A remark on Leray's inequality", Pacofic J. Math., 1993, Vol. 157, pp. 151-158. MR1197051 (93j:35135)

[42] R. Finn, "On steady-state solutions of the Navier–Stokes equations. III", Acta Math., 1961, Vol. 105, pp. 197-244. MR0166498 (29:3773)

[43] H. Morimoto and S. Ukai, "Perturbation of the Navier–Stokes equations in an annular domain with non-vanishing outflow condition", J. Math. Sci. Univ. Tokyo, 1995, Vol. 3, pp. 78-82. MR1414620 (97i:35139)

[44] W. Borchers and K. Pileckas, "Note on the flux problem for stationary Navier–Stokes equations in domains with multiply connected boundary", Acta Applicandae Mathematicae, 1994, Vol. 37, pp. 21-30. MR1308742 (95k:76026)

[45] Ch.J. Amick, "Existence of solutions to the non-homogeneous steady Navier–Stokes equations", Indiana Univ. Math. J., 1984, Vol. 33, pp. 817-830. MR763943 (86d:35116)

[46] L.I. Sazonov, "On the existence of a stationary symmetric solution of the two-dimensional fluid flow problem", Mathematical Notes, 1993, Vol. 54, pp. 1280–1283. MR1268380 (94m:76020)

[47] H. Fujita, "On stationary solutions to Navier–Stokes equations in symmetric plane domains under general outflow conditions", in: Proc. Int. Conf. on Navier–Stokes equations. Theory and Numerical Methods, 1997, Pitman Research Notes Math. Series, 1997, Vol. 388, pp. 16-30. MR1773581 (2001h:35142)

[48] H. Morimoto, "A remark on the existence of 2-D steady Navier–Stokes flow in bounded domain under general outflow condition", J. Math. Fluid Mech., 2007, Vol. 9, No. 4, pp. 411-418. MR2336076 (2008i:35182)

[49] V.V. Pukhnachev, "Viscous flows in domains with a multiply connected boundary", in: New Directions in Mathematical Fluid Mechanics. The Alexander V. Kazhikhov Memorial Volume, Eds. A.V. Fursikov, G.P. Galdi, and V.V. Pukhnachev (Birkhäuser, Basel, Boston, and Berlin, 2009, pp. 333–348). MR2732017 (2012c:76029)

[50] M.V. Korobkov, K. Pileckas, and R. Russo, "Steady Navier–Stokes system with non-homogeneous boundary conditions in the axially symmetric case", C. R. Acad. Sci. Paris, 2012, Vol. 340, Ser. I, pp. 115–119.

[51] G.V. Alekseev and V.V. Pukhnachev, "The axially symmetric flow-through problem for the Navier–Stokes equations in variables "vorticity–stream function", Doklady Physics, 2012, Vol. 57, No. 8, pp. 301–306. MR3050500

[52] M.V. Korobkov, K. Pileckas, and R. Russo, "On the flux problem in the theory of steady Navier–Stokes equations with non-homogeneous boundary conditions", Arch. Rational Mech. Anal., 2013, Vol. 207, pp. 185–213. MR3004771

[53] V.V. Pukhnachev, "Leray's problem and Yudovich's hypothesis," Izv. Vyssh. Uch. Zaved., Severo-Kavkazskii Raion, Estesv. Nauki, Special issue "Actual Problems of Mathematical Hydrodynamics", 2009, pp. 185–194 (in Russian).

[54] E. Hopf, "On nonlinear partial differential equations", Lecture Series of the Symposium on Partial Differential Equations, 1957, Univ. Kansas, pp. 1–31.

[55] H. Fujita, "On the existence and regularity of the steady-state solutions of the Navier–Stokes equations", J. Fac. Sci. Univ. Tokyo, 1961, Vol. 9, Part 1, pp. 59–102. MR0132307 (24:A2152)
[56] V.A. Kodratiev, "Boundary value problems for elliptic equations in domains with conical or angular points", Proceedings of the Moscow Mathematical Society, 1967, Vol. 16, pp. 227–313. MR0226187 (37:1777)
[57] H. Fujita, H. Morimoto, and H. Okamoto, "Stability analysis of the Navier–Stokes equations flows in annuli", Math. Methods in Appl. Sciences, 1997, Vol. 20, pp. 950–978. MR1458528 (98f:76025)
[58] V.A. Kondratiev, "Asymptotic of the solution of the Navier–Stokes equation near the corner point of the boundary", Prikl. Mat. Mekh., 1967, Vol. 31, No. 1, pp 119–123 (in Russian). MR0218732 (36:1816)
[59] Yu. B. Rumer, "Problem of a submerged jet", Prikl. Mat. Mekh. 1952, Vol. 16, No. 2, pp. 255–256 (in Russian). MR0046813 (13:791b)
[60] J. Happel and H. Brenner, Low Reynolds Number Hydrodynamics (Kluwer Academic Publisher, Dordrecht, 1983).
[61] A. Russo and A. Tartaglione, "On the signular solutions of the Navier–Stokes equations", Lithuanian Math. J. **53** (2013), no. 4, 423–437.

Lavrent'ev Institute of Hydrodynamics, Siberian Branch, Russian Academy of Sciences, Novosibirsk State University, Prosp. Akademika Lavrent'eva, 15, Novosibirsk 630090, Russia

E-mail address: pukhnachev@gmail.com

Amer. Math. Soc. Transl.
Volume **232**, 2014

The local regularity theory for the Navier–Stokes equations near the boundary

G. A. Seregin and T. N. Shilkin

Dedicated to the memory of our teacher Olga Alexandrovna Ladyzhenskaya

Abstract. This is an expository paper on the theory of local regularity for weak solutions to the non-stationary 3D Navier–Stokes equations near the boundary of a domain.

1. Introduction

The main problem of modern mathematical hydrodynamics is the global well-posedness of the 3D Navier–Stokes equations, i.e., the global existence of a unique solution, corresponding to a given smooth divergent-free initial data.

There are two main directions in the study of this problem. One can try to improve local well-posedness results (see classical papers [**17**] and [**15**] and many others on local well-posedness) or show that a global week solution, introduced essentially in [**17**] and [**11**] and called the weak Leray–Hopf solution, to the corresponding initial boundary value problem is in fact unique. On the other hand, the second aim can be achieved by proving regularity of weak solutions. Indeed, it is well known, since the celebrated paper [**17**] has been published, that smoothness of weak solutions implies their uniqueness in the class of weak Leray–Hopf solutions. Here, we are going to discuss regularity of weak solutions, keeping in mind that it is one of possible ways to attack the main problem on the global well-posedness. Our approach is quite typical for PDE theory and is in a sense local. The latter means that we have a solution to the Navier–Stokes system with a "finite energy" in a canonical parabolic domain, and we try to show that it is smoother in subdomains. The result depends on assumptions imposed on the pressure. Our choice of the class for pressure is motivated by the linear theory and reflects the fact that the whole Navier–Stokes (or Stokes) problem is not quite local because of the incompressibility condition.

Given a space-time point $z_0 = (x_0, t_0)$, the canonical domain is going to be a parabolic cylinder $Q(z_0, R) := B(x_0, R) \times]t_0 - R^2, t_0[$ if the local interior regularity

2010 *Mathematics Subject Classification.* Primary 35Q30, 76D05.
Key words and phrases. Boundary regularity, weak solutions, Navier–Stokes system.
This work was supported by RFBR grant 11-01-00324.

is studied or a parabolic half-cylinder $Q^+(z_0,R) := B^+(x_0,R)\times\,]t_0-R^2,t_0[$ if the local boundary regularity is under consideration. Here $B(x_0,R)$ denotes a ball in $\mathbb{R}^3$ of radius R centered at a point x_0, and $B^+(x_0,R)$ is a half-ball $B(x_0,R)\cap(x_0+\mathbb{R}^3_+)$, and $\mathbb{R}^3_+ := \{\, x=(x_1,x_2,x_3)\in\mathbb{R}^3 \mid x_3>0\,\}$.

Since the Navier–Stokes system is invariant under the scaling transformation

$$u^R(y,s) = Ru(x_0+Ry,t_0+R^2s), \qquad p^R(y,s) = R^2p(x_0+Ry,t_0+R^2s), \tag{1.1}$$

the problem of local regularity of weak solutions in a neighborhood of a point z_0 can be reduced to a model problem in some fixed domain, say, $Q = B\times\,]-1,0[$ in the internal case or $Q^+ = B^+\times\,]-1,0[$ in the boundary case. Here, B is the unit ball of $\mathbb{R}^3$ centered at the origin and $B^+ := B\cap\mathbb{R}^3_+$.

In contrast to equations of parabolic type, the local smoothing for the Navier–Stokes system has some special features. In particular, in the local setting, weak solutions may be not infinitely smooth in subdomains despite the right-hand side being infinitely smooth there. So, there might be a limiting smoothness which can be achieved locally. The roots of this phenomena lay in the linear theory, which we discuss in detail in Section 2.

Typical results in the known local regularity theory have the form of the so-called ε-regularity conditions. For the Navier–Stokes system in canonical domains, ε-regularity conditions ensure Hölder continuity of the velocity field around the origin, provided a certain integral quantity of a solution over the above domain is sufficiently small.

The first results on the local regularity for the 3D Navier–Stokes equations belong to Scheffer [**22**]. Scheffer considered a special class of weak solutions to the Cauchy problem that satisfy a *local energy inequality*. Motivated by this observation, later on, Caffarelli, Kohn, and Nirenberg introduced the so called *suitable weak solutions*, which are just solutions to the Navier–Stokes system with certain reasonable properties. That was a great step towards a complete local setting. By definition, a suitable weak solution u and p is such that the velocity u belongs to the energy class, pressure is an integrable function (with a certain exponent of integrability determined by the linear theory and by the integrability of the convective term), and u and p are assumed to satisfy the Navier–Stokes system in the sense of distributions and the local energy inequality. We will explore the definition of suitable weak solutions introduced in [**18**]; see also [**16**].

DEFINITION 1.1. *We say that a pair of functions u and p is a suitable weak solution to the Navier–Stokes system in Q if*

- $u\in L_{2,\infty}(Q)\cap W^{1,0}_2(Q)$, $p\in L_{\frac32}(Q)$,
- *u and p satisfy the Navier–Stokes system in Q in the sense of distributions,*
- *for a.a. $t\in\,]-1,0[$, the pair u and p satisfies the local energy inequality in Q*

$$\int\limits_B \zeta(x,t)|u(x,t)|^2\,dx \;+\; 2\int\limits_{-1}^{t}\int\limits_B \zeta|\nabla u|^2\,dxdt$$
$$\le \int\limits_{-1}^{t}\int\limits_B |u|^2\;(\partial_t\zeta+\Delta\zeta)\;dxdt \;+\; \int\limits_{-1}^{t}\int\limits_B u\cdot\nabla\zeta\;\big(|u|^2+2p\big)\;dxdt$$

for any non-negative test function $\zeta \in C^\infty(\mathbb{R}^3 \times \mathbb{R})$ *vanishing near the parabolic boundary* $\partial' Q := (\partial B\times\,]-1,0[) \cup (B \times \{t=-1\})$.

Here we denote by $L_s(Q)$ the Lesbegue space of functions integrable over Q with the exponent $s \in [1,+\infty]$; $W^{1,0}_s(Q) := \{u \in L_s(Q) \mid \nabla u \in L_s(Q)\}$, where ∇u denotes the gradient of u with respect to spatial variables; $L_{2,\infty}(Q) := L_\infty(-1,0;L_2(B))$.

This class of local solutions appears in the global setting. Indeed, it is not so difficult to show that, in a given space-time domain among all weak Leray–Hopf solutions corresponding to a given initial data, there exists at least one that is a suitable weak solution in any parabolic cylinder $Q(z_0,R)$ belonging to the space-time domain; see [**4**] and [**16**].

A significant contribution into the local regularity theory for the Navier–Stokes equation has been made by Caffarelli, Kohn, and Nirenberg. They showed that the set of all singular points is very small in the following sense: the one dimensional parabolic Hausdorff measure of this set is equal to zero. This is a consequence of the Caffarelli–Kohn–Nirenberg ε-regularity condition reading that there exists an absolute constant ε_0 such that, for any suitable weak solution u and p in Q with

$$\sup_{r<1} \frac{1}{r} \int\limits_{Q(r)} |\nabla u|^2\, dxdt \;<\; \varepsilon_0, \tag{1.2}$$

the velocity field u is essentially bounded near the origin (actually it is Hölder continuous, as it was shown later in [**16**]). Here and in what follows we denote $Q(r) := Q(0,r)$, $Q^+(r) := Q^+(0,r)$, etc.

Among various ε-regularity conditions (see, for example, papers [**4**], [**18**], [**16**], [**30**], and many others), we would like to point out the following one: there exists an absolute constant $\varepsilon_1 > 0$ such that, for any suitable weak solution u and p in Q, satisfying in addition the inequality

$$\int\limits_{Q} \left(|u|^3 + |p|^{\frac{3}{2}}\right) dxdt \;<\; \varepsilon_1, \tag{1.3}$$

the velocity field u is Hölder continuous in the completion of the set $Q(\frac{1}{2})$. In the present paper, condition (1.3) and its boundary analogue are called *the basic ε-regularity conditions*. The basic ε-regularity condition is remarkable as many other ε-regularity conditions, including the CKN-condition (1.2), can be derived from this one; see, for example, [**4**], [**18**], and [**16**]. In what follows, we are going to deal with the boundary analogue of condition (1.3).

Now, let us review known results on local regularity up to the spatial boundary for weak solutions to the Navier–Stokes system. In this case, we complement the Navier–Stokes equation with the non-slip boundary condition for the velocity field. We focus mostly on the explanation of what happens with weak solutions to the Navier–Stokes system in the canonical domain Q^+ with the boundary condition on the flat part of the semiball B^+, i.e.,

$$u|_{x_3=0} = 0.$$

An analogue of ε-regularity condition (1.2) for boundary points has been proven in [**24**]. The proof is typical for a PDE system and is based contradiction arguments. Later in [**25**], the same sufficient regularity condition was proved directly, which made it possible in principle to estimate the size of all constants in the corresponding assumptions.

The latest version of the definition of suitable weak solutions is as follows.

DEFINITION 1.2. *We say that a pair of functions u and p is a boundary suitable weak solution to the Navier–Stokes system in Q^+ if*

- $u \in L_{2,\infty}(Q^+) \cap W^{1,0}_2(Q^+)$, $p \in L_{\frac{3}{2}}(Q^+)$,
- $u|_{x_3=0} = 0$ *in the sense of traces,*
- u *and* p *satisfy the Navier–Stokes system in* Q^+ *in the sense of distributions,*
- *for a.a.* $t \in]-1, 0[$, *the pair* u *and* p *satisfies the local energy inequality in* Q^+

$$\int\limits_{B^+} \zeta(x,t)|u(x,t)|^2\,dx \;+\; 2\int\limits_{-1}^{t}\int\limits_{B^+} \zeta|\nabla u|^2\,dxdt$$
$$\le \int\limits_{-1}^{t}\int\limits_{B^+} |u|^2\ (\partial_t\zeta + \Delta\zeta)\ dxdt \;+\; \int\limits_{-1}^{t}\int\limits_{B^+} u\cdot\nabla\zeta\ \big(|u|^2 + 2p\big)\ dxdt$$

for any non-negative test function $\zeta \in C^\infty(\mathbb{R}^3 \times \mathbb{R})$ *vanishing near the parabolic boundary* $\partial' Q$.

However, in [**24**], [**25**], [**32**], the definition of suitable weak solutions is different. It is supposed there in addition that the second spatial derivatives and the first derivative in time of the velocity field and the gradient of the pressure must exist as integrable functions in Q^+:

$$\begin{gathered} u \in W^{2,1}_{s,l}(Q^+), \quad p \in W^{1,0}_{s,l}(Q^+) \\ \text{for some } \ s, l \in]1, +\infty[\ \text{ such that } \ \frac{3}{s} + \frac{2}{l} \ge 4. \end{gathered} \tag{1.4}$$

Here $L_{s,l}(Q^+)$ is the anisotropic Lebesgue space equipped with the norm

$$\|f\|_{L_{s,l}(Q^+)} := \Big(\int\limits_{-1}^{0}\Big(\int\limits_{B^+} |f(x,t)|^s\,dx\Big)^{l/s} dt\Big)^{1/l},$$

and we use the following notation for the functional spaces,

$$W^{1,0}_{s,l}(Q^+) \equiv L_l(-1,0; W^1_s(B^+)) = \{\ u \in L_{s,l}(Q^+) :\ \nabla u \in L_{s,l}(Q^+)\ \},$$
$$W^{2,1}_{s,l}(Q^+) = \{\ u \in W^{1,0}_{s,l}(Q^+) :\ \nabla^2 u,\ \partial_t u \in L_{s,l}(Q^+)\ \},$$

and the following notation for the norms,

$$\|u\|_{W^{1,0}_{s,l}(Q^+)} = \|u\|_{L_{s,l}(Q^+)} + \|\nabla u\|_{L_{s,l}(Q^+)},$$
$$\|u\|_{W^{2,1}_{s,l}(Q^+)} = \|u\|_{W^{1,0}_{s,l}(Q^+)} + \|\nabla^2 u\|_{L_{s,l}(Q^+)} + \|\partial_t u\|_{L_{s,l}(Q^+)}.$$

A particular choice of exponents s, l is a matter of convenience and can be made in various ways. For example, in [**24**] and in [**32**], it is assumed that $s = 9/8$, $l = 3/2$, in [**25**] $s = l = 5/4$ and $s = 15/14$, $l = 5/3$. It should be mentioned that the choice of s, l is always motivated by the following general idea: if we take the Leray–Hopf solution u of the initial boundary value problem for the Navier–Stokes equation in some domain and interpret the convective term $(u \cdot \nabla)u$ as a right-hand side of the initial boundary value problem for the Stokes system, then the extra condition (1.4) follows from the coercive estimates of the linear theory and from the uniqueness theorem for the Stokes problem in the class of weak solutions. This scheme works well in the case of interior regularity, while, for the boundary regularity case, its realization encounters some difficulties. Nevertheless, as is shown in [**26**], the scheme works up to the boundary as well, and extra assumptions (1.4) are simply superfluous.

As has been already mentioned, there is an essential difference between local interior regularity and local boundary regularity even for the Stokes system. One of the consequences of such an observation is that further smoothing in a neighborhood of regular points might happen differently. In a neighborhood of an interior regular point, a suitable weak solution has all the spatial derivatives that are Hölder continuous, while, in a neighborhood of boundary regular point, the spatial gradient is not necessarily even bounded; see [**13**] and [**33**].

Besides conditions (1.2) and (1.3), ε-regularity theory provides many other sufficient conditions of local regularity of suitable weak solutions to the Navier–Stokes equations; see, for example, papers [**12**], [**28**], [**29**], [**30**], and references in them. Most of these conditions are stated in terms of so-called *scale invariant functionals*. Some examples of such functionals (in the case of the flat part of the boundary) are as follows:

$$(1.5)\qquad \begin{aligned} A(u,r) &= \sup_{t\in\,]-r^2,0[} \Big(\frac{1}{r} \int\limits_{B^+(r)} |u(x,t)|^2\,dx \Big)^{1/2}, \\ C(u,r) &= \Big(\frac{1}{r^2} \int\limits_{Q^+(r)} |u(x,t)|^3\,dxdt \Big)^{1/3}, \\ E(u,r) &= \Big(\frac{1}{r} \int\limits_{Q^+(r)} |\nabla u(x,t)|^2\,dxdt \Big)^{1/2}. \end{aligned}$$

"Scale invariance" means that if $F(u,r)$ is one of these functionals and u^R and p^R are functions obtained from u and p by formulas (1.1) with $x_0 = 0$, $t_0 = 0$, then

$$F(u^R,1) \; = \; F(u,R) \qquad \forall\; R > 0.$$

The functionals $A(u,r)$, $C(u,r)$, $E(u,r)$ possess the following property: boundedness of one of them, i.e.,

$$\min\Big\{ \sup_{r<1} A(u,r),\; \sup_{r<1} C(u,r),\; \sup_{r<1} E(u,r) \Big\} \;<\; +\infty,$$

implies boundedness of all others,

$$(1.6)\qquad \max\Big\{ \sup_{r<1} A(u,r),\; \sup_{r<1} C(u,r),\; \sup_{r<1} E(u,r),\; \sup_{r<1} D(p,r) \Big\} \;<\; +\infty.$$

Here, $D(p,r)$ is a functional

$$D(p,r) = \Big(\frac{1}{r^2} \int_{Q^+(r)} |p(x,t)|^{3/2}\,dxdt \Big)^{2/3},$$

which is also invariant under the scaling transformation of p according to (1.1). Statement (1.6) has been proven in [**28**] in the internal case. Later on, in [**19**], it was generalized to the boundary case.

One of the basic principles in the ε-regularity theory for Navier–Stokes equations reads, If at least one of the scale invariant functionals is small uniformly with respect to all $r \in]0,1[$, i.e.,

$$\sup_{r<1} F(u,r) < \varepsilon_0, \tag{1.7}$$

then the origin is a regular point of the velocity field u (i.e., u is Hölder continuous near the origin). This statement has been rigorously proven by many authors for various types of scale invariant functionals; see, for example, references in [**12**], [**10**], [**29**], [**30**]. In our paper, we shall mention only those contributions that are concerned with the boundary case. In [**24**] and [**25**] the boundary regularity up to the flat part of the boundary has been proven if $F(u,r)$ is one of functionals in (1.5).

There is one more example of scale invariant functionals, which we call the Ladyzhenskaya–Prodi–Serrin-type functional (LPS functional),

$$M_{s,l}(u,r) := \|u\|_{L_{s,l}(Q^+(r))} = \Big(\int_{-r^2}^0 \Big(\int_{B^+(r)} |u(x,t)|^s\, dx\Big)^{l/s} dt\Big)^{1/l},$$
$$s,l \in]1,+\infty[.$$

Comparably with functionals defined in (1.5), LPS functionals possess two additional properties. First, it is monotone with respect to r, and hence if it is finite for some particular $r_0 > 0$, then it is uniformly bounded for all $r \in]0,r_0[$. And second, if s, $l > 1$, then LPS functionals are absolutely continuous functions of a domain, i.e.,

$$M_{s,l}(u,r) \to 0 \qquad \text{as} \qquad r \to 0. \tag{1.8}$$

So, ε-regularity theory developed above provides a simple proof of the following conditional result for a boundary suitable weak solution u to the Navier–Stokes equations in Q^+: if $s > 3$ and

$$\frac{3}{s} + \frac{2}{l} = 1 \qquad \text{and} \qquad M_{s,l}(u,1) < +\infty, \tag{1.9}$$

then u is regular near the origin. For a different approach, we refer to [**12**].

Note that formally the first inequality in (1.9) allows the following combination of parameters: $s = 3$, $l = +\infty$; i.e., one can ask the question about local regularity of solutions to the Navier–Stokes system belonging to the class $L_{3,\infty}(Q^+) := L_\infty(-1,0;L_3(B^+))$ which we call $L_{3,\infty}$-solutions. It is known (see [**14**]) that the initial boundary value problem for the Navier–Stokes equation is locally well-posed on L_3-space. But the method that we used to prove regularity weak solution with finite LPS functionals in the case of $s > 3$, $l < +\infty$, cannot be extended to the $L_{3,\infty}$-case as the functional $M_{s,l}(u,r)$ with $s = 3$, $l = +\infty$, in general does not possess property (1.8).

The proof of regularity of $L_{3,\infty}$-solutions to the Navier–Stokes system requires development of the completely new approach which is based on the backward-in-time uniqueness for the heat operator with lower order coefficients in the compliment to a ball or even in a half-space. This method has been introduced in [**6**] and then developed in [**5**] and in [**7**] in order to prove the interior regularity of

$L_{3,\infty}$-solutions. Later on, the same type of result was extended to the boundary case of $L_{3,\infty}$-solutions. It has been shown in [**31**] that

$$u \in L_{3,\infty}(Q^+) \qquad \Longrightarrow \qquad u \in C^{\alpha,\frac{\alpha}{2}}(\bar{Q}^+(\tfrac{1}{2})).$$

Here $C^{\alpha,\frac{\alpha}{2}}(\bar{Q}^+(r))$ denotes the space of functions that are Hölder continuous with the exponent $\alpha > 0$ with respect to the usual parabolic metric.

The paper is organized as follows. In Section 2 we discuss the local smoothness of weak solutions to the linear Stokes problem near the boundary. In Section 3 we present a proof of the basic ε-regularity condition for boundary suitable weak solutions to the Navier–Stokes equations near a flat part of the boundary. Finally, in Section 4 we give a brief overview of known results on the local regularity theory for the Navier–Stokes equations in a domain with a curvilinear boundary.

2. Linear theory

As is mentioned in the Introduction, weak solutions to the non-stationary Stokes system,

$$\left\{\begin{array}{c} \partial_t u - \Delta u + \nabla p = 0, \\ \operatorname{div} u = 0, \end{array}\right. \quad \text{in} \quad Q, \tag{2.1}$$

locally are not necessarily smooth. A simple example of a non-smooth solution to (2.1) is

$$u(x,t) = \varphi(t)\nabla h(x), \qquad p(x,t) = -\varphi'(t)h(x),$$

where h is a scalar harmonic function in spatial variables and φ is an arbitrary function of t having limited smoothness. The same effect takes place in the non-linear case and has been pointed out by J. Serrin in [**35**]. Nevertheless, the Stokes system (2.1) has the property of infinite smoothing of weak solutions with respect to spatial variables in internal points of a domain:

THEOREM 2.1. *Assume $s \in \,]1,+\infty[$, $l \in \,]1,2[$, and $u \in W^{1,0}_{s,l}(Q)$, $p \in L_{s,l}(Q)$ satisfy* (2.1) *in Q in the sense of distributions. Then for any $k = 0,1,\ldots,$ we have $\nabla^k u \in C^{\alpha,\frac{\alpha}{2}}(\bar{Q}(\frac{1}{2}))$ with $\alpha = 2 - \frac{2}{l}$.*

Surprisingly, the analog of Theorem 2.1 is not valid if we consider weak solutions to the Stokes system near the boundary

$$\left\{\begin{array}{c} \partial_t u - \Delta u + \nabla p = f, \\ \operatorname{div} u = 0, \\ u|_{x_3=0} = 0. \end{array}\right. \quad \text{in} \quad Q^+, \tag{2.2}$$

Actually, in contrast to the internal case, the first spatial gradient of a weak solution to system (2.2) is not necessarily bounded up to the boundary; i.e., there exist functions $u \in W^{1,0}_2(Q^+)$, $p \in L_{\frac{3}{2}}(Q^+)$ which satisfy system (2.2) with $f \equiv 0$ in Q^+ in the sense of distributions, u satisfies the boundary condition in the sense of traces but

$$\nabla u \notin L_\infty(Q^+(r)) \tag{2.3}$$

for any $0 < r \le 1/2$. The first counterexample of this kind has been constructed by Kang in [**13**]. Later Seregin and Sverak in [**33**] simplified his construction significantly. Here, we explain the counterexample, following [**33**]:

EXAMPLE 2.1. *Assume $\varphi(t)$ is an arbitrary function of t variables, and let $h : \mathbb{R}_+\times\ ]-2,0[\ \to \mathbb{R}$, $h = h(x,t)$, be a solution to the initial boundary value problem for the 1D heat equation in a half-line,*

$$\begin{cases} \dfrac{\partial h}{\partial t} - \dfrac{\partial^2 h}{\partial x^2} = \varphi(t), & \text{in } \ \mathbb{R}_+\times\ ]-2,0[, \\ h|_{t=-2} = 0, & h|_{x=0} = 0. \end{cases}$$

Let $u : \mathbb{R}^3_+\times\ ]-2,0[\ \to \mathbb{R}^3$, $p : \mathbb{R}^3_+\times\ ]-2,0[\ \to \mathbb{R}$ be functions representing the shear flow along the x_1-axe

$$u(x,t) := (h(x_3,t),0,0), \qquad p(x,t) := -\varphi(t)x_1.$$

Then functions u and p satisfy (formally) both the Stokes and the Navier–Stokes systems in Q^+. Moreover, if we assume that $\alpha \in\ ]\frac{1}{3},\frac{1}{2}[$ and take

$$\varphi(t) = \tfrac{1}{|t|^{1-\alpha}},$$

then $u \in W^{1,0}_2(Q^+)$ and $p \in L_{\frac{3}{2}}(Q^+)$. Functions u and p satisfy equations (2.2) with $f = 0$ in the sense of distributions and the boundary condition in the sense of traces as well. However, ∇u is unbounded in any neighborhood of the origin. In particular, (2.3) holds.

Example 2.1 shows that, in contrast to the internal case, the Stokes system does not possess the property of significant improvement of the regularity of weak solutions up to the boundary. This is a serious obstacle which makes the theory of boundary regularity for the Navier–Stokes equations different from the analogue theory in the internal case. The natural question that arises is what is the optimal regularity of weak solutions to the Stokes system up to the boundary in the local setup? A certain answer to this question has been given in [**23**], where Hölder continuity of the velocity field u up to the flat part of the boundary has been established for *strong solutions* to (2.2). But before we switch to the discussion of this result, let us introduce some terminology to explain the difference between weak and strong solutions to (2.2).

DEFINITION 2.1. *Assume $1 < s,l < +\infty$, and $f \in L_l(-1,0;W^{-1}_s(B^+))$. We say that functions u and p are a weak solution of system (2.2) if they belong to the spaces*

$$u \in W^{1,0}_{s,l}(Q^+), \qquad p \in L_{s,l}(Q^+),$$

u and p satisfy (2.2) in the sense of distributions, and u satisfies the boundary condition in the sense of traces.

Note that, for any weak solution u and p to system (2.2),

$$\partial_t u \in L_l(-1,0;W^{-1}_s(B^+))$$

and the following estimate holds:

$$\begin{aligned} &\|\partial_t u\|_{L_l(-1,0;W^{-1}_s(B^+))} \\ &\quad\le C\left(\|f\|_{L_l(-1,0;W^{-1}_s(B^+))} + \|u\|_{W^{1,0}_{s,l}(Q^+)} + \|p\|_{L_{s,l}(Q^+)}\right). \end{aligned} \tag{2.4}$$

Here $W^{-1}_s(B^+)$ is the space dual to $\overset{\circ}{W}{}^1_{s'}(B^+)$ and

$$\|u\|_{L_l(-1,0;W^{-1}_s(B^+))} = \left(\int_{-1}^0 \|u(\cdot,t)\|^l_{W^{-1}_s(B^+)}\,dt\right)^{1/l}.$$

DEFINITION 2.2. *Assume* $1 < s,l < +\infty$, *and* $f \in L_{s,l}(Q^+)$. *We say that functions* u *and* p *are a strong solution to* (2.2) *if they are a weak solution to* (2.2) *and*

$$u \in W^{2,1}_{s,l}(Q^+), \qquad p \in W^{1,0}_{s,l}(Q^+).$$

The idea of showing Hölder continuity of strong solutions proposed in [**23**] is as follows. We first show that strong solutions satisfy the usual local version of coercive estimate given by

THEOREM 2.2. *Suppose* $s,\ l \in]1,\infty[$. *For any* $f \in L_{s,l}(Q^+)$ *and for any strong solution* $u \in W^{2,1}_{s,l}(Q^+)$, $p \in W^{1,0}_{s,l}(Q^+)$ *to system* (2.2) *in* Q^+, *the following local estimate holds:*

$$\begin{aligned}(2.5)\qquad &\|u\|_{W^{2,1}_{s,l}(Q^+(\frac12))} + \|\nabla p\|_{L_{s,l}(Q^+(\frac12))}\\ &\leq C\left(\|f\|_{L_{s,l}(Q^+)} + \|\nabla u\|_{L_{s,l}(Q^+)} + \|p\|_{L_{s,l}(Q^+)}\right)\end{aligned}$$

with a positive constant C, *depending only on* s, l.

In contrast to the interior case, this is not a trivial statement. The first proof given in [**23**] has been based on duality arguments and inspired by paper [**37**]. In the present paper, we show that it can be deduced from a more general statement proved in [**8**]. But before explaining our approach, let us briefly described the main idea of getting Hölder continuity. Unlike the way it has been done in the internal case, in the boundary case we cannot get the result by gaining more derivatives in space variables. In fact, this is even impossible because of the counterexample in Example 2.1. But what we can do is gain more integrability in space and apply certain bootstrap arguments. So, we show

THEOREM 2.3. *Suppose* $s,\ l,\ m \in]1,\infty[$, $m \geq s$. *For any* $f \in L_{m,l}(Q^+)$ *and for any strong solution* $u \in W^{2,1}_{s,l}(Q^+)$, $p \in W^{1,0}_{s,l}(Q^+)$ *to* (2.2) *in* Q^+, *we have*

$$u \in W^{2,1}_{m,l}(Q^+(\tfrac12)), \qquad \nabla p \in L_{m,l}(Q^+(\tfrac12)),$$

and the local estimate

$$\begin{aligned}(2.6)\qquad &\|u\|_{W^{2,1}_{m,l}(Q^+(\frac12))} + \|\nabla p\|_{L_{m,l}(Q^+(\frac12))}\\ &\leq C\left(\|f\|_{L_{m,l}(Q^+)} + \|\nabla u\|_{L_{s,l}(Q^+)} + \|p\|_{L_{s,l}(Q^+)}\right)\end{aligned}$$

holds with a positive constant C, *depending only on* s, l, m.

If the exponent m in Theorem 2.3 is sufficiently large, then the Hölder continuity of u follows from the imbedding theorems for the anisotropic Sobolev spaces. Namely, the following result is true:

THEOREM 2.4. *Suppose* $s,\ l,\ m \in]1,\infty[$, $m > \frac{3l}{2(l-1)}$. *For any* $f \in L_{m,l}(Q^+)$ *and for any strong solution* $u \in W^{2,1}_{s,l}(Q^+)$, $p \in W^{1,0}_{s,l}(Q^+)$ *to system* (2.2), *we have* $u \in C^{\beta,\frac{\beta}{2}}(\bar Q^+(\frac12))$ *with* $\beta = 2 - \frac{3}{m} - \frac{2}{l}$ *and the local estimate*

$$(2.7)\qquad \|u\|_{C^{\beta,\frac{\beta}{2}}(\bar Q^+(\frac12))} \leq C\left(\|f\|_{L_{m,l}(Q^+)} + \|\nabla u\|_{L_{s,l}(Q^+)} + \|p\|_{L_{s,l}(Q^+)}\right),$$

holds, where a positive constant C *depends only on* s, l, m.

The important part is to show that any weak solution to (2.2) is actually a strong one, at least locally (here we use the terminology introduced in our Definitions 2.1 and 2.2). For the interior case, the corresponding claim is known (see details, for example, in [**16**]). For the boundary case, the analogous statement has been proven in [**26**] relatively recently. Here, our proof follows [**36**]. It is some modification of the approach in [**26**], which can be used in more general situations. So, the result is

THEOREM 2.5. *Assume $s,\, l\in\,]1,\infty[$. Then, for any $f\in L_{s,l}(Q^+)$ and for any weak solution $u\in W^{1,0}_{s,l}(Q^+)$, $p\in L_{s,l}(Q^+)$ to (2.2) in Q^+, the statements*

$$u\in W^{2,1}_{s,l}(Q^+(\tfrac12)),\qquad p\in W^{1,0}_{s,l}(Q^+(\tfrac12))$$

hold and functions u and p are a strong solution to system (2.2) in the half-cylinder $Q^+(\frac12)$.

As our example shows, regularity results up to the boundary described by the above statements are in a sense optimal.

Now we come to the detailed proofs of the results above.

PROOF OF THEOREM 2.2. The proof presented below is borrowed from [**8**]; we reproduce it here for the sake of completeness. Take arbitrary ρ, r such that $\frac12\le\rho<r\le\frac{9}{10}$. Consider a cut-off function $\zeta\in C^\infty(\bar Q^+)$ such that

$$0\le\zeta\le1\quad\text{in}\quad Q^+,\qquad \zeta\equiv1\quad\text{in}\quad Q^+(\rho),\qquad \zeta\equiv0\quad\text{in}\quad Q^+\setminus Q^+(r),$$
$$\|\nabla^k\zeta\|_{L_\infty(Q^+)}\le\frac{C}{(r-\rho)^k},\quad k=1,2,\quad \|\partial_t\nabla^k\zeta\|_{L_\infty(Q^+)}\le\frac{C}{(r-\rho)^k},\quad k=0,1.$$

Let u and p be a strong solution to system (2.2). Then functions $v:=\zeta u$, $q:=\zeta p$ satisfy the following initial boundary value problem

$$\left\{\begin{aligned} &\partial_t v-\Delta v+\nabla q=\tilde f,\\ &\operatorname{div} v=g,\\ &v|_{\partial\Omega\times\,]-1,0[}=0,\quad v|_{t=-1}=0,\end{aligned}\right.\qquad \text{in}\quad \Omega\times\,]-1,0[,\tag{2.8}$$

where

$$\tilde f=\zeta f+u(\partial_t\zeta-\Delta\zeta)-2(\nabla u)\nabla\zeta+p\nabla\zeta,\qquad g=u\cdot\nabla\zeta,\tag{2.9}$$

and Ω is some smooth canonical domain which is diffeomorphic to a ball and satisfies the inclusions $B^+_{9/10}\subset\Omega\subset B^+$. Applying Theorem 1.1 of [**8**], we obtain the estimate

$$\begin{aligned}&\|v\|_{W^{2,1}_{s,l}(\tilde Q)}+\|\nabla q\|_{L_{s,l}(\tilde Q)}\\ &\quad\le C\left(\|\tilde f\|_{L_{s,l}(\tilde Q)}+\|g\|_{W^{1,0}_{s,l}(\tilde Q)}+\|\partial_t g\|^{1/s}_{L_{s,l}(\tilde Q)}\|\partial_t g\|^{1/s'}_{L_l(-1,0;W^{-1}_s(\Omega))}\right),\end{aligned}\tag{2.10}$$

where we denote $\tilde Q:=\Omega\times\,]-1,0[$. Taking into account that $\frac{1}{r-\rho}\ge1$, after routine computations we obtain the estimate

$$\begin{aligned}\|u\|^s_{W^{2,1}_{s,l}(Q^+(\rho))}&\le C\|f\|^s_{L_{s,l}(Q^+)}\\ &+\frac{C}{(r-\rho)^{2s}}\Big(\|u\|^s_{W^{1,0}_{s,l}(Q^+)}+\|p\|^s_{L_{s,l}(Q^+)}+\|\partial_t u\|^s_{L_l(-1,0;W^{-1}_s(B^+))}\Big)\\ &+\frac{C}{(r-\rho)^{2s}}\|\partial_t u\|_{L_{s,l}(Q^+(r))}\Big(\|\partial_t u\|^{s-1}_{L_l(-1,0;W^{-1}_s(B^+))}+\|u\|^{s-1}_{L_{s,l}(Q^+)}\Big).\end{aligned}\tag{2.11}$$

Estimating the last term in the right-hand side of (2.11) via the Young inequality $ab \le \varepsilon a^s + C_\varepsilon b^{s'}$, we obtain

$$\|u\|^s_{W^{2,1}_{s,l}(Q^+(\rho))} \le C\|f\|^s_{L_{s,l}(Q^+)} + \varepsilon\|\partial_t u\|^s_{L_{s,l}(Q^+(r))} + \frac{C_\varepsilon}{(r-\rho)^{2ss'}}\Big(\|u\|^s_{W^{1,0}_{s,l}(Q^+)} + \|p\|^s_{L_{s,l}(Q^+)} + \|\partial_t u\|^s_{L_l(-1,0;W^{-1}_s(B^+))}\Big),$$

where a constant $\varepsilon > 0$ can be chosen arbitrarily small. By virtue of (2.4), we obtain

$$\|u\|^s_{W^{2,1}_{s,l}(Q^+(\rho))} \le \varepsilon\|\partial_t u\|^s_{L_{s,l}(Q^+(r))} + \frac{C_\varepsilon}{(r-\rho)^{2ss'}}\Big(\|f\|^s_{L_{s,l}(Q^+)} + \|u\|^s_{W^{1,0}_{s,l}(Q^+)} + \|p\|^s_{L_{s,l}(Q^+)}\Big). \tag{2.12}$$

Now, let us introduce the monotone function $\Psi(\rho) := \|u\|^s_{W^{2,1}_{s,l}(Q^+(\rho))}$ and the constant

$$A := C_\varepsilon\left(\|f\|^s_{L_{s,l}(Q^+)} + \|u\|^s_{W^{1,0}_{s,l}(Q^+)} + \|p\|^s_{L_{s,l}(Q^+)}\right).$$

The inequality (2.12) implies that

$$\Psi(\rho) \le \varepsilon\Psi(r) + \tfrac{A}{(r-\rho)^\alpha} \qquad \forall\ \rho,\ r: \quad R_1 \le \rho < r \le R_0, \tag{2.13}$$

for some $\alpha > 0$ depending only on s, and for $R_1 = \frac{1}{2}$, $R_0 = \frac{9}{10}$. Now we shall take advantage of the following lemma (which can be easily proved by iterations if one takes $r_k := R_0 - 2^{-k}(R_0 - R_1)$):

LEMMA 2.1. *Assume Ψ is a non-decreasing bounded function which satisfies inequality* (2.13) *for some $\alpha > 0$, $A > 0$, and $\varepsilon \in]0, 2^{-\alpha}[$. Then there exists a constant B depending only on ε and α such that*

$$\Psi(R_1) \le \frac{B\,A}{(R_0 - R_1)^\alpha}.$$

Fixing $\varepsilon = 2^{-3ss'}$ in (2.12), applying Lemma 2.1 to our function Ψ, and evaluating ∇p from equations (2.2) held a.e. in Q^+, we derive the estimate (2.5). Theorem 2.2 is proved. □

Theorems 2.2 together with results of [**8**] provide us the following proof of Theorem 2.5.

PROOF OF THEOREM 2.5. Let $\rho_m \to +0$ be an arbitrary sequence. Extend all functions u, p, f from Q^+ to the set $B^+ \times \mathbb{R}$ by zero. For any extended function u denote by u^m the mollification of the function u with respect to the t variable,

$$u^m(x,t) := (\omega_{\rho_m} * u)(x,t) \equiv \int_{\mathbb{R}} \omega_{\rho_m}(t-\tau)u(x,\tau)\,d\tau,$$

where $\omega_\rho(t) = \frac{1}{\rho}\omega(t/\rho)$, and $\omega \in C_0^\infty(-1,1)$ is a smooth kernel normalized by the identity $\int_0^1 \omega(t)dt = 1$.

As $u \in W^{1,0}_{s,l}(Q^+)$, $p \in L_{s,l}(Q^+)$, $f \in L_{s,l}(Q^+)$, we have

$$u^m \to u \quad \text{in } W^{1,0}_{s,l}(Q^+), \quad p^m \to p \quad \text{in } L_{s,l}(Q^+), \qquad f^m \to f \quad \text{in } L_{s,l}(Q^+). \tag{2.14}$$

Let us fix arbitrary $\delta \in]0, \frac{1}{12}[$. Then for any $\rho_m < \delta$ and for any $\eta \in C^\infty(\bar{Q}^+)$

$$\partial_t(\omega_{\rho_m} * \eta)(x,t) = (\omega_{\rho_m} * \partial_t \eta)(x,t), \quad \forall\, x \in B^+, \ t \in]-1+\delta, -\delta[.$$

A weak solution u and p to system (2.2) obeys the integral identity

$$-\int\limits_{Q^+} u \cdot (\partial_t \eta + \Delta \eta)\, dxdt \ = \ \int\limits_{Q^+} (f \cdot \eta + p \operatorname{div} \eta)\, dxdt$$

which holds for all $\eta \in C^\infty(\bar{Q}^+)$ satisfying conditions

(2.15)
$$\eta|_{\partial B^+ \times\,]-1,0[} = 0, \qquad \nabla \eta|_{\partial' B^+ \times\,]-1,0[} = 0, \qquad \eta|_{B^+ \times (]-1,-1+\delta[\cup]-\delta,0[)} = 0,$$

where $\partial' B^+ := \{x \in \mathbb{R}^n : |x| = 1, x_n > 0\}$. Take the test function $\eta = \omega_{\rho_m} * \tilde{\eta}$, where $\tilde{\eta} \in C^\infty(\bar{Q}^+)$ is an arbitrary function satisfying (2.15). Using properties of convolution, we find the identity

$$-\int\limits_{Q^+} u^m \cdot (\partial_t \tilde{\eta} + \Delta \tilde{\eta})\, dxdt \ = \ \int\limits_{Q^+} (f^m \cdot \tilde{\eta} + p^m \operatorname{div} \tilde{\eta})\, dxdt, \tag{2.16}$$

which holds for all $\tilde{\eta} \in C^\infty(\bar{Q}^+)$ satisfying (2.15).

Let $\zeta \in C^\infty(\bar{Q}^+)$ be a cut-off function vanishing in $Q^+ \setminus Q^+(\frac{5}{6})$ and such that $\zeta \equiv 1$ in $Q^+(\frac{2}{3})$. Denote $v^m := \zeta u^m$, $q^m := \zeta p^m$. Then from (2.16) we deduce that v^m and q^m obey the integral identity

$$-\int\limits_{B^+ \times\,]-1,-\delta[} v^m \cdot (\partial_t \eta + \Delta \eta)\, dxdt \ = \int\limits_{B^+ \times\,]-1,-\delta[} (\tilde{f}^m \cdot \eta + q^m \operatorname{div} \eta)\, dxdt$$

for any $\eta \in C^\infty(\bar{B}^+ \times [-1,-\delta])$ such that $\eta|_{\partial B^+ \times\,]-1,-\delta[} = 0$ and $\eta|_{B^+ \times \{t=-\delta\}} = 0$. Here $\tilde{f}^m$ and g^m are determined by formulas (2.9) with u, p, and f replaced by u^m, p^m, and f^m, respectively.

Assume $\Omega \subset \mathbb{R}^3$ is a smooth domain such that $B^+(\frac{5}{6}) \subset \Omega \subset B^+$, and denote $\tilde{Q} := \Omega \times\,]-1,0[$. As functions g^m are smooth with respect to t, we obtain from Theorem 1.1 of [**8**] that, for any $m \in \mathbb{N}$, there exists a strong solution $\tilde{v}^m \in W^{2,1}_{s,l}(\tilde{Q})$, $\tilde{q}^m \in W^{1,0}_{s,l}(\tilde{Q})$ to the problem

$$\left\{ \begin{array}{ll} \partial_t \tilde{v}^m - \Delta \tilde{v}^m + \nabla \tilde{q}^m = \tilde{f}^m, & \\ \operatorname{div} \tilde{v}^m = g^m, & \text{in } \ \tilde{Q}, \\ \tilde{v}^m|_{\partial\Omega \times\,]-1,0[} = 0, \qquad \tilde{v}^m|_{t=-1} = 0. & \end{array} \right. \tag{2.17}$$

Note that $\zeta \equiv 1$ in $Q^+(\frac{2}{3})$, and hence $g^m \equiv 0$ in $Q^+(\frac{2}{3})$. So, functions $\tilde{v}^m$ and $\tilde{q}^m$ satisfy all assumptions of Theorem 2.2 in $Q^+(\frac{2}{3})$ and, hence, by its obvious modification, we have the estimate

$$\begin{aligned} &\|\tilde{v}^m\|_{W^{2,1}_{s,l}(Q^+(\frac{1}{2}))} + \|\nabla \tilde{q}^m\|_{L_{s,l}(Q^+(\frac{1}{2}))} \\ &\quad \leq C \left(\|\tilde{f}^m\|_{L_{s,l}(Q^+(\frac{2}{3}))} + \|\tilde{v}^m\|_{W^{1,0}_{s,l}(Q^+(\frac{2}{3}))} + \|\tilde{q}^m\|_{L_{s,l}(Q^+(\frac{2}{3}))} \right), \end{aligned} \tag{2.18}$$

where a constant C depends neither on m nor on δ.

Since every strong solution of the Stokes system is a weak one, $\tilde{v}^m$ and $\tilde{q}^m$ satisfy the integral identity

$$-\int\limits_{\tilde{Q}} \tilde{v}^m \cdot (\partial_t \eta + \Delta \eta)\, dxdt \;=\; \int\limits_{\tilde{Q}} (\tilde{f}^m \cdot \eta + \tilde{q}^m \operatorname{div} \eta)\, dxdt$$

for all $\eta \in C^\infty(\overline{\tilde{Q}})$ such that $\eta|_{\partial\Omega\times\,]-1,0[} = 0$ and $\eta|_{\Omega\times\{t=0\}} = 0$. Hence, the differences $w^m := v^m - \tilde{v}^m$, $\pi^m := q^m - \tilde{q}^m$ are a weak solution to the Stokes system in $\Omega\times\,]-1,-\delta[$, satisfying the identities

$$\operatorname{div} w^m = 0 \quad \text{a.e. in} \quad \Omega\times\,]-1,-\delta[,$$

$$(2.19)\qquad -\int\limits_{\Omega\times\,]-1,-\delta[} w^m \cdot (\partial_t \eta + \Delta \eta)\, dxdt \;=\; \int\limits_{\Omega\times]-1,-\delta[} \pi^m \operatorname{div} \eta\, dxdt,$$

for any $\eta \in W^{2,1}_{s',l'}(\Omega\times\,]-1,-\delta[)$ such that $\eta|_{\partial\Omega\times\,]-1,-\delta[} = 0$ and $\eta|_{\Omega\times\{t=-\delta\}} = 0$. Denote $\varkappa = \min\{s,l\} > 1$. As u^m, $\tilde{u}^m \in L_{s,l}(\tilde{Q})$ and q^m, $\tilde{q}^m \in L_{s,l}(\tilde{Q})$, we have $w^m = v^m - \tilde{v}^m \in L_\varkappa(\tilde{Q})$ and $\pi^m = q^m - \tilde{q}^m \in L_\varkappa(\tilde{Q}^+)$. Hence $|w^m|^{\varkappa-2}w^m \in L_{\varkappa'}(\tilde{Q})$, and using results of [**37**] we can find functions $\eta \in W^{2,1}_{\varkappa'}(\Omega\times\,]-1,-\delta[)$ and $\kappa \in W^{1,0}_{\varkappa'}(\Omega\times\,]-1,-\delta[)$ such that

$$\begin{cases} \partial_t \eta + \Delta \eta + \nabla \kappa = |w^m|^{\varkappa-2}w^m, & \text{in} \quad \Omega\times\,]-1,-\delta[, \\ \operatorname{div} \eta = 0, & \\ \eta|_{\partial\Omega\times\,]-1,\delta[} = 0, \qquad \eta|_{t=-\delta} = 0. & \end{cases}$$

Substituting this η as a test function into identity (2.19), we obtain $w^m = 0$ in $\Omega\times\,]-1,-\delta[$. Hence $v^m = \tilde{v}^m \in W^{2,1}_{s,l}(\Omega\times\,]-1,-\delta[)$. From (2.19) we obtain

$$(2.20)\qquad \int\limits_{\Omega\times\,]-1,-\delta[} \pi^m \operatorname{div} \eta\, dxdt \;=\; 0 \quad \forall\, \eta \in L_{l'}(-1,-\delta; \overset{\circ}{W}{}^1_{s'}(\Omega)).$$

Correcting, if necessary, the function $\tilde{q}^m$ by a constant, we can assume that

$$\int\limits_{\Omega} \pi^m\, dx = 0$$

for a.e. $t \in\,]-1,-\delta[$. As $\pi^m \in L_\varkappa(\Omega)$ for a.e. $t \in\,]-1,-\delta[$, we have $|\pi^m|^{\varkappa-2}\pi^m \in L_{\varkappa'}(\Omega)$ for a.e. $t \in\,]-1,-\delta[$. Using results of [**2**] for a.e. t, we can find $\eta(\cdot,t) \in \overset{\circ}{W}{}^1_{\varkappa'}(\Omega)$ such that

$$\begin{cases} \operatorname{div} \eta = |\pi^m|^{\varkappa-2}\pi^m - (|\pi^m|^{\varkappa-2}\pi^m)_\Omega, & \text{a.e. } t \in\,]-1,-\delta[, \\ \|\eta\|_{W^1_{\varkappa'}(\Omega)} \le C\|\pi^m\|^{\varkappa-1}_{L_\varkappa(\Omega)}. & \end{cases}$$

From the latter estimate it follows that

$$\eta \in L_{\varkappa'}(-1,-\delta; \overset{\circ}{W}{}^1_{\varkappa'}(\Omega)) \subset L_{l'}(-1,-\delta; \overset{\circ}{W}{}^1_{s'}(\Omega)).$$

Substituting this η into identity (2.20), we obtain $\pi^m = 0$. This implies $q^m = \tilde{q}^m + \text{const}$ and we obtain the inclusion $q^m \in W^{1,0}_{s,l}(\Omega\times\,]-1,-\delta[)$. Moreover, from

(2.18), we find

$$\|v^m\|_{W^{2,1}_{s,l}(B^+(\frac12)\times\,]-\frac14,-\delta[)} + \|\nabla q^m\|_{L_{s,l}(B^+(\frac12)\times\,]-\frac14,-\delta[)}$$
$$\leq C\,\left(\|\tilde f^m\|_{L_{s,l}(Q^+(\frac23))} + \|v^m\|_{W^{1,0}_{s,l}(Q^+(\frac23))} + \|q^m - b\|_{L_{s,l}(Q^+(\frac23))}\right),$$

where C is independent on m and δ. Using identities $v^m = \zeta u^m$, $q^m = \zeta p^m$, $\zeta \equiv 1$ on $Q^+(\frac23)$ and expression (2.9) for $\tilde f^m$, we arrive at the estimate

$$\|u^m\|_{W^{2,1}_{s,l}(B^+(\frac12)\times\,]-\frac14,-\delta[)} + \|\nabla p^m\|_{L_{s,l}(B^+(\frac12)\times\,]-\frac14,-\delta[)}$$
$$\leq C\,\left(\|f^m\|_{L_{s,l}(Q^+(\frac23))} + \|u^m\|_{W^{1,0}_{s,l}(Q^+(\frac23))} + \|p^m\|_{L_{s,l}(Q^+(\frac23))}\right).$$

Making use of (2.14), we conclude that

$$u \in W^{2,1}_{s,l}\left(B^+(\tfrac12)\times\,]-\tfrac14,-\delta[\right),\quad p \in W^{1,0}_{s,l}\left(B^+(\tfrac12)\times\,]-\tfrac14,-\delta[\right),$$

and the estimate

$$\|u\|_{W^{2,1}_{s,l}(B^+(\frac12)\times\,]-\frac14,-\delta[)} + \|\nabla p\|_{L_{s,l}(B^+(\frac12)\times\,]-\frac14,-\delta[)}$$
$$\leq C\,\left(\|f\|_{L_{s,l}(Q^+(\frac23))} + \|u\|_{W^{1,0}_{s,l}(Q^+(\frac23))} + \|p\|_{L_{s,l}(Q^+(\frac23))}\right)$$

holds for any $\delta \in\,]0,\frac{1}{12}[$ with C independent on δ. The last inequality provides the required properties of u and p. Theorem 2.5 is proved. □

Now, we are able to prove Theorem 2.3:

PROOF OF THEOREM 2.3. For any $k = 0, 1, \dots$, denote $s_k = \frac{ns}{n-ks}$ if $n > ks$ and $\frac{ns}{n-ks} < m$ and $s_k = m$ otherwise. Denote also $N = \min\{k \in \mathbb{N} : s_k = m\}$ and $\rho_k = \frac12 + \frac{1}{2^{k+1}}$.

Using obvious modifications of Theorem 2.2 and Theorem 2.5, we observe that if $u \in W^{1,0}_{s_k,l}(Q^+(\rho_k))$ and $p \in L_{s_k,l}(Q^+(\rho_k))$ is a weak solution of problem (2.2) in $Q^+(\rho_k)$, then $u \in W^{2,1}_{s_k,l}(Q^+(\rho_{k+1}))$ and $p \in W^{1,0}_{s_k,l}(Q^+(\rho_{k+1}))$ and the following estimate holds:

$$\begin{aligned}&\|u\|_{W^{2,1}_{s_k,l}(Q^+(\rho_{k+1}))} + \|\nabla p\|_{L_{s_k,l}(Q^+(\rho_{k+1}))}\\ &\qquad\leq C\,\left(\|f\|_{L_{m,l}(Q^+)} + \|u\|_{W^{1,0}_{s_k,l}(Q^+(\rho_k))} + \|p\|_{L_{s_k,l}(Q^+(\rho_k))}\right).\end{aligned}\tag{2.21}$$

Moreover, due to the imbedding $W^1_{s_k}(B^+(\rho_{k+1})) \hookrightarrow L_{s_{k+1}}(B^+(\rho_{k+1}))$, we find the estimate

$$\begin{aligned}&\|u\|_{W^{1,0}_{s_{k+1},l}(Q^+(\rho_{k+1}))} + \|p\|_{L_{s_{k+1},l}(Q^+(\rho_{k+1}))}\\ &\qquad\leq C\,\left(\|u\|_{W^{2,1}_{s_k,l}(Q^+(\rho_{k+1}))} + \|p\|_{W^{1,0}_{s_k,l}(Q^+(\rho_{k+1}))}\right).\end{aligned}\tag{2.22}$$

Iterating (2.21) and (2.22) from $k = 0$ to $k = N$, we finally obtain the bound

$$\|u\|_{W^{2,1}_{s_N,l}(Q^+(\frac12))} + \|\nabla p\|_{L_{s_N,l}(Q^+(\frac12))}$$
$$\leq C^N\,\left(\|f\|_{L_{m,l}(Q^+)} + \|u\|_{W^{1,0}_{s_0,l}(Q^+)} + \|p\|_{L_{s_0,l}(Q^+)}\right).$$

This estimate is equivalent to (2.6). Theorem 2.3 is proved. □

PROOF OF THEOREM 2.4. Theorem 2.4 follows from Theorem 2.3 and the following imbedding theorem for anisotropic Sobolev spaces (see [**1**]):

$$W^{2,1}_{m,l}(Q^+(\tfrac12)) \hookrightarrow C^{\beta,\frac{\beta}{2}}(\bar Q^+(\tfrac12)), \qquad \text{if} \quad m > \tfrac{3l}{2(l-1)} \quad \text{and} \quad \beta = 2 - \tfrac{3}{m} - \tfrac{2}{l}$$
$$\|u\|_{C^{\beta,\frac{\beta}{2}}(\bar Q^+(\frac12))} \le C\|u\|_{W^{2,1}_{m,l}(Q^+(\frac12))} \qquad \forall\, u \in W^{2,1}_{m,l}(Q^+(\tfrac12)).$$

Theorem 2.4 is proved. □

3. Proof of the basic ε-regularity condition

In this section we consider the Navier–Stokes system in a half-cylinder Q^+

$$\left\{\begin{aligned} \partial_t u + (u\cdot\nabla)u - \Delta u + \nabla p = 0,\\ \operatorname{div} u = 0,\\ u|_{x_3=0} = 0.\end{aligned}\right. \qquad \text{in} \quad Q^+, \tag{3.1}$$

The aim of this section is to provide a proof of the following theorem, which is the boundary analogue of ε-regularity condition (1.3):

THEOREM 3.1. *For any given $\alpha \in]0,\frac23[$ there exists a constant $\varepsilon_* > 0$ depending only on α such that for any boundary suitable weak solution u and p in Q^+ subject to the condition*

$$\int\limits_{Q^+} \Big(|u|^3 + |p|^{\frac32} \Big) dxdt \; < \; \varepsilon_*, \tag{3.2}$$

the velocity field $u \in C^{\alpha,\frac{\alpha}{2}}(\bar Q^+(\frac12))$.

We prove Theorem 3.1 following the method developed in [**24**]. This method is based on the indirect approach in the regularity theory (see terminology, for example, in [**9**]), and its crucial step is the decay estimate of Theorem 3.2. For a direct proof of partial regularity in Navier–Stokes theory, we refer to [**25**].

THEOREM 3.2. *For any $\theta \in]0,\frac12[$, $\beta \in]0,\frac23[$, there exists a constant $\varepsilon_0(\theta,\beta) > 0$ such that, for any boundary suitable weak solution u and p to system* (3.1) *in Q^+, the following implication holds:*

$$\textit{if} \qquad Y_1(u,p) < \varepsilon_0, \qquad \textit{then} \qquad Y_\theta(u,p) \le C_*\, \theta^\beta\, Y_1(u,p).$$

Here $C_* > 0$ is some absolute constant.

Here we denote

$$Y_\theta(u,p) := \Big(\fint_{Q^+(\theta)} |u|^3\, dxdt \Big)^{1/3} + \theta \Big(\fint_{Q^+(\theta)} |p - [p]_{B^+(\theta)}|^{3/2}\, dxdt \Big)^{2/3},$$

where for any $f \in L_1(Q^+(\theta))$ we denote

$$\fint\limits_{Q^+(\theta)} f(x,t)\, dxdt = \frac{1}{|Q^+(\theta)|} \int\limits_{Q^+(\theta)} f(x,t)\, dxdt,$$
$$[f]_{B^+(\theta)} = \frac{1}{|B^+(\theta)|} \int\limits_{B^+(\theta)} f(x,t)\, dx.$$

PROOF OF THEOREM 3.2. Arguing by contradiction, we assume there is a number $\theta \in]0, \frac{1}{2}[$, the sequence $\varepsilon_h \to 0$, and functions u^h and p^h, which are the boundary suitable weak solutions in the sense of Definition 1.2 satisfying relations

$$Y_1(u^h, p^h) = \varepsilon_h \to 0, \qquad Y_\theta(u^h, p^h) \geq C_* \theta^\beta \varepsilon_h.$$

We introduce new functions

$$v^h = \frac{1}{\varepsilon_h} u^h, \qquad q^h = \frac{1}{\varepsilon_h}(p^h - [p^h]_{B+}).$$

They meet relations

$$Y_1(v^h, q^h) = 1, \qquad Y_\theta(v^h, q^h) \geq C_* \theta^\beta, \tag{3.3}$$

as well as the system

$$\left\{ \begin{array}{c} \partial_t v^h + \varepsilon_h \operatorname{div}(v^h \otimes v^h) - \Delta v^h + \nabla q^h = 0, \\ \operatorname{div} v^h = 0, \\ v^h|_{x_3=0} = 0, \end{array} \right. \quad \text{in } Q^+, \tag{3.4}$$

which holds in the sense of distributions, and the boundary condition is understood in the sense of traces. Moreover, functions v^h and q^h satisfy the local energy inequality

$$\begin{aligned} &\int_{B+} \zeta(x,t)|v^h(x,t)|^2\, dx + 2\int_{-1}^t \int_{B+} \zeta |\nabla v^h|^2\, dxdt \\ &\leq \int_{-1}^t \int_{B+} \left\{ \, |v^h|^2 \left(\partial_t \zeta + \Delta \zeta\right) + v^h \cdot \nabla \zeta \left(\varepsilon_h |v^h|^2 + 2q^h\right) \, \right\} \, dxdt \end{aligned} \tag{3.5}$$

for a.e. $t \in]-1, 0[$ and all non-negative $\zeta \in C^\infty(\bar{Q})$ which vanish near $\partial' Q$.

From (3.3), we derive the estimate

$$\|v^h\|_{L_3(Q^+)} + \|q^h\|_{L_{\frac{3}{2}}(Q^+)} \leq C. \tag{3.6}$$

Picking up a cut-off function ζ so that $\zeta \equiv 1$ on $Q^+(\frac{3}{4})$ and taking into account (3.6), we find

$$\|v^h\|_{L_{2,\infty}(Q^+(\frac{3}{4}))} + \|v^h\|_{W_2^{1,0}(Q^+(\frac{3}{4}))} \leq C. \tag{3.7}$$

The known multiplicative inequality allows us to conclude that

$$\|v^h\|_{L_{\frac{10}{3}}(Q^+(\frac{3}{4}))} \leq C. \tag{3.8}$$

Another bound easily follows from (3.4) and has the form

$$\|\partial_t v^h\|_{L_{\frac{3}{2}}(-(\frac{3}{4})^2, 0; W_{\frac{3}{2}}^{-1}(B^+(\frac{3}{4})))} \leq C. \tag{3.9}$$

Estimates (3.6) and (3.7) provide the existence of subsequences $\{v^h\}$ and $\{q^h\}$ with the following properties

$$v^h \rightharpoonup v^0 \quad \text{in} \quad W_2^{1,0}(Q^+(\tfrac{3}{4})) \cap L_3(Q^+), \tag{3.10}$$

$$q^h \rightharpoonup q^0 \quad \text{in} \quad L_{\frac{3}{2}}(Q^+). \tag{3.11}$$

Routine compactness arguments imply

$$v^h \to v^0 \quad \text{in} \quad L_3(Q^+(\tfrac{3}{4})). \tag{3.12}$$

Convergence (3.10)–(3.11) allow us to pass to the limit in equations (3.4) (if we take these equations in the weak form). So, v^0 and q^0 is a weak solution to the system

$$\left\{\begin{aligned} \partial_t v^0 - \Delta v^0 + \nabla q^0 = 0, \\ \operatorname{div} v^0 = 0, \\ v^0|_{x_3=0} = 0. \end{aligned}\right. \quad \text{in} \quad Q^+(\tfrac{3}{4}), \tag{3.13}$$

Moreover, from the second relation in (3.3), we deduce the estimate

$$\liminf_{h\to\infty} Y_\theta(v^h, q^h) \geq C_* \theta^\beta. \tag{3.14}$$

On the other hand, below we will show that

$$\limsup_{h\to\infty} Y_\theta(v^h, q^h) \leq C_{**} \theta^\beta, \tag{3.15}$$

with a constant $C_{**} > 0$. Taking in (3.14) a constant $C_* > C_{**}$, we arrive at a contradiction between (3.14) and (3.15). This will complete our proof Theorem 3.2.

To prove (3.15), we split $Y_\theta(v^h, q^h)$ into two parts:

$$Y_\theta(v^h, q^h) = Y_\theta^1(v^h) + Y_\theta^2(q^h),$$

where

$$Y_\theta^1(v^h) \equiv \Big(\fint_{Q^+(\theta)} |v^h|^3 \, dxdt \Big)^{\frac{1}{3}}, \qquad Y_\theta^2(q^h) \equiv \theta \Big(\fint_{Q^+(\theta)} |q^h|^{\frac{3}{2}} \, dxdt \Big)^{\frac{2}{3}}.$$

As $\theta \in]0, \frac{1}{2}[$, the strong convergence (3.12) gives us

$$\lim_{h\to\infty} Y_\theta^1(v^h) = Y_\theta^1(v^0). \tag{3.16}$$

Since $v^0 \in W_2^{1,0}(Q^+(\frac{3}{4}))$, $q^0 \in L_{\frac{3}{2}}(Q^+(\frac{3}{4}))$ are a weak solution to the Stokes system (3.13) in $Q^+(\frac{3}{4})$ (in the sense of Definition 2.1), one can apply Theorems 2.5 and 2.4 to find that $v^0 \in C^{\beta,\frac{\beta}{2}}(\bar{Q}^+(\frac{1}{2}))$ with the estimate

$$\|v^0\|_{C^{\beta,\frac{\beta}{2}}(\bar{Q}^+(\frac{1}{2}))} \leq C \left(\|\nabla v^0\|_{L_{\frac{3}{2}}(Q^+(\frac{3}{4}))} + \|q^0\|_{L_{\frac{3}{2}}(Q^+)} \right). \tag{3.17}$$

Thanks to estimates (3.6), (3.7), and the lower semicontinuity of the corresponding norms with respect to weak convergence (3.10), (3.11), the right-hand side of (3.17) can be estimated by some absolute constant C. As $v^0|_{x_3=0} = 0$, relation (3.17) yields the estimate

$$Y_\theta^1(v^0) \leq C\theta^\beta \|v^0\|_{C^{\beta,\frac{\beta}{2}}(\bar{Q}^+(\frac{1}{2}))} \leq C_0 \theta^\beta. \tag{3.18}$$

Combining (3.16) and (3.18), we find the estimate

$$\limsup_{h\to\infty} Y_\theta(v^h, q^h) \leq C_0 \theta^\beta + \limsup_{h\to\infty} Y_\theta^2(q^h). \tag{3.19}$$

So, to show (3.15), we need to estimate the second term in the right-hand side of (3.19). For this purpose, let us define

$$f^h = -\varepsilon_h \operatorname{div}(v^h \otimes v^h) \quad \text{in} \quad Q^+(\tfrac{3}{4}).$$

From the Hölder inequality, it follows that $f^h \in L_{\frac98,\frac32}(Q^+(\frac34))$, and using (3.7), we obtain

(3.20)
$$\|f^h\|_{L_{\frac98,\frac32}(Q^+(\frac34))} \leq C\,\varepsilon_h\,\|v^h\|^{\frac23}_{L_{2,\infty}(Q^+(\frac34))}\|\nabla v^h\|^{\frac43}_{L_2(Q^+(\frac34))} \to 0 \quad \text{as} \quad h\to\infty.$$

Assume $\Omega \subset \mathbb{R}^3$ is a smooth canonical domain such that $B^+(\frac58)\subset\Omega\subset B^+(\frac34)$, and denote $\tilde Q := \Omega\times\,]-\frac{9}{16},0[$. It is known (see, for example, [**39**]) that there exist a unique pair of functions $\hat v^h \in W^{2,1}_{\frac98,\frac32}(\tilde Q)$, $q_1^h \in W^{1,0}_{\frac98,\frac32}(\tilde Q)$, $[q_1^h]_\Omega = 0$ a.e. $t\in\,]-\frac{9}{16},0[$, which obey the following initial boundary value problem

$$\begin{cases} \partial_t \hat v^h - \Delta \hat v^h + \nabla q_1^h = f^h, & \\ \qquad \operatorname{div}\hat v^h = 0, & \text{in} \quad \tilde Q, \\ \hat v^h|_{t=-\frac{9}{16}} = 0, \qquad \hat v^h|_{\partial\Omega\times\,]-\frac{9}{16},0[} = 0, & \end{cases}$$

and is subject to the estimate

$$\|\hat v^h\|_{W^{2,1}_{\frac98,\frac32}(\tilde Q)} + \|q_1^h\|_{W^{1,0}_{\frac98,\frac32}(\tilde Q)} \leq C\|f^h\|_{L_{\frac98,\frac32}(\tilde Q)}. \tag{3.21}$$

From (3.20), (3.21), and the imbedding $W^1_{\frac98}(\Omega)\hookrightarrow L_{\frac32}(\Omega)$, we can conclude

$$\begin{aligned} Y_\theta^2(q_1^h) &\leq C\theta^{-2}\|\nabla q_1^h\|_{L_{\frac98,\frac32}(\tilde Q)} \\ &\leq C\,\theta^{-2}\|f^h\|_{L_{\frac98,\frac32}(Q^+(\frac34))} \to 0 \quad \text{as} \quad h\to\infty. \end{aligned} \tag{3.22}$$

Now, consider the functions $\tilde v^h \equiv v^h - \hat v^h$, $q_2^h = q^h - q_1^h$. Note that $\tilde v^h \in W^{1,0}_{\frac98,\frac32}(\tilde Q)$, $q_2^h \in L_{\frac98,\frac32}(\tilde Q)$ and, hence, $\tilde v^h$ and q_2^h is a weak solution (in the sense of Definition 2.1) of the homogeneous Stokes system in $\tilde Q$

$$\begin{cases} \partial_t \tilde v^h - \Delta \tilde v^h + \nabla q_2^h = 0, & \\ \qquad \operatorname{div}\tilde v^h = 0, & \text{in} \quad \tilde Q, \\ \qquad \tilde v^h|_{x_3=0} = 0. & \end{cases}$$

Let us take $m := \frac{9}{2-3\beta}$, $m\in\,]\frac92,+\infty[$. By an obvious modification of Theorem 2.3 with $f\equiv 0$, we obtain inclusions $\tilde v^h \in W^{2,1}_{m,\frac32}(Q^+(\frac12))$, $q_2^h \in W^{2,1}_{m,\frac32}(Q^+(\frac12))$, and the estimate

$$\|\nabla q_2^h\|_{L_{m,\frac32}(Q^+(\frac12))} \leq C\left(\|\tilde v^h\|_{W^{1,0}_{\frac98,\frac32}(\tilde Q)} + \|q_2^h\|_{L_{\frac98,\frac32}(\tilde Q)}\right).$$

The right-hand side of the last inequality can be controlled by majorants which do not depend on h. Indeed, as $\tilde v^h = v^h - \hat v^h$, from (3.7), (3.21), we find

$$\|\nabla \tilde v^h\|_{L_{\frac98,\frac32}(Q^+(\frac34))} \leq \|\nabla v^h\|_{L_2(Q^+(\frac34))} + \|\nabla \hat v^h\|_{L_{\frac98,\frac32}(Q^+(\frac34))} \leq C.$$

As $q_2^h = q^h - q_1^h$, using (3.6), (3.20), and (3.21), we get

$$\|q_2^h\|_{L_{\frac98,\frac32}(\tilde Q)} \leq C\left(\|q^h\|_{L_{\frac32}(Q^+)} + \|q_1^h\|_{L_{\frac98,\frac32}(\tilde Q)}\right) \leq C\left(1+\|f^h\|_{L_{\frac98,\frac32}(Q^+(\frac34))}\right) \leq C.$$

Hence, it has been shown that

$$\|\nabla q_2^h\|_{L_{m,\frac32}(Q^+(\frac12))} \leq C. \tag{3.23}$$

Using (3.23) and the Poincaré and Hölder inequalities, we find

$$Y_\theta^2(q_2^h) \le C\theta^\beta \|\nabla q_2^h\|_{L_{m,\frac{3}{2}}(Q^+(\frac{1}{2}))} \le C_1\theta^\beta. \tag{3.24}$$

Combining (3.24) with (3.22), we show

$$\limsup_{h\to\infty} Y_\theta^2(q^h) \le \limsup_{h\to\infty} Y_\theta^2(q_1^h) + \limsup_{h\to\infty} Y_\theta^2(q_2^h) \le 0 + C_1\theta^\beta = C_1\theta^\beta.$$

Hence from (3.19) we obtain (3.15) with $C_{**} = C_0 + C_1$, which contradicts (3.14) if we take $C_* > C_{**}$. Theorem 3.2 is proved. □

THEOREM 3.3. *Let $C_* > 1$ be the absolute constant defined by Theorem 3.2. Assume $\beta \in \,]0, \frac{2}{3}[$, $\theta \in \,]0, \frac{1}{2}[$ are arbitrary. Denote by $\varepsilon_0 > 0$, $\varepsilon_0 = \varepsilon_0(\theta,\beta)$ a constant determined by Theorem 3.2. Then, for any boundary suitable weak solution u and p to the Navier–Stokes system in Q^+ and for any $k = 0, 1, 2, \ldots$, the following is true:*

$$\text{if} \quad Y_{\theta^k}(u,p) < \varepsilon_0, \qquad \text{then } Y_{\theta^{k+1}}(u,p) \le C_*\theta^\beta \, Y_{\theta^k}(u,p). \tag{3.25}$$

PROOF. Define functions u^{θ^k} and p^{θ^k} by formulas

$$\begin{aligned} u^{\theta^k}(x,t) &:= \theta^k u(\theta^k x, \theta^{2k} t), \\ p^{\theta^k}(x,t) &:= \theta^{2k} p(\theta^k x, \theta^{2k} t), \end{aligned} \qquad (x,t) \in Q^+.$$

As u and p are a boundary suitable weak solution of the Navier–Stokes equations in $Q^+(\theta^k)$, the functions u^{θ^k} and p^{θ^k} are a boundary suitable weak solution of the Navier–Stokes system in Q^+. Moreover,

$$Y_1(u^{\theta^k}, p^{\theta^k}) \;=\; Y_{\theta^k}(u,p) \;<\; \varepsilon_0,$$

and, hence, by Theorem 3.2,

$$Y_\theta(u^{\theta^k}, p^{\theta^k}) \;\le\; C_*\theta^\alpha \; Y_1(u^{\theta^k}, p^{\theta^k}).$$

From this inequality, the conclusion of the implication (3.25) follows by change of variables. Theorem 3.3 is proved. □

THEOREM 3.4. *Let $C_* > 1$ be the absolute constant defined by Theorem 3.2, and let $\alpha \in \,]0, \frac{2}{3}[$, $\beta \in \,]\alpha, \frac{2}{3}[$ be arbitrary. Assume a number $\theta \in \,]0, \frac{1}{2}[$ is fixed in such a way that*

$$C_*\theta^{\beta-\alpha} < 1, \tag{3.26}$$

and let $\varepsilon > 0$, $\varepsilon_0 = \varepsilon_0(\theta,\beta)$ be a constant determined by Theorem 3.2. Then, for any boundary suitable weak solution u and p of the Navier–Stokes system in Q^+ and for any $k = 0, 1, 2, \ldots$, the following is valid:

$$\text{if } Y_1(u,p) < \varepsilon_0, \quad \text{then for any } k = 0,1,2,\ldots \quad \left\{ \begin{array}{c} Y_{\theta^k}(u,p) < \varepsilon_0, \\ Y_{\theta^{k+1}}(u,p) < \theta^{\alpha(k+1)} Y_1(u,p). \end{array} \right.$$

PROOF. The proof follows easily from Theorem 3.3 by induction in k. □

THEOREM 3.5. *Assume $\alpha \in \,]0, \frac{2}{3}[$ is arbitrary, and take $\beta = \frac{1}{2}(\alpha + \frac{2}{3})$. Let us fix $\theta \in \,]0, \frac{1}{2}[$, $\theta = \theta(\alpha)$ so that (3.26) holds, and let $\varepsilon_0 > 0$, $\varepsilon_0 = \varepsilon_0(\theta,\beta)$ be a constant determined by Theorem 3.2. Denote $\varepsilon'_* := \varepsilon_0(\theta,\beta)$, and note that $\varepsilon'_* > 0$ actually depends only on α. Then for any $z_0 = (x_0,t_0)$, $x_0 \in \partial\mathbb{R}^3_+$, and for any boundary* *suitable weak solution u and p of the Navier–Stokes equations in $Q^+(z_0,R)$, the following is valid:*

$$\text{if } R\, Y_{z_0,R}(u,p) < \varepsilon'_*,$$

$$\text{then for any } 0<r<R \qquad Y_{z_0,r}(u,p) \;\le\; C(\alpha)\left(\frac{r}{R}\right)^{\alpha} Y_{z_0,R}(u,p).$$

Here,

$$Y_{z_0,R}(u,p) := \Big(\fint_{Q^+(z_0,R)} |u|^3\,dxdt\Big)^{\frac13} + R\Big(\fint_{Q^+(z_0,R)} |p-[p]_{B^+(x_0,R)}|^{\frac32}\,dxdt\Big)^{\frac23}.$$

Proof. The proof follows easily from Theorem 3.4 by the scaling transformation (1.1), if we fix $k\in\mathbb{N}\cup\{0\}$ in such a way that $\theta^{k+1}R\le r<\theta^k R$. □

Proof of Theorem 3.1. Let

$$\begin{aligned}\bar Y_{z_0,R}(u,p) := \Big(\fint_{Q(z_0,R)\cap Q^+} |u-(u)_{Q(z_0,R)\cap Q^+}|^3\,dxdt\Big)^{\frac13}\\ + R\Big(\fint_{Q(z_0,R)\cap Q^+} |p-[p]_{B(x_0,R)\cap B^+}|^{\frac32}\,dxdt\Big)^{\frac23},\end{aligned}$$

where $(u)_{Q(z_0,R)\cap Q^+}$ denotes the space-time average of u over the set $Q(z_0,R)\cap Q^+$, and $[p]_{B(x_0,R)\cap B^+}$ denotes the spatial average of p over the set $B(x_0,R)\cap B^+$.

For any $x_0\in\mathbb{R}^3_+$, $x_0=(x_1^0,x_2^0,x_3^0)$, $z_0=(x_0,t_0)$, we denote by x_0' the point with coordinates $(x_1^0,x_2^0,0)$, $z_0':=(x_0',t_0)$, and $d(x_0):=\operatorname{dist}\{x_0,\partial\mathbb{R}^3_+\}=x_3^0$.

From the internal regularity theory (see, for example, [**28**]), we know there is a constant $\varepsilon''_*>0$ depending only on α such that for any $z_0\in Q^+$, $R>0$ the following implication holds:

$$\begin{aligned}&Q(z_0,R)\subset Q^+,\ R\,\bar Y_{z_0,R}(u,p)<\varepsilon''_* \implies\\ &\implies \bar Y_{z_0,r}(u,p)\le c\left(\frac{r}{R}\right)^{\alpha}\bar Y_{z_0,R}(u,p).\end{aligned}\tag{3.27}$$

On the other hand, Theorem 3.5 reads that there is a constant $\varepsilon'_*>0$ depending only on α such that for any $z_0'\in\partial\mathbb{R}^3_+$, $R>0$, we have

$$\begin{aligned}&Q^+(z_0',R)\subset Q^+,\ R\,Y_{z_0',R}(u,p)<\varepsilon'_* \implies\\ &\implies Y_{z_0',r}(u,p)\le c\left(\frac{r}{R}\right)^{\alpha} Y_{z_0',R}(u,p).\end{aligned}\tag{3.28}$$

Besides, it is easy to see that there is an absolute constant c_1 such that for any $z_0\in B^+(\frac12)$, $z_0'\in\bar B^+(\frac12)\cap\partial\mathbb{R}^3_+$,

$$\bar Y_{z_0,\frac14}(u,p)\le c_1 Y_1(u,p),\qquad Y_{z_0',\frac14}(u,p)\le c_1Y_1(u,p)\tag{3.29}$$

Assume $z_0=(x_0,t_0)\in\bar Q^+(\frac12)$ is arbitrary and $0<r<\frac18$. Denote $d:=d(x_0)$. There are three possible cases:

Case 1. $0\le d<r<\frac18$.

Case 2. $0<r\le d<\frac18$.

Case 3. $\frac18\le d\le\frac12$.

In Case 1 we have $\bar Y_{z_0,r}(u,p)\le 2\,Y_{z_0',2r}(u,p)$. Let us fix $\varepsilon_1:=\frac{\varepsilon'_*}{c_1}$, where c_1 is fixed in (3.29). Then if we assume $Y_1(u,p)<\varepsilon_1$ from (3.29), we obtain

$Y_{z_0',\frac14}(u,p) < \varepsilon_*'$, and hence with the help of (3.28) we obtain

$$\bar Y_{z_0,r}(u,p) \;\le\; 2\,Y_{z_0',2r}(u,p) \;\le\; 2\,c\,\left(\frac{2r}{\frac14}\right)^\alpha\, Y_{z_0',\frac14}(u,p) \;\le\; c\varepsilon_*' r^\alpha.$$

In Case 2 we have $\bar Y_{z_0,d}(u,p) \;\le\; 2\,Y_{z_0',2d}(u,p)$. If we assume $Y_1(u,p) < \varepsilon_1$, then (3.29) yields $Y_{z_0',\frac14}(u,p) < \varepsilon_*'$, and hence with the help of (3.28), (3.29) we state that

$$\bar Y_{z_0,d}(u,p) \;\le\; 2\,Y_{z_0',2d}(u,p) \;\le\; 2\,c\,\left(\frac{2d}{\frac14}\right)^\alpha\, Y_{z_0',\frac14}(u,p) \;\le\; c_2 Y_1(u,p) d^\alpha.$$

Taking into account that $d < \frac18$, we can conclude that $d\,\bar Y_{z_0,d}(u,p) \le c_2 Y_1(u,p)$. Hence, if we fix $\varepsilon_2 := \min\{\frac{\varepsilon_*'}{c_1}, \frac{\varepsilon_*''}{c_2}\}$ and assume $Y_1(u,p) < \varepsilon_2$, we can apply (3.27) with $R = d$ and show

$$\bar Y_{z_0,r}(u,p) \;\le\; c\,\left(\frac{r}{d}\right)^\alpha\, \bar Y_{z_0,d}(u,p) \;\le\; c\,\varepsilon_*''\,r^\alpha.$$

In Case 3 we have $d\,\bar Y_{z_0,d}(u,p) \le c_3 Y_1(u,p)$ with an absolute constant $c_3 > 0$. Hence if we take $\varepsilon_3 := \frac{\varepsilon_*''}{c_3}$ and assume $Y_1(u,p) < \varepsilon_3$, we observe that $d\,Y_{z_0,d}(u,p) < \varepsilon_*''$, and hence we can apply (3.27) with $R = d$. Taking into account $\frac18 \le d \le \frac12$, we find

$$\bar Y_{z_0,r}(u,p) \;\le\; c\,\left(\frac{r}{d}\right)^\alpha\, \bar Y_{z_0,d}(u,p) \le c\,\varepsilon_*''\,r^\alpha.$$

Finally, if we fix $\varepsilon_* := \min\{\varepsilon_1,\varepsilon_2,\varepsilon_3\}$ and assume $Y_1(u,p) < \varepsilon_*$, then in all cases we get the estimate

$$\forall\, z_0 \in \bar Q^+(\tfrac12), \quad \forall\, r \in (0,\tfrac18) \qquad \bar Y_{z_0,r}(u,p) \le K r^\alpha,$$

with some $K > 0$ depending only on α, ε_*' and ε_*''. From this estimate, we deduce Hölder continuity of u in the set $\bar Q^+(\frac12)$ via the Campanato criterion; see, for example, [**3**]. Theorem 3.1 is proved. □

4. Further results

In this section we discuss further results of ε-regularity theory for the Navier–Stokes system. We start with the Navier–Stokes equations in the neighborhood of a point x_0 belonging to the smooth curvilinear part of the boundary $\partial\Omega$ of a domain $\Omega \subset \mathbb{R}^3$. Namely, assume $x_0 \in \partial\Omega$, denote by $\Omega(x_0,R)$ the intersection of some neighborhood of x_0 with Ω, and consider the system

$$\text{(4.1)}\qquad \left\{\begin{array}{l} \partial_t u - \Delta u + (u\cdot\nabla)u + \nabla p \;=\; f, \\ \operatorname{div} u = 0, \\ u|_{\partial\Omega\times]-R^2,0[} = 0. \end{array}\right. \qquad \text{in}\quad \Omega(x_0,R)\times\,]-R^2,0[,$$

Without loss of generality we can assume that our Cartesian coordinate system is chosen in such a way that x_0 coincides with its origin (i.e., $x_0 = 0$), and the set $\Omega(x_0,R)$ is described by relations

$$\text{(4.2)}\qquad \Omega(x_0,R) \;=\; \Big\{\, x = (x_1,x_2,x_3) \in \mathbb{R}^3 \mid x' \in S_R, \\ \varphi(x') < x_3 < \varphi(x') + \sqrt{R^2 - |x'|^2}\,\Big\}.$$

Here we denote $x' := (x_1, x_2)$ and $S_R := \{ x' \in \mathbb{R}^2 \mid \sqrt{x_1^2 + x_2^2} < R \}$. With this assumption the boundary condition in (4.1) is equivalent to the relation

$$u|_{x_3 = \varphi(x')} = 0.$$

We assume φ is of class W^3_∞ (i.e., its second derivatives are Lipschitz continuous) and the Cartesian coordinate system is chosen in such a way that the following relations hold:

$$\varphi(0) = 0, \qquad \nabla\varphi(0) = 0, \qquad \|\varphi\|_{W^3_\infty(S_R)} \le \mu. \tag{4.3}$$

Now we apply the diffeomorphism flattering the boundary, or, in other words, we introduce new coordinates $y = \psi(x)$ by formulas

$$\psi : \Omega_R \to B_R^+, \qquad y = \psi(x) \;=\; \begin{pmatrix} x' \\ x_3 - \varphi(x') \end{pmatrix}, \tag{4.4}$$

$$x \in \Omega(x_0, R) \quad \Longleftrightarrow \quad y \in B_R^+.$$

Denote

$$v := u \circ \psi^{-1}, \qquad q := p \circ \psi^{-1}, \qquad \tilde{f} := f \circ \psi^{-1}.$$

Then for $y = \psi(x)$, we have relations

$$\nabla p(x) = \hat{\nabla}_\varphi q(y), \quad \Delta u(x) = \hat{\Delta}_\varphi v(y), \quad \operatorname{div} u(x) = (\hat{\nabla}_\varphi \cdot v)(y),$$

where $\hat{\Delta}_\varphi$ and $\hat{\nabla}_\varphi$ are the differential operators with variable coefficients defined via a function φ by the formulas

$$\begin{aligned} \hat{\Delta}_\varphi v \;&:=\; \Delta v - 2v_{,\alpha 3}\varphi_{,\alpha} + v_{,33}|\nabla'\varphi|^2 - v_{,3}\Delta'\varphi, \\ \hat{\nabla}_\varphi \cdot v \;&:=\; \operatorname{div} v - v_{\alpha,3}\varphi_{,\alpha}, \\ \hat{\nabla}_\varphi q \;&:=\; \nabla q - q_{,3}\begin{pmatrix} \nabla'\varphi \\ 0 \end{pmatrix}. \end{aligned}$$

Here we assume summation from 1 to 2 over repeated Greek indexes, and ∇' and Δ' denote the gradient and Laplacian with respect to (y_1, y_2) variables.

The Navier–Stokes system (4.1) in $\Omega(x_0, R) \times (-R^2, 0)$ in x-variables is transformed to the following system with variable coefficients depending on the y-variables:

$$\left\{ \begin{aligned} \partial_t v \;-\; \hat{\Delta}_\varphi v \;+\; (v \cdot \hat{\nabla}_\varphi)v + \; \hat{\nabla}_\varphi q \;&=\; \tilde{f}, \\ \hat{\nabla}_\varphi \cdot v \;&=\; 0, \\ v|_{y_3 = 0} &= 0. \end{aligned} \right. \qquad \text{in} \quad Q^+(R), \tag{4.5}$$

We call this system the *Perturbed Navier–Stokes system.* Note that if the boundary $\partial\Omega$ is smooth in the neighborhood of x_0, then the coefficients of system (4.5) are also smooth.

The Perturbed Navier–Stokes system possesses the following scaling property: if functions v, q, f, φ satisfy (4.5) in the cylinder $Q^+(R)$ with φ satisfying (4.3), then the functions

$$\begin{aligned} v^R(x,t) &= Rv(Rx, R^2t), \qquad & q^R(x,t) &= R^2 q(Rx, R^2t), \\ f^R(x,t) &= R^3 f(Rx, R^2t), \qquad & \varphi^R(x') &= \frac{1}{R}\varphi(Rx') \end{aligned} \tag{4.6}$$

satisfy the Perturbed Navier–Stokes system (4.5) in Q^+, and from Taylor decomposition of the function φ^R, one can obtain for $R \le 1$

$$\varphi^R(0) = 0, \qquad \nabla'\varphi^R(0) = 0, \qquad \|\varphi^R\|_{W^3_\infty(S_1)} \le \mu R. \tag{4.7}$$

Hence, if assumptions (4.3) hold with some constant μ whose value can be arbitrarily large (in particular, this implies that the curvature of the boundary of Ω in the neighborhood of $x_0 \in \partial\Omega$ can be arbitrary), then applying diffeomorphism (4.4) and the scaling transformation (4.6), we can reduce the study of regularity of weak solutions to the Navier–Stokes system (4.1) in domain $\Omega(x_0, R) \times]-R^2, 0[$ to the study of the Perturbed Navier–Stokes system (4.5) in the canonical domain Q^+. Moreover, for any given $\mu_* > 0$, choosing the radius $R = R(\mu_*) > 0$ sufficiently small thanks to (4.7), we can assume that condition

$$\mu R \le \mu_* \tag{4.8}$$

holds with some absolute constant $\mu_* > 0$. If the value μ_* in (4.8) is chosen sufficiently small, then variable coefficients in the Perturbed Navier–Stokes system (4.5) for functions v^R, q^R, f^R, φ^R in Q^+ can be interpreted as small perturbations of the "constant coefficients" in the usual Navier–Stokes system (3.1) for functions v^R, q^R, f^R in Q^+.

The linear theory of strong solutions to the Stokes system extending results of [**23**] to the case of the curvilinear boundary was developed in [**38**]. Later, in [**36**] a similar theory was developed for the linear Perturbed Stokes system, which is the linearization of (4.5). Moreover, in contrast to [**38**], in [**36**] the local estimates were obtained not for strong but for weak solutions. In particular, the analogs of our Theorems 2.2–2.5 were proved in [**36**] for the Perturbed Stokes system under the assumption that its coefficients are small perturbations of the constant coefficients of the usual Stokes system.

The linear theory developed in [**38**] and [**36**] allows us to prove the analogue of the basic ε-regularity condition (3.2) at the neighborhood of point x_0 belonging to a curvilinear part of the boundary. To formulate a result, we need to define boundary suitable weak solution to the Navier–Stokes equation near a curvilinear part of the boundary. The definition of this class of solutions is analogous to one given in our Definition 1.2 for the case of flat boundaries; see details in [**32**]. So, we obtain the following result:

THEOREM 4.1. *Let $\Omega(x_0, R)$ be defined by (4.2), where $\varphi \in W^3_\infty(S_R)$ satisfies (4.3), and assume $z_0 = (x_0, t_0)$. There exist an absolute constant $\varepsilon_* > 0$ and a constant $R_* \in (0, R)$ depending only on μ and R such that for any boundary suitable weak solution u and p of the Navier–Stokes equations (4.1) in $\Omega(x_0, R) \times]t_0 - R^2, t_0[$ the following is true: if there exists $r \le R_*$ such that*

$$\frac{1}{r^2} \int\limits_{t_0 - r^2}^{t_0} \int\limits_{\Omega(x_0, r)} \left(|u|^3 + |p|^{\frac{3}{2}}\right)\, dxdt < \varepsilon_*, \tag{4.9}$$

then u is Hölder continuous on $\bar{\Omega}(x_0, \frac{r}{2}) \times [t_0 - \frac{r^2}{4}, t_0]$.

Theorem 4.1 is proved in [**32**]. Using this theorem one can obtain the following result, which is a boundary version of the Caffarelli–Kohn–Nirenberg theorem (see [**4**]):

THEOREM 4.2. *Let all assumptions of Theorem 4.1 be satisfied. There exists an absolute constant* $\varepsilon_{**} > 0$ *such that, for any boundary suitable weak solution* u *and* p *of the Navier–Stokes equations* (4.1) *in* $\Omega(x_0, R) \times]t_0 - R^2, t_0[$*, the velocity field* u *is Hölder continuous in some neighborhood of* z_0 *provided*

$$\limsup_{r\to 0} \frac{1}{r}\int_{t_0-r^2}^{t_0}\int_{\Omega(x_0,r)} |\nabla u|^2\, dxdt \; < \; \varepsilon_{**}. \tag{4.10}$$

Theorem 4.2 provides the following estimate of the parabolic Hausdorff measure of a singular set:

THEOREM 4.3. *Let* $\Omega \subset \mathbb{R}^3$ *be a domain whose boundary* $\partial\Omega$ *is of class* W^3_∞*, and assume there is a constant* $\mu > 0$ *such that for any* $x_0 \in \partial\Omega$ *there exists a neighborhood* $\Omega(x_0, R)$ *which can be described in an appropriate Cartesian coordinate system by formulas* (4.2) *with some function* $\varphi = \varphi_{x_0}$ *satisfying conditions* (4.3)*. Then for any boundary suitable weak solution* u *and* p *of the Navier–Stokes system in* $\Omega\times\,]0, T[$*, there exists a closed set* $\Sigma \subset \partial\Omega\times\,]0, T]$ *such that for any point* $z_0 \in (\partial\Omega \times]0, T]) \setminus \Sigma$*, the function* u *is Hölder continuous in some neighborhood of* z_0 *and, moreover,*

$$\mathcal{P}^1(\Sigma) = 0,$$

where $\mathcal{P}^1(\Sigma)$ *is the one-dimensional parabolic Hausdorff measure of* Σ.

It is worthy to notice that there is an essential difference between ε-regularity conditions (4.9) and (4.10). Condition (4.10) requires smallness of functional

$$E(u, r) = \Big(\frac{1}{r}\int_{t_0-r^2}^{t_0}\int_{\Omega(x_0,r)} |\nabla u|^2\, dydt\Big)^{1/2}$$

uniformly with respect to all $r \in (0, R_1)$ with some $R_1 > 0$; i.e.,

$$\limsup_{r\to 0} E(u, r) < \varepsilon_0. \tag{4.11}$$

The question is whether (4.11) could be weakened to

$$\liminf_{r\to 0} E(u, r) < \varepsilon_0. \tag{4.12}$$

In general, an answer to this question is unknown. But if we assume *a priori* boundedness of scale invariant functionals of type (1.5), then it is possible to improve most of ε-regularity conditions, replacing in (4.11) lim sup with lim inf. Namely, the boundary suitable weak solution u and p to the Navier–Stokes system in Q^+ is regular near the origin if (1.6) holds, and besides one of the following conditions is valid:

- $\min\Big\{ \liminf\limits_{r\to 0} A(u, r),\ \liminf\limits_{r\to 0} C(u, r),\ \liminf\limits_{r\to 0} E(u, r),\ \liminf\limits_{r\to 0} D(p, r) \Big\} < \varepsilon_0$
- $\liminf\limits_{r\to 0} C_2(u, r) < \varepsilon_0$, where

$$C_2(u, r) := \Big(\frac{1}{r^3} \int\limits_{Q^+(r)} |u(x, t)|^2\, dxdt \Big)^{1/2}.$$

This statement has been proven in [**28**] in the internal case. Later on, it was extended to the boundary case in [**19**] and [**20**].

References

[1] O.V. Besov, V.P. Il'in, S.M. Nikolskii, Integral representations of functions and imbedding theorems. Nauka, Moscow, 1975. Translation: Wiley&Sons, 1978. MR0430771 (55:3776)

[2] M.E. Bogovskii, *On solution of some problems of vectoral analysis related to* div *and* grad *operators*, Proc. of S.L. Sobolev Seminar **1** (1980), 5-40. MR631691 (82m:26014)

[3] S. Campanato, *Proprieta di Hölderianita di alcune classi di funzioni*, Ann. Scuola Norm. Sup. Pisa **17 (3)** (1963), 175-188. MR0156188 (27:6119)

[4] L. Caffarelli, R.V. Kohn, L. Nirenberg, *Partial regularity of suitable weak solutions of the Navier-Stokes equations*, Comm. Pure Appl. Math. **35** (1982), 771-831. MR673830 (84m:35097)

[5] L. Escauriaza, G.A. Seregin, V. Sverak, *On backward uniquness for parabolic equations*, Arch. Rational Mech. Anal., **169** (2003), 145-157. MR2005639 (2005j:35097)

[6] L. Escauriaza, G.A. Seregin, V. Sverak, *$L_{3,\infty}$-solutions to the Navier-Stokes equations and backward uniqueness*, Uspekhi Matematicheskih Nauk, **58** (2003) no. 2 (350), 3-44. English translation in Russian Mathematical Surveys, **58** (2003) no. 2, 211-250. MR1992563 (2004m:35204)

[7] L. Escauriaza, G.A. Seregin, V. Sverak, *Backward uniquness for the heat operator in a half-space*, St.-Petersburg Mathematical Journal, **15** (2004), no. 1, 139-148. MR1979722 (2004c:35153)

[8] N.D. Filonov, T.N. Shilkin, *On the Stokes problem with non-zero divergence*, Zap. Nauchn. Semin. POMI **370** (2009), 184-202.

[9] M. Giaquinta, Multiple integrals in the calculus of variations and nonlinear elliptic systems, Princeton, 1983. MR717034 (86b:49003)

[10] S. Gustafson, K. Kang, T.-P. Tsai, *Interiour regularity criteria for suitable weak solutions to he Navier-Stokes equations*, Commun. Math. Phys., **273** (2007), 161-176. MR2308753 (2008j:35142)

[11] E. Hopf, *Uber die Anfangswertaufgabe fur die hydrodynamischen Grundgleichungen.* Math. Nachr. **4**, (1950), 213-231. MR0050423 (14:327b)

[12] K. Kang, *On boundary regularity of the Navier-Stokes equations*, Communications in Partial Differential Equations, **29** (7-8) (2004), 955-987. MR2097573 (2005j:35168)

[13] K. Kang, *Unbounded normal derivative for the Stokes system near boundary*, Mathemtische Annalen **331** (2005), 87-109. MR2107441 (2005h:35271)

[14] T. Kato, *Strong L_p-solutions of the Navier-Stokes equations in $\mathbb{R}^n$, with applications to weak solutions*, Math. Z., **197** (1984), 471-480. MR760047 (86b:35171)

[15] A.A. Kiselev, O.A. Ladyzhenskaya, *On the existence and uniqueness of the solution of the nonstationary problem for a viscous, incompressible fluid*, Izv. Akad. Nauk SSSR Ser. Mat., **21:5** (1957), 655680. MR0100448 (20:6881)

[16] O.A. Ladyzhenskaya, G.A. Seregin, *On partial regularity of suitable weak solutions to the three-dimensional Navier-Stokes equations*, Journal of Mathematical Fluid Mechanics, **1** (1999), 356-387. MR1738171 (2001b:35242)

[17] J. Leray, *Sur le mouvement d'un liquide visqueux emplissant l'espace*, Acta Math., **63** (1934), 193-248. MR1555394

[18] F.-H. Lin, *A new proof of the Caffarelli-Kohn-Nirenberg theorem*, Comm. Pure Appl. Math. **51** (1998), no. 3, 241-257. MR1488514 (98k:35151)

[19] A.S. Mikhaylov, *Local regularity for suitable weak solutions of the Navier–Stokes equations near the boundary,* Zap. Nauchn. Semin. POMI **370** (2009), 73-93. MR2749212 (2011m:35274)

[20] A.S. Mikhaylov, *On local regularity for suitable weak solutions of the Navier-Stokes equations near the boundary,* Zap. Nauchn. Semin. POMI **385** (2010), 83-97. MR2749371 (2012b:35252)

[21] A.S. Mikhaylov, T.N Shilkin, *$L_{3,\infty}$ solutions to the Navier-Stokes equations near the boundary*, Zap. Nauchn. Semin. POMI **336** (2006), 133-152.

[22] V. Scheffer, *Hausdorff measure and the Navier-Stokes equations*, Comm. Math. Phys. **55** (1977), 97-112. MR0510154 (58:23176)

[23] G.A. Seregin, *Some estimates near the boundary for solutions to the non-stationary linearized Navier-Stokes equations*, Zap. Nauchn. Semin. POMI **271** (2000), 204-223. MR1810618 (2001k:35245)

[24] G.A. Seregin, Local regularity of suitable weak solutions to the Navier-Stokes equations near the boundary, Journal of Mathematical Fluid Mechanics **4** (2002) no.1, 1-29. MR1891072 (2003a:35152)
[25] G.A. Seregin, *Differentiability properties of weak solutions of the Navier-Stokes equations*, St.-Petersburg Math. Journal **14** (2003), no.1, 147-178. MR1893325 (2003f:35234)
[26] G.A. Seregin, *A note on local boundary regularity for the Stokes system*, Zap. Nauchn. Semin. POMI **370** (2009), 151-159. MR2749216 (2012a:35251)
[27] G.A. Seregin, *A certain necessary condition of potential blow up for Navier-Stokes equations*, Communications in Math. Phys. **312 (3)** (2012), 833-845. MR2925135
[28] G.A. Seregin, *Local regularity for suitable weak solutions of the Navier-Stokes equations*, Russian Mathematical Surveys **62 (3)** (2007), 595-614. MR2355422 (2008f:76054)
[29] G.A. Seregin, *Local regularity theory of the Navier-Stokes equations*, Handbook of Mathematical Fluid Dynamics, Vol. 4, (2007), 159-200.
[30] G.A. Seregin, *Weak solutions to the Navier-Stokes equations with bounded scale-invariant quantities*, Proceedings of the International Congress of Mathematics, 2010 (Hyderabad, India, 2010). MR2827878 (2012m:35250)
[31] G.A. Seregin, *On smoothness of $L_{3,\infty}$-solutions to the Navier-Stokes equations up to boundary*, Mathematische Annalen, **332** (2005), 219-238. MR2139258 (2005m:76051)
[32] G.A. Seregin, T.N. Shilkin, V.A. Solonnikov, *Boundary patial regularity for the Navier-Stokes equations*, Zap. Nauchn. Semin. POMI **310** (2004), 158-190. MR2120190 (2006d:35186)
[33] G.A. Seregin, V. Sverak, *On a bounded shear flow in half-space*, Zap. Nauchn. Semin. POMI, **385** (2010), 200-205. MR2749375 (2012b:35253)
[34] G. Seregin, V. Sverak, *On type I singularities of the local axi-symmentric solutions of the Navier-Stokes equations*, Communications in PDE's, **34** (2009), 171-201. MR2512858 (2010k:35356)
[35] J. Serrin, *On the interior regularity of weak solutions of the Navier-Stokes equations*, Arch. Ration. Mech. Anal., **9** (1962), 187-195. MR0136885 (25:346)
[36] T.N. Shilkin, V.A. Vyalov, *Estimates of solutions to the perturbed Stokes system*, Zap. Nauchn. Semin. POMI **410** (2013), 5-24. MR3048259
[37] V.A. Solonnikov, *Estimates of solutions to the non-stationary Navier-Stokes system*, Zap. Nauchn. Semin. POMI **28** (1973), 153-231. MR0415097 (54:3188)
[38] V.A. Solonnikov, *Estimates of solutions of the Stokes equations in Sobolev spaces with a mixed norm*, Zap. Nauchn. Semin. POMI **288** (2002), 204-231. MR1923551 (2003f:35235)
[39] V.A. Solonnikov, *On the estimates of solutions of nonstationary Stokes problem in anisotropic S.L. Sobolev spaces and on the estimate of resolvent of the Stokes problem*, Uspekhi Matematicheskih Nauk, **58** (2003) 2 (350) 123-156. MR1992567 (2004h:35184)

Mathematical Institute, University of Oxford, Oxford OX2 6GG, United Kingdom and St.-Petersburg Department of Steklov Mathematical Institute, Fontanka, 27, St.-Petersburg, 191023, Russia
E-mail address: Gregory.Seregin@maths.ox.ac.uk

St.-Petersburg Department of Steklov Mathematical Institute, Fontanka, 27, St.-Petersburg, 191023, Russia
E-mail address: shilkin@pdmi.ras.ru

Amer. Math. Soc. Transl.
Volume **232**, 2014

L_p-theory of free boundary problems of magnetohydrodynamics in simply connected domains

V. A. Solonnikov

Dedicated to the memory of O. A. Ladyzhenskaya

ABSTRACT. We present L_p-theory of solvability of free boundary problems of magnetohydrodynamics for a viscous incompressible fluid constructed in a 2001 paper by Padula and Solonnikov for $p = 2$. We consider the simplest problem studied in this paper.

1. Introduction

This paper is concerned with the problem of finding a bounded variable domain $\Omega_{1t} \subset \mathbb{R}^3$ with the boundary Γ_t, $t > 0$, that is a strictly interior subdomain of another bounded fixed domain $\Omega \subset \mathbb{R}^3$ with the boundary S. In addition, it is required to find the vector fields $\boldsymbol{H}(x,t) = (H_1, H_2, H_3)$, $x \in \Omega_{1t} \cup \Omega_{2t}$, $\boldsymbol{v}(x,t) = (v_1, v_2, v_3)$, and a scalar function $p(x,t)$, $x \in \Omega_{1t}$, satisfying the relations

$$(1.1)\quad \begin{cases} \boldsymbol{v}_t + (\boldsymbol{v}\cdot\nabla)\boldsymbol{v} - \nabla\cdot T(\boldsymbol{v},p) - \nabla\cdot T_M(H) = 0, \\ \nabla\cdot\boldsymbol{v}(x,t) = 0, \quad x\in\Omega_{1t},\ \ t>0, \\ \mu_1\boldsymbol{H}_t + \alpha^{-1}\mathrm{rot}\,\mathrm{rot}\,\boldsymbol{H} - \mu_1\mathrm{rot}\,(\boldsymbol{v}\times\boldsymbol{H}) = 0, \\ \nabla\cdot\boldsymbol{H}(x,t) = 0, \quad x\in\Omega_{1t}, \\ \mathrm{rot}\,\boldsymbol{H} = 0, \quad \nabla\cdot\boldsymbol{H}(\mathrm{x},\mathrm{t}) = 0, \quad \mathrm{x}\in\Omega_{2t}, \\ \big(T(\boldsymbol{v},p) + [T_M(\boldsymbol{H})]\big)\boldsymbol{n} = \sigma\boldsymbol{n}H, \quad V_n = \boldsymbol{v}\cdot\boldsymbol{n}, \quad x\in\Gamma_t, \\ [\mu\boldsymbol{H}\cdot\boldsymbol{n}] = 0, \quad [\boldsymbol{H}_\tau] = 0, \quad x\in\Gamma_t, \\ \boldsymbol{H}\cdot\boldsymbol{n} = 0, \quad x\in S, \\ \boldsymbol{v}(x,0) = \boldsymbol{v}_0(x), \quad x\in\Omega_{10}, \quad \boldsymbol{H}(x,0) = \boldsymbol{H}_0(x), \quad x\in\Omega_{10}\cup\Omega_{20}, \end{cases}$$

where $\Omega_{2t} = \Omega\backslash\overline{\Omega}_{1t}$, $T(\boldsymbol{v},p) = -pI + \nu S(\boldsymbol{v})$ is the viscous stress tensor; $S(\boldsymbol{v}) = \nabla\boldsymbol{v} + (\nabla\boldsymbol{v})^T$ is the doubled rate-of-strain tensor; $T_M(\boldsymbol{H}) = \mu(\boldsymbol{H}\otimes\boldsymbol{H} - \frac{1}{2}|\boldsymbol{H}|^2 I)$, $x \in \Omega_{it}$, is the magnetic stress tensor; μ is a piecewise constant function equal to μ_i in Ω_{it}; $\boldsymbol{n}$ is the exterior normal to Γ_t and to S; V_n is the velocity of evolution of Γ_t in the direction $\boldsymbol{n}$; $\boldsymbol{H}_\tau = \boldsymbol{H} - \boldsymbol{n}(\boldsymbol{n}\cdot\boldsymbol{H})$ is the tangential component of $\boldsymbol{H}$; and H is the doubled mean curvature of Γ_t negative for convex domains. The parameters ν, μ_i,

α, σ are positive constants. By $[u]$ we mean the jump on Γ_t of the function defined in Ω_{it}: $[u] = u^{(1)} - u^{(2)}$, and $u^{(i)} = u(x,t)|_{x\in\Omega_{it}}$.

Both Ω_{1t} and Ω are assumed to be simply connected.

The problem (1.1) is studied in the paper [1] where the local-in-time existence theorem of the problem is proved in the Sobolev–Slobodetskii spaces $W_2^{2+l,1+l/2}$, $l \in (1/2, 1)$. In the present paper the solution belonging to $W_p^{2,1}$ with $p > 3$ is obtained. Working in these spaces allows one to get rid of some technical difficulties, in particular, in the estimates of non-linear terms.

In the case of a fixed domain Ω_1 filled with liquid, the problem has been studied in the joint papers [2, 3] of the author and his teacher O. A. Ladyzhenskaya, to whose memory the present article is dedicated.

We assume that Γ_0 is located in the neighborhood of a smooth connected surface $\mathcal{G}$ of arbitrary shape (but such that the domain $\mathcal{F}_1$ bounded by $\mathcal{G}$ is simply connected). Then Γ_0 can be regarded as a normal perturbation of $\mathcal{G}$:

$$\Gamma_0 = \{x = y + \boldsymbol{N}(y)\rho_0(y), \quad y \in \mathcal{G}\},$$

where ρ_0 is a given small function and $\boldsymbol{N}(y)$ is the exterior normal to $\mathcal{G}$. Moreover, we assume that also for $t > 0$

$$\Gamma_t = \{x = y + \boldsymbol{N}(y)\rho(y,t), \quad y \in \mathcal{G}\},$$

with an unknown function $\rho(y,t)$ such that $\rho(y,0) = \rho_0(y)$. We extend $\boldsymbol{N}(y)$ and $\rho(y,t)$ from $\mathcal{G}$ into Ω in such a way that the extension $\boldsymbol{N}^*$ of $\boldsymbol{N}$ is a smooth non-zero regular function in Ω, and ρ^* vanishes near S and satisfies the inequalities (5.7) and (5.8) (hence ρ^* is small for small ρ).

The transformation

$$x \equiv e_\rho(y,t) = y + \boldsymbol{N}^*(y)\rho^*(y,t), \quad y \in \Omega, \tag{1.2}$$

maps $\mathcal{F}_1$ on Ω_{1t}, $\mathcal{F}_2 = \Omega \setminus \overline{\mathcal{F}_1}$ on Ω_{2t} and, as shown in [1], it converts (1.1) into (1.3)

$$\begin{cases} \boldsymbol{u}_t - \rho_t^*(\mathcal{L}^{-1}\boldsymbol{N}^*(y)\cdot\nabla)\boldsymbol{u} + (\mathcal{L}^{-1}\boldsymbol{u}\cdot\nabla)\boldsymbol{u} - \widetilde{\nabla}\cdot\widetilde{T}(\boldsymbol{u},q) - \widetilde{\nabla}\cdot T_M(\frac{\mathcal{L}}{L}\boldsymbol{h}) = 0, \\ \nabla\cdot\widehat{\mathcal{L}}\boldsymbol{u} = 0, \quad y \in \mathcal{F}_1, \quad t > 0, \\ \mu_1\Big(\boldsymbol{h}_t - \frac{1}{L}\widehat{\mathcal{L}}_t\mathcal{L}\boldsymbol{h} - \rho_t^*\widehat{\mathcal{L}}(\mathcal{L}^{-1}\boldsymbol{N}^*(y)\cdot\nabla)\frac{1}{L}\mathcal{L}\boldsymbol{h}\Big) \\ \quad + \alpha^{-1}\mathrm{rot}\,\mathcal{P}\mathrm{rot}\,\mathcal{P}\boldsymbol{h} - \mu_1\mathrm{rot}\,(\mathcal{L}^{-1}\boldsymbol{u}\times\boldsymbol{h}) = 0, \\ \nabla\cdot\boldsymbol{h} = 0, \quad y \in \mathcal{F}_1, \\ \mathrm{rot}\,\mathcal{P}\boldsymbol{h} = 0, \quad \nabla\cdot\boldsymbol{h} = 0, \quad y \in \mathcal{F}_2, \\ \widetilde{T}(\boldsymbol{u},q)\boldsymbol{n}(e_\rho) + [T_M(\frac{1}{L}\mathcal{L}h)]\boldsymbol{n}(e_\rho) = \sigma H(e_\rho)\boldsymbol{n}(e_\rho), \\ \rho_{\mathrm{t}} = \dfrac{\boldsymbol{u}\cdot\widehat{\mathcal{L}}^{\mathrm{T}}\boldsymbol{N}}{\boldsymbol{N}\cdot\widehat{\mathcal{L}}^{\mathrm{T}}\boldsymbol{N}}, \quad \mathrm{y} \in \mathcal{G}, \\ [\mu\boldsymbol{h}\cdot\boldsymbol{N}] = 0, \quad [\boldsymbol{h}_\tau] = \Big(\dfrac{\widehat{\mathcal{L}}\widehat{\mathcal{L}}^T\boldsymbol{N}}{|\widehat{\mathcal{L}}^T\boldsymbol{N}|^2} - \boldsymbol{N}\Big)[\boldsymbol{h}\cdot\boldsymbol{N}], \quad y \in \mathcal{G}, \\ \boldsymbol{h}\cdot\boldsymbol{n} = 0, \quad y \in S, \\ \boldsymbol{u}(y,0) = \boldsymbol{u}_0(y), \quad y \in \mathcal{F}_1, \quad \boldsymbol{h}(y,0) = \boldsymbol{h}_0(y), \quad y \in \mathcal{F}_1\cup\mathcal{F}_2, \\ \rho(y,0) = \rho_0(y), \quad y \in \mathcal{G}, \end{cases}$$

where $\mathcal{L} = \mathcal{L}(y, \rho^*) = (\delta_{ij} + \frac{\partial}{\partial x_j} N_i^*(y)\rho^*(y,t))_{i,j=1,2,3}$ is the Jacobi matrix of the transformation (1.2), $L = \det\mathcal{L}$, $\widehat{\mathcal{L}} = L\mathcal{L}^{-1}$ is the cofactors matrix of $\mathcal{L}$, $\mathcal{P}(y,\rho) = L^{-1}\mathcal{L}^T\mathcal{L}$,

$\boldsymbol{u}(y,t) = \boldsymbol{v}(e_\rho(y,t),t)$, $q(y,t) = p(e_\rho, t)$, $\boldsymbol{h}(y,t) = \widehat{\mathcal{L}}(y,\rho^*)\boldsymbol{H}(e_\rho, t)$,

$\widetilde{\nabla} = \mathcal{L}^{-T}(y,\rho^*)\nabla_y$ is the transformed gradient $\nabla_x = \left(\frac{\partial}{\partial x_1}, \frac{\partial}{\partial x_2}, \frac{\partial}{\partial x_3}\right)$,

$\mathcal{L}^{-T} = (\mathcal{L}^{-1})^T$ (the sign T means transposition),

$\widetilde{T}(\boldsymbol{u}, q)$ is the transformed stress tensor: $\widetilde{T} = -qI + \nu\widetilde{S}(\boldsymbol{u})$,

$\widetilde{S}(\boldsymbol{u}) = (\widetilde{\nabla}\boldsymbol{u}) + (\widetilde{\nabla}\boldsymbol{u})^T$ is the transformed rate-of-strain tensor,

ρ is an additional unknown function,

$\boldsymbol{u}_0(y) = \boldsymbol{v}_0(e_{\rho_0}(y))$, $\boldsymbol{h}_0(y) = \widehat{\mathcal{L}}(y,\rho_0^*)\boldsymbol{H}_0(e_{\rho_0}(y))$.

The vectors $\boldsymbol{n}(e_\rho)$ and $\boldsymbol{N}(y)$ are connected by

$$\boldsymbol{n}(e_\rho) = \frac{\widehat{\mathcal{L}}^T\boldsymbol{N}(y)}{|\widehat{\mathcal{L}}^T\boldsymbol{N}|}, \quad y \in \mathcal{G}.$$

We consider the problem (1.3) in the spaces

$$W_p^{2,1}(\mathcal{D}_T) = W_p^{2,0}(\mathcal{D}_T) \cap W_p^{0,1}(\mathcal{D}_T), \quad p > 3, \quad \mathcal{D}_T = \mathcal{D} \times (0,T), \quad \mathcal{D} \subset \mathbb{R}^n,$$

where $W_p^{2,0}(\mathcal{D}_T) = L_p(0,T; W_p^2(\mathcal{D}))$, $W_2^{0,1}(\mathcal{D}_T) = W_p^1(0,T; L_p(\mathcal{D}))$. The norm in $W_p^{2,1}(\mathcal{D}_T)$ is defined by

$$(1.4)\quad \|u\|^p_{W_p^{2,1}(\mathcal{D}_T)} = \int_0^T \|u(\cdot,t)\|^p_{W_p^2(\mathcal{D})}dt + \int_0^T (\|u_t(\cdot,t)\|^p_{L_p(\mathcal{D})} + \|u(\cdot,t)\|^p_{L_p(\mathcal{D})})dt,$$

and the integrals in the right-hand side represent the norms in $W_p^{2,0}(\mathcal{D}_T)$ and $W_p^{0,1}(\mathcal{D}_T)$, respectively. By $W_p^l(\mathcal{D})$ we mean the space of functions $u(x)$, $x \in \mathcal{D}$, with finite norm

$$\|u\|_{W_p^l(\mathcal{D})} = \Big(\sum_{|j|\le l} \|D^j u\|^p_{L_p(\mathcal{D})}\Big)^{1/p},$$

if l is an integer, and

$$\|u\|_{W_p^l(\mathcal{D})} = \Big(\|u\|^p_{W_p^{[l]}(\mathcal{D})} + \sum_{|j|=[l]} \int_{\mathcal{D}}\int_{\mathcal{D}} \frac{|D^j u(x) - D^j u(y)|^p}{|x-y|^{n+p\lambda}} dxdy\Big)^{1/p},$$

if $l = [l] + \lambda$, $0 < \lambda < 1$. The anisotropic space $W_p^{l,l/2}(\mathcal{D}_T)$ is defined as $W_p^{l,0}(\mathcal{D}_T) \cap W_p^{0,l/2}(\mathcal{D}_T) = L_p(0,T; W_p^l(\mathcal{D})) \cap W_p^{l/2}(0,T; L_p(\mathcal{D}))$.

The W_p^l-spaces on manifolds, in particular, on the surfaces $\mathcal{G}$ and S, are introduced with the help of local maps and the partition of unity.

We recall some imbedding theorems for the space $W_p^{2,1}(\mathcal{D}_T)$:

1. *If* $u \in W_p^{2,1}(\mathcal{D}_T)$, *then* $u|_{t=t_0} \in W_p^{2-2/p}(\mathcal{D})$, $u|_{x\in\Sigma} \in W_p^{2-1/p,1-1/(2p)}(\Sigma_T)$, $\Sigma_T = \Sigma \times (0,T)$, *where* $\Sigma = \partial\mathcal{D}$, *and*

$$\|u(\cdot,t_0)\|_{W_p^{2-2/p}(\mathcal{D})} \le c\|u\|_{W_p^{2,1}(\mathcal{D}_T)},$$

$$\|u\|_{W_p^{2-1/p,1-1/(2p)}(\Sigma_T)} \le c\|u\|_{W_p^{2,1}(\mathcal{D}_T)}.$$

2. *For arbitrary* $\varphi \in W_p^{2-2/p}(\mathcal{D})$, *there exists such* $u \in W_p^{2,1}(\mathcal{D}_T)$ *that* $u(x,0) = \varphi(x)$, *and*

$$\|u\|_{W_p^{2,1}(\mathcal{D}_T)} \le c\|\varphi\|_{W_p^{2-2/p}(\mathcal{D})}.$$

For arbitrary $\psi \in W_p^{2-1/p,1-1/(2p)}(\Sigma_T)$ *there exists such* $u \in W_p^{2,1}(\mathcal{D}_T)$ *that* $u|_{x\in\Sigma} = \psi(x,t)$ *and*

$$\|u\|_{W_p^{2,1}(\mathcal{D}_T)} \leq c\|\psi\|_{W_p^{2-1/p,1-1/(2p)}(\Sigma_T)}.$$

3. *If* $\rho \in W_p^{3-1/p,0}(\Sigma_T)\cap W_p^1(0,T;W_p^{2-1/p}(\Sigma))$, *then* $\rho(\cdot,t_0) \in W_p^{3-2/p}(\Sigma)$, *and*

$$\|\rho(\cdot,t_0)\|_{W_p^{3-2/p}(\Sigma)} \leq c(\|\rho\|_{W_p^{3-1/p,0}(\Sigma_T)} + \|\rho_t\|_{W_p^{2-1/p,0}(\Sigma_T)}).$$

For arbitrary $\rho_0 \in W_p^{3-2/p}(\Sigma)$ *there exists* $\rho \in W_p^{3-1/p,0}(\Sigma_T)$ *with* $\rho_t \in W_p^{2-1/p,1-1/2p}(\Sigma_T)$ *such that* $\rho(x,0) = \rho_0(x)$, *and*

$$\|\rho\|_{W_p^{3-1/p,0}(\Sigma_T)} + \|\rho_t\|_{W_p^{2-1/p,1-1/2p}(\Sigma_T)} \leq c\|\rho_0\|_{W_p^{3-2/p}(\Sigma)}$$

(*cf.* [4, *Proposition* 4.1] *for* $p = 2$ *and* [10]).

4. *If* $lp > n$, *then*

$$\sup_{\mathcal{D}} |u(x)| \leq c\|u\|_{W_p^l(\mathcal{D})}.$$

Now we present the main result of the paper.

THEOREM 1. *Let* $\boldsymbol{u}_0 \in W_p^{2-2/p}(\mathcal{F}_1)$, $\rho_0 \in W_p^{3-2/p}(\mathcal{G})$, $\boldsymbol{h}_0 \in W_r^{2-2/r}(\mathcal{F}_i)$, $i = 1,2$, *with*

$$3 < r < p, \quad 1/r - 1/p \leq 1/5(1-2/p), \tag{1.5}$$

and let the compatibility conditions

(1.6)

$$\begin{aligned}
&\nabla\cdot\widehat{\mathcal{L}}(y,\rho_0^*)\boldsymbol{u}_0 = 0, \quad y\in\mathcal{F}_1, \quad \widetilde{S}(\boldsymbol{u}_0)\boldsymbol{n}_0(e_{\rho_0}) - \boldsymbol{n}_0(\boldsymbol{n}_0\cdot\widetilde{S}(\boldsymbol{u}_0)\boldsymbol{n}_0) = 0, \quad y\in\mathcal{G},\\
&\nabla\cdot\boldsymbol{h}_0 = 0, \quad y\in\mathcal{F}_i, \quad i=1,2,\\
&\operatorname{rot}\mathcal{P}(y,\rho_0^*)\boldsymbol{h}_0 = 0, \quad y\in\mathcal{F}_2,\\
&[\mu\boldsymbol{h}_0\cdot\boldsymbol{N}] = 0, \quad [\boldsymbol{h}_{0\tau}] = \Big(\frac{\widehat{\mathcal{L}}(y,\rho_0)\widehat{\mathcal{L}}^T\boldsymbol{N}}{|\widehat{\mathcal{L}}^T\boldsymbol{N}|^2} - \boldsymbol{N}\Big)[\boldsymbol{h}_0\cdot\boldsymbol{N}], \quad y\in\mathcal{G},\\
&\boldsymbol{h}_0\cdot\boldsymbol{n} = 0, \quad y\in S,
\end{aligned}$$

where $\boldsymbol{n}_0$ *is the normal to* Γ_0, *and the smallness condition*

$$\|\rho_0\|_{W_p^{2-1/p}(\mathcal{G})} \leq \epsilon \ll 1 \tag{1.7}$$

is satisfied. Then the problem (1.3) *has a unique solution defined in a certain (small) time interval* $(0,T)$ *with the regularity properties*

$$\begin{aligned}
&\boldsymbol{u}\in W_p^{2,1}(Q_T^1), \quad \nabla q\in L_p(Q_T^1), \quad q\in W_p^{l-1/p,0}(G_T), \quad \rho\in W_p^{3-1/p,0}(G_T),\\
&\rho_t\in W_p^{2-1/p,1-1/2p}(G_T), \quad \boldsymbol{h}^{(i)}\in W_r^{2,1}(Q_T^i), \quad i=1,2,
\end{aligned}$$

where $Q_T^i = \mathcal{F}_i \times (0,T)$, $G_T = \mathcal{G} \times (0,T)$, $\boldsymbol{h}^{(i)} = \boldsymbol{h}|_{Q_T^i}$, $i = 1,2$. *The solution satisfies the inequality*

$$
\begin{aligned}
&\|\boldsymbol{u}\|_{W_p^{2,1}(Q_T^1)} + \|\nabla q\|_{L_p(Q_T^1)} + \|q\|_{W_p^{1-1/p,0}(G_T)} + \sup_{t<T} \|\boldsymbol{u}\|_{W_p^{2-2/p}(\mathcal{F}_1)} \\
&\quad + \|\rho\|_{W_p^{3-1/p,0}(G_T)} + \|\rho_t\|_{W_p^{2-1/p,1-1/2p}(G_T)} + \sup_{t<T} \|\rho\|_{W_p^{3-2/p}(\mathcal{G})} \\
&\quad + \sum_{i=1}^{2} \big(\|\boldsymbol{h}^{(i)}\|_{W_r^{2,1}(Q_T^i)} + \sup_{t<T} \|\boldsymbol{h}^{(i)}\|_{W_r^{2-2/r}(\mathcal{F}_i)}\big) \\
&\quad \le c\Big(\|\boldsymbol{f}\|_{L_p(Q_T^1)} + \|\mathcal{H}\|_{W_p^{1-1/p}(\mathcal{G})} \\
&\quad + \|\boldsymbol{u}_0\|_{W_p^{2-2/p}(\mathcal{F}_1)} + \|\rho_0\|_{W_p^{3-2/p}(\mathcal{G})} + \sum_{i=1}^{2} \|\boldsymbol{h}_0\|_{W_r^{2-2/r}(\mathcal{F}_i)}\Big),
\end{aligned}
\tag{1.8}
$$

where $\mathcal{H}$ *is the doubled mean curvature of* $\mathcal{G}$.

For the problem (1.1) this means that it is solvable in the time interval $(0,T)$ and

$$
\begin{aligned}
\boldsymbol{v} \circ e_\rho &\in W_p^{2.1}(Q_T^1), \\
\nabla p \circ e_\rho &\in L_p(Q_T^1), \\
p \circ e_\rho &\in W_p^{1-1/p,0}(G_T), \\
\rho &\in W_p^{3-1/p,0}(G_T), \\
\rho_t &\in W_p^{2-1/p,1-1/2p}(G_T), \\
\boldsymbol{H}^{(i)} \circ e_\rho &\in W_r^{2,1}(Q_T^i), \quad i = 1,2.
\end{aligned}
$$

Once the solution with the above-mentioned properties is obtained, it can be shown that

$$\boldsymbol{h}^{(i)} \in W_p^{2,1}(Q_T^i), \quad i = 1,2,$$

provided $\boldsymbol{h}_0 \in W_p^{2-2/p}(\Omega_{i0})$, $i = 1,2$ (see Section 5).

2. Linear problems

The proof of Theorem 1 is based on the analysis of the following non-homogeneous linear problems.

1. Find $(\boldsymbol{v}, p, \rho)$ such that

$$
\begin{cases}
\boldsymbol{v}_t - \nu\nabla^2\boldsymbol{v} + \nabla p = \boldsymbol{f}(y,t), \\
\nabla \cdot \boldsymbol{v} = f(y,t) = \nabla \cdot \boldsymbol{F}(y,t), \quad y \in \mathcal{F}_1, \ t > 0, \\
T(\boldsymbol{v}, p)\boldsymbol{N}(y) + \sigma \boldsymbol{N}(y)\mathfrak{B}\rho = \boldsymbol{d}(y,t), \\
\rho_t + \boldsymbol{V}(y) \cdot \nabla_\tau \rho - \boldsymbol{v} \cdot \boldsymbol{N}(y) = g(y,t), \quad y \in \mathcal{G}, \\
\boldsymbol{v}(y,0) = \boldsymbol{v}_0(y), \quad y \in \mathcal{F}_1, \quad \rho(y,0) = \rho_0(y), \quad y \in \mathcal{G},
\end{cases}
\tag{2.1}
$$

where $\mathfrak{B}\rho = -\Delta_{\mathcal{G}}\rho - b(y)\rho$, $b = (\mathcal{H}^2 - 2\mathcal{K})$, $\Delta_{\mathcal{G}}$ is the Laplace–Beltrami operator on $\mathcal{G}$, $\mathcal{H}$ and $\mathcal{K}$ are the doubled mean curvature and the Gaussian curvature of $\mathcal{G}$, respectively, and $\boldsymbol{V}(x)$ is a given vector field from $W_p^{2-1/p}(\mathcal{G})$.

2. Find the vector field $\boldsymbol{H}(y,t)$, satisfying the equations

$$(2.2)\qquad \begin{cases} \mu_1 \boldsymbol{H}_t + \alpha^{-1}\mathrm{rot}\,\mathrm{rot}\,\boldsymbol{H} = \boldsymbol{G}(y,t), \\ \nabla\cdot\boldsymbol{H} = 0, \quad y\in\mathcal{F}_1, \\ \mathrm{rot}\,\boldsymbol{H} = \mathrm{rot}\,\boldsymbol{\ell}(y,t), \quad \nabla\cdot\boldsymbol{H}=0, \quad y\in\mathcal{F}_2, \\ [\mu\boldsymbol{H}\cdot\boldsymbol{N}] = 0, \quad [\boldsymbol{H}_\tau] = \boldsymbol{a}(y,t), \quad y\in\mathcal{G}, \quad \boldsymbol{H}\cdot\boldsymbol{n}(y) = 0, \quad y\in S, \\ \boldsymbol{H}(y,0) = \boldsymbol{H}_0(y), \quad y\in\mathcal{F}_1\cup\mathcal{F}_2. \end{cases}$$

3. In addition, we need to consider the auxiliary problem

$$(2.3)\qquad \begin{cases} \mathrm{rot}\,\boldsymbol{h}(y,t) = \boldsymbol{k}(y,t), \quad \nabla\cdot\boldsymbol{h} = 0, \quad y\in\mathcal{F}_1\cup\mathcal{F}_2, \\ [\mu\boldsymbol{h}\cdot\boldsymbol{N}] = 0, \quad [\boldsymbol{h}_\tau] = \boldsymbol{a}, \quad y\in\mathcal{G}, \\ \boldsymbol{h}\cdot\boldsymbol{n}(y) = 0, \quad y\in S. \end{cases}$$

THEOREM 2 ([5]). *Let $p>3$. Assume that $\boldsymbol{f}\in L_p(Q_T^1)$, $f\in W_p^{1,0}(Q_T^1)$, $f=\nabla\boldsymbol{F}$, $\boldsymbol{F}\in W_p^{0,1}(Q_T^1)$, $\boldsymbol{d}\cdot\boldsymbol{N}\in W_2^{1-1/p,0}(G_T)$, $\boldsymbol{d}-\boldsymbol{N}(\boldsymbol{d}\cdot\boldsymbol{N})\equiv\boldsymbol{d}_\tau\in W_p^{1-1/p,1/2-1/(2p)}(G_T)$, $g\in W_p^{2-1/p,1-1/2p}(G_T)$, $\boldsymbol{v}_0\in W_p^{2-2/p}(\mathcal{F}_1)$, $\rho_0\in W_p^{3-2/p}(\mathcal{G})$, $\boldsymbol{V}\in W_p^{2-1/p}(\mathcal{G})$, and let the compatibility conditions*

$$(2.4)\qquad \nabla\cdot\boldsymbol{v}_0(x) = f(x,0), \quad x\in\mathcal{F}_1, \quad \nu(S(\boldsymbol{v}_0)\boldsymbol{N})_\tau = \boldsymbol{d}_\tau(x,0), \quad x\in\mathcal{G},$$

be satisfied. Then problem (2.1) has a unique solution $\boldsymbol{v},p,\rho$ such that $\boldsymbol{v}\in W_p^{2,1}(Q_T^1)$, $\nabla p\in L_p(Q_T^1)$, $p\in W_p^{1-1/p,0}(G_T)$, $\rho\in W_p^{3-1/p,0}(G_T)$, $\rho_t\in W_p^{2-1/p,1-1/2p}(G_T)$, $\rho(\cdot,t)\in W_p^{3-2/p}(\mathcal{G})$, $\forall t\in(0,T)$, and the solution satisfies the inequality

$$(2.5)\qquad \begin{aligned} &\|\boldsymbol{v}\|_{W_p^{2,1}(Q_T^1)} + \|\nabla p\|_{L_p(Q_T^1)} + \|p\|_{W_p^{1-1/p,0}(G_T)} \\ &\quad + \|\rho\|_{W_p^{3-1/p,0}(G_T)} + \|\rho_t\|_{W_2^{2-1/p,1-1/2p}(G_T)} + \sup_{t<T}\|\rho(\cdot,t)\|_{W_p^{3-2/p}(\mathcal{G})} \\ &\le c(T)\Big(\|\boldsymbol{f}\|_{L_p(Q_T^1)} + \|f\|_{W_p^{1,0}(Q_T^1)} + \|\boldsymbol{F}\|_{W_p^{0,1}(Q_T^1)} \\ &\quad + \|\boldsymbol{d}_\tau\|_{W_2^{1-1/p,l/2-1/(2p)}(G_T)} + \|\boldsymbol{d}\cdot\boldsymbol{N}\|_{W_2^{1-1/p,0}(G_T)} \\ &\quad + \|g\|_{W_p^{2-1/p,1-1/2p}(G_T)} + \|\boldsymbol{v}_0\|_{W_p^{2-2/p}(\mathcal{F})} + \|\rho_0\|_{W_p^{3-2/p}(\mathcal{G})}\Big). \end{aligned}$$

The compatibility condition $\nabla\cdot\boldsymbol{v}_0(x) = f(x,0)$, $x\in\mathcal{F}_1$ can be understood in a weak sense as

$$\int_{\mathcal{F}_1}(\boldsymbol{v}_0(x) - \boldsymbol{F}(x,0))\cdot\nabla\eta(x)dx = 0$$

for an arbitrary smooth $\eta(x)$ vanishing on $\mathcal{G}$.

THEOREM 3. *Assume that $\boldsymbol{k} = \mathrm{rot}\,\boldsymbol{K}(y,t)$, $\boldsymbol{a} = [\boldsymbol{A}]$, $\boldsymbol{K},\boldsymbol{A}\in W_p^{2,1}(Q_T^i)$, $i=1,2$,*

$$(2.6)\quad \boldsymbol{K} = \boldsymbol{A} = 0, \quad y\in S, \quad [\boldsymbol{K}_\tau] = \boldsymbol{a}, \quad \boldsymbol{A}^{(1)}\cdot\boldsymbol{N}(y) = \boldsymbol{A}^{(2)}\cdot\boldsymbol{N}(y) = 0, \quad y\in\mathcal{G}.$$

Then the problem (2.3) has a unique solution $\boldsymbol{h}\in W_p^{2,1}(\mathcal{F}_i)$, $i=1,2$, that satisfies the inequality

$$(2.7)\qquad \sum_{i=1}^2\|\boldsymbol{h}\|_{W_p^{2,1}(\mathcal{F}_i)} \le c\sum_{i=1}^2(\|\boldsymbol{K}\|_{W_p^{2,1}(Q_T^i)} + \|\boldsymbol{A}\|_{W_p^{2,1}(Q_T)^i}).$$

PROOF. Following [1, Theorem 3], we seek the solution in the form

$$\boldsymbol{h} = \boldsymbol{A} + \nabla\phi + \boldsymbol{X},$$

where ϕ and $\boldsymbol{X}$ are solutions to the problems

$$\begin{cases} \nabla^2\phi(x,t) = -\nabla\cdot\boldsymbol{A}(x,t), \quad x\in\mathcal{F}_1\cup\mathcal{F}_2, \quad \dfrac{\partial\phi}{\partial n} = 0, \quad x\in S, \\ [\phi] = 0, \quad [\mu\dfrac{\partial\phi}{\partial N}] = 0, \quad x\in\mathcal{G}, \end{cases} \tag{2.8}$$

$$\begin{cases} \operatorname{rot}\boldsymbol{X} = \operatorname{rot}(\boldsymbol{K}(x,t) - \boldsymbol{A}(x,t)), \quad \nabla\cdot\boldsymbol{X} = 0, \quad x\in\mathcal{F}_1\cup\mathcal{F}_2, \\ [\mu\boldsymbol{X}\cdot\boldsymbol{N}] = 0, \quad [\boldsymbol{X}_\tau] = 0, \quad x\in\mathcal{G}, \quad \boldsymbol{X}\cdot\boldsymbol{n} = 0, \quad x\in S. \end{cases}$$

The problem (2.8) can be written in a weak form as

$$\int_\Omega \mu\nabla\phi\cdot\nabla\eta dx = -\int_\Omega \mu\boldsymbol{A}\cdot\nabla\eta dx$$

with an arbitrary smooth $\eta(x)$. It is uniquely solvable, up to a constant, and

$$\|\nabla\phi\|_{L_p(\Omega)} \le c\|\boldsymbol{A}\|_{L_p(\Omega)}, \quad \sum_{i=1}^2 \|\nabla\phi\|_{W_p^2(\mathcal{F}_i)} \le c\sum_{i=1}^2 \|\boldsymbol{A}\|_{W_p^2(\mathcal{F}_i)}, \tag{2.9}$$

which implies

$$\sum_{i=1}^2 \|\nabla\phi\|_{W_p^{2,1}(Q_T^i)} \le c\sum_{i=1}^2 \|\boldsymbol{A}\|_{W_p^{2,1}(Q_T^i)}. \tag{2.10}$$

As for $\boldsymbol{X}$, we represent it as the sum

$$\boldsymbol{X}(x,t) = \boldsymbol{X}_1(x,t) + \nabla U(x,t),$$
$$\boldsymbol{X}_1(x,t) = \frac{1}{4\pi}\operatorname{rot}\int_\Omega \frac{\operatorname{rot}(\boldsymbol{K}(y,t) - \boldsymbol{A}(y,t))}{|x-y|}dy,$$

$$\begin{cases} \nabla^2 U(x,t) = 0, \quad x\in\mathcal{F}_1\cup\mathcal{F}_2, \quad [U(x,t)] = 0, \quad [\mu\dfrac{\partial U}{\partial N}] = -[\mu]\boldsymbol{X}_1\cdot\boldsymbol{N}, \quad x\in\mathcal{G}, \\ \dfrac{\partial U}{\partial n} = -\boldsymbol{X}_1\cdot\boldsymbol{n}, \quad x\in S, \end{cases}$$

or, in a weak form,

$$\int_\Omega \mu\nabla U\cdot\nabla\eta dx = \int_{\mathcal{G}} [\mu]\boldsymbol{X}_1\cdot\boldsymbol{N}\eta dS + \int_S \mu_2\boldsymbol{X}_1\cdot\boldsymbol{n}\eta dS = \int_\Omega \mu\boldsymbol{X}_1\cdot\nabla\eta dx.$$

Since $[\boldsymbol{N}\times(\boldsymbol{K}-\boldsymbol{A})]|_{x\in\mathcal{G}} = 0$, $\boldsymbol{K}_\tau = \boldsymbol{A}_\tau|_{x\in S} = 0$, it holds $\operatorname{rot}\boldsymbol{X}_1 = \boldsymbol{K} - \boldsymbol{A}$,

$$\boldsymbol{X}_1(x,t) = \frac{1}{4\pi}\operatorname{rot}\int_\Omega \nabla\frac{1}{|x-y|}\times(\boldsymbol{K}(y,t) - \boldsymbol{A}(y,t))dy,$$

$$\|\boldsymbol{X}_1(\cdot,t)\|_{L_p(\Omega)} \le c(\|\boldsymbol{K}(\cdot,t)\|_{L_p(\Omega)} + \|\boldsymbol{A}(\cdot,t)\|_{L_p(\Omega)}),$$

$$\sum_{i=1}^2 \|\boldsymbol{X}_1(\cdot,t)\|_{W_p^2(\mathcal{F}_i)} \le c\sum_{i=1}^2(\|\boldsymbol{K}(\cdot,t)\|_{W_p^2(\mathcal{F}_i)} + \|\boldsymbol{A}(\cdot,t)\|_{W_p^2(\mathcal{F}_i)}),$$

and, as a consequence,

$$\sum_{i=1}^2 \|\boldsymbol{X}_1\|_{W_p^{2,1}(Q_T^i)} \le c\sum_{i=1}^2(\|\boldsymbol{K}\|_{W_p^{2,1}(Q_T^i)} + \|\boldsymbol{A}\|_{W_p^{2,1}(\mathcal{F}_i)}). \tag{2.11}$$

In addition, we have

$$(2.12)\qquad \begin{aligned} &\|\nabla U(\cdot,t)\|_{L_p(\Omega)} \le c\|\boldsymbol{X}_1(\cdot,t)\|_{L_p(\Omega)},\\ &\|\nabla U(\cdot,t)\|_{W^2_p(\Omega)} \le c\sum_{i=1}^{2}\|\boldsymbol{X}_1(\cdot,t)\|_{W^2_p(\mathcal{F}_i)},\\ &\sum_{i=1}^{2}\|\nabla U\|_{W^{2,1}_p(Q^i_T)} \le c\sum_{i=1}^{2}\|\boldsymbol{X}_1\|_{W^{2,1}_p(Q^i_T)}. \end{aligned}$$

Inequality (2.7) follows from (2.9)–(2.12). If $\boldsymbol{k}=0$, $\boldsymbol{a}=0$, then $\boldsymbol{h}=0$, hence the solution of (2.3) constructed above is unique. The theorem is proved.

THEOREM 4. *Assume that the data of the problem* (2.2) *possess the following properties:* $\boldsymbol{G}\in L_p(Q^1_T)$, $\boldsymbol{H}_0\in W_p^{2-2/p}(\mathcal{F}_1)\cap W_p^{2-2/p}(\mathcal{F}_2)$, $\boldsymbol{a}\in W_p^{2-1/p,1-1/2p}(G_T)$, $\boldsymbol{\ell}\in W^{2,1}_p(Q^2_T)$, $\boldsymbol{\ell}|_{x\in S}=0$, *moreover,* $\boldsymbol{a}=[\boldsymbol{A}]$ *with* $\boldsymbol{A}^{(i)}\in W^{2,1}_p(Q^i_T)$, $\boldsymbol{A}^{(i)}\cdot\boldsymbol{N}(x)|_{x\in\mathcal{G}}=0$, $i=1,2$, *and the compatibility conditions*

$$(2.13)\qquad \begin{aligned} &\nabla\cdot\boldsymbol{G}(x,t)=0,\quad x\in\mathcal{F}_1,\quad \nabla\cdot\boldsymbol{H}_0(x)=0,\quad x\in\mathcal{F}_1\cup\mathcal{F}_2,\\ &\operatorname{rot}\boldsymbol{H}_0(\mathrm{x})=\operatorname{rot}\boldsymbol{\ell}(\mathrm{x},0),\quad \mathrm{x}\in\mathcal{F}_2,\\ &[\mu\boldsymbol{H}_0\cdot\boldsymbol{N}]=0,\quad [\boldsymbol{H}_{0\tau}]=\boldsymbol{a}(x,0)=[\boldsymbol{A}(x,0)],\quad x\in\mathcal{G},\\ &\boldsymbol{H}_0\cdot\boldsymbol{n}=0,\quad x\in S \end{aligned}$$

are satisfied. Then problem (2.2) *has a unique solution* $\boldsymbol{H}\in W^{2,1}_p(Q^1_T)\cap W^{2,1}_p(Q^2_T)$, *and*

$$(2.14)\qquad \begin{aligned} \sum_{i=1}^{2}\|\boldsymbol{H}^{(i)}\|_{W^{2,1}_p(Q^i_T)} \le c(T)\Big(&\|\boldsymbol{G}\|_{L_p(Q^1_T)}+\|\boldsymbol{\ell}\|_{W^{2,1}_p(Q^2_T)}\\ &+\sum_{i=1}^{2}(\|\boldsymbol{H}_0\|_{W_p^{2-2/p}(\mathcal{F}_i)}+\|\boldsymbol{A}^{(i)}\|_{W^{2,1}_p(Q^i_T)})\Big). \end{aligned}$$

PROOF. We reduce the problem to the particular case $\boldsymbol{\ell}=0$, $\boldsymbol{a}=0$. We extend $\boldsymbol{\ell}$ into Ω with the preservation of class, i.e., so that the extended function $\boldsymbol{\ell}^*$ satisfies

$$(2.15)\qquad \|\boldsymbol{\ell}^*\|_{W^{2,1}_p(Q_T)} \le c\|\boldsymbol{\ell}\|_{W^{2,1}_p(Q^2_T)},$$

and we define

$$\begin{aligned} &\boldsymbol{a}^*(x,t)=\boldsymbol{A}^{(1)}(x,t)-\boldsymbol{A}^{(2)*}(x,t),\quad x\in\mathcal{F}_1,\\ &\boldsymbol{a}^*(x,t)=0,\quad x\in\mathcal{F}_2, \end{aligned}$$

where $\boldsymbol{A}^{(2)*}$ is the extension of $\boldsymbol{A}^{(2)}$ such that

$$\|\boldsymbol{A}^{(2)*}\|_{W^{2,1}_p(Q_T)} \le c\|\boldsymbol{A}^{(2)}\|_{W^{2,1}_p(Q^2_T)}.$$

It is easily verified that

$$[\boldsymbol{a}^*]=[\boldsymbol{A}]=\boldsymbol{a},\quad [\boldsymbol{\ell}^*+\boldsymbol{a}^*]=\boldsymbol{a},\quad x\in\mathcal{G}.$$

Now we define $\boldsymbol{h}_1(x,t)$ as a solution of the problem (2.3) with $\boldsymbol{k}(x,t) = \operatorname{rot}(\boldsymbol{\ell}^* + \boldsymbol{a}^*)$. By (2.7),

$$
\begin{aligned}
\sum_{i=1}^{2} \|\boldsymbol{h}_1\|_{W_p^{2,1}(Q_T^i)} &\le c \sum_{i=1}^{2} (\|\boldsymbol{\ell}^*\|_{W_p^{2,1}(Q_T^i)} + \|\boldsymbol{a}^*\|_{W_p^{2,1}(Q_T^i)}) \\
&\le c(\|\boldsymbol{\ell}\|_{W_p^{2,1}(Q_T^2)} + \sum_{i=1}^{2} \|\boldsymbol{A}\|_{W_p^{2,1}(Q_T^i)}).
\end{aligned} \tag{2.16}
$$

For $\boldsymbol{h} = \boldsymbol{H} - \boldsymbol{h}_1$, we obtain the problem

$$
\begin{cases}
\mu_1 \boldsymbol{h}_t + \alpha^{-1} \operatorname{rot}\operatorname{rot} \boldsymbol{h} = \boldsymbol{G}'(y,t), \\
\nabla \cdot \boldsymbol{h} = 0, \quad y \in \mathcal{F}_1, \\
\operatorname{rot} \boldsymbol{h} = 0, \quad \nabla \cdot \boldsymbol{h} = 0, \quad y \in \mathcal{F}_2, \\
[\mu \boldsymbol{h} \cdot \boldsymbol{N}] = 0, \quad [\boldsymbol{h}_\tau] = 0, \quad y \in \mathcal{G}, \quad \boldsymbol{h} \cdot \boldsymbol{n} = 0, \quad y \in S, \\
\boldsymbol{h}(y,0) = \boldsymbol{h}_0(y) = \boldsymbol{H}_0(x) - \boldsymbol{h}_1(y,0), \quad y \in \mathcal{F}_1 \cup \mathcal{F}_2,
\end{cases} \tag{2.17}
$$

where $\boldsymbol{G}' = \boldsymbol{G} - \mu_1 \boldsymbol{h}_{1t} - \alpha^{-1} \operatorname{rot}\operatorname{rot} \boldsymbol{h}_1$.

The equations $\operatorname{rot} \boldsymbol{h}(y,t) = 0$, $\nabla \cdot \boldsymbol{h}(y,t) = 0$ in $\mathcal{F}_2$ imply

$$
\begin{aligned}
&\boldsymbol{h}(y,t) = \nabla \varphi(y,t), \\
&\nabla^2 \varphi(y,t) = 0, \quad y \in \mathcal{F}_2, \\
&\frac{\partial \varphi}{\partial n}\Big|_{y \in S} = 0, \quad \mu_2 \frac{\partial \varphi}{\partial n} - \mu_1 \boldsymbol{h}^{(1)} \cdot \boldsymbol{N}\Big|_{y \in \mathcal{G}} = 0, \\
&\boldsymbol{h}_\tau^{(1)} - \nabla_\tau \varphi|_{y \in \mathcal{G}} = 0,
\end{aligned} \tag{2.18}
$$

so (2.17) can be written in the form

$$
\begin{cases}
\mu_1 \boldsymbol{h}_t + \alpha^{-1} \operatorname{rot}\operatorname{rot} \boldsymbol{h} = \boldsymbol{G}'(y,t), \quad \nabla \cdot \boldsymbol{h} = 0, \quad y \in \mathcal{F}_1, \\
\nabla^2 \varphi(y,t) = 0, \quad y \in \mathcal{F}_2, \quad \dfrac{\partial \varphi}{\partial n}\Big|_{y \in S} = 0, \\
\mu_2 \dfrac{\partial \varphi}{\partial N} - \mu_1 \boldsymbol{h}^{(1)} \cdot \boldsymbol{N}\Big|_{y \in \mathcal{G}} = 0, \quad \boldsymbol{h}_\tau^{(1)} - \nabla_\tau \varphi|_{y \in \mathcal{G}} = 0, \\
\boldsymbol{h}(y,0) = \boldsymbol{h}_0(y), \quad y \in \mathcal{F}_1 \cup \mathcal{F}_2.
\end{cases} \tag{2.19}
$$

As shown in [6], this problem is uniquely solvable, and

$$
\|\boldsymbol{h}\|_{W_p^{2,1}(Q_T^1)} + \|\nabla \varphi\|_{W_p^{2,1}(Q_T^2)} \le c(\|\boldsymbol{G}'\|_{L_p(Q_T^1)} + \sum_{i=1}^{2} \|\boldsymbol{h}_0\|_{W_2^{2-2/p}(\mathcal{F}_i)}).
$$

Together with (2.16), this inequality implies (2.14). The theorem is proved.

Equations (2.18) show that $\boldsymbol{h}^{(2)} = \nabla \varphi$ is completely determined by $\boldsymbol{h}^{(1)} \cdot \boldsymbol{N}|_{y \in \mathcal{G}}$ and

$$
\begin{aligned}
&\|\boldsymbol{h}^{(2)}\|_{W_p^{2,1}(Q_T^2)} \le c \|\boldsymbol{h}^{(1)}\|_{W_p^{2,1}(Q_T^1)}, \\
&\|\boldsymbol{h}_0^{(2)}\|_{W_p^{2-2/p}(\mathcal{F}_2)} \le c \|\boldsymbol{h}_0^{(1)}\|_{W_p^{2-2/p}(\mathcal{F}_1)}.
\end{aligned}
$$

3. Non-linear problem

In this section we outline main ideas of the proof of Theorem 1. Following [1], we write the problem (1.3) in the form

$$
(3.1)\quad \begin{cases}
\boldsymbol{u}_t(y,t) - \nu\nabla^2\boldsymbol{u} + \nabla q = \boldsymbol{l}_1(\boldsymbol{u},q,\boldsymbol{h},\rho),\\
\nabla\cdot\boldsymbol{u} = l_2(\boldsymbol{u},\rho), \quad y\in\mathcal{F}_1,\ t>0,\\
\Pi_{\mathcal{G}}S(\boldsymbol{u})\boldsymbol{N} = \boldsymbol{l}_3(\boldsymbol{u},\rho),\\
-q + \nu\boldsymbol{N}\cdot S(\boldsymbol{u})\boldsymbol{N}(y) + \sigma\mathfrak{B}\rho = l_4(\boldsymbol{u},\boldsymbol{h},\rho) + l_5(\rho) + \sigma\mathcal{H}(y),\\
\rho_t + \boldsymbol{V}(x)\cdot\nabla_\tau\rho - \boldsymbol{u}\cdot\boldsymbol{N}(y) = l_6(\boldsymbol{u},\rho), \quad y\in\mathcal{G},\\
\mu_1\boldsymbol{h}_t + \alpha^{-1}\mathrm{rot\,rot}\,\boldsymbol{h} = \boldsymbol{l}_7(\boldsymbol{h},\boldsymbol{u},\rho), \quad \nabla\cdot\boldsymbol{h} = 0, \quad y\in\mathcal{F}_1,\\
\mathrm{rot}\,\boldsymbol{h} = \mathrm{rot}\,\boldsymbol{l}_8(\boldsymbol{h},\rho), \quad \nabla\cdot\boldsymbol{h}=0, \quad y\in\mathcal{F}_2,\\
[\mu\boldsymbol{h}\cdot\boldsymbol{N}] = 0, \quad [\boldsymbol{h}_\tau] = \boldsymbol{l}_9(\boldsymbol{h},\rho), \quad y\in\mathcal{G}, \quad \boldsymbol{h}\cdot\boldsymbol{n} = 0, \quad y\in S,\\
\boldsymbol{u}(y,0) = \boldsymbol{u}_0(y), \quad y\in\mathcal{F}_1, \quad \boldsymbol{h}(y,0) = \boldsymbol{h}_0(y), \quad y\in\mathcal{F}_1\cup\mathcal{F}_2,\\
\rho(y,0) = \rho_0(y), \quad y\in\mathcal{G},
\end{cases}
$$

where

$$
(3.2)\quad \begin{cases}
\boldsymbol{l}_1(\boldsymbol{u},q,\boldsymbol{h},\rho) = \nu(\widetilde{\nabla}^2 - \nabla^2)\boldsymbol{u} + (\nabla - \widetilde{\nabla})q + \rho_t^*(\mathcal{L}^{-1}\boldsymbol{N}^*(y)\cdot\nabla)\boldsymbol{u}\\
\quad -(\mathcal{L}^{-1}\boldsymbol{u}\cdot\nabla)\boldsymbol{u} + \widetilde{\nabla}\cdot T_M(\frac{\mathcal{L}}{L}\boldsymbol{h}),\\
l_2(\boldsymbol{u},\rho) = (I-\widehat{\mathcal{L}}^T)\nabla\cdot\boldsymbol{u} = \nabla\cdot\boldsymbol{L}(\boldsymbol{u},\rho),\\
\boldsymbol{L}(\boldsymbol{u},\rho) = (I-\widehat{\mathcal{L}})\boldsymbol{u}, \quad y\in\mathcal{F}_1,\\
\boldsymbol{l}_3(\boldsymbol{u},\rho) = \Pi_{\mathcal{G}}(\Pi_{\mathcal{G}}S(\boldsymbol{u})\boldsymbol{N}(y)) - \Pi\widetilde{S}(\boldsymbol{u})\boldsymbol{n}(e_\rho(y)),\\
l_4(\boldsymbol{u},\boldsymbol{h},\rho) = \nu(\boldsymbol{N}\cdot S(\boldsymbol{u})\boldsymbol{N} - \boldsymbol{n}\cdot\widetilde{S}(\boldsymbol{u})\boldsymbol{n}) - [\boldsymbol{n}\cdot T_M(\frac{\mathcal{L}}{L}\boldsymbol{h})\boldsymbol{n}],\\
l_5(\rho) = -\int_0^1 (1-s)\frac{d^2}{ds^2}\mathcal{L}^{-T}(y,s\rho)\nabla\cdot\frac{\mathcal{L}^T(y,s\rho)\boldsymbol{N}}{|\mathcal{L}^T(y,s\rho)\boldsymbol{N}|}ds,\\
l_6(\boldsymbol{u},\boldsymbol{h},\rho) = (\frac{\widehat{\mathcal{L}}^T\boldsymbol{N}}{\boldsymbol{N}\cdot\widehat{\mathcal{L}}^T\boldsymbol{N}} + \nabla_\tau\rho - \boldsymbol{N})\cdot\boldsymbol{u} + (\boldsymbol{V}-\boldsymbol{u})\cdot\nabla_\tau\rho, \quad y\in\mathcal{G},\\
\boldsymbol{l}_7(\boldsymbol{h},\boldsymbol{u},\rho) = \alpha^{-1}\mathrm{rot}\,(\mathrm{rot}\,\boldsymbol{h} - \mathcal{P}\mathrm{rot}\,\mathcal{P}\boldsymbol{h})\\
\quad + \mu_1(\frac{1}{L}\widehat{\mathcal{L}}_t\mathcal{L}\boldsymbol{h} + \rho_t^*\widehat{\mathcal{L}}(\mathcal{L}^{-1}\boldsymbol{N}\cdot\nabla)\frac{1}{L}\mathcal{L}\boldsymbol{h}) + \mu_1\mathrm{rot}\,(\mathcal{L}^{-1}\boldsymbol{u}\times\boldsymbol{h}), \quad \mathrm{y}\in\mathcal{F}_1,\\
\boldsymbol{l}_8(\boldsymbol{h},\rho) = (I-\mathcal{P})\boldsymbol{h}, \quad y\in\mathcal{F}_2,\\
\boldsymbol{l}_9(\boldsymbol{h},\rho) = (\frac{\widehat{\mathcal{L}}\widehat{\mathcal{L}}^T\boldsymbol{N}}{|\widehat{\mathcal{L}}^T\boldsymbol{N}|^2} - \boldsymbol{N})[\boldsymbol{h}\cdot\boldsymbol{N}] = [\boldsymbol{A}(\boldsymbol{h},\rho)], \quad y\in\mathcal{G},\\
\boldsymbol{A}^{(i)}(\boldsymbol{h},\rho^*) = \Big(\frac{\widehat{\mathcal{L}}(y,\rho^*)\widehat{\mathcal{L}}^T(y,\rho^*)\boldsymbol{N}^*}{|\widehat{\mathcal{L}}^T(y,\rho^*)\boldsymbol{N}^*|^2} - \frac{\boldsymbol{N}^*}{|\boldsymbol{N}^*|^2}\Big)\boldsymbol{h}^{(i)}\cdot\boldsymbol{N}^*,
\end{cases}
$$

$$y\in\mathcal{F}_i, \quad i=1,2, \quad \Pi\boldsymbol{f} = \boldsymbol{f} - \boldsymbol{n}(\boldsymbol{n}\cdot\boldsymbol{f}), \quad \Pi_{\mathcal{G}}\boldsymbol{g} = \boldsymbol{g} - \boldsymbol{N}(\boldsymbol{g}\cdot\boldsymbol{N}).$$

We have used the formula for the variation of the mean curvature under the normal perturbation of the surface:

$$H(e_\rho) - \mathcal{H}(y) = -\mathfrak{B}\rho - \int_0^1 (1-s)\frac{d^2}{ds^2}\mathcal{L}^{-T}(y,s\rho)\nabla\cdot\frac{\mathcal{L}^T(y,s\rho)\boldsymbol{N}}{|\mathcal{L}^T(y,s\rho)\boldsymbol{N}|}ds.$$

The vector field $\boldsymbol{V} \in W_p^{2-1/p}(\mathcal{G})$ should be close to $\boldsymbol{u}_0$ in the norm $W_p^{1-1/p}(\mathcal{G})$ (see (3.6)) and provide good estimates for the non-linear expression $l_6(\boldsymbol{u}, \boldsymbol{h}, \rho)$.

The solvability of the problems (3.1), (3.2) is proved by successive approximations, according to a standard scheme:

(3.3)
$$\begin{cases} \boldsymbol{u}_{m+1,t}(y,t) - \nu\nabla^2\boldsymbol{u}_{m+1} + \nabla q_{m+1} = \boldsymbol{l}_1(\boldsymbol{u}_m, q_m, \boldsymbol{h}_m, \rho_m), \\ \nabla \cdot \boldsymbol{u}_{m+1} = l_2(\boldsymbol{u}_M, \rho_M) + \nabla \cdot \boldsymbol{L}(\boldsymbol{u}_m, \rho_m), \quad y \in \mathcal{F}_1,\ t > 0, \\ \Pi_{\mathcal{G}} S(\boldsymbol{u}_{m+1})\boldsymbol{N} = \boldsymbol{l}_3(\boldsymbol{u}_m, \rho_m), \\ -q_{m+1} + \nu \boldsymbol{N} \cdot S(\boldsymbol{u}_{m+1})\boldsymbol{N}(y) + \sigma\mathfrak{B}\rho_{m+1} = l_4(\boldsymbol{u}_m, \boldsymbol{h}_m, \rho_m) + l_5(\rho_m) + \sigma\mathcal{H}(y), \\ \rho_{m+1,t} + \boldsymbol{V}(x)\cdot\nabla_\tau\rho_{m+1} - \boldsymbol{u}_{m+1}\cdot\boldsymbol{N}(y) = l_6(\boldsymbol{u}_m, \rho_m), \quad y \in \mathcal{G}, \\ \mu_1\boldsymbol{h}_{m+1,t} + \alpha^{-1}\mathrm{rot}\,\mathrm{rot}\,\boldsymbol{h}_{m+1} = \boldsymbol{l}_7(\boldsymbol{h}_m, \boldsymbol{u}_m, \rho_m), \quad \nabla\cdot\boldsymbol{h}_{m+1} = 0, \quad y \in \mathcal{F}_1, \\ \mathrm{rot}\,\boldsymbol{h}_{m+1} = \mathrm{rot}\,\boldsymbol{l}_8(\boldsymbol{h}_m, \rho_m), \quad \nabla\cdot\boldsymbol{h}_{m+1} = 0, \quad y \in \mathcal{F}_2, \\ [\mu\boldsymbol{h}_{m+1}\cdot\boldsymbol{N}] = 0, \quad [\boldsymbol{h}_{m+1,\tau}] = \boldsymbol{l}_9(\boldsymbol{h}_m, \rho_m), \quad y \in \mathcal{G}, \quad \boldsymbol{h}_{m+1}\cdot\boldsymbol{n} = 0, \quad y \in S, \\ \boldsymbol{u}_{m+1}(y,0) = \boldsymbol{u}_0(y), \quad y \in \mathcal{F}_1, \quad \boldsymbol{h}_{m+1}(y,0) = \boldsymbol{h}_0(y), \quad y \in \mathcal{F}_1 \cup \mathcal{F}_2, \\ \rho_{m+1}(y,0) = \rho_0(y), \quad y \in \mathcal{G}, \quad m = 1, \dots. \end{cases}$$

The first approximation, $(\boldsymbol{u}_1, q_1, \rho_1, \boldsymbol{h}_1)$, is defined in the following way: $q_1 = 0$, $\boldsymbol{u}_1$ and ρ_1 satisfy the initial conditions

$$\boldsymbol{u}_1(y,0) = \boldsymbol{u}_0(y), \quad y \in \mathcal{F}_1, \quad \rho_1(y,0) = \rho_0(y), \quad y \in \mathcal{G},$$

and the inequalities

$$\begin{aligned} &\|\boldsymbol{u}_1\|_{W_p^{2,1}(Q_T^1)} \le c\|\boldsymbol{u}_1\|_{W_p^{2,1}(Q_\infty^1)} \le c\|\boldsymbol{u}_0\|_{W_p^{2-2/p}(\mathcal{F}_1)}, \\ &\|\rho_1\|_{W_p^{3-1/p,0}(G_\infty)} + \|\rho_{1,t}\|_{W_p^{2-1/p,0}(G_\infty)} \le c\|\rho_0\|_{W_p^{3-2/p}(\mathcal{G})}, \end{aligned} \tag{3.4}$$

and $\boldsymbol{h}_1$ is defined as a divergence-free vector field satisfying the equations (2.18), initial conditions

$$\boldsymbol{h}_1(y,0) = \boldsymbol{h}_0(y), \quad y \in \mathcal{F}_1 \cup \mathcal{F}_2,$$

and the inequality

$$\sum_{i=1}^{2} \|\boldsymbol{h}_1^{(i)}\|_{W_r^{2,1}(Q_\infty^i)} \le c\|\boldsymbol{h}_0^{(1)}\|_{W_r^{2-2/r}(\mathcal{F}_1)}. \tag{3.5}$$

The existence of such $\boldsymbol{u}_1$, ρ_1 follows from the inverse trace theorems (see Section 1) and $\boldsymbol{h}_1$ can be constructed exactly as in the proof of Theorem 3 in [1].

The proof of the solvability of (3.3) is based on Theorems 2 and 4 (with the exponent r instead of p), and on the estimates of nonlinear terms (3.2) obtained in Section 5.

THEOREM 5. *Assume that* $\boldsymbol{u}_m \in W_p^{2,1}(Q_T^1)$, $\nabla q_m \in L_p(Q_T^1)$, $\rho_m \in W_p^{3-1/p,0}(G_T)$, $\rho_{m,t} \in W_p^{2-1/p,1-1/2p}(G_T)$, $\boldsymbol{h}_m \in W_r^{2,0}(Q_T^i)$, $i = 1, 2$, *and the conditions* (1.5), *as well as*

$$\begin{aligned} &\|\rho_m(\cdot,t)\|_{W_p^{1-1/p}(\mathcal{G})} \le \delta \ll 1, \\ &\|\boldsymbol{V} - \boldsymbol{u}_0\|_{W_p^{1-1/p}(\mathcal{G})} \le \delta, \\ &T^{1/2-1/p}\|\boldsymbol{V}\|_{W_p^{2-1/p}(\mathcal{G})} \le \delta \end{aligned} \tag{3.6}$$

are satisfied. Then

$$Z_m(T) \leq \vartheta(\delta,T)(Y_m(T)+Y_m^2(T)), \tag{3.7}$$

where

$$\begin{aligned} Z_m(T) =& \|\boldsymbol{l}_1(\boldsymbol{u}_m, q_m, \boldsymbol{h}_m, \rho)\|_{L_p(Q_T^1)} + \|l_2(\boldsymbol{u}_m,\rho_m)\|_{W_p^{1,0}(Q_T^1)} \\ &+ \|\boldsymbol{L}(\boldsymbol{u}_m,\rho_m)\|_{W_p^{0,1}(Q_T^1)} \\ &+ \|\boldsymbol{l}_3(\boldsymbol{u}_m,\rho_m)\|_{W_p^{1-1/p,1/2-1/2p}(G_T)} \\ &+ \|l_4(\boldsymbol{u}_m,\boldsymbol{h}_m,\rho_m)\|_{W_p^{1-1/p,0}(G_T)} + \|l_5(\rho_m)\|_{W_p^{1-1/p,0}(G_T)} \\ &+ \|l_6(\boldsymbol{u}_m,\rho_m)\|_{W_p^{2-1/p,1-1/2p}(G_T)} + \|\boldsymbol{l}_7(\boldsymbol{h}_m,\boldsymbol{u}_m,\rho_m)\|_{L_r(Q_T^1)} \\ &+ \|\boldsymbol{l}_8(\boldsymbol{h}_m,\rho_m)\|_{W_r^{2,1}(Q_T^2)} + \sum_{i=1}^2 \|\boldsymbol{A}^{(i)}\|_{W_r^{2,1}(Q_T^i)}, \end{aligned}$$

where $\vartheta(\delta,T)$ is a constant that can be made arbitrarily small by the choice of small δ and T.

Estimate (3.7) is a direct consequence of the inequalities (5.15)–(5.20), (5.22), (5.23), (5.26), (5.27), (5.32), (5.34)-(5.38) obtained below in Section 5.

If $(\boldsymbol{u}_m, q_m, \rho_m, \boldsymbol{h}_m)$ are found, then, by Theorems 2 and 4, the problem (3.3) is uniquely solvable and

$$Y_{m+1}(T) \leq c\theta(\delta,T)(Y_m(T)+Y_m^2(T)) + c_1 N, \tag{3.8}$$

where

(3.9)

$$\begin{aligned} Y_m(T) =& \|\boldsymbol{u}_m\|_{W_p^{2,1}(Q_T^1)} + \sup_{t<T}\|\boldsymbol{u}_m\|_{W_p^{2-2/p}(\mathcal{F}_1)} + \|\nabla q_m\|_{L_p(Q_T^1)} + \|q_m\|_{W_p^{1-1/p,0}(G_T)} \\ &+ \|\rho_m\|_{W_p^{3-1/p,0}(G_T)} + \|\rho_{m,t}\|_{W_p^{2-1/p,1-1/2p}(G_T)} + \sup_{t<T}\|\rho_m(\cdot,t)\|_{W_p^{3-2/p}(\mathcal{G})} \\ &+ \sum_{i=1}^2 (\|\boldsymbol{h}_m\|_{W_r^{2,1}(Q_T^i)} + \sup_{t<T}\|\boldsymbol{h}_m\|_{W_r^{2-2/r}(\mathcal{F}_i)}) \\ \equiv& Y(\boldsymbol{u}_m, q_m, \rho_m, \boldsymbol{h}_m), \end{aligned}$$

$$N = \|\boldsymbol{u}_0\|_{W_p^{2-2/p}(\mathcal{F}_1)} + \|\rho_0\|_{W_p^{3-2/p}(\mathcal{G})} + \|\mathcal{H}\|_{W_p^{1-1/p}(\mathcal{G})} + \sum_{i=1}^2 \|\boldsymbol{h}_0\|_{W_r^{2-2/r}(\mathcal{F}_i)}. \tag{3.10}$$

It follows that

$$Y_{m+1}(T) \leq 2c_1 N, \tag{3.11}$$

if $Y_m(T) \leq 2c_1 N$ and ϑ is so small that

$$c\vartheta(\delta,T)(2c_1 N + 4(c_1 N)^2) \leq c_1 N,$$

i.e.,

$$\vartheta(\delta,T) \leq \frac{1}{2c(1+2c_1 N)}; \tag{3.12}$$

in addition,

$$\begin{aligned} \|\rho_{m+1}(\cdot,t)\|_{W_p^{2-1/p}(\mathcal{G})} &\leq \|\rho_0\|_{W_p^{2-1/p}(\mathcal{G})} + \int_0^t \|\rho_{m+1,\tau}(\cdot,\tau)\|_{W_p^{2-1/p}(\mathcal{G})} d\tau \\ &\leq \epsilon + 2c_1 T^{1-1/p} N \leq \delta. \end{aligned} \tag{3.13}$$

Hence if c_1 is chosen in such a way that (3.11) holds for $m = 0$, then (3.11) and (3.13) are satisfied for all m.

In order to prove the convergence of the sequence $(\boldsymbol{u}_m, q_m, \rho_m, \boldsymbol{h}_m)$, we estimate the differences

$$\boldsymbol{u}_{m+1} - \boldsymbol{u}_m, \quad q_{m+1} - q_m, \quad \rho_{m+1} - \rho_m, \quad \boldsymbol{h}_{m+1} - \boldsymbol{h}_m.$$

We show that

$$Y(\boldsymbol{u}_{m+1} - \boldsymbol{u}_m, q_{m+1} - q_m, \rho_{m+1} - \rho_m, \boldsymbol{h}_{m+1} - \boldsymbol{h}_m)$$
$$\leq \vartheta_1 Y(\boldsymbol{u}_m - \boldsymbol{u}_{m-1}, q_m - q_{m-1}, \rho_m - \rho_{m-1}, \boldsymbol{h}_m - \boldsymbol{h}_{m-1}),$$

$m > 1$, with a small ϑ_1, which guarantees the boundedness of

$$\sum_{m=1}^{\infty} Y(\boldsymbol{u}_{m+1} - \boldsymbol{u}_m, q_{m+1} - q_m, \rho_{m+1} - \rho_m, \boldsymbol{h}_{m+1} - \boldsymbol{h}_m)$$

and the convergence of $(\boldsymbol{u}_m, q_m, \rho_m, \boldsymbol{h}_m)$. This is made in the same way as in the papers [1, 7] in the case $p, r = 2$, and we omit the details.

It is clear that $(\boldsymbol{u}_m, q_m, \rho_m, \boldsymbol{h}_m)$ tend to the solution of the problem (3.1), as $m \to \infty$ and the solution satisfies (3.11):

$$Y(T) \leq 2c_1 N. \tag{3.14}$$

Remark. In the proof of (3.11) we have used the fact that the constants $c(T)$ in (2.5) and (2.14) are bounded for small T. This follows from the assumption $p > 3$, which implies $1/2 - 1/2p > 1/p$. As a consequence, the functions $d \in W_p^{k-1/p,k/2-1/2p}(G_T)$, $k = 1, 2$, vanishing for $t = 0$ (such as $\boldsymbol{d}_\tau$ in (2.1) in the case of zero initial data) admit zero extension in the domain $t < 0$, and the extended functions $d^{(0)}$ satisfy the inequality

$$\|d^{(0)}\|_{W_p^{k-1/p,k/2-1/2p}(\mathcal{G}\times(-\infty,T))} \leq c\|d\|_{W_p^{k-1/p,k/2-1/2p}(G_T)}, \quad \forall t < T$$

(see [8, Sec.4, §4]) without any additional terms of the type

$$\frac{1}{T^\alpha}\|d\|_{L_p(\mathcal{G})}, \quad \alpha > 0,$$

as was in the case $p = 2$ (see details in [7]).

4. Construction of the electric field

The complete system of equations of magnetohydrodynamics contains the electric field $\boldsymbol{E}(x, t)$. In particular, the problem (1.1) consists of finding the free boundary Γ_t and the functions $\boldsymbol{v}, p, \boldsymbol{H}$ and $\boldsymbol{E}$ such that

$$\begin{cases} \boldsymbol{v}_t + (\boldsymbol{v}\cdot\nabla)\boldsymbol{v} - \nabla\cdot T(\boldsymbol{v}, p) - \nabla\cdot T_M(\boldsymbol{H}) = 0, \\ \nabla\cdot\boldsymbol{v}(x,t) = 0, \quad x \in \Omega_{1t}, \ t > 0, \end{cases} \tag{4.1}$$

$$\begin{cases} \mu\boldsymbol{H}_t = -\operatorname{rot}\boldsymbol{E}, \quad \nabla\cdot\boldsymbol{H} = 0, \qquad x \in \Omega_{1t}\cup\Omega_{2t}, \\ \operatorname{rot}\boldsymbol{H} = \alpha(\boldsymbol{E} + \mu(\boldsymbol{v}\times\boldsymbol{H})), \quad x \in \Omega_{1t}, \ t > 0, \\ \operatorname{rot}\boldsymbol{H} = 0, \quad \nabla\cdot\boldsymbol{H} = 0, \quad \nabla\cdot\boldsymbol{E} = 0, \quad x \in \Omega_{2t}. \end{cases} \tag{4.2}$$

$$\begin{cases} \boldsymbol{H}\cdot\boldsymbol{n} = 0, \quad \boldsymbol{E}_\tau = 0, \quad x \in S, \\ [\mu\boldsymbol{H}\cdot\boldsymbol{n}] = 0, \quad [\boldsymbol{H}_\tau] = 0, \\ (T(\boldsymbol{v}, p) + [T_M(\boldsymbol{H})])\boldsymbol{n} = \sigma\boldsymbol{n}H, \quad V_n = \boldsymbol{v}\cdot\boldsymbol{n}, \quad x \in \Gamma_t, \end{cases} \tag{4.3}$$

$$\mathfrak{n}_t[\mu \boldsymbol{H}] + [\mathfrak{n}_x \times \boldsymbol{E}] = 0, \quad x \in \Gamma_t, \tag{4.4}$$

$$\boldsymbol{v}(x,0) = \boldsymbol{v}_0(x), \quad x \in \Omega_{10}, \quad \boldsymbol{H}(x,0) = \boldsymbol{H}_0(x), \quad x \in \Omega_{10} \cup \Omega_{20}, \tag{4.5}$$

where $\mathfrak{n}_x = (\mathfrak{n}_1, \mathfrak{n}_2, \mathfrak{n}_3)$ and $\mathfrak{n}_t$ are components of the normal vector $\mathfrak{n}$ to the surface $\mathfrak{G} = \{x \in \Gamma_t, t > 0\}$ in $\mathbb{R}^4$. In particular, if $\mu_1 = \mu_2$, then condition (4.4) takes a standard form $[\mathfrak{n}_x \times \boldsymbol{E}] = 0$, i.e., $[\boldsymbol{E}_\tau] = 0$.

Condition (4.4) can be deduced from the assumption that the first equation in (4.2) is satisfied in the sense of the distributions theory, i.e.,

$$\int_{t_0}^{t_0+\tau} \int_K (-\boldsymbol{B} \cdot \boldsymbol{\varphi}_t + \boldsymbol{E} \cdot \operatorname{rot} \boldsymbol{\varphi}) dx dt = 0, \tag{4.6}$$

where $\boldsymbol{B} = \mu \boldsymbol{H}$, $\boldsymbol{\varphi} \in C_0^\infty(K)$, $K \subset \Omega$, $K \cap \Gamma_t \neq \emptyset$. From (4.6) it follows that

$$\boldsymbol{B}_t + \operatorname{rot} \boldsymbol{E} = 0, \quad x \in K,$$
$$\mathfrak{n}_t[\boldsymbol{B}] + [\mathfrak{n}_x \times \boldsymbol{E}] = 0, \quad x \in K \cup \Gamma_t, \quad t \in (t_0, t_0 + \tau).$$

The following proposition is important for the construction of $\boldsymbol{E}$.

THEOREM 6. *If the relations*

$$\begin{aligned} &\nabla \cdot \boldsymbol{B}(x,t) = 0, \quad x \in K \cap (\Omega_1 \cup \Omega_2), \quad [\boldsymbol{B} \cdot \boldsymbol{n}] = 0, \quad x \in K \cap \Gamma_t, \\ &\mathfrak{n}_t[\boldsymbol{B}] + [\boldsymbol{n}_x \times \boldsymbol{E}] = 0, \quad x \in K \cap \Gamma_t \end{aligned} \tag{4.7}$$

are satisfied, then

$$[\boldsymbol{n} \cdot \operatorname{rot} \boldsymbol{E}] = -[\boldsymbol{n} \cdot \boldsymbol{B}_t], \quad x \in \Gamma_t \cap K. \tag{4.8}$$

PROOF. Suppose the surface $\mathfrak{G}_\tau = \mathfrak{G} \cap (K \times (t_0, t_0 + \tau))$ is given by the equation $\xi_3 = z(\xi_1, \xi_2, t)$, where (ξ_1, ξ_2, ξ_3) are Cartesian coordinates in $\mathbb{R}^3$. The functions given on $\mathfrak{G}_\tau$ can be considered as functions of ξ_1, ξ_2, t. The normal $\mathfrak{n}$ to $\mathfrak{G}_\tau$ is parallel to the vector $(\widetilde{\boldsymbol{n}}, z_t)$, where $\widetilde{\boldsymbol{n}} = (z_{\xi_1}, z_{\xi_2}, -1)$.

It is easily verified that

$$\begin{aligned} -[\widetilde{\boldsymbol{n}} \cdot \operatorname{rot} \boldsymbol{E}] &= [z_{\xi_1}(E_{2,3} - E_{3,2}) + z_{\xi_2}(E_{3,1} - E_{1,3}) - (E_{1,2} - E_{2,1})] \\ &= \Big[\frac{\partial}{\partial \xi_1}(z_{\xi_2} E_3 + E_2) + \frac{\partial}{\partial \xi_2}(-E_1 - z_{\xi_1} E_3)\Big] \\ &= \frac{\partial}{\partial \xi_1}[\widetilde{\boldsymbol{n}} \times \boldsymbol{E}]_1 + \frac{\partial}{\partial \xi_2}[\widetilde{\boldsymbol{n}} \times \boldsymbol{E}]_2, \end{aligned} \tag{4.9}$$

where $E_{i,j}$ is the partial derivative of E_i with respect to ξ_j and

$$\frac{\partial \boldsymbol{E}}{\partial \xi_\alpha} = \boldsymbol{E}_{,\alpha} + z_{\xi_\alpha} \boldsymbol{E}_{,3}, \quad \alpha = 1, 2,$$

are derivatives calculated taking the dependence of $\boldsymbol{E}$ on ξ_3 into account. By (4.7), equation (4.9) is equivalent to

$$\begin{aligned} -[\widetilde{\boldsymbol{n}} \cdot \operatorname{rot} \boldsymbol{E}] &= -\frac{\partial}{\partial \xi_1} z_t[B_1] - \frac{\partial}{\partial \xi_2} z_t[B_2] \\ &= -[z_{\xi_1,t} B_1 + z_t B_{1,\xi_1} + z_{\xi_1} z_t B_{1,3} + z_{\xi_2,t} B_2 + z_t B_{2,\xi_2} + z_{\xi_2} z_t B_{2,3}]. \end{aligned} \tag{4.10}$$

Now we differentiate $[\widetilde{\boldsymbol{n}} \cdot \boldsymbol{B}] = 0$ with respect to t, which leads to

$$[z_{\xi_1,t} B_1 + z_{\xi_1} B_{1,t} + z_t z_{\xi_1} B_{1,3} + z_{\xi_2,t} B_2 + z_{\xi_2} B_{2,t} + z_{\xi_2} z_t B_{2,3} - B_{3,t} - z_t B_{3,t}] = 0. \tag{4.11}$$

When we add (4.11) to (4.10) and make use of the equation $\nabla \cdot \boldsymbol{B} = 0$, we obtain

$$-[\widetilde{\boldsymbol{n}} \cdot \operatorname{rot} \boldsymbol{E}] = [z_{\xi_1} B_{1,t} + z_{\xi_2} B_{2,t} - B_{3,t}] = [\widetilde{\boldsymbol{n}} \cdot \boldsymbol{B}_t].$$

The proof of the theorem is due to Dr. N. Filonov.

The transformation (1.2) converts (4.1)–(4.4) into

$$\begin{cases} \boldsymbol{u}_t - \rho_t^*(\mathcal{L}^{-1}\boldsymbol{N}^*(y)\cdot\nabla)\boldsymbol{u} + (\mathcal{L}^{-1}\boldsymbol{u}\cdot\nabla)\boldsymbol{u} \\ \qquad -\widetilde{\nabla}\cdot\widetilde{T}(\boldsymbol{u},q) - \widetilde{\nabla}\cdot T_M(\frac{\mathcal{L}}{L}\boldsymbol{h}) = 0, \\ \nabla\cdot\widehat{\mathcal{L}}\boldsymbol{u} = 0, \quad y\in\mathcal{F}_1, \ t>0, \end{cases} \tag{4.12}$$

$$\begin{cases} \mu(\boldsymbol{h}_t - \frac{1}{L}\widehat{\mathcal{L}}_t\mathcal{L}\boldsymbol{h} - \rho_t^*\widehat{\mathcal{L}}(\mathcal{L}^{-1}\boldsymbol{N}^*(y)\cdot\nabla)\frac{1}{L}\mathcal{L}\boldsymbol{h}) = -\mathrm{rot}\,\mathcal{P}\boldsymbol{e}, \quad y\in\mathcal{F}_1\cup\mathcal{F}_2, \\ \mathcal{P}\mathrm{rot}\,\mathcal{P}\boldsymbol{h} = \alpha(\mathcal{P}\boldsymbol{e} + \mu(\mathcal{L}^{-1}\boldsymbol{u}\times\boldsymbol{h})), \quad \nabla\cdot\boldsymbol{h} = 0, \quad y\in\mathcal{F}_1, \\ \mathrm{rot}\,\mathcal{P}\boldsymbol{h} = 0, \quad \nabla\cdot\boldsymbol{h} = 0, \quad \nabla\cdot\boldsymbol{e} = 0, \quad x\in\mathcal{F}_2, \end{cases} \tag{4.13}$$

$$\begin{cases} \boldsymbol{h}\cdot\boldsymbol{n} = 0, \quad \boldsymbol{e}_\tau = 0, \quad y\in S, \\ [\mu\boldsymbol{h}\cdot\boldsymbol{N}] = 0, \quad [\boldsymbol{h} - \dfrac{\widetilde{\mathcal{L}}\widetilde{\mathcal{L}}^T\boldsymbol{N}}{|\widetilde{\mathcal{L}}^T\boldsymbol{N}|^2}(\boldsymbol{h}\cdot\boldsymbol{N})] = 0, \\ \widetilde{T}(\boldsymbol{u},q)\boldsymbol{n}(e_\rho) + [T_M(\frac{\mathcal{L}}{L}\boldsymbol{h})\boldsymbol{n}(e_\rho)] = \sigma H\boldsymbol{n}, \quad \rho_t = \dfrac{\boldsymbol{u}\cdot\widehat{\mathcal{L}}^T\boldsymbol{N}}{\Lambda(y,\rho)}, \\ -\Lambda\rho_t[\mu\boldsymbol{h}] + L[\boldsymbol{N}\times\mathcal{P}\boldsymbol{e}] = 0, \quad y\in\mathcal{G}, \end{cases} \tag{4.14}$$

and (4.7), (4.8) (with $K=\Omega$) into

$$\begin{gathered} \nabla\cdot\boldsymbol{b} = 0, \quad y\in\mathcal{F}_1\cup\mathcal{F}_2, \quad [\boldsymbol{b}\cdot\boldsymbol{N}] = 0, \quad y\in\mathcal{G}, \\ [\boldsymbol{N}\times\mathcal{P}\boldsymbol{e}] = \boldsymbol{\Psi}(\boldsymbol{b},\rho), \quad y\in\mathcal{G}, \end{gathered} \tag{4.15}$$

$$[\boldsymbol{N}\times\mathrm{rot}\,\mathcal{P}\boldsymbol{e}] = \boldsymbol{\Phi}(\boldsymbol{b},\rho), \tag{4.16}$$

where $\boldsymbol{e} = \widehat{\mathcal{L}}\boldsymbol{E}(e_\rho,t)$, $\boldsymbol{b} = \widehat{\mathcal{L}}\boldsymbol{B}(e_\rho,t)$, $\boldsymbol{h} = \widehat{\mathcal{L}}\boldsymbol{H}(e_\rho,t)$, $y\in\mathcal{F}_1\cup\mathcal{F}_2$,

$$\begin{gathered} \boldsymbol{\Phi}(\boldsymbol{b},\rho) = \frac{1}{L}\widehat{\mathcal{L}}_t^T\mathcal{L}\boldsymbol{b} + \rho_t^*\widehat{\mathcal{L}}(\mathcal{L}^{-1}\boldsymbol{N}\cdot\nabla)\frac{1}{L}\mathcal{L}\boldsymbol{b}, \quad \boldsymbol{\Psi}(\boldsymbol{b},\rho) = \frac{\Lambda\rho_t}{L}[\boldsymbol{b}], \\ \Lambda(y,\rho) = \boldsymbol{N}(y)\cdot\widehat{\mathcal{L}}^T(y,\rho)\boldsymbol{N}(y) = 1 - \rho\mathcal{H} + \rho^2\mathcal{K}, \quad y\in\mathcal{G}. \end{gathered} \tag{4.17}$$

Hence (4.15) implies (4.16). We observe that the vector field $\boldsymbol{\Phi}(\boldsymbol{h},\rho)$ is divergence free in $\mathcal{F}_1\cup\mathcal{F}_2$ [1].

Now we pass to the construction of $\boldsymbol{e}$ satisfying (4.13), (4.14) together with $\boldsymbol{u},p,\rho,\boldsymbol{h}$ whose existence is established above. We introduce the auxiliary vector field $\boldsymbol{e}_1(y,t)$ such that

$$\begin{gathered} \nabla\cdot\boldsymbol{e}_1(y,t) = 0, \quad y\in\mathcal{F}_1\cup\mathcal{F}_2, \\ [\boldsymbol{N}(y)\times\mathcal{P}\boldsymbol{e}_1] = \boldsymbol{\Psi}(\mu\boldsymbol{h},\rho), \quad [\boldsymbol{N}\cdot\boldsymbol{e}_1] = 0, \quad y\in\mathcal{G}, \quad \boldsymbol{e}_1(y,t) = 0, \quad y\in S, \\ \sum_{i=1}^2 \|\boldsymbol{e}_1\|_{W_r^{1,0}(Q_T^i)} \le c\|\boldsymbol{\Psi}\|_{W_r^{1-1/r,0}(G_T)}, \end{gathered} \tag{4.18}$$

and we solve the problem

$$\begin{gathered} \mathrm{rot}\,\boldsymbol{\mathcal{E}}(y,t) = -\mu\boldsymbol{h}_t + \boldsymbol{\Phi}(\mu\boldsymbol{h},\rho) - \mathrm{rot}\,\mathcal{P}\boldsymbol{e}_1(y,t), \\ \nabla\cdot\mathcal{P}^{-1}\cdot\boldsymbol{\mathcal{E}}(y,t) = 0, \quad y\in\Omega, \\ \boldsymbol{\mathcal{E}}_\tau = 0, \quad y\in S. \end{gathered} \tag{4.19}$$

In view of Theorem 6, we have

$$[\boldsymbol{N}(y)\cdot(\mu(\boldsymbol{h}_t-\boldsymbol{\Phi})+\operatorname{rot}\mathcal{P}\boldsymbol{e}_1)]=0,\quad [\mu\boldsymbol{h}_t\cdot\boldsymbol{N}]=0,\quad y\in\mathcal{G},$$
$$\boldsymbol{n}\cdot(\mu(\boldsymbol{h}_t-\boldsymbol{\Phi})+\operatorname{rot}\mathcal{P}\boldsymbol{e}_1(y,t))=0,\quad y\in S;$$

hence problem (4.19) is solvable. The solution has the form $\boldsymbol{\mathcal{E}}=\boldsymbol{\mathcal{E}}_1+\nabla Z(y,t)$, where

$$\boldsymbol{\mathcal{E}}_1(y,t)=\frac{1}{4\pi}\operatorname{rot}\int_\Omega\frac{-\mu\boldsymbol{h}_{\rm t}+\mu\boldsymbol{\Phi}-\operatorname{rot}\mathcal{P}\boldsymbol{e}_1(\rm z,t)}{|\rm y-z|}\rm dz,$$
$$\nabla\cdot\mathcal{P}^{-1}\nabla Z=-\nabla\mathcal{P}^{-1}\boldsymbol{\mathcal{E}}_1,\quad y\in\Omega,\quad Z(y,t)=-g(y,t),\quad y\in S.$$

It is easily seen that

$$\operatorname{rot}\boldsymbol{\mathcal{E}}_1=-\mu\boldsymbol{h}_t+\mu\boldsymbol{\Phi}-\operatorname{rot}\mathcal{P}\boldsymbol{e}_1(y,t),\quad y\in\Omega.$$

The function g is defined as follows. Since $(\mu\boldsymbol{h}_t-\mu\boldsymbol{\Phi}+\operatorname{rot}\mathcal{P}\boldsymbol{e}_1)\cdot\boldsymbol{n}|_S=0$, we have $\int_\Sigma\operatorname{rot}\boldsymbol{\mathcal{E}}_1\cdot\boldsymbol{n}\mathrm{dS}=0$ for arbitrary $\Sigma\subset S$. Hence by the Stokes formula,

$$\int_\gamma\boldsymbol{\mathcal{E}}_1\cdot d\boldsymbol{l}=0$$

for arbitrary closed contour $\gamma=\partial\Sigma\subset S$, which implies $\boldsymbol{\mathcal{E}}_1|_S=\nabla_\tau g(y,t)$ with a certain single-valued g. Now equations (4.19) are easily verified.

By the Calderon–Zygmund theorem,

$$\begin{aligned}\|\boldsymbol{\mathcal{E}}_1\|_{W_r^{1,0}(Q_T)}&\le c\|\mu\boldsymbol{h}_t-\mu\boldsymbol{\Phi}+\operatorname{rot}\mathcal{P}\boldsymbol{e}_1\|_{L_r(Q_T)}\\&\le c(\|\boldsymbol{h}_t\|_{L_r(Q_T)}+\|\boldsymbol{\Phi}\|_{L_r(Q_T)}+\|\boldsymbol{\Psi}\|_{W_r^{1-1/r,0}(G_T)}),\end{aligned}\tag{4.20}$$

in addition,

$$\|\nabla Z\|_{W_r^{1,0}(Q_T)}\le c\|\boldsymbol{\mathcal{E}}_1\|_{W_r^{1,0}(Q_T)},$$

hence

$$\|\boldsymbol{\mathcal{E}}\|_{W_r^{1,0}(Q_T)}\le c(\|\boldsymbol{h}_t\|_{L_r(Q_T)}+\|\boldsymbol{\Phi}\|_{L_r(Q_T)}+\|\boldsymbol{\Psi}\|_{W_r^{1-1/r,0}(G_T)}).\tag{4.21}$$

The norms of $\boldsymbol{\Phi}$ and $\boldsymbol{\Psi}$ are estimated below in (5.37) and (5.38).

Now we define $\boldsymbol{e}^{(1)}$ by

$$\mathcal{P}\boldsymbol{e}^{(1)}=\alpha_1^{-1}\mathcal{P}\operatorname{rot}\mathcal{P}\boldsymbol{h}-\mu(\mathcal{L}^{-1}\boldsymbol{u}\times\boldsymbol{h}).\tag{4.22}$$

Then

$$\begin{aligned}\operatorname{rot}\mathcal{P}\boldsymbol{e}^{(1)}&=\operatorname{rot}(\alpha_1^{-1}\mathcal{P}\operatorname{rot}\mathcal{P}\boldsymbol{h}-\mu(\mathcal{L}^{-1}\boldsymbol{u}\times\boldsymbol{h}))\\&=-\mu\boldsymbol{h}_t+\mu\boldsymbol{\Phi}=\operatorname{rot}\boldsymbol{\mathcal{E}}+\operatorname{rot}\mathcal{P}\boldsymbol{e}_1,\quad y\in\mathcal{F}_1,\end{aligned}$$

which implies

$$\mathcal{P}\boldsymbol{e}^{(1)}=\boldsymbol{\mathcal{E}}^{(1)}+\mathcal{P}\boldsymbol{e}_1^{(1)}+\nabla\chi_1(y,t),\quad y\in\mathcal{F}_1\tag{4.23}$$

with a certain single-valued $\chi_1(y,t)$. Finally, we set

$$\mathcal{P}\boldsymbol{e}^{(2)}=\boldsymbol{\mathcal{E}}^{(2)}+\mathcal{P}\boldsymbol{e}_1^{(2)}+\nabla\chi_2(y,t)+C(t)\nabla w(y),\quad y\in\mathcal{F}_2,\tag{4.24}$$

where χ_2 and $\boldsymbol{w}$ are solutions to the problems

$$\begin{aligned}&\nabla\cdot\mathcal{P}^{-1}\nabla\chi_2(y,t)=0,\quad y\in\mathcal{F}_2,\\&\chi_2(y,t)=\chi_1(y,t),\quad y\in\mathcal{G},\quad \chi_2(y,t)=0,\quad y\in S,\\&\nabla\cdot\mathcal{P}^{-1}\nabla w(y)=0,\quad y\in\mathcal{F}_2,\quad w(y)=1,\quad y\in\mathcal{G},\quad w(y,t)=0,\quad y\in S.\end{aligned}\tag{4.25}$$

The vector field $\boldsymbol{e}$ defined in this way satisfies (4.13) and (4.14). Indeed,

$$\mu \boldsymbol{h}_t - \boldsymbol{\Phi} + \operatorname{rot} \mathcal{P}\boldsymbol{e}^{(i)} = \mu \boldsymbol{h}_t - \boldsymbol{\Phi} + \operatorname{rot} \boldsymbol{\mathcal{E}}^{(i)} + \operatorname{rot} \mathcal{P}\boldsymbol{e}_1^{(i)} = 0, \quad y \in \mathcal{F}_i, \quad i = 1, 2,$$
$$\nabla \cdot \boldsymbol{e}^{(2)} = \nabla \cdot \mathcal{P}^{-1}\boldsymbol{\mathcal{E}}^{(2)} + \nabla \cdot \boldsymbol{e}_1^{(2)} = 0, \quad y \in \mathcal{F}_2,$$
$$[\boldsymbol{N} \times \mathcal{P}\boldsymbol{e}] = [\boldsymbol{N} \times \boldsymbol{\mathcal{E}}] + [\boldsymbol{N} \times \mathcal{P}\boldsymbol{e}_1] = \boldsymbol{\Psi}(\mu \boldsymbol{h}, \rho), \quad y \in \mathcal{G}, \quad \boldsymbol{e}_\tau|_S = \mathcal{E}_\tau + \boldsymbol{e}_{1\tau}|_S = 0.$$

In order to find $C(t)$, we need to impose on $\boldsymbol{e}^{(2)}$ a normalization restriction, for instance of the form

$$\int_S \mathcal{P}\boldsymbol{e}^{(2)}(y,t) \cdot \boldsymbol{n}(y) dS - 0, \tag{4.26}$$

which coincides with

$$\int_S \boldsymbol{E}^{(2)}(x,t) \cdot \boldsymbol{n}(x) dS = 0.$$

Since $I = \int_S \frac{\partial w}{\partial n} dS \neq 0$, we have

$$C(t) = -I^{-1} \int_S (\boldsymbol{\mathcal{E}}^{(2)} + \mathcal{P}\boldsymbol{e}_1^{(2)} + \nabla \chi_2(y,t)) \cdot \boldsymbol{n}(y) dS. \tag{4.27}$$

As for the estimates of $\boldsymbol{e}$, we have, by (4.21)–(4.24),

$$\begin{aligned}
&\|\boldsymbol{e}^{(1)}\|_{W_r^{1,0}(Q_T^1)} \le c(\|\mathcal{P}\boldsymbol{h}\|_{W_r^{2,0}(Q_T^1)} + \|(\mathcal{L}^{-1}\boldsymbol{u} \times \boldsymbol{h})\|_{W_r^{1,0}(Q_T^1)}),\\
&\|\boldsymbol{e}^{(2)}\|_{W_r^{1,0}(Q_T^2)} \le c(\|\boldsymbol{\mathcal{E}}^{(2)}\|_{W_r^{1,0}(Q_T^2)} + \|\boldsymbol{e}_1\|_{W_r^{1,0}(Q_T^2)}\\
&\qquad\qquad + \|\nabla \chi_2\|_{W_r^{1,0}(Q_T^2)} + \|C\|_{L_r(0,T)}),\\
&\|\nabla \chi_2\|_{W_r^{1,0}(Q_T^2)} \le c\|\nabla \chi_1\|_{W_r^{1,0}(Q_T^2)},\\
&\|\nabla \chi_1\|_{W_r^{1,0}(Q_T^1)} \le c(\|\boldsymbol{e}^{(1)}\|_{W_r^{1,0}(Q_T^1)} + \|\boldsymbol{\mathcal{E}}^{(1)}\|_{W_r^{1,0}(Q_T^1)} + \|\boldsymbol{e}_1\|_{W_r^{1,0}(Q_T^1)}),\\
&\|C\|_{L_p(0,T)} \le c(\|\boldsymbol{e}_1\|_{W_r^{1,0}(Q_T^2)} + \|\boldsymbol{\mathcal{E}}^{(2)}\|_{W_r^{1,0}(Q_T^2)} + \|\nabla \chi_2\|_{W_r^{1,0}(Q_T^2)}).
\end{aligned} \tag{4.28}$$

Taking (4.21), (4.18), (2.5), (2.15) into account, we conclude that $\boldsymbol{e} \in W_r^{1,0}(Q_T^i)$, $i = 1, 2$.

5. Estimates of non-linear terms

In this section we estimate the expressions (3.2). For this we need some auxiliary propositions.

PROPOSITION 1. *Arbitrary functions $u(x), v(x)$ given in a domain $\mathcal{D} \subset \mathbb{R}^n$ satisfy the following inequalities:*

$$\begin{aligned}
\|uv\|_{W_p^1(\mathcal{D})} &\le c(\sup_{\mathcal{D}} |v(x)| \|u\|_{W_p^1(\mathcal{D})} + \sup_{\mathcal{D}} |u(x)| \|v\|_{W_p^1(\mathcal{D})})\\
&\le c\|u\|_{W_p^1(\mathcal{D})} \|v\|_{W_p^1(\mathcal{D})},
\end{aligned} \tag{5.1}$$

$$\|uv\|_{W_p^2(\mathcal{D})} \le c(\|u\|_{W_p^2(\mathcal{D})} \|v\|_{W_p^1(\mathcal{D})} + \|v\|_{W_p^2(\mathcal{D})} \|u\|_{W_p^1(\mathcal{D})}), \tag{5.2}$$

if $p > n$,

$$\|uv\|_{W_p^l(\mathcal{D})} \le c(\|u\|_{W_p^l(\mathcal{D})} \sup_{\mathcal{D}} |v(x)| + \|v\|_{W_p^l(\mathcal{D})} \sup_{\mathcal{D}} |u(x)|), \tag{5.3}$$

if $pl > n$ and l is not an integer.

PROOF. The inequality (5.1) is obvious, and (5.2) follows from (5.1) and

$$\begin{aligned}
\|\nabla(uv)\|_{W_p^1(\mathcal{D})} &\le \|v \nabla u\|_{W_p^1(\mathcal{D})} + \|u \nabla v\|_{W_p^1(\mathcal{D})}\\
&\le c(\|u\|_{W_p^2(\mathcal{D})} \|v\|_{W_p^1(\mathcal{D})} + \|v\|_{W_p^2(\mathcal{D})} \|u\|_{W_p^1(\mathcal{D})}).
\end{aligned} \tag{5.4}$$

To prove (5.3), we estimate the semi-norm

$$\|w\|_{\dot W^l_p(\mathbb{R}^n)} = \Big(\int_{\mathbb{R}^n} \|\Delta^m(z)w\|^p_{L_p(\mathbb{R}^n)} \frac{dz}{|z|^{n+pl}}\Big)^{1/p}, \quad m > l, \tag{5.5}$$

of the product $w(x) = u(x)v(x)$, assuming that $u, v \in C_0^\infty(\mathbb{R}^n)$. By $\Delta^m(z)w(x)$, we mean a finite difference

$$\Delta^m(z)w(x) = \sum_{j=0}^m (-1)^{m-j} C^j_m w(x + jz),$$

and we notice that the semi-norms (5.5) are equivalent to each other for different $m > l$; see [9]. We assume that $m > 2l$; then in every term of the sum

$$\Delta^m(z)u(x)v(x) = \sum_{j=0}^m C^j_m (\Delta^j(z)u(x))\Delta^{m-j}(z)v(x + (m - j)z),$$

one of the numbers j or $m - j$ is greater than l. Hence

$$\|\Delta^m(z)uv\|_{L_p(\mathbb{R}^n)} \le c(\sup_{\mathbb{R}^n} |v(x)| \|\Delta^k(z)u\|_{L_p(\mathbb{R}^n)} + \sup_{\mathbb{R}^n} |u(x)| \|\Delta^k(z)v\|_{L_p(\mathbb{R}^n)}),$$

where $k > l$, and

$$\|uv\|_{\dot W^l_p(\mathbb{R}^n)} \le c(\sup_{\mathbb{R}^n} |v(x)| \|u\|_{\dot W^l_p(\mathbb{R}^n)} + \sup_{\mathbb{R}^n} |u(x)| \|v\|_{\dot W^l_p(\mathbb{R}^n)}).$$

From this inequality and

$$\|uv\|_{L_p(\mathbb{R}^n)} \le c(\sup_{\mathbb{R}^n} |v(x)| \|u\|_{L_p(\mathbb{R}^n)} + \sup_{\mathbb{R}^n} |u(x)| \|v\|_{L_p(\mathbb{R}^n)}),$$

it follows that

$$\|uv\|_{\widetilde W^l_p(\mathbb{R}^n)} \le c(\sup_{\mathbb{R}^n} |v(x)| \|u\|_{\widetilde W^l_p(\mathbb{R}^n)} + \sup_{\mathbb{R}^n} |u(x)| \|v\|_{\widetilde W^l_p(\mathbb{R}^n)}), \tag{5.6}$$

where

$$\|u\|_{\widetilde W^l_p(\mathbb{R}^n)} = \|u\|_{L_p(\mathbb{R}^n)} + \|u\|_{\dot W^l_p(\mathbb{R}^n)}.$$

This norm is equivalent to the norm in $W^l_p(\mathbb{R}^n)$ defined in Section 1, if l is not an integer. Since every function from $W^l_p(\mathcal{D})$ can be extended in $\mathbb{R}^n$ with preservation of class, (5.6) implies (5.3). The proposition is proved.

Now we formulate our assumptions concerning the extensions $\boldsymbol{N}^*$ and ρ^*. We assume that $\boldsymbol{N}^*(y)$ is a sufficiently regular non-zero vector field in Ω and $\rho^* = E\rho$, where E is a linear extension operator with the properties

$$\begin{gathered} \frac{\partial \rho^*(x,t)}{\partial N}\Big|_{\mathcal{G}} = 0, \quad \rho^*(y,t) = 0 \quad \text{near} \quad \mathrm{S}, \\ \|\rho^*(\cdot,t)\|_{W^r_p(\Omega)} \le c\|\rho\|_{W^{r-1/p}_p(\mathcal{G})}, \quad r \in (1/p, 3]. \end{gathered} \tag{5.7}$$

It follows that the time derivative of ρ^* satisfies similar inequalities

$$\|\rho^*_t(\cdot,t)\|_{W^r_p(\Omega)} \le c\|\rho_t\|_{W^{r-1/p}_p(\mathcal{G})}, \quad r \in (1/2, 2]. \tag{5.8}$$

Proposition 2. *Let* $\boldsymbol{R}(x,t) = \Big(\frac{\partial N^*_i \rho^*}{\partial x_j}\Big)_{i,j=1,2,3}$, *and let the conditions* $p > 3$ *and*

$$\|\rho\|_{W^{2-1/p}_p(\mathcal{G})} \le \delta \ll 1 \tag{5.9}$$

be satisfied. For arbitrary smooth function $f(\widetilde{\boldsymbol{R}})$ defined for $|\widetilde{\boldsymbol{R}}| \le \delta_0$, $\delta_0 > \delta$, the inequalities

$$
\begin{aligned}
&\|\boldsymbol{R} f(\boldsymbol{R})\|_{L_p(\Omega)} \le c\|\boldsymbol{R}\|_{L_p(\Omega)},\\
&\|\boldsymbol{R} f(\boldsymbol{R})\|_{W^1_p(\Omega)} \le c\|\boldsymbol{R}\|_{W^1_p(\Omega)} \le c\|\rho\|_{W^{2-1/p}_p(\mathcal{G})},\\
&\|f(\boldsymbol{R})\|_{W^1_p(\Omega)} \le c,
\end{aligned} \tag{5.10}
$$

$$
\|\boldsymbol{R} f(\boldsymbol{R})\|_{W^{1-1/p}_p(\mathcal{G})} \le c\|\rho\|_{W^{1-1/p}_p(\mathcal{G})} \tag{5.11}
$$

hold.

PROOF. Inequalities (5.10) are obvious, and (5.11) follows from (5.10):

$$
\|\boldsymbol{R} f(\boldsymbol{R})\|_{W^{1-1/p}_p(\mathcal{G})} \le c\|\boldsymbol{R} f(\boldsymbol{R})\|_{W^1_p(\Omega)} \le c\|\boldsymbol{R}\|_{W^1_p(\Omega)} \le c\|\rho\|_{W^{2-1/p}_p(\mathcal{G})}.
$$

We also have

$$
\sup_{\Omega} |\boldsymbol{R}(x)| \le c\|\boldsymbol{R}\|_{W^1_p(\Omega)} \le c\|\rho\|_{W^{2-1/p}_p(\Omega)} \le c\delta. \tag{5.12}
$$

Examples of functions satisfying the assumptions of Proposition 2 are provided by the elements of the matrices $\mathcal{L}(x,\rho^*)$, $\mathcal{L}^{-1}(x,\rho^*)$, $\widehat{\mathcal{L}}(x,\rho^*)$, whereas $\boldsymbol{n}(e_\rho)$ and $\widehat{\mathcal{L}}(x,\rho)\boldsymbol{N}(x)(\Lambda(x,\rho)^{-1}$ are smooth functions of $\boldsymbol{R}$, ρ and x. They satisfy the inequality

$$
\begin{aligned}
&\|\boldsymbol{R} f(\boldsymbol{R},\rho,\cdot)\|_{W^1_p(\Omega)} + \|\rho^* f(\boldsymbol{R},\rho,\cdot)\|_{W^1_p(\Omega)}\\
&\qquad\le c\Big(\|\boldsymbol{R}\|_{W^1_p(\Omega)} + \|\rho^*\|_{W^1_p(\mathcal{F})}\Big) \le c\delta.
\end{aligned} \tag{5.13}
$$

It is obtained in the same way as (5.10).

We shall use a similar estimate in the two-dimensional case:

$$
\begin{aligned}
&\|\boldsymbol{R} f(\boldsymbol{R},\rho,\cdot)\|_{W^{1-1/p}_p(\mathcal{G})} + \|\rho f(\boldsymbol{R},\rho,\cdot)\|_{W^{1-1/p}_p(\mathcal{G})}\\
&\le c\Big(\|\boldsymbol{R}\|_{W^{1-1/p}_p(\mathcal{G})} + \|\rho\|_{W^{1-1/p}_p(\mathcal{G})}\Big) \le c\delta.
\end{aligned} \tag{5.14}
$$

Now we estimate all the terms in (3.2), step by step. We always assume that $p > 3$.

1. *Estimate of* $(\nabla - \widetilde{\nabla})q$. Since

$$
(\nabla - \widetilde{\nabla})q = (I - \mathcal{L}^{-T})\nabla q
$$

and $|I - \mathcal{L}^{-T}| \le c|\boldsymbol{R}|$, we have

$$
\|(\nabla - \widetilde{\nabla})q\|_{L_p(Q^1_T)} \le c \sup_{Q^1_T} |\boldsymbol{R}(x,t)| \|\nabla q\|_{L_p(Q_T)} \le c\delta\|\nabla q\|_{L_p(Q_T)}. \tag{5.15}
$$

2. *Estimate of* $(\widetilde{\nabla}^2 - \nabla^2)\boldsymbol{u}$. It is easily seen that

$$
\begin{aligned}
(\widetilde{\nabla}^2 - \nabla^2)\boldsymbol{u} &= \widetilde{\nabla}\cdot(\widetilde{\nabla} - \nabla)\boldsymbol{u} + (\widetilde{\nabla} - \nabla)\cdot\nabla\boldsymbol{u}\\
&= (\mathcal{L}^{-T}\nabla\cdot(\mathcal{L}^{-T} - I)\nabla)\boldsymbol{u} + (\mathcal{L}^{-T} - I)\nabla\cdot\nabla\boldsymbol{u}\\
&= (\mathcal{L}^{-1}(\mathcal{L}^{-T} - I) : D^2)\boldsymbol{u} + (\mathcal{L}^{-T} - I)\nabla\cdot\nabla\boldsymbol{u} + (\boldsymbol{M}\cdot\nabla)\boldsymbol{u},
\end{aligned}
$$

where $\boldsymbol{M}$ is the vector field whose components $M_j = \sum_{k,i=1}^3 l^{ki}\frac{\partial}{\partial x_k} l^{ji}$ are linear combinations of $\nabla\boldsymbol{R} g(\boldsymbol{R})$ (l^{ij} are elements of $\mathcal{L}^{-1}$). The first two terms on the right-hand side are estimated in the same way as $(\widehat{\nabla} - \nabla)q$:

$$
\|(\mathcal{L}^{-T} - I)\nabla\cdot\nabla\boldsymbol{u}\|_{L_p(Q^1_T)} + \|(\mathcal{L}^{-1}(\mathcal{L}^{-T} - I) : D^2)\boldsymbol{u}\|_{L_p(Q^1_T)} \le c\delta\|\boldsymbol{u}\|_{W^{2,0}_p(Q^1_T)},
$$

and

$$\begin{aligned}\|(\boldsymbol{M}\cdot\nabla)\boldsymbol{u}\|_{L_p(Q_T^1)} &\le c\Big(\int_0^T \sup_{x\in\mathcal{F}_1}|\nabla\boldsymbol{u}|^p dt\Big)^{1/p}\sup_{t<T}\|\boldsymbol{M}\|_{L_p(\mathcal{F}_1)}\\ &\le c\sup_{t<T}\|\rho^*\|_{W_p^2(\mathcal{F}_1)}\|\boldsymbol{u}\|_{W_p^{2,0}(Q_T^1)}\\ &\le c\sup_{t<T}\|\rho\|_{W_p^{2-1/p}(\mathcal{G})}\|\boldsymbol{u}\|_{W_p^{2,0}(Q_T^1)}\le c\delta\|\boldsymbol{u}\|_{W_p^{2,0}(Q_T^1)}.\end{aligned}\tag{5.16}$$

3. *Estimate of* $(\mathcal{L}^{-1}\boldsymbol{u}\cdot\nabla)\boldsymbol{u}$.

$$\begin{aligned}\|(\mathcal{L}^{-1}\boldsymbol{u}\cdot\nabla)\boldsymbol{u}\|_{L_p(Q_T^1)} &\le c\sup_{Q_T^1}|\boldsymbol{u}(x,t)|\|\nabla\boldsymbol{u}\|_{L_p(Q_T^1)}\\ &\le cT^{1/p}(\sup_{t<T}\|\boldsymbol{u}\|_{W_p^1(\mathcal{F}_1)})^2.\end{aligned}\tag{5.17}$$

4. *Estimate of* $\rho_t^*(\mathcal{L}^{-1}\boldsymbol{N}^*\cdot\nabla)\boldsymbol{u}$.

$$\begin{aligned}\|\rho_t^*(\mathcal{L}^{-1}\boldsymbol{N}^*\cdot\nabla)\boldsymbol{u}\|_{L_p(Q_T^1)} &\le c\sup_{Q_T^1}|\rho_t^*(x,t)|\|\nabla\boldsymbol{u}\|_{L_p(Q_T^1)}\\ &\le c\sup_{t<T}\|\rho_t^*\|_{W_p^1(\mathcal{F}_1)}\|\boldsymbol{u}\|_{W_p^{1,0}(Q_T^1)}\\ &\le cT^{1/p}\sup_{t<T}\|\boldsymbol{u}\|_{W_p^1(\mathcal{F}_1)}\sup_{t<T}\|\rho_t\|_{W_p^{1-1/p}(\mathcal{G})}.\end{aligned}\tag{5.18}$$

5. *Estimate of* $(I-\widehat{\mathcal{L}}^T)\nabla\cdot\boldsymbol{u}=\nabla\cdot(I-\widehat{\mathcal{L}})\boldsymbol{u}$.

$$\begin{aligned}\|(I-\widehat{\mathcal{L}}^T)\nabla\boldsymbol{u}\|_{W_p^{1,0}(Q_T^1)} &\le c\sup_{t<T}\|I-\widehat{\mathcal{L}}^T\|_{W_p^1(\mathcal{F}_1)}\|\nabla\boldsymbol{u}\|_{W_p^{1,0}(Q_T^1)}\\ &\le c\sup_{t<T}\|\rho\|_{W_p^{2-1/p}(\mathcal{G})}\|\boldsymbol{u}\|_{W_p^{2,0}(Q_T^1)}\\ &\le c\delta\|\boldsymbol{u}\|_{W_p^{2,0}(Q_T^1)},\end{aligned}\tag{5.19}$$

$$\begin{aligned}\|(I-\widehat{\mathcal{L}}^T)\boldsymbol{u}\|_{L_p(Q_T^1)}+\|(I-\widehat{\mathcal{L}}^T)\boldsymbol{u}_t\|_{L_p(Q_T^1)} &\le c\delta\|\boldsymbol{u}\|_{W_p^{0,1}(Q_T)},\\ \|\widehat{\mathcal{L}}_t^T\boldsymbol{u}\|_{L_p(Q_T^1)} &\le c\sup_{Q_T^1}|\boldsymbol{u}|\|\rho_t\|_{W_p^{1-1/p}(G_T)}\\ &\le cT^{1/p}\sup_{t<T}\|\boldsymbol{u}\|_{W_p^1(\mathcal{F}_1)}\sup_{t<T}\|\rho_t\|_{W_p^{1-1/p}(\mathcal{G})}.\end{aligned}\tag{5.20}$$

6. *Estimate of* $\nabla\cdot T_M(\frac{\mathcal{L}}{L}\boldsymbol{h})$. If (1.5) holds, then

$$\|\boldsymbol{h}\|_{W_p^i(\mathcal{F}_1)}\le c\|\boldsymbol{h}\|_{W_r^{2-2/r}(\mathcal{F}_i)},\tag{5.21}$$

hence

$$\begin{aligned}\|\nabla\cdot T_M(\frac{\mathcal{L}}{L}\boldsymbol{h})\|_{L_p(Q_T^i)} \le& c\big(\|\boldsymbol{h}\|_{W_p^{1,0}(Q_T^i)}\\ &+\|\rho^*\|_{W_p^{2,0}(Q_T^i)}\sup_{Q_T^i}|\boldsymbol{h}(y,t)|\big)\sup_{Q_T^i}|\boldsymbol{h}(y,t)|)\\ \le& cT^{1/p}\big(\sup_{t<T}\|\boldsymbol{h}\|_{W_r^{2-2/r}(\mathcal{F}_i)}\\ &+\sup_{t<T}\|\rho\|_{W_p^{2-1/p}(\mathcal{G})}\sup_{t<T}\|\boldsymbol{h}\|_{W_r^1(\mathcal{F}_i)}\big)\sup_{t<T}\|\boldsymbol{h}\|_{W_r^1(\mathcal{F}_i)}).\end{aligned}\tag{5.22}$$

It follows that

$$
\begin{aligned}
\|[\boldsymbol{n}\cdot T_M(\frac{\mathcal{L}}{L}\boldsymbol{h})\boldsymbol{n}]\|_{W_p^{1-1/p,0}(G_T)} &\le c\sum_{i=1}^{2}\big(\|\boldsymbol{h}\|_{W_p^{1,0}(Q_T^i)}\\
&\quad+\|\rho^*\|_{W_p^{2,0}(Q_T^i)}\sup_{Q_T^i}|\boldsymbol{h}(y,t)|\big)\sup_{Q_T^i}|\boldsymbol{h}(y,t)|\\
&\le c\sum_{i=1}^{2}T^{1/p}\big(\sup_{t<T}\|\boldsymbol{h}\|_{W_r^{2-2/r}(\mathcal{F}_i)}\\
&\quad+\sup_{t<T}\|\rho\|_{W_p^{2-1/p}(\mathcal{G})}\sup_{t<T}\|\boldsymbol{h}\|_{W_r^1(\mathcal{F}_i)}\big)\sup_{t<T}\|\boldsymbol{h}\|_{W_r^1(\mathcal{F}_i)}.
\end{aligned}
\tag{5.23}
$$

7. *Estimates of* $\widetilde{S}(\boldsymbol{u})\boldsymbol{n}-S(\boldsymbol{u})\boldsymbol{N}$. Since $\boldsymbol{n}(e_\rho)=\frac{\widetilde{\mathcal{L}}^T\boldsymbol{N}(y)}{|\widetilde{\mathcal{L}}^T\boldsymbol{N}(y)|}$, we have

$$
\sup_{\mathcal{G}}|\boldsymbol{n}-\boldsymbol{N}|\le c\|\boldsymbol{n}-\boldsymbol{N}\|_{W_p^{1-1/p}(\mathcal{G})}\le c\|\rho\|_{W_p^{2-1/p}(\mathcal{G})}\le c\delta,
$$

hence

$$
\begin{aligned}
\|\widetilde{S}(\boldsymbol{u})\boldsymbol{n}-S(\boldsymbol{u})\boldsymbol{N}\|_{W_p^{1-1/p,0}(G_T)} &\le \|(\widetilde{S}(\boldsymbol{u})-S(\boldsymbol{u}))\boldsymbol{N}\|_{W_p^{1-1/p,0}(G_T)}\\
&\quad+\|\widetilde{S}(\boldsymbol{u})(\boldsymbol{n}-\boldsymbol{N})\|_{W_p^{1-1/p,0}(G_T)}\le c\delta\|\boldsymbol{u}\|_{W_p^{2-1/p,0}(G_T)}.
\end{aligned}
\tag{5.24}
$$

Now we pass to the estimate of

$$
J=\int_0^T\frac{dh}{h^{1/2+p/2}}\int_h^T\|\Delta_t(-h)(\widetilde{S}(\boldsymbol{u})\boldsymbol{n}-S(\boldsymbol{u})\boldsymbol{N})\|_{L_p(\mathcal{G})}^p dt,
$$

where $\Delta_t(-h)f(t)=f(t-h)-f(t)$. We have

$$
\begin{aligned}
&\|\Delta_t(-h)(\widetilde{S}(\boldsymbol{u})\boldsymbol{n}-S(\boldsymbol{u})\boldsymbol{N})\|_{L_p(\mathcal{G})}\le\|\Delta_t(-h)(\widetilde{S}(\boldsymbol{u})-S(\boldsymbol{u}))\boldsymbol{N}\|_{L_p(\mathcal{G})}\\
&+\|(\Delta_t(-h)\widetilde{S}(\boldsymbol{u}))(\boldsymbol{n}-\boldsymbol{N})\|_{L_p(\mathcal{G})}+\|S(\boldsymbol{u})\Delta_t(-h)\boldsymbol{n}\|_{L_p(\mathcal{G})}\\
&\le c\big(\delta\|\Delta_t(-h)\nabla\boldsymbol{u}\|_{L_p(\mathcal{G})}+\|\nabla\boldsymbol{u}\|_{L_p(\mathcal{G})}\int_0^h\|\rho_t(\cdot,t-\xi)\|_{W_p^{2-1/p}(\mathcal{G})}d\xi\big),
\end{aligned}
$$

which implies

$$
J^{1/p}\le c\delta\|\nabla\boldsymbol{u}\|_{W_p^{0,1/2-1/2p}(G_T)}+cT^{1/2-1/2p}\sup_{t<T}\|\nabla\boldsymbol{u}\|_{L_p(\mathcal{G})}\|\rho_t\|_{W_p^{2-1/p,0}(G_T)},
$$

hence

$$
\begin{aligned}
&\|\widetilde{S}(\boldsymbol{u})\boldsymbol{n}-S(\boldsymbol{u})\boldsymbol{N}\|_{W_p^{1-1/p,1/2-1/2p}(G_T)}\\
&\le c\delta\|\boldsymbol{u}\|_{W_p^{2,1}(Q_T^1)}+cT^{1/2-1/2p}\sup_{t<T}\|\nabla\boldsymbol{u}\|_{L_p(\mathcal{G})}\|\rho_t\|_{W_p^{2-1/p,0}(G_T)}.
\end{aligned}
\tag{5.25}
$$

It follows that

$$(5.26)\quad \begin{aligned} &\|\boldsymbol{l}_3(\boldsymbol{u},\rho)\|_{W_p^{1-1/p,1/2-1/2p}(G_T)} \le c\big(\delta\|\boldsymbol{u}\|_{W_p^{2,1}(Q_T^1)} \\ &\quad + T^{1/2-1/2p}\sup_{t<T}\|\nabla\boldsymbol{u}\|_{L_p(\mathcal{G})}\|\rho_t\|_{W_p^{2-1/p,0}(G_T)}\big), \\ &\|l_4(\boldsymbol{u},\boldsymbol{h},\rho)\|_{W_p^{1-1/p,0}(G_T)} \le c\big(\delta\|\boldsymbol{u}\|_{W_p^{2,1}(Q_T^1)} \\ &\quad + \sum_{i=1}^2 (T^{1/p}\sup_{t<T}\|\boldsymbol{h}\|_{W_r^{2-2/r}(\mathcal{F}_i)} \\ &\quad + T^{1/p}\sup_{t<T}\|\rho\|_{W_p^{2-1/p}(\mathcal{G})}\sup_{t<T}\|\boldsymbol{h}\|_{W_r^1(\mathcal{F}_i)})\sup_{t<T}\|\boldsymbol{h}\|_{W_r^1(\mathcal{F}_i)}\big). \end{aligned}$$

The next step is the estimate of $l_5(\rho)$ and $l_6(\boldsymbol{u},\rho)$.

8. *Estimate of* $\frac{d^2}{ds^2}\mathcal{L}^{-T}(x,s\rho)\nabla\cdot\frac{\widehat{\mathcal{L}}^T(x,s\rho)\boldsymbol{N}(x)}{|\widehat{\mathcal{L}}^T(x,s\rho)\boldsymbol{N}(x)|}$ *and* $\frac{d^2}{ds^2}\frac{\widehat{\mathcal{L}}^T(x,s\rho)\boldsymbol{N}}{\Lambda(x,s\rho)}$. The function $\frac{d^2}{ds^2}\mathcal{L}^{-T}(x,s\rho)\nabla\cdot\frac{\widehat{\mathcal{L}}^T(x,s\rho)\boldsymbol{N}(x)}{|\widehat{\mathcal{L}}^T(x,s\rho)\boldsymbol{N}(x)|}$ is a linear combination of the expressions $\nabla\boldsymbol{R}\boldsymbol{R}f(s\boldsymbol{R},x)$, where $\boldsymbol{R}=\Big(\frac{\partial N_i(x)\rho(x,t)}{\partial x_j}\Big)_{i,j=1,2,3}$, $s\in(0,1)$ and f is the function satisfying the assumptions of Proposition 2 but depending also on x. Making use of this proposition (in the two-dimensional case), we obtain

$$\|\nabla\boldsymbol{R}\boldsymbol{R}f(s\boldsymbol{R},\cdot)\|_{W_p^{1-1/p}(\mathcal{G})} \le c\|\nabla\boldsymbol{R}\|_{W_p^{1-1/p}(\mathcal{G})}\|\boldsymbol{R}\|_{W_p^{1-1/p}(\mathcal{G})}.$$

This inequality implies

$$\begin{aligned} &\|\nabla\boldsymbol{R}\boldsymbol{R}f(s\boldsymbol{R},\cdot)\|_{W_p^{1-1/p,0}(G_T)} \\ &\qquad \le c\sup_{t<T}\|\rho\|_{W_p^{2-1/p}(\mathcal{G})}\|\rho\|_{W_p^{3-1/p,0}(G_T)} \le c\delta\|\rho\|_{W_p^{3-1/p,0}(G_T)}, \end{aligned}$$

hence

$$(5.27)\qquad \|l_5\|_{W_p^{1-1/p,0}(G_T)} \le c\delta\|\rho\|_{W_p^{3-1/p,0}(G_T)}.$$

The function $\frac{d^2}{ds^2}\frac{\widehat{\mathcal{L}}^T(x,s\rho)\boldsymbol{N}}{\Lambda(x,s\rho)}$ is a linear combination of the expressions

$$\boldsymbol{R}\boldsymbol{R}f_1(s\boldsymbol{R},s\rho,x),\quad \rho\boldsymbol{R}f_2(s\boldsymbol{R},s\rho,x),\quad \rho^2f_3(s\boldsymbol{R},s\rho,x)$$

with f_i as in Proposition 2, but depending also on ρ and x. We estimate the leading term $\boldsymbol{R}\boldsymbol{R}f_1$, since two other functions are estimated in a similar way, as ρ^* is more regular than $\mathcal{R}$. As above, we have

$$\begin{aligned} &\|\boldsymbol{R}\boldsymbol{R}f_1(s\boldsymbol{R},s\rho,\cdot)\|_{W_p^{2-1/p,0}(G_T)} \le c\delta\|\rho\|_{W_p^{3-1/p,0}(G_T)}, \\ &\Delta_t(-h)|\boldsymbol{R}\boldsymbol{R}f_1(s\boldsymbol{R},s\rho,\cdot)\|_{L_p(\mathcal{G})} \le c\sup_{\mathcal{G}}|\mathcal{R}|\int_0^h\|\rho_t(\cdot,t-\xi)\|_{W_p^{2-1/p}(\mathcal{G})}d\xi, \\ &\Big(\int_0^T\frac{dh}{h^{1/2+p}}\int_h^T\|\Delta_t(-h)\boldsymbol{R}\boldsymbol{R}f_1\|^p_{L_p(\mathcal{G})}dt\Big)^{1/p} \\ &\qquad \le cT^{1/2p}\sup_{t<T}\|\rho\|_{W_p^{2-1/p}(\mathcal{G})}\|\rho_t\|_{W_p^{2-1/p,0}(G_T)}. \end{aligned}$$

The same inequalities hold for $\frac{d^2}{ds^2}\frac{\widehat{\mathcal{L}}^T(x,s\rho)\boldsymbol{N}}{\Lambda(x,s\rho)}$.

It follows that

$$\boldsymbol{D}=\frac{\widehat{\mathcal{L}}^T\boldsymbol{N}}{\Lambda(y,\rho)}-\boldsymbol{N}+\nabla_\tau\rho=\int_0^1(1-s)\frac{d^2}{ds^2}\frac{\widehat{\mathcal{L}}^T(x,s\rho)\boldsymbol{N}}{\Lambda(x,s\rho)}ds$$

satisfies

$$\|\boldsymbol{D}\|_{W_p^{2-1/p,1-1/2p}(G_T)} \leq c\delta(\|\rho\|_{W_p^{3-1/p,0}(G_T)} + T^{1/2p}\|\rho_t\|_{W_p^{2-1/p,0}(G_T)}),$$
$$\sup_{\mathcal{G}} |\boldsymbol{D}(y,t)| \leq c \sup_{\mathcal{G}} |\mathcal{R}\mathcal{R}| \leq c\delta\|\rho\|_{W_p^{2-1/p}(\mathcal{G})},$$

hence by (5.3)

$$\begin{aligned}\|\boldsymbol{D}\cdot\boldsymbol{u}\|_{W_p^{2-1/p,0}(G_T)} &\leq c(\|\boldsymbol{D}\|_{W_p^{2-1/p,0}(G_T)} \sup_{t<T}\|\boldsymbol{u}(\cdot,t)\|_{W_p^1(\mathcal{F}_1)} \\ &\quad + \delta\|\boldsymbol{u}\|_{W_p^{2-1/p,0}(G_T)} \sup_{t<T}\|\rho\|_{W_p^{2-1/p}(\mathcal{G})}) \\ &\leq c\delta((\|\rho\|_{W_p^{3-1/p,0}(G_T)} + T^{1/2p}\|\rho_t\|_{W_p^{2-1/p,0}(G_T)}) \sup_{t<T}\|\boldsymbol{u}\|_{W_p^1(\mathcal{F}_1)} \\ &\quad + \|\boldsymbol{u}\|_{W_p^{2-1/p,0}(G_T)} \sup_{t<T}\|\rho\|_{W_p^{2-1/p}(\mathcal{G})}), \\ \|\boldsymbol{D}\cdot\boldsymbol{u}\|_{W_p^{0,1-1/2p}(G_T)} &\leq c\delta(T^{1/2p}\|\rho_t\|_{W_p^{2-1/p,0}(G_T)} \sup_{t<T}\|\boldsymbol{u}\|_{W_p^1(\mathcal{F}_1)} \\ &\quad + \|\boldsymbol{u}\|_{W_p^{0,1-1/2p}(G_T)} \sup_{t<T}\|\rho\|_{W_p^{2-2/p}(\mathcal{G})}).\end{aligned} \tag{5.28}$$

9. *Estimate of* $(\boldsymbol{V}-\boldsymbol{u})\cdot\nabla_\tau\rho$. We have, by (5.3),

$$\begin{aligned}\|(\boldsymbol{V}-\boldsymbol{u})\cdot\nabla_\tau\rho\|_{W_p^{2-1/p,0}(G_T)} &\leq c\|\nabla_\tau\rho\|_{W_p^{2-1/p,0}(G_T)} \sup_{t<T}\|\boldsymbol{V}-\boldsymbol{u}\|_{W_p^{1-1/p}(\mathcal{G})} \\ &\quad + \Big(\int_0^T \|\boldsymbol{V}-\boldsymbol{u}\|^p_{W_p^{2-1/p}(\mathcal{G})}\|\nabla_\tau\rho\|^p_{W_p^{1-1/p}(\mathcal{G})}dt\Big)^{1/p}.\end{aligned} \tag{5.29}$$

We require that

$$\|\boldsymbol{V}-\boldsymbol{u}_0\|_{W_p^{1-1/p}(\mathcal{G})} \leq \delta, \tag{5.30}$$

then

$$\begin{aligned}\sup_{t<T}\|\boldsymbol{V}-\boldsymbol{u}\|_{W_p^{1-1/p}(\mathcal{G})} &\leq \delta + \sup_{t<T}\|\boldsymbol{u}-\boldsymbol{u}_0\|_{W_p^{1-1/p}(\mathcal{G})} \\ &\leq (\delta + cT^\beta\|\boldsymbol{u}\|_{W_p^{2,1}(Q_T^1)}), \quad \beta = 1/2 - 1/p,\end{aligned}$$

because $\boldsymbol{u} \in W_p^{1/2}(0,T;W_p^1(\mathcal{F}_1))$. Hence the first term in (5.29) does not exceed

$$c(\delta + T^\beta\|\boldsymbol{u}\|_{W_p^{2,1}(Q_T)})\|\rho\|_{W_p^{3-1/p,0}(G_T)},$$

and the second term is not greater than

$$\begin{aligned}&\|\boldsymbol{V}\|_{W_p^{2-1/p}(\mathcal{G})}\|\nabla_\tau\rho\|_{W_p^{1-1/p,0}(G_T)} + \|\boldsymbol{u}\|_{W_p^{2,0}(Q_T^1)} \sup_{t<T}\|\nabla_\tau\rho\|_{W_p^{1-1/p}(\mathcal{G})} \\ &\leq c(T^{1/p}\sup_{t<T}\|\nabla_\tau\rho\|_{W_p^{1-1/p}(\mathcal{G})}\|\boldsymbol{V}\|_{W_p^{2-1/p}(\mathcal{G})} + \delta\|\boldsymbol{u}\|_{W_p^{2,0}(Q_T^1)}) \\ &\leq c\delta(\sup_{t<T}\|\rho\|_{W_p^{2-1/p}(\mathcal{G})} + \|\boldsymbol{u}\|_{W_p^{2,0}(Q_T^1)}),\end{aligned}$$

if

$$T^{1/p}\|\boldsymbol{V}\|_{W_p^{2-1/p}(\mathcal{G})} \leq \delta. \tag{5.31}$$

We also have

$$\begin{aligned}&\|(\Delta_t(-h)\boldsymbol{u}))\cdot\nabla_\tau\rho\|_{L_p(\mathcal{G})} + \|(\boldsymbol{V}-\boldsymbol{u})\cdot\Delta_t(-h)\nabla_\tau\rho\|_{L_p(\mathcal{G})} \\ &\leq c\|\Delta_t(-h)\boldsymbol{u}\|_{L_p(\mathcal{G})}\|\rho\|_{W_p^{2-1/p}(\mathcal{G})} + c\|\boldsymbol{V}-\boldsymbol{u}\|_{W_p^{1-1/p}(\mathcal{G})}\|\Delta_t(-h)\nabla_\tau\rho\|_{L_p(\mathcal{G})},\end{aligned}$$

$$\Big(\int_0^T \frac{dh}{h^{1/2+p}}\int_h^T \|\Delta_t(-h)\boldsymbol{u}\|^p_{L_p(\mathcal{G})}dt\Big)^{1/p}\sup_{t<T}\|\rho\|_{W_p^{2-1/p}(\mathcal{G})}$$
$$+c\sup_{t<T}\|\boldsymbol{V}-\boldsymbol{u}\|_{W_p^{1-1/p}(\mathcal{G})}\Big(\int_0^T \frac{dh}{h^{1/2+p}}\int_h^T \|\Delta_t(-h)\nabla_\tau\rho\|^p_{L_p(\mathcal{G})}dt\Big)^{1/p}\le$$
$$c\sup_{t<T}\|\rho\|_{W_p^{2-1/p}(\mathcal{G})}\|\boldsymbol{u}\|_{W_p^{0,1-1/2p}(G_T)}$$
$$+cT^{1/2p}(\delta+cT^\beta\|\boldsymbol{u}\|_{W_p^{2,1}(Q_T^1)})\|\rho_t\|_{W_p^{2-1/p,0}(G_T)}$$

It follows that

(5.32)
$$\|l_6\|_{W_p^{2-1/p,1-1/2p}(G_T)}\le c\delta(\|\rho\|_{W_p^{3-1/p,0}(G_T)}+\sup_{t<T}\|\rho\|_{W_p^{2-1/p}(\mathcal{G})}+\|\boldsymbol{u}\|_{W_p^{2,1}(Q_T^1)})$$
$$+cT^{1/2p}(\delta+cT^\beta\|\boldsymbol{u}\|_{W_p^{2,1}(Q_T^1)})\|\rho_t\|_{W_p^{2-1/p,0}(G_T)}.$$

Now we pass to the estimate of $\boldsymbol{l}_8$, $\boldsymbol{l}_9$, $\boldsymbol{A}$.

10. *Estimate of* $(I-\mathcal{P})\boldsymbol{h}$. We have

$$\|(I-\mathcal{P})\boldsymbol{h}\|_{W_r^{2,0}(Q_T^i)}\le c\big(\delta\|\boldsymbol{h}\|_{W_r^{2,0}(Q_T^i)}+\|\rho^*\|_{W_r^{3,0}(Q_T^i)}\sup_{t<T}\|\boldsymbol{h}\|_{W_r^1(\mathcal{F}_i)}\big),$$
$$\|\rho^*\|_{W_r^{3,0}(Q_T^i)}=\Big(\int_0^T\|\rho^*\|^r_{W_r^3(\mathcal{F}_i)}dt\Big)^{1/r}\le T^{1/r-1/p}\Big(\int_0^T\|\rho^*\|^p_{W_r^3(\mathcal{F}_i)}dt\Big)^{1/p}$$
$$\le cT^{1/r-1/p}\|\rho\|_{W_p^{3-1/p,0}(Q_T^i)};$$

hence

$$\|(I-\mathcal{P})\boldsymbol{h}\|_{W_r^{2,0}(Q_T^i)}\le c\big(\delta\|\boldsymbol{h}\|_{W_r^{2,0}(Q_T^i)}+T^{1/r-1/p}\|\rho\|_{W_p^{3-1/p,0}(Q_T^i)}\sup_{t<T}\|\boldsymbol{h}\|_{W_r^1(\mathcal{F}_i)}\big).$$

Moreover,

$$\|((I-\mathcal{P})\boldsymbol{h})_t\|_{L_r(Q_T^i)}\le c\big(\delta\|\boldsymbol{h}_t\|_{L_r(Q_T^i)}+T^{1/r}\sup_{t<T}\|\rho_t\|_{W_p^{1-1/p}(\mathcal{G})}\sup_{t<T}\|\boldsymbol{h}\|_{W_r^1(\mathcal{F}_i)}\big),$$

and finally

$$\begin{aligned}&\|(I-\mathcal{P})\boldsymbol{h}\|_{W_r^{2,1}(Q_T^i)}\\ &\le c\big(\delta\|\boldsymbol{h}\|_{W_r^{2,1}(Q_T^i)}+T^{1/r-1/p}\|\rho\|_{W_p^{3-1/p,0}(Q_T^i)}\sup_{t<T}\|\boldsymbol{h}\|_{W_r^1(\mathcal{F}_i)}\\ &+T^{1/r}\sup_{t<T}\|\rho_t\|_{W_p^{1-1/p}(\mathcal{G})}\sup_{t<T}\|\boldsymbol{h}\|_{W_r^1(\mathcal{F}_i)}\big),\quad i=1,2,\end{aligned}\tag{5.33}$$

The same inequalities hold for $\boldsymbol{l}_8(\boldsymbol{h},\rho)$. Moreover, since the elements of the matrix

$$\frac{\widehat{\mathcal{L}}\widehat{\mathcal{L}}^T}{|\widehat{\mathcal{L}}^T\boldsymbol{N}^*(y)|^2}-\frac{I}{|\boldsymbol{N}^*|^2}\equiv f(y,\boldsymbol{R})$$

in (3.2) satisfy the assumptions of Proposition 2 (just as the elements of $I-\mathcal{P}$), (5.33) holds also for $\|\boldsymbol{A}^{(i)}\|_{W_r^{2,1}(Q_T^i)}$.

11. *Estimate of* $\boldsymbol{l}_7(\boldsymbol{u},\boldsymbol{h},\rho)$ *and* $\boldsymbol{\Phi}(\boldsymbol{h},\rho)=\frac{1}{L}\widehat{\mathcal{L}}_t\mathcal{L}\boldsymbol{h}+\rho^*\widehat{\mathcal{L}}(\mathcal{L}^{-1}\boldsymbol{N}\cdot\nabla)\frac{1}{L}\mathcal{L}\boldsymbol{h}$. We estimate the expression

$$\operatorname{rot}\boldsymbol{h}-\mathcal{P}\operatorname{rot}\mathcal{P}\boldsymbol{h}=\mathcal{P}\operatorname{rot}(\boldsymbol{h}-\mathcal{P}\boldsymbol{h})+(1-\mathcal{P})\operatorname{rot}\boldsymbol{h}.$$

We have

$$\|(I-\mathcal{P})\operatorname{rot}\boldsymbol{h}\|_{W_r^{1,0}(Q_T^1)}\le c\delta\|\boldsymbol{h}\|_{W_r^{2,0}(Q_T^1)},$$
$$\|\mathcal{P}\operatorname{rot}(I-\mathcal{P})\boldsymbol{h}\|_{W_r^{1,0}(Q_T^1)}\le c\|\operatorname{rot}(I-\mathcal{P})\boldsymbol{h}\|_{W_r^{1,0}(Q_T^1)},$$

and this norm has been estimated above in (5.33). Hence

$$\begin{aligned}&\|\operatorname{rot}\boldsymbol{h}-\mathcal{P}\operatorname{rot}\mathcal{P}\boldsymbol{h}\|_{W_r^{1,0}(Q_T^1)}\\ &\leq c\big(\delta\|\boldsymbol{h}\|_{W_r^{2,0}(Q_T^1)}+T^{1/r-1/p}\|\rho\|_{W_p^{3-1/p,0}(Q_T^1)}\sup_{t<T}\|\boldsymbol{h}\|_{W_r^1(\mathcal{F}_1)}\big).\end{aligned}\tag{5.34}$$

Other terms in $\boldsymbol{l}_7$ are estimated as follows:

$$\begin{aligned}\|\rho_t^*\widehat{\mathcal{L}}(\mathcal{L}^{-1}\boldsymbol{N}\cdot\nabla)\frac{1}{L}\mathcal{L}\boldsymbol{h}\|_{L_r(Q_T^1)}&\leq c\sup_{t<T}\|\rho_t^*\|_{W_p^1(\mathcal{F}_1)}\|\frac{\mathcal{L}}{L}\boldsymbol{h}\|_{W_r^{1,0}(\mathcal{F}_1)}\\ &\leq cT^{1/r}\sup_{t<T}\|\rho_t\|_{W_p^{1-1/p}(\mathcal{G})}\sup_{t<T}\|\boldsymbol{h}\|_{W_r^1(\mathcal{F}_1)},\\ \|\frac{1}{L}\widehat{\mathcal{L}}_t\mathcal{L}\boldsymbol{h}\|_{L_r(Q_T^1)}&\leq c\sup_{Q_T^1}|\boldsymbol{h}(x,t)|\|\nabla\rho_t^*\|_{L_r(Q_T^1)}\\ &\leq cT^{1/r}\sup_{t<T}\|\boldsymbol{h}\|_{W_r^1(\mathcal{F}_1)}\sup_{t<T}\|\rho_t\|_{W_p^{1-1/p}(\mathcal{G})},\end{aligned}\tag{5.35}$$

$$\begin{aligned}\|\mathcal{L}^{-1}\boldsymbol{u}\times\boldsymbol{h}\|_{W_r^{1,0}(Q_T^1)}&\leq c(\|\boldsymbol{u}\|_{W_r^{1,0}(Q_T^1)}\sup_{t<T}\|\boldsymbol{h}\|_{W_r^1(\mathcal{F}_1)}\\ &\quad+\|\boldsymbol{h}\|_{W_r^{1,0}(Q_T^1)}\sup_{t<T}\|\boldsymbol{u}\|_{W_r^1(\mathcal{F}_1)})\\ &\leq cT^{1/r}\sup_{t<T}\|\boldsymbol{u}\|_{W_r^1(\mathcal{F}_1)}\sup_{t<T}\|\boldsymbol{h}\|_{W_r^1(\mathcal{F}_1)},\end{aligned}\tag{5.36}$$

hence

$$\|\boldsymbol{\Phi}(\boldsymbol{h},\rho)\|_{L_r(Q_T)}\leq cT^{1/r}\sum_{i=1}^{2}\sup_{t<T}\|\boldsymbol{h}\|_{W_r^1(\mathcal{F}_i)}\sup_{t<T}\|\rho_t\|_{W_p^{1-1/p}(\mathcal{G})}.\tag{5.37}$$

12. *Estimate of* $\boldsymbol{\Psi}(\boldsymbol{h},\rho)$.

$$\begin{aligned}\|\boldsymbol{\Psi}\|_{W_r^{1-1/r,0}(G_T)}&=\|\frac{\Lambda(\rho)\rho_t}{L}[\mu\boldsymbol{h}]\|_{W_r^{1-1/r,0}(G_T)}\\ &\leq c\sum_{i=1}^{2}\|\frac{\Lambda(\rho^*)\rho_t^*}{L(\cdot,\rho^*)}\mu\boldsymbol{h}\|_{W_r^{1,0}(Q_T^i)}\\ &\leq cT^{1/r}\|\rho_t\|_{W_p^{1-1/p}(\mathcal{G})}\sum_{i=1}^{2}\sup_{t<T}\|\boldsymbol{h}\|_{W_r^1(\mathcal{F}_i)}.\end{aligned}\tag{5.38}$$

In conclusion, we show that Theorem 1 holds for $r=p$, i.e., that the solution of the problem under consideration possesses the property $\boldsymbol{h}\in W_p^{2,1}(Q_T)$, if $\boldsymbol{h}_0\in W_p^{2-1/p}(\mathcal{F}_i)$, $i=1,2$. We restrict ourselves to obtaining the a priori estimate. It suffices to estimate the norms

$$\|l_7(\boldsymbol{u},\boldsymbol{h},\rho)\|_{L_p(Q_T^i)},\quad \|\boldsymbol{l}_8\|_{W_p^{2,1}(Q_T^i)},\quad \|\boldsymbol{A}\|_{W_p^{2,1}(Q_T^i)},\quad \|\boldsymbol{\Psi}\|_{W_p^{1-1/p,0}(G_T)}.$$

Instead of (5.33), we have

$$\begin{aligned}&\|(I-\mathcal{P})\boldsymbol{h}\|_{W_p^{2,1}(Q_T^i)}\leq c\big(\delta\|\boldsymbol{h}\|_{W_p^{2,1}(Q_T^i)}\\ &+(\|\rho\|_{W_p^{3-1/p,0}(Q_T^i)}+T^{1/p}\sup_{t<T}\|\rho_t\|_{W_p^{1-1/p}(\mathcal{G})})\sup_{t<T}\|\boldsymbol{h}\|_{W_p^1(\mathcal{F}_i)}\big).\end{aligned}\tag{5.39}$$

With the help of this inequality and the estimate (5.21) we obtain

$$\begin{aligned}&\|\boldsymbol{l}_8\|_{W_p^{2,1}(Q_T^2)}+\sum_{i=1}^2\|\boldsymbol{A}\|_{W_p^{2,1}(Q_T^i)}\\ &\leq c\sum_{i=1}^2(\delta\|\boldsymbol{h}\|_{W_p^{2,1}(Q_T^i)}+C_1(\rho)\sup_{t<T}\|\boldsymbol{h}\|_{W_r^{2-2/r}(\mathcal{F}_i)})\end{aligned}\tag{5.40}$$

with $C_1(\rho)$ depending on the norms of ρ estimated in (3.14).

In view of (5.35), (5.36), (5.37), $\|\boldsymbol{l}_7\|_{L_p(Q_T^i)}$ also satisfies (5.40), and

$$\|\boldsymbol{\Psi}\|_{W_p^{1-1/p,0}(G_T)}\leq cT^{1/p}\|\rho_t\|_{W_p^{1-1/p}(\mathcal{G})}\sum_{i=1}^2\sup_{t<T}\|\boldsymbol{h}\|_{W_r^{2-2/r}(\mathcal{F}_i)}.$$

Hence by Theorem 4

$$\begin{aligned}&\sum_{i=1}^2\|\boldsymbol{h}\|_{W_p^{2,1}(Q_T^1)}\\ &\leq c\sum_{i=1}^3(\delta\|\boldsymbol{h}\|_{W_p^{2,1}(Q_T^i)}+C_2(\rho,\boldsymbol{u})\sup_{t<T}\|\boldsymbol{h}\|_{W_r^{2-2/r}(\mathcal{F}_i)}+c\|\boldsymbol{h}_0\|_{W_p^{2-2/p}(\mathcal{F}_i)})\end{aligned}$$

with a certain $C_2(\rho,\boldsymbol{u})<\infty$, which yields the desired estimate in the case of small δ.

Inequalities (4.18), (4.20), (4.21), (4.28) with p instead of r show that $\boldsymbol{e}\in W_p^{1,0}(Q_T^i)$, $i=1,2$.

References

[1] M. Padula, V. A. Solonnikov, *On the free boundary problem of magnetohydrodynamics,* J. Math. Sci. **178** No 3 (2011), 313-344.

[2] O. A. Ladyzhenskaya, V. A. Solonnikov. *Solution of some non-stationary problems of magnetohydrodynamics for a viscous incompressible fluid,* Proc. Mat. Inst. Steklov, **59** (1960), 115–173.

[3] O. A. Ladyzhenskaya, V. A. Solonnikov. *The linearization principle and invariant manifolds for problems of magnetohydrodynamics,* J. Math. Sci., **8** (1977), 384–422.

[4] V. A. Solonnikov, *On the linear problem arising in the study of free boundary problem for the Navier-Stokes equations*, St. Petersburg Math. J. **22** No 6 (2011), 1023–1039.

[5] V. A. Solonnikov, *L_p-estimates for the linear problem arising in the study of the motion of an isolated liquid mass.* J. Math. Sci. 189, No 4 (2013), p. 699.

[6] V. A. Solonnikov, *L_p-estimates for the solution of a linear problem that arises in magnetohydrodynamics.* (Russian) Algebra i Analiz 23 (2011), no. 1, 232-254.

[7] M. Padula, V. A. Solonnikov, *On the local solvability of free boundary problem for the Navier-Stokes equations*, J. Math. Sci. **170** No 4 (2010), 522-553.

[8] O. A. Ladyzhenskaya, V. A. Solonnikov, N. N. Uraltseva, *Linear and quasilinear equations of parabolic type*, AMS, Providence, Rhode Island, 1968.

[9] K. K. Golovkin, *On the equivalent normalizations of fractional spaces*, Proc. Math. Inst. Steklov, **66** (1962), 364-383.

[10] V. A. Solonnikov, E. V. Frolova, *L_p-theory for the Stefan problem,* Zapiski Nauchn. Semin. POMI **243** (1997), 299-323.

St.-Petersburg Department of Steklov Mathematical Institute, Fontanka, 27, St.-Petersburg, 191023, Russia

E-mail address: solonnik@pdmi.ras.ru